PERGAMON INTERNATIONAL LIBRARY
of Science, Technology, Engineering and Social Studies

The 1000-volume original paperback library in aid of education,
industrial training and the enjoyment of leisure

Publisher: Robert Maxwell, M.C.

ELECTRICAL MACHINES
AND THEIR APPLICATIONS
FOURTH EDITION

||||||| ||||||||||| ||| |||| ||||||||||||| ||| |||

☑ **W9-AQG-099**

Price: US $42/UK £26 (Hard)
US $16 / UK £9.50 (Paper)

THE PERGAMON TEXTBOOK
INSPECTION COPY SERVICE

An inspection copy of any book published in the Pergamon International Library will gladly be sent to academic staff without obligation for their consideration for course adoption or recommendation. Copies may be retained for a period of 60 days from receipt and returned if not suitable. When a particular title is adopted or recommended for adoption for class use and the recommendation results in a sale of 12 or more copies, the inspection copy may be retained with our compliments. The Publishers will be pleased to receive suggestions for revised editions and new titles to be published in this important International Library.

APPLIED ELECTRICITY AND ELECTRONICS

General Editor: P. HAMMOND

Other titles of interest in the

PERGAMON INTERNATIONAL LIBRARY

JOURNALS OF RELATED INTEREST—FREE SPECIMEN COPY GLADLY SENT ON REQUEST

Computers & Electrical Engineering
Electric Technology USSR
Microelectronics & Reliability
Solid State Electronics

ELECTRICAL MACHINES
AND THEIR APPLICATIONS
FOURTH EDITION

J. Hindmarsh, B.Sc.(Eng.), C.Eng., M.I.E.E.

Formerly Senior Lecturer at the University of Manchester Institute of Science and Technology, U.K.

PERGAMON PRESS

OXFORD · NEW YORK · TORONTO · SYDNEY · PARIS · FRANKFURT

U.K.	Pergamon Press Ltd., Headington Hill Hall, Oxford OX3 0BW, England
U.S.A.	Pergamon Press Inc., Maxwell House, Fairview Park, Elmsford, New York 10523, U.S.A.
CANADA	Pergamon Press Canada Ltd., Suite 104, 150 Consumers Rd., Willowdale, Ontario M2J 1P9, Canada
AUSTRALIA	Pergamon Press (Aust.) Pty. Ltd., P.O. Box 544, Potts Point, N.S.W. 2011, Australia
FRANCE	Pergamon Press SARL, 24 rue des Ecoles, 75240 Paris, Cedex 05, France
FEDERAL REPUBLIC OF GERMANY	Pergamon Press GmbH, Hammerweg 6, D-6242 Kronberg-Taunus, Federal Republic of Germany

Copyright © 1977 J. Hindmarsh

All Rights Reserved. No part of this publication may be reproduced, stored in a retrieval system or transmitted in any form or by any means: electronic, electrostatic, magnetic tape, mechanical, photocopying, recording or otherwise, without permission in writing from the publishers.

First edition 1965
Second edition 1970
Third edition 1977
Reprinted 1977
Reprinted (with corrections) 1979
Reprinted (with modifications) 1981
Reprinted 1982, 1983
Fourth edition 1984

Library of Congress Cataloging in Publication Data

Hindmarsh, John, 1922-
Electrical machines and their applications.
(Pergamon international library: Electrical engineering division)
First ed. published in 1965 under title: Electrical machines.
Bibliography: p.
Includes index.
1. Electrical machinery. I. Title.
TK2182.H48 1976   621.31'042   76-20595

ISBN 0 08 030572 5 hardcover
ISBN 0 08 030573 3 flexicover

Printed in Great Britain by A. Wheaton & Co. Ltd., Exeter

DEDICATED TO
THE AUTHOR OF LIFE
WHO MADE IT POSSIBLE
AND TO MY LONG-SUFFERING FAMILY
WHO PATIENTLY ENDURED LONG PERIODS
OF DEPRIVATION AND NEGLECT DURING THE
PREPARATION AND REVISIONS OF THIS BOOK

PREFACE TO THE FOURTH EDITION

ALTHOUGH the various reprints since the third edition was first published have included some modifications, it has seemed appropriate in the present instance to undertake a more thorough revision, combing the text carefully, ensuring that all references conform to the latest national (BS) and international (IEC) recommendations and recording latest developments. The objective of the book remains, to provide understanding and instruction in the essential, unchanging principles governing electrical-machine performance. After substantial changes in the earlier editions, one might hope that the correct balance is being approached. But many minor textual alterations have been made in an attempt to clarify further any incomplete explanations and add material reflecting current thinking and practice. The old Appendix A dealt with alternative phasor-diagram conventions but these discussions have subsided somewhat and the conventions adopted since the second edition have become more generally accepted, as are the terms *time-phasor* and *space-phasor*, which designate the representation of sinusoidal time and space variations. This practice, which is now followed, avoids any possible confusion with true vector terminology, even though the methods of combination and presentation share some similarities. The new Appendix A covers a different topic. It deals more fully with permanent-magnet theory, recognising the growing importance of permanent-magnet machines, following the extensive developments of new materials.

The reaction of modern electrical machines to the continued proliferation of power-electronic circuits remains stoical, dignified and co-operative; not changing their basic electromechanical behaviour appreciably, except in terms of the more rapid and precise response such circuits have made possible. Taking a broad view, these developing systems are, for the most part, a convenient means of subjecting machines to changes of voltage and/or frequency, inevitably involving somewhat distorted waveforms. The electromechanical result is an "averaged out" behaviour, attenuating the high supply-harmonic content but with the possibility of increased noise and vibration. New designs proceed, with the aim of producing integrated machine/ electronic-systems for optimum performance. The general performance with the above features is covered in the text, but the interaction of machines and solid-state circuitry is relatively complex, and it receives greater consideration in the author's companion volume, *Worked Examples in Electrical Machines and Drives.*

J. H.

Sale, Cheshire 1984

PREFACE TO THE THIRD EDITION

THE changes to this edition are not so extensive as those made in the second edition. The purpose of the book is to establish a sound understanding of the basic principles of electrical machine operation. The method is first to delineate the common ground, before considering alternative possibilities for providing the electromagnetic reaction between the two elements, the one fixed and the other moving, normally. From this it emerges that there are four main types of machine, each with its own special characteristics in terms of performance, convenience and cost. One of the interesting consequences of the control and semiconductor revolutions has been the attempt to produce an economic, robust and flexibly controlled machine-drive system which combines the desirable features of all types. The main changes in the book are concerned with discussing more fully, the practice and influence of variable-frequency control which have made these attempts possible—especially in the further developments of "brushless" machines.

Although these developments are instructive, even fascinating, the bulk of the text must still be concerned with explaining how these desirable (and less desirable) characteristics arise in the first place, and further, how they can be applied to the appropriate engineering situations. It is not easy to decide how much engineering detail to include in a book of this kind; students' requirements differ. Much importance is given, not only to providing a physical understanding, but also to answering the practical questions which arise and exploring theoretical notions which present themselves in the study of electrical machines. But students must also learn to be selective and for example, the chapter on Windings, which is there for those who want it or need it, does not require study in depth for an adequate understanding of machine performance and equations. Some guidance on selecting a short course for study is given in the Preface for the second edition.

PREFACE TO THIRD EDITION

The diagram on the front cover of the book indicates in a schematic way, the external and internal viewpoints of an electrical machine. "Access" to the machine is gained via external "terminals", through which it is connected as a link between the electrical and mechanical systems, and transfer of power between the systems is regulated. Internally, the required reactions are developed electromagnetically and inevitably power losses are involved in the conversion process. These must be minimised. Since the majority of students will eventually be involved with electrical machines only as devices to be used when required, some emphasis is given to the terminal characteristics which show for which application they are suited. An additional Appendix has been included to present, concisely, the factors which govern the choice of machine for a particular drive and the drive requirements which may have to be met.

The text is updated generally, new developments are put in context and a few more illustrations, problems and references for further reading are added. Opportunity has been taken to scrutinise the text carefully so that any ambiguities and errors could be removed. The author is grateful both to those who have spoken or written to him to point these out, and to his colleagues and students at UMIST who patiently allow themselves to be used as sounding boards in trying out new explanations and attempts to elucidate obscure points.

The University of Manchester Institute of Science and Technology J. H.

PREFACE TO THE SECOND EDITION

SEVERAL changes have been made in this revised edition though the general pattern remains the same. More emphasis is given to considering the machine as an element in an electrical circuit and as a link between electrical and mechanical systems. In this way it is intended that the special terminal characteristics of a machine as an active circuit element will follow naturally from an extension of simple circuit theory. The phasor diagrams have been redrawn, so that together with the equivalent circuits they can be seen clearly as alternative methods of presenting and remembering the machine equations. Other conventions used in the drawing of phasor diagrams are reviewed briefly in Appendix A.

To emphasise further the common principles of machine operation, two new approaches are explained which derive the equivalent circuits from the basic equations of the transformer. The first method, given in Section 3.9, is closely related to the electrical circuit, electromagnetic and physical concepts. The alternative method,[1] explained in Appendix B, is based on feedback-control principles which are first outlined briefly and then used to consider the machine as a simple closed-loop system with e.m.f. and m.m.f. feedback. For the reader with some prior knowledge of electrical machines, or one who wishes to follow a short, concentrated course, either of these two approaches can be used as a starting point since the first three chapters are of a revisory nature. Subsequently, the remainder of such a short course could be based on the derivation and application of the machine equations given in Sections 4.2, 6.7, 7.3, 8.3 and 8.4, supplemented if necessary by the earlier sections of these chapters and the later sections of Chapter 5. This procedure might well be suitable for those who are not specialising in electrical machines and applications, but who are taking one of the many Electronic Engineering courses now available. For those who wish to go further and acquaint themselves with a more comprehensive analytical approach, a new chapter has been added to introduce generalised theory. In Chapter 10, this powerful method, which requires a sound mathematical background for successful application, is explained in sufficient detail to relate it to the equivalent-circuit approach as adopted in the main text and to indicate its appropriate fields of application.

The full text covers the needs of those whose training requires a deep understanding of all aspects of machine behaviour, as in studies of power systems, control systems, machine design and general industrial applications. The text reviews the latest developments, including the new modes of operation made available by the use of combined semiconductor/machine circuits and the first practical application of superconducting techniques to electrical machines. The work on electromechanical transients has been extended to show some simple practical examples involving the solution of linear and non-linear differential equations. Finally, more information relevant to the electrical machines laboratory course is included, to complete the comprehensive treatment of the subject which is the object of the book.

Much of the new material and modifications to the presentation have resulted from the author's discussions with his colleagues. To Dr. N. N. Hancock in particular, for his co-operation during the preparation of Chapter 10, and to Dr. B. J. Chalmers, the author records his acknowledgements and appreciation.

J. H.

Sale, Cheshire

PREFACE TO THE FIRST EDITION

THE subject of Electrical Machinery, which once formed the core of the Electrical Engineering syllabus, must now constitute a more modest portion of the course so that more time may be devoted to covering the vast range of developments in other fields. The situation is not without advantages. It has stimulated thought on means of rationalising the theory to emphasise the unity underlying the behaviour of the numerous different types of machine. For those with a flair for mathematics, generalised treatments have been and are being developed which are very satisfying in their results. By considering the electrical machine as a combination of inductively coupled coils, the interconnection of which determines the characteristic machine behaviour, an all-embracing theory has been approached. This has been made possible by the work of Gabriel Kron who first developed and applied the techniques of Tensor Analysis to electrical circuits and machines. One book on this basis has already been published in this series.[2]

However, the problem still remains of conveying in a manner more concise than previously, the essential details concerning the physical nature of machines. Further, there are many practical problems which will have to be faced by the electrical student in the laboratory, by the applications engineer and by the machines specialist for which the physical approach is more suitable.

In this volume an attempt has been made to survey practically the whole field of Electrical Machinery, without recourse to advanced mathematics but with the intention of supporting any mathematical generalised treatment. The method adopted is to give most space to a careful explanation of four main electromagnetic devices in their basic form, viz. the two-coil transformer, the d.c. machine, the induction machine and the synchronous machine. A sound knowledge of these four will permit the ready understanding of all other machines, which for the most part are modifications of one or other type. Although

xiii

there are distinguishing features between the main kinds of machine there are many similarities too; these are brought out wherever possible, and the treatment of each machine follows similar lines where this is practicable. The theory is extended to introduce other modes of operation in a manner adequate for general understanding but without detailed explanation in all cases. The text is supported by nearly fifty comprehensive worked examples which cover most of the theoretical points raised.

The first chapter is a brief review of the common principles underlying machine operation and the application of these principles to produce the different machine types. A minimal knowledge of electrical machinery is assumed. However, Chapters 2 and 3 revise the appropriate magnetic and circuit theory so that to some extent the book is self-contained. These two chapters can be skipped by the student who feels secure enough on their subject matter, but they will be referred to frequently either implicitly or explicitly in the later chapters. Towards the end of Chapter 3, a concentrated review of more advanced work is given which necessitates the use of simple differential equations. For example, transient behaviour is discussed and a few elementary problems are considered in the later text. The ideas which form the basis of generalised treatments are also explained briefly in order that reference may be made in Chapter 9 to certain possible uses of the Generalised Laboratory Machine.

Since machine windings are the means by which theory is translated into practice, it has been considered necessary to devote one short chapter to them. In themselves they form a unifying link between different machines. Although the treatment has been simplified, it is adequate for straightforward winding problems, and the material is useful support for the remaining chapters.

Following the bulk of the text which is devoted to the main machine types, the final chapter serves to consider the general relationships obtaining when both commutator and slip rings are provided on the armature. Practical applications of these principles to various commutator machines are described, together with a discussion of the recently available generalised laboratory machines.

Although work for the final year of Electrical Engineering Degree courses is introduced, the main intention of the book is to meet the requirements of all courses dealing with Electrical Machinery up to this stage, where students take up a few specialised studies. However, the Machines portions of the Electrical Power and Machines syllabus for the External B.Sc.(Eng.) degree of London University and for Higher National Certificate and equivalent courses are all covered.

The author is grateful to his colleagues in the Department of Electrical Engineering for the assistance received in various ways when preparing the text. He is particularly indebted to Dr. N. N. Hancock, F.I.E.E., who gave generously of his time to discuss and clarify many aspects of machine theory. Much helpful advice was received from Professor P. Hammond, M.A., F.I.E.E., the consulting editor, and Mr. G. E. Middleton, M.A., F.I.E.E., read through the manuscript and made many useful suggestions. Various firms have taken considerable trouble to find suitable photographs and these are acknowledged where they appear in the book. Figure 2.7 was copied from a flux plot kindly loaned by Mr. F. J. Pepworth, M.I.E.E., who had spent many hours in its preparation.

J. H.

*Manchester College of Science
and Technology*

CONTENTS

CONTENTS

LIST OF SYMBOLS

THE following list comprises those symbols which are used fairly frequently throughout the text. Other symbols which are confined to certain sections of the book and those which are in general use are not included, e.g. the circuit symbols like R for resistance and the use of A, B and C for 3-phase quantities. Some symbols are used for more than one quantity as indicated in the list. With few exceptions, the symbols conform to those recommended by the British Standards Institution BS 1991.

Instantaneous values are given small letters, e.g. e, i, for e.m.f. and current respectively.

R.M.S. and steady d.c. values are given capital letters, e.g. E, I.

Maximum values are written thus: $\hat{E}$, $\hat{I}$.

Bold face type is used for phasor and vector quantities, e.g. **E, I.**

In general, the symbol E (e) is used for induced e.m.f.s due to mutual flux and the symbol V (v) is used for terminal voltages.

a	Number of pairs of parallel circuits in machine winding.
At	Ampere turns—equivalent to amperes enclosed.
B	Flux density, in teslas (webers/metre^2).
C	Number of coils or commutator bars.
d	Symbol for direct-axis quantities.
d	Armature diameter, in metres.
e	Base of natural logarithms.
E_f	Induced e.m.f. due to field m.m.f. F_f.
f	Frequency, in hertz (Hz) (cycles per second).
F	Magnetomotive force (m.m.f.) in ampere turns. Peak m.m.f. per pole per phase.
F_a'	Effective d.c. armature-winding magnetising m.m.f. per pole.
F_a	Peak armature-winding m.m.f. per pole.
F_f	Peak field-winding m.m.f. per pole.
	(Note that the suffices a and f are also used with the symbols for currents, fluxes and resistances of armature and field respectively.)

F_r Peak resultant m.m.f. per pole.

$I_{f.l.}$ Full-load current.

I_0 Current in magnetising branch.

I_p Power component of I_0.

I_m Reactive or magnetising component of I_0.

J Polar moment of inertia (rotational inertia), in kg m^2.

k Coefficient of coupling. A constant.

k_{pn} Coil-pitch factor for the nth harmonic.

k_{dn} Distribution factor for the nth harmonic.

k_f Generated volts per field ampere or per unit of m.m.f.

k_{fs} Saturated value of k_f.

k_ϕ Flux factor; generated volts per radian/sec or torque per ampere.

l Conductor length. Magnetic path length.

l or l_1, l_2, etc., leakage inductance.

L General inductance symbol; e.g. L_{11} = self-inductance of coil 1; L_{12}, L_{13}, etc., for mutual inductances.

m Number of phases.

M Alternative mutual-inductance symbol for two coils.

n Rev/sec. n_s Rev/sec synchronous   = f/p.

N Number of turns. Rev/min. N_s Rev/min synchronous = $60f/p$.

p   Operator d/dt.

p Number of pole pairs.

p.u.   Suffix for per-unit quantities.

P Power per phase.

P_g (or P) Air-gap power per phase.

q Slots per pole per phase.

q   Symbol for quadrature-axis quantities.

Q Slots per pole.

R_m Magnetising resistance, representing iron losses.

s Fractional slip = $(n_s - n)/n_s$.

S Per-unit relative motion n/n_s (= $1-s$). Number of slots.

$T_{coupling}$ Torque acting at mechanical shaft coupling.

T_e Torque developed electromagnetically, in newton metres.

T_{loss} Sum of all mechanical, internal loss-torques.

T_m Torque arising mechanically = T_e in steady state.

v Velocity, in metres per second.

V Voltage measured at the terminals of a circuit or machine.

x (or x_1, x_2, x_{a1} etc.) Leakage reactance.

X General reactance symbol.

X_m Magnetising reactance.

X_{ms} Saturated value of X_m. X_{mu} unsaturated value of X_m.

X_s Synchronous reactance $= X_m + x_{a1}$.

y_c Coil-end pitch, measured in commutator bars where appropriate.

z_s Number of series-connected conductors per phase or per parallel path of a winding.

Z Total number of armature conductors.

Z_s Synchronous impedance.

α General angle. Slot angle. Impedance angle $\tan^{-1} R/X$.

β General angle. Chording angle.

δ Load angle. δ_{fa} torque angle.

η Efficiency.

λ Flux linkage, in weber turns.

Λ Magnetic permeance, webers/ampere-turn.

μ_0 Magnetic constant $= 4\pi/10^7$.

μ_r Relative permeability.

μ Absolute permeability $= B/H = \mu_0 \mu_r$.

φ Power-factor angle. *N.B. This must be distinguished from the symbol for flux ϕ below.

ϕ Instantaneous value of flux. Flux per pole, in webers.

ϕ_m Mutual flux, in webers, due to resultant m.m.f.

$\boldsymbol{\Phi'}$ Flux space-phasor.

$\boldsymbol{\Phi}$ Flux time-phasor.

θ Shaft angular position. Temperature rise. General variable.

τ Time constant.

ω Angular velocity of rotating time-phasors $= 2\pi f$ radians/sec.

ω_m Mechanical angular rotational velocity $= 2\pi n$ radians/sec.

ω_s Synchronous angular velocity $= 2\pi n_s = 2\pi f/p$ radians/sec.

Note: SI units (Système International d'Unités) are used in the text unless specifically stated otherwise.

CHAPTER 1

INTRODUCTION AND BASIC IDEAS

1.1 AIM OF THE BOOK

This book is intended to explain the behaviour of certain electro-magnetic devices which are able to convert power from one form to another. Electrical, mechanical or both forms of energy may appear in either input or output. Energy may be stored and recovered during transient processes, in both the magnetic field and in the mechanical inertia. An appreciable amount of power will be converted to heat and lost; this will be discussed when dealing with losses. Trivial amounts of power will be converted to other forms but will be neglected here, not-withstanding their importance in other contexts, e.g. acoustic noise and radio noise. Rotation will be involved in the conversion process with the exception of the transformer, but the title Electrical Machines will be taken to cover this also.

A remarkable number of different machines has been devised, and as many different methods of analysis have grown up around them. Until quite recently it was the custom to discuss each machine separately as if it had a unique existence, but the whole subject has become so vast in content that to teach it on this basis is unsatisfactory. It is possible with some thought, to see in Electrical Machines many common features, and while it is erroneous to pretend that all machines are really the same, much can be done to economise in general ideas. The present chapter is an attempt to bring out certain points of similarity.

At this stage, certain liberties as to the extent of the reader's know-ledge will be taken. For example, it will be assumed that the electro-magnetic principles governing the operation of Electrical Machines are not entirely unfamiliar. They will be discussed however, in Chapters 2

1

and 3 with a view to emphasising their relative importance and consolidating a suitable foundation. Certain general ideas which are also discussed in these earlier chapters will be elaborated in the later text, so it is suggested that there should be no undue delay at this stage if a difficulty is encountered. The abilities of students to absorb generalities are best judged by the individual lecturer concerned and he may therefore wish either to defer some of this material or to extend it.

1.2 GENERATION OF ELECTRO-MOTIVE FORCE

A voltage may be generated in various ways, but for present purposes, only the e.m.f. of electromagnetic induction is of importance. One way of expressing the result of Faraday's well-known experiment is by means of the equation:

$$e = N(\mathrm{d}\phi/\mathrm{d}t) \text{ volts.} \qquad (1.1)$$

Here is introduced the concept of magnetic flux lines, ϕ webers, and their linkage with the N turns of a coil. Not all of the flux lines link all of the turns and the meaning assigned to ϕ is given by the equation $\phi = \lambda/N$ where λ is termed the flux linkage. λ is obtained by considering the number of turns encircled by each flux line or group of lines having the same turn linkage. Examples will be considered in Chapter 2, but the point is demonstrated for a simple case in Fig. 1.1. The total flux linkage is $4 \times 3 + 2 \times 1 = 14$. Hence ϕ, the mean flux per turn, is 14/3.

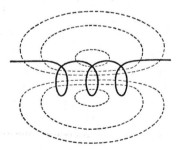

FIG. 1.1. Flux linkage with 3-turn coil.

2

The direction of the induced e.m.f. is such that if the coil circuit was closed, it would tend to drive a current giving a magnetic effect opposing the change of flux. This is sometimes referred to as Lenz's law, but in fact the alternative would be unthinkable in a stable universe; if the induced current supported the flux change, both e.m.f. and current would increase indefinitely. The negative sign often used in eqn. (1.1) indicates this opposition but it will be found more convenient when writing down the circuit equations, to use a positive sign and treat this voltage as a *back* e.m.f. opposing the increase of current, not as a *forward* e.m.f. This avoids confusion when drawing phasor diagrams.

Electrical machines are provided with coils, and flux changes are provoked by one means or another to produce an e.m.f. Consider, for example, a single-turn coil placed as in Fig. 1.2a. In this position, it embraces the whole of the flux $+\phi$ entering the south pole. If, with the coil stationary, the poles are magnetised by alternating current such that after a time t the flux reverses to $-\phi$, the coil will experience a total change of 2ϕ webers and there will be an induced e.m.f. of average value, $1 \times 2\phi/t$ volts. The actual e.m.f. at any particular instant will depend on the time variation of ϕ. If, for example, $\phi = \hat{\phi} \sin 2\pi ft$, where $2\pi f = \omega$ radians per second, then the instantaneous e.m.f. $e = N.d\phi/dt = 1.(2\pi f\hat{\phi} \cos 2\pi ft)$, which is another sinusoidal function advanced in time by $1/4f$ sec. Since this e.m.f. could provide electrical power in the coil circuit, there could be an electrical/electrical power "conversion" through the medium of the magnetic field, the input power coming from the a.c. source magnetising the poles. This is called *transformer* action, and the resulting e.m.f. a *transformer e.m.f.*

If the poles are permanent magnets or d.c. excited, they require no electrical input other than that to sustain the I^2R loss in the exciting coils if provided. However, it is still possible, even though the flux is constant in time, to induce an e.m.f. If the coil is moved to the position of Fig. 1.2b in a time t seconds, it will now embrace a flux $-\phi$. The flux has changed, as far as the coil is concerned, from $+\phi$, through zero when the coil sides are midway under the poles and therefore there is no net flux linkage, to $-\phi$. As before, the average e.m.f. during the interval will be $2\phi/t$ volts, though it has been produced by different

3

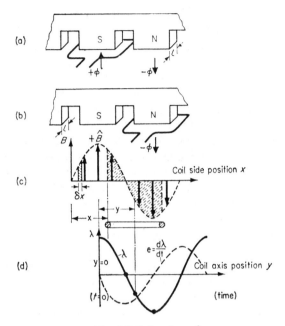

FIG. 1.2. Induced e.m.f.

means. It is a *motional e.m.f.* and motion involves mechanical energy in the conversion process.

The instantaneous e.m.f. this time will depend on the way in which the flux is distributed in space since this stationary distribution is "seen" by the moving coil as a time variation. Figure 1.2c shows a possible case where the spatial distribution of flux density B is assumed to be sinusoidal. Considering an incremental area bounded by the pole length l and a small distance δx, the flux passing through the area $l.\delta x$ will be $B.l.\delta x$ which is proportional to the area under the B curve over distance δx. For any coil position then, the total flux through the coil is proportional to the area under the B curve within the coil sides. This is indicated by shading for a particular case. When the coil axis is midway between the poles the area will have equal

4

positive and negative halves, flux entering into, and emerging from the top of the coil without linking it. With the coil centrally placed about any pole, the whole of the pole flux passes through, thus giving maximum linkage. Salient points on a flux linkage curve can therefore be plotted. Since they are derived from the area under the sinusoidal flux-density curve, they will themselves lie on a sine curve as shown by Fig. 1.2d. Although this is a spatial variation, it can also represent the time variation of flux linkage when "viewed" from a coil moving with uniform velocity. On Fig. 1.2d, $t = 0$ has been taken to correspond with the instant when the coil axis is under the middle of a south pole, i.e. $x = 0$ and $y = 0$. The coil e.m.f. can therefore be obtained by differentiating this curve, giving another sine wave which rises and falls in phase with the B variation as "seen" by the coil sides, e.g. at $t = 0$ the coil sides are under zero flux density and e is also zero.

The same average e.m.f. $2\phi/t$, could have been produced if, with the coil stationary, the whole pole system had been moved one pole pitch to the left in a time t sec. Viewed from the coil, the flux changes would have been identical. One could think of many interesting combinations of pole motion and coil motion which would give the same e.m.f. The important thing to realise is that as far as the coil is concerned, if at one instant of time t_1, it is linked by a flux ϕ_1 webers, and at a later instant t_2 it is linked by ϕ_2 webers, due account being taken of the sense or sign of the flux, then an average e.m.f. of value $N(\phi_2 - \phi_1)/(t_2 - t_1)$ volts will be induced during the time interval. Whether the coil moves is not the issue, it must just experience a change of flux. If, for example, the coil was moving, but the poles too were moving at the same speed and in the same direction, no e.m.f. would be induced; it is just as if they were both stationary, as in fact they are, *relative to one another*.

Although it may be useful to distinguish between a *transformer e.m.f.* and a *motional e.m.f.* because of power conversion considerations, the general statement is still true, that the total e.m.f. induced is due to the total flux change viewed from the coil. On some a.c. machines, both e.m.f. components are present since the coils are moving through a pulsating flux.

It is convenient to use an alternative expression for induced e.m.f. when relative motion is involved. In the normal machine the conductors of the coils are constrained so that they move at right angles to the field. Figure 1.3 shows the plan view of a coil, the two sides of which are moving in flux densities B_1 and B_2 teslas respectively. If in a time δt seconds, a coil side of length l metres moves a distance δx metres, in a

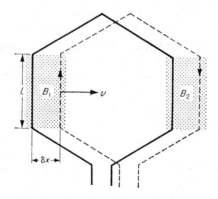

FIG. 1.3. *Blv* formula.

direction perpendicular to a field of B teslas, the change of flux linkage due to the movement of this coil side is:

$$\text{flux density} \times \text{area change} = B.l.\delta x \text{ webers.}$$

The e.m.f. induced is therefore $B.l.dx/dt = Blv$ where v is the velocity in metres/sec. The total coil e.m.f. due to the movement of the two coil sides will be $B_1lv + B_2lv$. For a full pitch coil as in Fig. 1.2, $B_2 = -B_1$, giving e.m.f.s in opposite senses. However, the end connections result in the two conductor voltages being additive round the coil. The total e.m.f. is $2Blv$ where $B = B_1 = B_2$ in magnitude. The expression $N(d\phi/dt)$ applied to the whole coil linkage would yield the same result. All that has been done here is to express the flux change in terms of flux density and velocity. The e.m.f. is thus given at any instant of time corresponding to a particular position in space and is in

6

phase with the B variation, e.g., see Fig. 1.2d; when $x = 0$, $B = 0$, $y = 0$ and $e = 0$.

An alternative derivation of the Blv formula can be obtained from consideration of the magnetic force on the loosely bound conductor electrons moving with velocity v through the field B. The electrons are displaced along the conductor until the electrostatic attractive force following from the charge separation balances the magnetic force. An electric field is therefore established along the conductor having a magnitude of Bv volts per metre. This expression is convenient when explaining the theory underlying the attempts to exploit Magneto-hydrodynamic Power Generation (MHD). Here, a hot conductive gas is forced at high velocity v, through a magnetic field B. A transverse electric field, Bv volts per metre is established due to the charge separation. Electrical power can thus be extracted from electrodes suitably placed in this electric field, the power being converted directly from the heat in the gas, thus avoiding the wasteful losses in the turbine.

The generation of e.m.f. is not just associated with a generator and an electrical output, it occurs whenever there is a change of flux linked with a coil. In a motor, it is usually referred to as a back e.m.f. since it opposes the flow of current.

1.3 PRODUCTION OF TORQUE

A second electromagnetic effect will now be considered with reference to Fig. 1.4. Although this figure could also be used to explain transformer and motional e.m.f.s induced in the rotor coil, as for Fig. 1.2, the emphasis now will be on the forces produced. The centrally pivoted rotor magnet positioned as in Fig. 1.4a, will not have any tendency to turn. If it was rotated through 180°, this statement would still be true but only theoretically, the equilibrium being unstable. A slight vibration would be sufficient to swing the magnet back to the original position. Suppose now that there is an angular displacement δ (delta), between the two magnetic axes, Fig. 1.4b. As indicated by the arrows, and also by the field lines round the conductors, forces will be exerted both attractive and repulsive tending to bring the magnets into align-

7

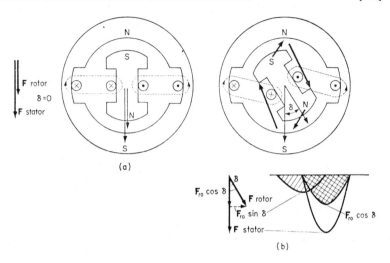

FIG. 1.4. Production of torque.

ment. The total force and the corresponding turning moment or torque will vary with the angular position, being a maximum when $\delta = 90°$. Coil excited magnets have been indicated but the same effects would be experienced with permanent magnets. In either case, it is possible to consider the force as due to the mutual reaction of currents; a fundamental approach, or due to the mutual reactions of the poles, which is perhaps simpler to understand; see also Fig. 2.13. Normal machines are provided with conductors and coils on at least one member, the armature, in which the motional voltage is produced. The force is then easier to calculate using the concept of magnetic flux. The force on a current-carrying conductor in, and perpendicular to a magnetic field B is found by experiment to be Bli newtons, i being the current in amperes.

It is customary to indicate the magnetic axis of each member by an arrow like a vector. The two "vectors" are in the same sense for the position of stable equilibrium and displaced by angle δ for any other position, see Fig. 1.4. If the spatial flux density due to either member acting alone is distributed sinusoidally, cf. Section 5.5, the "vector" for the inner member say, can be resolved along and in quadrature with the other "vector". The quadrature component, $F_{\text{rotor}} \sin \delta_1$ gives rise to a tangential force and torque. The in-phase component, $F_{\text{rotor}} \cos \delta_2$, will give rise to opposing radial forces, equal in magnitude on opposite sides of the rotor if this is perfectly central, otherwise unequal and causing an unbalanced magnetic pull. When $\delta = 0$, there is no tangential component.

If the inner magnet were to be released from an angular position δ, the resulting motion would be short lived. A revolving machine must be arranged to preserve the angular displacement between the two axes of magnetisation even when rotation takes place; thus the force will be continually exerted and the movement will persist. There are several ways of achieving this arrangement practically.

The Direct-current Machine

Consider Fig. 1.5; wire is wound on the iron ring, which is free to rotate within the field of the stationary magnet. The rotor here

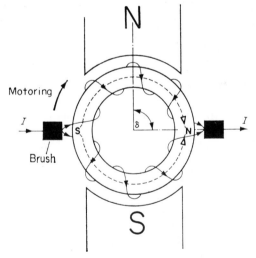

Fig. 1.5. Simple d.c. machine.

is the armature and the stator is the field system. Small stationary conducting blocks (brushes) make sliding contact with each turn of wire in sequence. Direct current fed into one brush and out of the other is divided in parallel paths on either side of the brush axis. This division occurs whether or not the coils are moving. The magnetic field established by these rotor currents gives rise to polarities on the surface as shown, irrespective of the rotational speed. The torque and motion will be clockwise as can be seen by considering the attractive and repulsive forces on the rotor poles. Miniature battery operated motors with many more turns of wire have in fact been made this way. It is one of the simplest forms of *d.c. machine*, the characteristic feature of which is the unidirectional rotor magnetomotive force. On the normal machine the axis of this m.m.f. is in line with the brush axis, not being disturbed substantially by rotor movement nor by load changes. The brushes are usually set for maximum torque, i.e. on the *quadrature axis* which is at right-angles to the field axis (*direct axis*). δ is then equal to 90°.

The Synchronous Machine

It is not essential for torque production, that the stator and rotor magnetic axes are stationary in space. For example, considering Fig. 1.5, if the stator magnets and the brushes, still d.c. supplied, were rotated together at say *n* revolutions per second, the torque would be unchanged; between the two magnetic axes there would still be no *relative* motion. Although this arrangement has no practical application, a similar situation arises in the operation of the *synchronous machine*. As will be explained more fully in Chapter 5, when polyphase currents are supplied to a polyphase winding, a travelling magnetic field is produced due to time variations of flux density at different points in space. Figure 1.6a shows the **B** vectors at various points around the developed periphery of such a winding. Figure 1.6b shows the vectors at a later instant. The envelope is taken to be sinusoidal, a close approximation to the practical case. It can be seen that the **B** vectors pulsate in such a manner as to cause movement of the envelope relative to the winding. With the coils arranged on the stator say, for 2*p* poles,

10

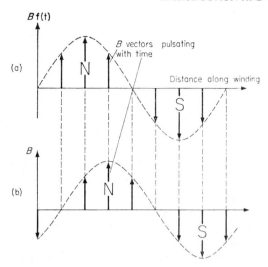

FIG. 1.6. Travelling flux wave.

and supplied at a frequency of f cycles per second, there will be rotation of the pole- and resultant m.m.f.-axis at a speed $n_s = f/p$ rev/sec (cf. p. 258). n_s is called the synchronous speed, and the arrangement is virtually equivalent to rotating the stator magnets of Fig. 1.5.

On the rotor is now provided a field system of north and south poles, $2p$ altogether, Fig. 1.7. If, and only if the rotor is moving at the same speed and in the same direction as the rotating field of the stator, a unidirectional force is developed between the stator and rotor. There will be no relative motion between the magnetic axes which are therefore at some fixed angular displacement δ'. The tangential component of the force between the poles could be sufficient to sustain the rotation once the rotor had been brought up to synchronous speed. The characteristic of the synchronous machine is the fixed speed determined by the supply frequency and the number of poles. Unlike the d.c. machine, the angular displacement between the m.m.f. axes is not constant but must increase with load from $\delta = 0°$ when unloaded, to $\delta = 90°$ approximately, when developing maximum force, beyond which synchronism can be lost.

11

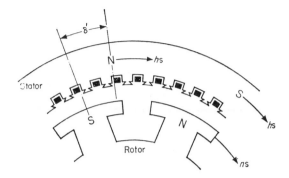

FIG. 1.7. Synchronous machine (motoring).

It can be noted, that if the rotor poles were initially unmagnetised but were brought up to synchronous speed, they would then be magnetised by induction and the rotation could still be maintained. This action is the basis of the *reluctance motor*, used for example, on electric clocks.

The Asynchronous or Induction Machine

The third main type of machine also utilises this property of polyphase windings and currents, the stator being basically the same as for the synchronous machine. Now, however, the rotor too is provided with a polyphase winding supplied with polyphase currents either inductively, conductively or by both means. There are thus two rotating fields, each produced by polyphase currents. In the case of the rotor, normally moving, its field speed in space is made up of its field speed relative to the rotor winding, plus the physical speed of the rotor and winding with respect to the stator, Fig. 1.8. If the machine is to be self-maintaining in rotation, the condition already stated must be fulfilled, i.e. the stator and rotor magnetic axes must move at the same speed and in the same direction. For example, if the stator field speed relative to a fixed point in space is n_s and the physical speed of the rotor with respect to the same point is n, then the motion of the rotor field with

12

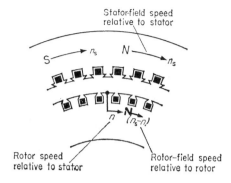

Stator-field speed
relative to stator

Rotor speed
relative to stator

Rotor–field speed
relative to rotor

FIG. 1.8. Induction machine (motoring).

respect to the rotor, must make up the difference between n_s and n, i.e. $(n_s - n)$ so that the two magnetic axes are kept in step. Under these circumstances, it can be appreciated that a unidirectional force will exist between stator and rotor which may be large enough to overcome the resisting forces and maintain the rotation at say n rev/sec. This action is the basis of the *induction motor*. The rotor winding must, of course, be arranged for the same number of poles as the stator winding and for the correct field speed, the rotor currents must be of a frequency $(n_s - n)/n_s$ times the stator frequency. In fact, this occurs automatically as will be seen (p. 373) so that the rotor speed in space, n, may have any value in simple theory and yet its m.m.f. axis will still rotate in space at synchronous speed n_s. This independence of the rotor field speed from the speed of the rotor itself is shared with the d.c. machine though in fact the vast majority of induction machines operate at a substantially constant speed which is very nearly equal to n_s. The angular displacement of the m.m.f. axes varies with load as on the synchronous machine.

The distinguishing feature of the induction machine is the dual nature of power conversion associated with the a.c. magnetisation of the field. For the d.c. and synchronous machines on steady state, the rotor motion causes only electrical/mechanical or mechanical/electrical power conversion. On the induction machine, the effect of rotor motion is to divide the total power converted, partly into electrical and partly

13

into mechanical power. For example there can be no motional e.m.f. at standstill with the stator and rotor coils at rest with respect to one another so there is no electrical/mechanical conversion. The pulsation of B causes a transformer e.m.f. and electrical/electrical "conversion". As the rotor speed increases from standstill in the same direction as the rotating-field axis, the pulsations of B in particular spatial positions at synchronous frequency are experienced by any rotor coil at a rate reduced in proportion to the increase of speed n. The standstill transformer e.m.f. due to pulsation is thus reduced by a factor $(n_s - n)/n_s$. This factor is called the slip s and has a value of unity at standstill when $n = 0$ and is zero when $n = n_s$. Thus at synchronous speed the induced e.m.f. would be zero, the rotor motion coinciding with the pulsation and the rotor coils being stationary with respect to the flux wave. It is perhaps worth noting, that if a d.c. current was now fed to the rotor, establishing a rotor field, stationary with respect to the rotor, the machine would operate in a synchronous mode; see pp. 373 and 537.

The changes resulting from rotor movement can be conceived as being due to the presence of a motional e.m.f. superimposed on and reducing the induced e.m.f. and current from the standstill values. With motion, the rotor conductors are as it were, cutting the pulsating **B** vectors at a rate proportional to the speed $n = (1-s)n_s$. This motional e.m.f. component, proportional to $(1-s)$, in conjunction with the rotor current due to the transformer e.m.f. which is proportional to s, gives rise to electromechanical conversion at all speeds other than synchronous and standstill where this motional e.m.f. × current product is zero. This viewpoint is discussed again on p. 379.

Although torque has been discussed with reference to motors, it occurs whenever a current-carrying conducting material is suitably arranged in a magnetic field. Electromagnetically produced torque therefore, arises in a generator too. In this case it provides the resisting force against which the mechanically applied force, developed by the so-called *prime mover*, does work, to convert mechanical to electrical power.

It should be pointed out that torque can be produced whether or not the machine is rotating but of course no power is developed until the associated force is allowed to do work by acting through a distance. The horse-power using mechanical engineers' units is given by the expression:

$$\text{hp} = (2\pi \times \text{rev/sec} \times \text{torque in lbf ft})/550 = \omega T'/550.$$

The power in watts would be $746 \times \text{hp} = \omega_m T'$. $746/550 = \omega_m T$. The SI unit T is the newton metre (Nm), and this expression shows that the lbf ft is $1 \cdot 36$ times the newton metre. Throughout the book, SI units must be presumed unless specifically stated otherwise. However, since machines are associated with mechanical energy, occasional use will be made, as in engineering practice, of the following units: horse-power, pound, foot-pound, pound-foot, and rev/sec or rev/min.

1.4 TORQUE ANGLE (LOAD ANGLE)

In the following discussion the stator and rotor m.m.f.s will be assumed sinusoidally distributed in space as for Fig. 1.6, though this is only a rough approximation for the normal d.c. machine. If the air gap between stator and rotor is assumed uniform round the periphery and if in addition the iron paths are taken to have negligible reluctance, either m.m.f. component if acting alone would produce a sinusoidal flux density distribution in the gap. With both windings excited, the field m.m.f. F_f must be combined with the armature m.m.f. F_a at every point in the air gap, to give the resultant m.m.f. F_r which determines the air-gap flux and distribution. For the conditions stated, this will also be sinusoidal. Figure 1.9 shows how the resultant can be obtained using a vector-type summation; the "vectors" representing peak values of m.m.f. Neglecting the effect of iron losses, the peak of gap flux density B_r is proportional to and in space phase with F_r but in general is out of phase with and not proportional to either F_f or F_a. Under the stated conditions, the gap flux density could have been obtained by vector summation of the component densities B_f and B_a. Only B_r actually exists in the gap but superposition of fluxes

15

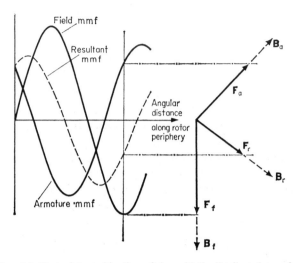

FIG. 1.9. Vectorial combination of sinusoidally distributed m.m.f.s.

may be more convenient and is a valid procedure if the magnetic circuit is linear, i.e. magnetic permeance constant; see Section 3.3. For most machines there is more or less distortion from pure sinusoidal form, in which case the waves could be resolved into fundamentals and harmonics, though this is not necessarily the simplest solution. If for particular studies, these space harmonics can be neglected, then vector combination can be applied to the fundamentals only.

For the three main types of machine discussed in Section 1.3, the torque is proportional to the sum of all the *Bli* forces. *l* is constant, but *B* and *i* could vary for each conductor. A convenient way of taking this variation into account is to consider the overall effect of the whole winding in terms of either the m.m.f. or flux distribution produced. This method, already used in effect when dealing with Fig. 1.4, brings out the concept of the angular displacement δ between the peaks of either the m.m.f. or flux-density waves. With the idealised assumptions made at the beginning of this section, the different machines can now be compared on the same basis. Remembering that it is the component

16

of one "vector" resolved in space quadrature with the other which is responsible for the tangential force, and with reference to Figs. 1.4 and 1.10,

$$\text{torque } T \propto F_{\text{stator}} . F_{\text{rotor}} . \sin \delta.$$

With regard to the angular displacement, there are now three m.m.f. axes to consider if that of the resultant $\mathbf{F_r}$ is included. For convenience and by convention, different pairs of axes are chosen for the different machines.

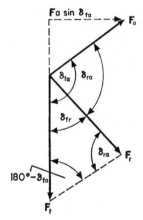

FIG. 1.10. Torque angle.

For the d.c. machine, the obvious choice (see p. 317) is to consider the field-pole axis and the brush axis, i.e. using F_f and F_a so that:

$$T \propto F_f . F_a . \sin \delta_{fa}. \tag{1.2a}$$

But from the sine rule applied to Fig. 1.10:

$$\frac{F_f}{\sin \delta_{ra}} = \frac{F_a}{\sin \delta_{fr}} = \frac{F_r}{\sin(180° - \delta_{fa})} = \frac{F_r}{\sin \delta_{fa}},$$

so that $F_a \sin \delta_{fa} = F_r \sin \delta_{fr}$. Hence the torque is also given as

$$T \propto F_f . F_r . \sin \delta_{fr} \tag{1.2b}$$

which is the form preferred for the synchronous machine as will be seen in Chapter 8 (p. 503).

Substituting now for $F_f . \sin \delta_{fr}$ from the sine rule we have:

$$T \propto F_r . F_a . \sin \delta_{ra} \qquad (1.2c)$$

which, considering the rotor winding as the armature, is the form preferred for the induction machine as will be seen in Chapter 7 (p. 383).

The angle δ_{fa} is called the torque angle but really there is a choice; the torque being proportional to the product of *any* two space "vectors" of m.m.f. and the sine of the angle between them. For a.c. machines, the angle is a function of load and the term *load angle* is used to describe the physical shift in the rotor alignment, which can in fact be measured. In terms of eqns. (1.2b) and (1.2c) the load angle is very nearly equal to δ_{fr} or δ_{ra} respectively, though modified slightly by the effect of leakage impedance. It is closely related to the angular displacement between certain time phasors.

The constants of proportionality in eqns. (1.2a), (1.2b) and (1.2c) will appear in eqns. (6.4a), (8.7a) and (7.3) respectively, though it will then be found more convenient to replace the m.m.f.s by related functions of flux, current or voltage.

1.5　MULTIPOLAR MACHINES

The three basic machine types discussed will form the subject matter in rotating machinery and subsequent chapters will deal with their theory and development in greater detail. Although particular windings have been assigned to stator and rotor, this order could be inverted since it is relative motion which determines the electromagnetic performance. However, the arrangements described are the usual ones and Figs. 1.11a, b and c show simplified cross-sectional outlines of the three types in their common form.

It will be noticed that all coils apart from those on salient poles are placed in slots. This gives support to the conductors and relieves pressure on the insulation since most of the tangential force is exerted on the tips of the teeth. The coils span a pole pitch approximately so

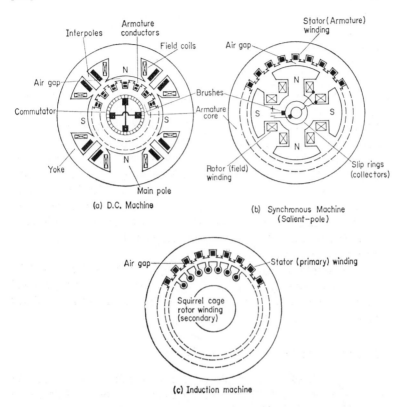

FIG. 1.11. Outline of three main machine types.

that when one coil side is under a north pole say, the other side is under a south pole and their induced e.m.f.s are additive round the coil. Often, each slot has two layers of conductors which makes for convenience in winding, as will be discussed in Chapter 5.

It will also be noticed that a multipolar construction is shown, four poles in this case. For a.c. synchronous and induction machines, the required number of poles is determined by the supply frequency and the operating speed since $n_s = f/p$. For d.c. machines, the number of

19

poles is fixed from other design considerations which lead, in general, to an increase in the number of poles as the armature diameter is increased.

In one revolution of a machine with $2p$ poles there are p complete cycles of flux change, two in the cases shown on Fig. 1.11. If n revolutions take place in one second, there will be $pn = f$ complete cycles per second. Now one cycle corresponds to 360° from the electrical point of view. Therefore in general, the number of electrical degrees is p times the number of mechanical degrees. Electrical angles will always be meant unless otherwise stated; and so a pole pitch corresponds to 180 electrical degrees or π electrical radians.

1.6 HOMOPOLAR MACHINES

Figure 1.12 shows in a schematic fashion, the essential difference between the normal multipolar (heteropolar) construction and the special, homopolar construction; a modification of the Faraday disc machine. It will be noticed that the flux crossing from stator to rotor radially, has the same direction at all points round the periphery for the homopolar machine. This type of magnetic circuit is used occasionally

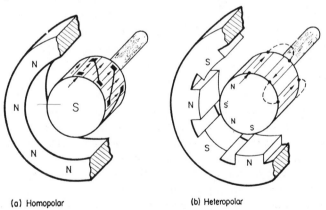

(a) Homopolar   (b) Heteropolar

FIG. 1.12. Alternative machine magnetic circuits.

20

for a.c. high-frequency generation, see p. 546. For d.c. operation, since the e.m.f.s are in the same axial direction at all points round the periphery, a solid, smooth rotor could be used to generate a ripple-free voltage of magnitude Blv, by using pick-up brushes at either end of the rotor. In practice, within the restrictions of normal design limits, this e.m.f. could only be a few volts, though the current rating could be very high.

To increase the voltage, several rotor conductors could be brought out to slip rings and connected in series, externally. Figure 1.12a shows the principle of a recent development,[3, 4] in which the rotor has several segments with one pair of brushes per *alternate* segment, cross connected to put the various e.m.f.s in series. The brushes must be separated circumferentially in this way, to avoid the short circuit which would occur during the period when the brushes bridged two segments, if cross connections were taken to adjacent segments. In practice, to make sure that the intersegment voltage does not exceed Blv at any point round the periphery, there would be a second path, similar to and in parallel with the one shown, but picking up the voltage from the segments on the left-hand half of the periphery.

The value of B for electromagnetic machines is typically of the order of 1 tesla but in this case it can be increased to perhaps 5 teslas if the co-axial field coil on the stator is made of a suitable superconducting alloy and enclosed in a cryogenic (refrigerated) environment. Near the absolute zero of temperature at about $4°K$, the coil has zero resistance so that very large magnetising currents, and, even without iron, very large fluxes can be produced, with zero I^2R loss. Such a superconducting machine, with newly developed, low-loss and low-wear brushes, shows considerable savings in volume over a conventional machine and may prove economically feasible for the larger ratings. Although these machines are very different in appearance and construction from conventional d.c. machines, the performance equations developed in Chapter 6 apply, particularly for the non-cryogenic homopolar machines. However, for superconducting machines, reductions of field current for control purposes need to take into account the very large value of stored field-energy which must be dissipated. A.C. machines are also being developed but at present there are unsolved problems with superconductors carrying alternating current. The d.c.-excited member of such a machine is not of course restricted in this way.

1.7 LINEAR MACHINES

Rotating machines can be built for very high peripheral speeds; some turbo-alternator rotors, for example, have surface speeds of more than 400 m.p.h. This leads to high power-efficiencies, nearly 99%, because voltage and power increase with speed whereas the losses can be made to increase at a lower rate. The electromagnetic principles outlined in this chapter are applicable also to restricted motion or to linear motion, but the efficiencies of such devices are usually low because of the low speed. The power-output/power-input efficiency may not be of prime importance in special applications, however, where the convenience of a linear drive may justify its use.

The common case of linear motion is the solenoid and plunger, a reluctance type of machine, but there are several other linear devices which, in general, give a better, uniform-force characteristic. If a conducting liquid has to be pumped, as in nuclear power plant, using sodium as a coolant, the liquid can form the "rotor" to a developed, or "unrolled" stator, across which the flux wave travels linearly. Sometimes the liquid is used as the "armature" in a d.c.-type conduction pump. In the M.H.D. generator, p. 7, the hot gas moves linearly across the field. The first large-scale linear motor was made for aeroplane launching,[5] but modern practice includes a wide variety of applications, many of them rather unusual. Actuators and conveyors are sometimes powered by linear motors, which would also be considered if it was necessary to produce motion where a mechanical coupling was undesirable. Linear drives have relevance in the developing field of high-speed traction and much experimental work is being conducted in this area. Reference 6 is a useful survey of linear machine applications.

Linear machines are really special cases and require an outlook rather different from rotating machines when considering their design. The concept of "goodness" has been introduced recently,[7] to provide a basis for assessing the best arrangement of such devices to achieve their purpose. It is measured by a "goodness" factor which is essentially the ability to produce maximum current and maximum flux at maximum speed. Consequently, the factor is proportional to the product of electrical conductance, magnetic permeance and supply frequency. Reference 7 also has instructive information on other unusual electromagnetic devices.

The student should not be discouraged if the unifying ideas considered in this chapter are not at first fully understood. They will form the basis of subsequent chapters in which the ideas will be developed more slowly. It should be possible however, to explain why the following statements are true:

The synchronous motor, as such, is not self-starting.

The induction motor, as such, cannot run at synchronous speed.

The d.c. motor must have a switching device to reverse the current in the armature coils in accordance with their position relative to the field system.

CHAPTER 2

THE MAGNETIC ASPECT

2.1 THE MAGNETISATION CURVE

For a point in a magnetic circuit, the relationship between the flux density B in teslas (wb/m^2) and the magnetising force H in amperes per metre is given by:

$$B = \mu_0 \mu_r H \tag{2.1}$$

where $\mu_0 = 4\pi.10^{-7}$ and μ_r is the relative permeability. In an electrical machine the flux paths are either in air, hydrogen, water, insulation or other non-magnetic material, or else in iron for which μ_r varies with H and may have a value of a thousand or more.

The different behaviour in iron is due to the uncompensated spin of the electrons within the iron atoms. The effect is just as if the iron contained many current loops. In the unmagnetised state the loops are so oriented that their net magnetising action is zero. The iron crystals are divided into small domains, less than one millimetre in size, for each of which the effective current loops are aligned, giving a definite direction to the domain magnetic axis. This will correspond to one of the six easily magnetised directions of the iron crystal, all of which follow the edge of a cube. When no external magnetism is evident, the various domain axes are along one or other of these directions but with a random distribution. The application of an external magnetic field tends to bring the domain axes into line to reinforce the field with their own magnetic action so that the flux becomes very much more intense.

Rewriting eqn. (2.1) as:

$$B = \mu_0 H + (\mu_0 \mu_r H - \mu_0 H) = \mu_0 H + \mu_0 (\mu_r - 1) H,$$

24

the first term is seen to be the flux density which would exist in a non-magnetic medium. The second term is the intrinsic contribution to the flux density made by the iron itself, and is denoted by the symbol B_i and called the *magnetic polarisation*. B_i is a function of H and varies with different ferromagnetic materials. When all the domain axes are in line with the external field direction, the material is said to be saturated. For pure iron, B_i has a saturation value of about 2 teslas. The typical magnetisation curve of Fig. 2.1 shows that the relationship

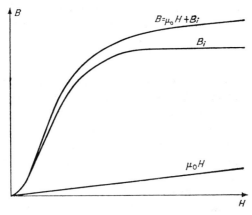

FIG. 2.1. Typical magnetisation curve for iron.

between B and H is highly non-linear. The domain changes take place in three definite stages. In the early, slowly-rising part of the curve, those domains nearest in line with the external field grow at the expense of the others. The sharp rise of the curve corresponds to sudden, successive, inelastic rotations of these other domain axes in turn, so that they are aligned with those of the enlarged domains. The final slow rise of the curve is due to gradual rotation of all the domain axes together so that at saturation they are all in line with the externally applied field.

25

The Hysteresis Loop

If H is decreased and reversed, the curve does not retrace the same path. Energy expended in domain changes is not all recoverable and the domains retain their new magnetic axes which followed the sudden rotations until a powerful enough reversing field is applied. Consequently, as Fig. 2.2 shows, when H is zero there is a residual value of flux density the magnitude of which, B_{res}, depends on the material, its crystal structure and the value of $\hat{B}$. To bring the flux density down to zero requires a *coercive field intensity*, or coercivity, H_c, which is negative.

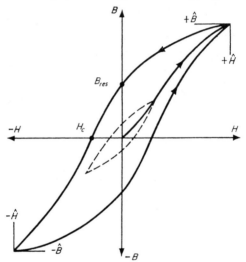

FIG. 2.2. Hysteresis loop.

This quantity is sometimes regarded as the residual magnetising force contributed by the iron alone, and producing B_{res} at a point on a magnetisation curve starting from H_c. On some machines, B_{res} is deliberately increased by choice of material to permit self-excitation processes to start.

As H reverses to $-\hat{H}$ and back again to $+\hat{H}$, a complete *hysteresis loop* is traced out. The loop area is a function of $\hat{B}$ and is found experimentally to vary as $\hat{B}^x$ up to moderate values of flux density. The index x is named after Steinmetz and is about 1·6 though it may be higher. Whenever a simple cycle of magnetisation is carried out it follows a closed curve. A minor loop is also indicated on Fig. 2.2. The magnetisation characteristic of a material is often shown as a single curve taken upwards from a residual value or from a demagnetised state. In the latter case, as indicated on Fig. 2.2, it nearly follows the locus of $\hat{B}$ for a series of hysteresis loops taken up to various values of peak flux density. Typical magnetisation curves for various materials used on electrical machines are shown on Fig. 2.3. Permanent-magnet materials are discussed in Appendix A.

It is interesting to note, that since any change of B lags the change of magnetising force H producing it, there will be an angular displacement between the rotating m.m.f. wave of a stator winding say, and the alternating field thereby induced in the rotor iron. Consequently, there will be a hysteretic torque whenever the iron is moving relative to the inducing m.m.f. wave. Small *hysteresis motors* operate on this principle, see p. 546.

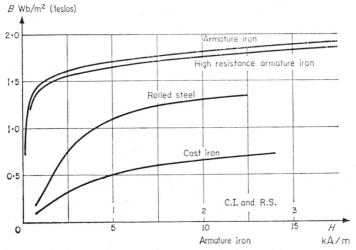

FIG. 2.3. B/H curves for different materials.

2.2 IRON LOSSES

Hysteresis Loss

Only a part of the energy expended in domain growth and rotation is recoverable. The loss can be calculated.

If a ring of iron of cross-sectional area A and mean circumferential length l is magnetised by a coil carrying i amperes through N turns, the supply must provide, apart from the iR drop and associated resistance loss, a component of voltage $e = N(\mathrm{d}\phi/\mathrm{d}t)$ to sustain a changing current and flux. The work done against the e.m.f. in a time $\mathrm{d}t$ is:

$$e.i.\mathrm{d}t = N\,\frac{\mathrm{d}(BA)}{\mathrm{d}t}.\frac{Hl}{N}.\mathrm{d}t$$

where B is the flux density and $H = Ni/l$ amperes per metre. On simplifying and dividing throughout by the volume Al, most of the terms cancel, giving:

$$\text{total work done per unit volume} = \frac{\int e.i.\mathrm{d}t}{Al}$$

$$= \int H.\mathrm{d}B \text{ joules/m}^3.$$

Consider first a flux change corresponding to a variation of H from 0 to $\hat{H}$. The energy required from this equation is represented by the area between the B/H curve and the y axis; see Fig. 2.4a.

On decreasing H to zero again, some energy is returned to the supply, $\mathrm{d}B$ being negative, and this again is represented by the area between the B/H curve and the B axis, see Fig. 2.4b. It can be seen that the net expenditure of energy is represented by the area of the half hysteresis loop, shown shaded. A complete cycle of magnetisation will clearly require energy represented by the whole loop area, and for a cyclic frequency of f hertz the rate of energy flow in joules/sec or watts will be:

$$\text{(loop area)} \times f = k_h \hat{B}^x.f \text{ watts/m}^3. \tag{2.2}$$

k_h is a constant following from the proportionality of area to $\hat{B}^x$.

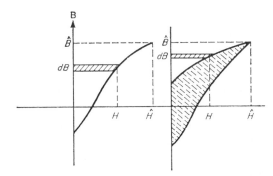

Fig. 2.4. Hysteresis loss.

Equation (2.2) should be applied to the *static* hysteresis loop taken, for example, by means of a ballistic galvanometer. The *dynamic* hysteresis loop, which can be displayed oscillographically, will be of different width if the frequency is more than a few cycles per second. This is due mainly to the eddy-current loss.

In rotating machinery, magnetisation is not in the simple forward and reverse manner implied in the previous discussion. The magnetic axis revolves due to the relative movement of the iron and the field axis. Under these circumstances the hysteresis loss can be larger than as calculated from eqn. (2.2) since magnetisation is forced to take place along difficult directions in the iron crystals.

Eddy-current Loss

Since iron is a conductor, a changing flux induces e.m.f.s and currents within the iron mass. These eddy currents, as they are called, produce losses, heating and demagnetisation. If the current paths are assumed wholly resistive and the redistribution of flux due to demagnetisation is neglected so that the flux density may be taken as uniform, a simple expression for the loss can be obtained.

In practice the iron is used in thin laminations so that the resistance is almost entirely due to such vertical paths within the section of iron

29

as indicated in Fig. 2.5. Denoting iron resistivity by ρ, the resistance to i_e is therefore $r_e = 2\rho h/l \cdot dx$ and the elemental path is linked by a flux $\phi = B \cdot 2hx$ producing an e.m.f. $d\phi/dt$.

Hence loss in path is equal to:

$$\frac{(d\phi/dt)^2}{r_e} = \frac{(dB/dt)^2 (2xh)^2}{2h\rho/l \cdot dx} = (dB/dt)^2 \cdot \frac{2hl}{\rho} x^2 \, dx.$$

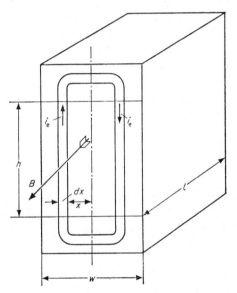

FIG. 2.5. Eddy-current loss.

The loss over the whole section is:

$$(dB/dt)^2 \frac{2hl}{\rho} \int_0^{w/2} x^2 \, dx.$$

Integrating and dividing by lhw to get the loss per unit volume gives:

$$\text{eddy-current loss} = \left(\frac{dB}{dt}\right)^2 \cdot \frac{w^2}{12\rho} \text{ watts/m}^3.$$

30

It is now clear why iron is laminated if it is to be subjected to alternating magnetisation, the loss being reduced in proportion to (thickness)^2.
It is economical at 50 Hz for example, to go to the trouble and expense
of dividing the iron into laminations about 0·4 mm thick and insulating
them from one another by means of thin paper or varnish. High resistance silicon iron is used in many cases even though the magnetising
m.m.f. is thereby increased, see Fig. 2.3. In a built up packet of laminations, the iron may represent only about 90–95 % of the measured overall
thickness. A *building factor* of this order must be allowed for in any
calculations of iron cross-section and volume.

For the special case of sinusoidal flux variations, the mean power
loss can be obtained in terms of frequency and maximum flux density.

Let $B = \hat{B} \sin \omega t$, then $(dB/dt)^2 = (\omega \hat{B} \cos \omega t)^2$.

The mean value of $\cos^2 \theta$ over a cycle is 0·5 so the loss is:

$$(2\pi f \, \hat{B})^2 \times 0\cdot 5 \times w^2/12\rho = \pi^2 f^2 \hat{B}^2 w^2/6\rho \text{ watts/m}^3. \qquad (2.3)$$

For any other wave form of form factor k_{ff} the (r.m.s. current)^2 and
the loss are changed in the ratio $(k_{ff}/1\cdot 11)^2$.

In practice the loss is much greater than given by eqn. (2.3) due
mainly to the local high rates of dB/dt associated with domain changes,
but the equation shows the main factors upon which the loss is dependent.

It should be noted from Fig. 2.5 that the directions of the eddy-
currents are such as to oppose by their magnetic effects any increase
in the value of B out of the paper. In fact, any change of flux is delayed,
in accordance with Lenz's law. In magnetic circuits where the flux,
though not varying cyclically, is required to have rapid response,
lamination is necessary to reduce the eddy-currents even though the
eddy-current loss itself is negligible. Figure 3.32 illustrates how the
resultant m.m.f. would be modified by induced currents.

In a given sample of iron, it is possible to measure the iron loss
and separate the hysteresis and eddy-current loss components. A
discussion on this aspect will be deferred until Chapter 4.

31

2.3 CALCULATION OF EXCITATION M.M.F. FOR A MACHINE

Calculations of excitation are usually based on a closed line of force for which Ampere's circuital equation applies:

$$\int H.dl = \text{current enclosed.} \qquad (2.4)$$

To understand the meaning of this equation consider Fig. 2.6 which shows a part cross-section of a salient-pole machine. Two typical lines of force are drawn. B, and the corresponding value of H, depend on the material and vary along the length of such a line. It will be possible, however, to divide the total length into sections l_a, l_b, etc., over which

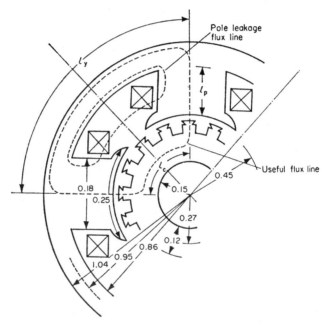

FIG. 2.6. Typical magnetic circuit. Diameters and other dimensions in metres.

H will be sensibly constant. The sum of the products $H_a l_a$, $H_b l_b$, etc., around any closed line will be equal to the resultant sum of all the currents flowing through the area bounded by this closed line. It is common practice, as in the remainder of the text, to measure the current enclosed in *ampere-turns* (At), since this may be due to a large current through a few turns, a small current through a large number of turns, or the summed effect of several coils having different numbers of turns and carrying different currents. If the armature currents are zero, i.e. on no load conditions, it can be seen that the longer line of force shown on Fig. 2.6 encloses the field current multiplied by the turns of two poles, whereas the shorter line only encloses a fraction of the ampere-turns per pair of poles. Since the position of the lines of force between the poles is not known until an elaborate flux plot has been made, see Fig. 2.7, a line through the middle of the pole for which the ampere-turns are clearly defined, is much more convenient as a basis for calculation. In addition, the flux density, and the value of H along the path are more easily computed. Nevertheless, for machines with different magnetic circuits, e.g. induction machines, a path away from the pole centre may offer a simpler solution.

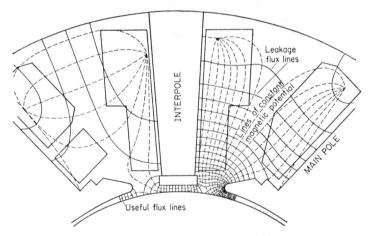

FIG. 2.7. Flux plot for loaded d.c. machine.

Figure 2.7 shows a complete flux plot for a loaded d.c. machine and includes the effects of currents in the field coils, interpole coils and armature coils. Allowance is also made for saturation of the iron although the flux lines through the iron are not shown. The flux pattern is only obtained after much labour and by trial and error until the laws of the magnetic circuit are satisfied. Although expensive Computer-aided-design (CAD) packages, with interactive graphics are now available for this purpose, a simpler method is desirable in most cases to get a reasonably close answer in a shorter time and at less cost. The usual design procedure will be reviewed briefly to bring out certain features of the magnetic circuit and indicate the approximations involved. The field coils only, will be excited, allowance for load currents will be discussed in the appropriate chapter referring to each particular machine.

Leakage Flux

The air-gap flux crossing the clearance between stator and rotor and linking both windings, is called the mutual or useful flux since it produces torque. The flux which avoids the armature and crosses between the poles is called the pole leakage flux and may be 10–20% of the useful flux under normal conditions. A leakage factor equal to the ratio (useful flux + leakage flux)/useful flux, and typically 1·2, expresses the relationship between the two components. There is also a leakage flux linking the armature alone when the armature carries current, and this will be discussed under Section 2.6.

In design, the required useful flux will be known and thus the flux density in the various parts of the magnetic circuit can be found. The m.m.f. required to magnetise each part of the path follows as below. Note, from Fig. 2.6, that since there is a plane of symmetry between the poles, it is usual to work out the m.m.f. for one pole by considering the path through the middle of one pole, one air gap, one length through the armature teeth, and half the length between a pole pitch in both the yoke and the armature core. This procedure corresponds to a division of the whole magnetic circuit into $2p$ equal parts, where $2p$ is the number of poles.

Pole Ampere-turns

Neglecting the fact that the density at the pole face is a little less than in the pole body, the flux density is:

$$B_p = \frac{\text{useful flux} \times \text{leakage factor}}{\text{cross-sectional area of pole body}}.$$

From the magnetisation curve for the pole iron, H_p follows and the m.m.f. required for the pole iron is $H_p l_p$ ampere turns.

Air Gap Ampere-turns

For the arrangement of Fig. 2.8, the flux density under the pole face consists of a constant component and a superimposed ripple due to the presence of slots and teeth. The effective air gap l'_g, is slightly longer than that measured between pole and teeth l_g, by a factor depending on the gap and slot dimensions. An alternative method of allowing for the increased magnetic reluctance would be to contract the area of the air gap (pole arc × axial pole length), by a related factor.

Beyond the pole tips, in the fringe field, the flux density falls off in a manner dependent upon the air-gap dimensions and the distance between adjacent pole tips. In calculations, the flux density at a particular position in the air gap is required, e.g. B_{max}, the mean value of the peak flux density under the centre of the pole. This value is related to a known quantity such as the average density B_{av} or the peak of the fundamental wave B_1, in a ratio depending on the field form.

Hence, for example, using B_{max}, the air-gap m.m.f. required is:

$$H_g l'_g = (B_{max}/\mu_0) \times \text{effective gap length} \quad \text{At.}$$

Tooth Ampere-turns

For parallel-sided rotor slots, the flux density increases towards the root of the tooth and because of saturation and the high values of H, an appreciable amount of flux is carried in the slot, particularly near

35

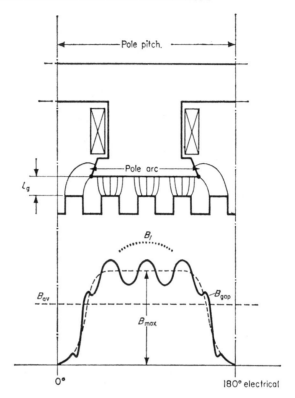

FIG. 2.8. Field form of salient-pole machine.

the root. The real density B_{real} is less than the apparent density B_{app} calculated from the useful flux divided by the tooth-iron area under the pole arc. For accurate computation of the tooth ampere-turns, H should be found at several positions down the tooth and $\Sigma H.\Delta l$ taken over the tooth length. Quite a close answer is obtained if B_{real} and the corresponding value of H are found at a position two-thirds of the length from the wider end.

Hence tooth m.m.f. $= H_{2/3}.l_t$ At. (see Fig. 2.9).

36

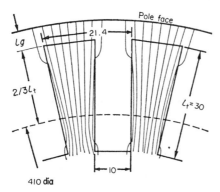

FIG. 2.9. Flux distribution in tapered tooth.
Dimensions in millimetres.

Yoke Ampere-turns

The flux entering the yoke divides into two parallel paths so that:

$$B_y = \frac{\text{useful flux} \times \text{leakage factor}}{2 \times \text{yoke cross-sectional area}}.$$

Reading off the corresponding value of H_y from the magnetisation curve for the yoke iron, the ampere-turns per pole required to magnetise the yoke $= H_y l_y/2$ At.

Armature Core Ampere-turns

The calculation of B_c is similar to that for the yoke but includes only the useful flux;

$$\therefore \text{core m.m.f.} = H_c l_c/2 \text{ At.}$$

Open-circuit Curve

Every portion of the path has now been dealt with for one particular value of useful flux. The summation of the Hl products gives the total m.m.f. to be provided on each pole to produce this flux, with the armature open circuited. By repeating this calculation for several values

37

of useful flux and plotting a curve against field ampere-turns, or field current if the field turns are known, the magnetisation characteristic of the machine is obtained (Fig. 2.10). When a machine is run at constant speed, it will be found that the generated e.m.f. is directly proportional to the useful flux (cf. Section 5.5). The magnetic calculations are therefore checked by running the machine at constant speed; the e.m.f./field current relationship, which is called the open-circuit

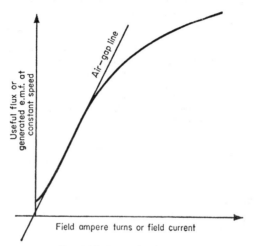

FIG. 2.10. Open-circuit curve.

(o.c.) curve, is really the magnetisation characteristic to a different scale. In Fig. 2.10 it will be noticed that the initial part of the curve is nearly straight, the so-called unsaturated region where the majority of the m.m.f. is being absorbed in the air gap. Much of machine analysis is based on this region where the performance equations are linear. Specialised techniques must be used to deal with operation in the saturated region, a term used to describe conditions beyond the air-gap line; see Section 3.3. Note that for a.c. excited machines, the open-circuit curve refers to the relationship between the induced e.m.f. and the magnetising current (r.m.s. value), at constant frequency.

Although it is not the intention here to cover magnetic circuit calculations in detail, it will help to clarify the points raised in this section if a numerical example is worked out.

EXAMPLE E.2.1

The various dimensions of a magnetic circuit are shown on Fig. 2.6, some of them having been calculated from the given inside and outside diameters of yoke and armature. There are 66 teeth and slots, only indicated schematically, the full dimensions being given on Fig. 2.9. The axial length of pole and armature is 0·35 m and the corresponding dimension for the yoke is 0·45 m. The effective air-gap length corrected for slotting effects may be taken as 5 mm. The ratio of B_{av}/B_{max} may be based on a rectangular field form as in Fig. 2.12a, i.e. equal to the pole-arc/pole-pitch ratio. Pole and armature laminations are of armature iron and a 95% building factor can be used. (In practice, an armature of this length would have two or three radial ventilating ducts along the core length but these will be neglected for the purpose of the problem.) The yoke is made of rolled steel. See Fig. 2.3 for B/H curves.

Assuming that $B_{real} = 0.95 \times B_{app}$ and that the leakage coefficient is 1·2, calculate the excitation m.m.f. required for a flux per pole of 0·075 webers.

Pole

Length $= 0.2$ m

Cross-sectional area $= 0.18 \times 0.35 \times 0.95 = 0.06$ m^2

$$B_p = \frac{0.075 \times 1.2}{0.06} = 1.5 \text{ teslas}$$

$H_p = 1.15$ kA/m

$H_p l_p = \underline{230 \text{ At.}}$

Air Gap

Length $= 5$ mm effective

Area over pole pitch $= \dfrac{0.45 \times \pi}{4} \times 0.35 = 0.124$ m^2

$B_{av} = 0.075/0.124 = 0.605$ tesla

$\dfrac{\text{pole arc}}{\text{pole pitch}} = \dfrac{25}{45\pi/4} = 0.705$

39

hence $B_{max} = 0.605/0.705 = 0.86$ tesla

$$H_g l_g = \frac{0.86}{4\pi \times 10^{-7}} \times \frac{5}{10^3} = \underline{3420 \text{ At.}}$$

Teeth

Length $= 3$ cm

The cross-sectional area of the tooth iron is calculated 2/3 the way down the slot. Assuming that all the gap flux over a tooth pitch is carried by this reduced area, the apparent tooth density is obtained on increasing B_{max} by the area ratio:

$$\frac{\text{area over tooth pitch}}{\text{area of one tooth}} = \frac{21.4 \times 350}{[(\pi \times 410)/66 - 10] \times 350 \times 0.95} = 2.36$$

Apparent tooth density $= 2.36 \times 0.86 = 2.03$ teslas

Real tooth density $= 2.03 \times 0.95 = 1.93$ teslas

for which $H_t = 17.3$ kA/m

Tooth ampere-turns $= 30 \times 10^{-3} \times 17.3 \times 10^3 = \underline{520 \text{ At.}}$

Yoke

Length $= l_y/2 = \dfrac{0.95 \times \pi}{4 \times 2} = 0.374$ m

Area $= 0.09 \times 0.45 = 0.0405$ m^2

$B_y = \dfrac{0.075 \times 1.2}{0.0405 \times 2} = 1.11$ teslas

$H_y = 1.05$ kA/m

$H_y l_y = \underline{390 \text{ At.}}$

Armature Core

Length $= l_c/2 = \dfrac{0.27 \times \pi}{4 \times 2} = 0.106$ m

Area $= 0.12 \times 0.35 \times 0.95 = 0.0398$ m^2

$B_c = \dfrac{0.075}{0.0398 \times 2} = 0.94$ tesla

$H_c = 0.2$ kA/m

$H_c l_c = \underline{20 \text{ At.}}$

Summing the m.m.f.s. for the five parts, gives a total excitation of $\underline{4580 \text{ At}}$ per pole.

40

2.4 THE ELECTRICAL ANALOGUE OF THE MAGNETIC CIRCUIT

Certain features of the magnetic circuit can be conveniently illustrated by drawing an analogy with the electrical circuit.

From eqn. (2.1), $H = B/\mu_0\mu_r$ ampere-turns/metre, and therefore:

$$H.\mathrm{d}l = \frac{\phi}{A}\cdot\frac{1}{\mu_0\mu_r}.\mathrm{d}l$$

ϕ being the flux across area A, giving flux density B.

If H is constant over a length l then:

$$Hl = \phi.\frac{l}{\mu_0\mu_r A} \text{ ampere-turns.} \qquad (2.5)$$

Hl is the *magnetomotive force* (m.m.f.) in ampere-turns, absorbed across a length l in establishing a flux ϕ across area A.

Equation (2.5) is analogous to the electrical circuit equation:

potential drop = current × resistance.

For the magnetic circuit:

magnetomotive force = flux × reluctance.   (2.5)

The reluctance $\mathscr{R} = l/\mu_0\mu_r A$ bears the same relationship to the length and area of the magnetic circuit as resistance bears to the length and area of a conductor. The coefficient $1/\mu_0\mu_r$, analogous to the specific resistance, varies widely with the relative permeability and so the equation is only readily applicable in air, or where the permeability variations are only over a limited range. The reciprocal of reluctance $\mu_0\mu_r A/l$ is called the *permeance* Λ, having units Wb/At. It is analogous to conductance in the electrical circuit.

Figure 2.11 shows the magnetic circuit of Fig. 2.6 arranged in an analogous manner to an electrical circuit. For simplicity only one leakage path per pole is shown, in a position where it is influenced by

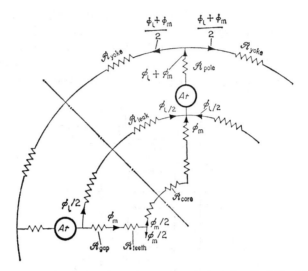

FIG. 2.11. Electrical analogue (magnetic equivalent circuit) of Fig. 2.6.

the whole of the pole m.m.f. Really, leakage is a distributed effect and several parallel reluctances should be included as a closer approximation, paths away from the air gap being tapped across only a portion of the pole m.m.f.

Determination of Flux Distribution

When flux is the unknown, eqn. (2.5) can be rearranged as:

$$\text{flux} = \text{m.m.f.} \times \text{permeance}$$

$$\phi = F \times \Lambda. \tag{2.6}$$

This equation is not readily applicable for manual calculation, because the iron permeance is not known until the flux density is known. Assumptions and approximations are therefore necessary, but the results are still of value in that the general effects may be clearly indicated even though the flux magnitudes are in error. For a rotating machine, both the m.m.f.

42

and permeance may vary with position round the periphery, and if the variations can be expressed graphically or analytically, the corresponding space variation of flux can be obtained by multiplication.

Consider Fig. 2.12a, for example, which shows a rectangular pole over a uniform air gap and magnetised by a concentrated field coil. If

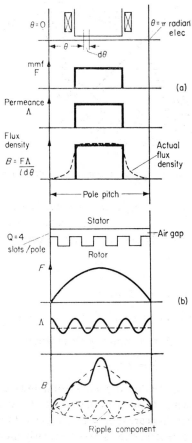

FIG. 2.12. Flux distribution from eqn. (2.6).

43

the reluctance of the iron parts is negligible, the m.m.f. across the air gap is substantially constant and equal to the field At per pole. Under the pole arc, the permeance at any angle θ, for a small area bounded by $d\theta$ and the axial gap length perpendicular to the paper, is substantially constant. The product of permeance and m.m.f. waves represents the flux in constant incremental areas of width $d\theta$. Under the pole arc, another rectangle results in this case, the ordinates of which are therefore proportional to flux density in the uniform portion of the air gap. Beyond the edges of the pole, the m.m.f. and permeance variations are less easy to define, and either a flux plot or mathematically deduced fringing curves are necessary to predict the remainder of the field form. The actual density distribution over the pole pitch would be as indicated by the broken line which shows that the error is not very serious for some purposes and an approximate estimate of the field form can be made.

A further, less idealised example is shown in Fig. 2.12b. Here, the fundamental of the m.m.f. space wave, moving relative to the slots, is acting on a gap permeance which consists of a constant component and a superimposed ripple due to the slots. The fundamental of the permeance ripple has a period corresponding to a slot pitch. When m.m.f. and permeance waves are multiplied, a flux ripple will result which will vary in amplitude over the pole pitch. This ripple, shown at one instant, is mathematically equivalent to two oppositely travelling waves of constant amplitude but slightly different angular period. Under suitable conditions, these space harmonics can give rise to time harmonics in the induced e.m.f. of a winding, the frequency being $(2Q \pm 1)f$ where Q is the number of slots per pole and f is the fundamental frequency of the m.m.f. wave. With slots on both sides of the air gap, there will be stator and rotor space harmonics which may interact at certain speeds, and interfere, usually in an undesirable fashion, with the speed/torque characteristic. Although the assumptions in the method militate against the calculation accuracy of the harmonic magnitudes, the frequencies and speeds at which interference can take place can be calculated correctly.

2.5 MOTORING AND GENERATING OPERATION

Direction of Force

A conductor of length *l* carrying a current *i*, in and perpendicular to a field where the flux density is *B*, experiences a force of magnitude *Bli* newtons in a direction mutually perpendicular to both conductor and field direction. There are many rules for remembering the sense of this force and it is well to be familiar with one of them. The simple pictures of Fig. 2.13a and 2.13b are easy to remember. With the conductor carrying current flowing into the paper, indicated by the rear end of a receding arrow, a clockwise field is produced around the con-

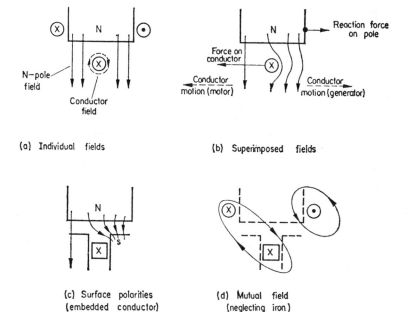

(a) Individual fields   (b) Superimposed fields

(c) Surface polarities   (d) Mutual field
(embedded conductor)   (neglecting iron)

FIG. 2.13. Force on current-carrying conductor in a magnetic field.

45

ductor (corkscrew rule). This field would react with the downward field lines emerging from the north pole to make the resultant field stronger on the right-hand side and weaker on the left-hand side. The electromagnetic picture itself suggests the correct sense of the force; towards the left on the conductor, the reaction force on the pole being towards the right.

Confirmation of the force direction can be obtained using the alternative physical concepts mentioned in connection with Fig. 1.4. It is almost universal practice to embed the conductors in slots and the superimposed fields are drawn for this condition in Fig. 2.13c. It is clearly seen that the surface polarities established and the distortion and concentration of the field on the right-hand side of the conductor give rise to an attractive force to the left, on the rotor structure. The third method of showing the force direction is demonstrated on Fig. 2.13d in which the resultant field of stator and rotor currents is indicated. With the left-hand conductor of the stator coil, the rotor conductor establishes a sympathetic field pattern, with low flux density between these two conductors and a strong field outside them giving rise to an attractive force to the left. With the right-hand conductor of the stator coil, the converse is true in that there is a strong field between the conductors causing mutual repulsion. Thus, either of the three concepts for explaining the nature of the force give rise to the same result, but in general, Fig. 2.13b is simpler to draw, in the various situations to be encountered in the text.

Direction of Induced E.M.F.

In a practical rotating machine, the conductor of Fig. 2.13 will generally be part of a coil, the other side of which is under the field of a south pole and carrying current in the opposite sense. There will be several such coils distributed round the periphery developing a gross electromagnetic torque T_e, equal to the summed effect of all the Bli forces multiplied by the radius of rotation. If the machine is a motor, the motion will be caused by, and be in the same direction as the force. If the machine is a generator, an external mechanical force must do work against the electromagnetic force to convert mechanical power into electrically generated power and the motion must therefore be in opposition to the electromagnetic torque as in Fig. 2.14a. The necessary current sense follows from the field polarity assumed and the sense of the induced e.m.f. must be such as to maintain the current direction shown. The sense of this generated e.m.f. can thus be ascertained by considering the generated current it would drive through the conductor. If the magnetic field lines are now considered to have a physical

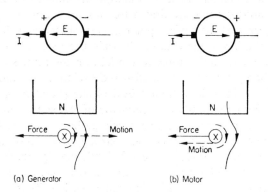

(a) Generator   (b) Motor

FIG. 2.14. Motoring and generating operation.

existence, then the movement of the conductor through the field will cause a distortion which, with a little thought, is found to be of the same general form as produced by the generated current and its field reacting with the stator field. Therefore, with the stator field polarity and field/conductor relative motion known, a simple rule (the *whiplash* rule), for determining the sense of the induced e.m.f. for either a generator or a motor, can be stated as follows:

If the lines of force emanating from a north pole (or entering a south pole), are imagined free to wind themselves round the conductor as a result of relative motion, they will be in the same direction round the conductor as the field which a current driven by the induced e.m.f. would establish.

47

For the d.c. and synchronous motors, the external voltage applied to the armature, drives the current in opposition to the induced e.m.f. but the rule still applies of course because it refers to the current which *would* flow if it was determined by the *induced e.m.f.* sense. Figure 2.14a shows the generator condition, with force and motion opposed but with current and e.m.f. in the same sense in the circuit. Figure 2.14b is for the motor, with force and motion in the same sense. The current must be as shown to give the correct electromagnetic force sense but since the relative motion is reversed from Fig. 2.14a, so is the induced e.m.f. which now opposes current flow in the circuit. The application of the rule is best illustrated by a few examples which also show the torque direction and the machine function.

Referring to Fig. 2.15, the *n* vectors represent motion relative to space. The field-pole motion *relative to the conductors* is also shown, since this determines the way in which the field lines wind round the

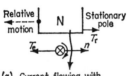

(a) Current flowing with generated e.m.f. Torque T_e opposing motion (generator)

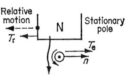

(b) Current flowing in opposition to generated e.m.f. Torque assisting motion (motor)

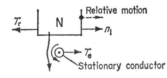

(c) Current flowing with generated e.m.f. Pole reaction torque opposing pole motion (generator)

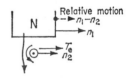

(d) Pole and conductor moving; current flowing with generated e.m.f. Torque assisting conductor motion (motor)

FIG. 2.15. Electromagnetic pictures for torque, motion and function.

conductor and hence the sign of the induced e.m.f. The electromagnetic torque on the conductor is T_e and T_r is the corresponding reaction torque on the pole.

FIG. 2.15a   Pole stationary; d.c. generator.

FIG. 2.15b   Pole stationary; d.c. motor; induced e.m.f. in opposite sense to current direction shown, i.e. same sense as Fig. 2.15a since relative motion and polarity unchanged.

FIG. 2.15c   Conductor stationary; current flowing with induced e.m.f.; poles moving in opposition to reaction torque exerted on them; synchronous generator.

FIG. 2.15d   Field and conductor moving; current flowing with induced e.m.f.; torque tending to reduce field/conductor relative motion but assisting conductor motion in space; i.e. motoring. This is the condition in the induction motor.

The rules demonstrated in Fig. 2.15 are very convenient when both field and conductors are moving, or indeed for all the various modes of operation to be discussed and so these electromagnetic pictures will be used frequently in the book. With a little thought to gain familiarity, they will be found less confusing and more informative than the left-hand and right-hand rules, since torque, e.m.f. and function can easily be visualised with one simple drawing.

Interaction of Electromagnetic and Mechanical Speed/Torque Characteristics

It is one of the features of electromagnetic machines that they can operate either as motors or generators, and under suitable conditions, a machine will change over smoothly from one function to the other. Consider the situation at standstill for a machine which can develop a starting torque. The behaviour depends on the nature of the mechanical torques T_m applied at the shaft. If they are less than T_e, the difference $T_e - T_m$ will be absorbed in the inertia torque $J\, d\omega_m/dt$, in accelerating the machine and coupled mechanical load. J must include the whole of the coupled inertia, the units being kilogramme (metres)^2. The speed will continue to rise until $T_e - T_m$ falls to zero as a result of changes to T_e, T_m or to both of them as the speed changes. The curves of Fig. 2.16 show these variations for a d.c. series motor driving a mechanical load such as a fan. Both torques in this case change with speed. T_m is drawn with the same sign as T_e for convenience in showing $T_e - T_m$, the net torque available for acceleration. T_m is actually opposing T_e. The

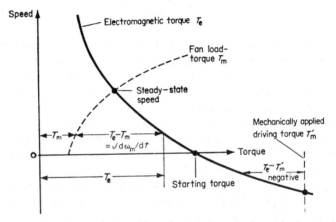

FIG. 2.16. Machine and mechanical-load speed/torque characteristics.

direction of motion has been decided by the electromagnetic torque and the machine would be converting electrical power supplied to the coils into mechanical power absorbed by the load; i.e. motoring.

On the other hand, if a mechanical driving torque T_m' had been applied, sufficient to overcome both T_e and any mechanical resisting torque, motion would have been in the opposite direction, $J\, d\omega_m/dt$ being negative. When the speed settled down all the mechanical power supplied, less the mechanical losses, would then be converted to electrical power; i.e. the machine would be generating in a mode discussed later on p. 345.

50

2.6   LEAKAGE FLUX AND REACTANCE

Inductance

A coil of N turns carrying current i establishes an average flux per turn of $\phi = iN.\Lambda$ webers, eqn (2.6). If the current, and therefore the flux change with time, the e.m.f. induced is $e = N \, \mathrm{d}\phi/\mathrm{d}t$. This relationship can be re-formed as:

$$e = N \frac{\mathrm{d}\phi}{\mathrm{d}i} \cdot \frac{\mathrm{d}i}{\mathrm{d}t} = L \cdot \frac{\mathrm{d}i}{\mathrm{d}t} \qquad (2.7)$$

where L, the flux turns per ampere or flux linkage per ampere is termed the *self-inductance*, the unit being the *henry*. L is constant if the flux is proportional to current, otherwise, as in ferromagnetic materials, it is proportional to the slope of the flux/current curve. This is roughly constant up to moderate values of flux density.

The circuit approach would give eqn. (2.7) directly in the final form by observing that quite apart from the IR drop, an additional voltage appears across a coil in which the current is changing with time. Under certain conditions this voltage is proportional to the rate of current change, L being the constant of proportionality. However, the concept of flux is useful in explaining the variation of L due to saturation, in the calculation of inductance and in the general understanding of machine behaviour.

It is also observed that another coil, in the vicinity of, but not connected to the first coil, may have a voltage induced when current in the first coil changes. The two coils are then said to be inductively coupled. Again the voltage is proportional to the rate of current change in the first coil. The constant of proportionality is called the *mutual inductance M* henrys, and will be discussed further in Section 3.7.

Reactance

For the special case where a coil current changes sinusoidally with time, it can be expressed as $i = \hat{I} \sin \omega t$. Consequently $\mathrm{d}i/\mathrm{d}t = \omega \hat{I} \cos \omega t$

51

and $e = L\,di/dt = \omega L\hat{I}\cos\omega t$. From this relationship it can be seen that ωL must be measured in ohms. It is the *self-inductive reactance* and is given the symbol X. If there is *mutual inductance*, the reactance symbol used will be X_m, which is proportional to ωM, see eqn. (3.17).

Leakage Reactance

The leakage flux between salient poles has already been discussed, its main effect being to bring on saturation earlier, due to the extra flux carried by certain parts of the iron circuit. It is also effective in limiting the current in any coupled winding which is subjected to sudden changes in terminal impedance; see Section 3.10. When leakage flux is associated with a winding carrying "steady-state" alternating current, a reactive voltage arises. The magnetic reluctance offered to leakage flux is predominantly due to the air path under normal conditions and so this flux is directly proportional to current, giving a constant inductance. Consequently the reactive voltage may be regarded as due to a constant *leakage reactance* and will be given the symbol x. It is a determining factor in machine performance.

Typical examples of leakage flux paths are shown in Fig. 2.17. Figure 2.17a shows a partial cross-section of a transformer core with two concentric coils which carry currents magnetising in opposite senses for mcst of the time. The majority of the flux established, ϕ_m, links both the coils. There is, however, a certain amount of flux passing between, and therefore not common to the two coils. This is the leakage flux, conveniently considered in two components ϕ_1 and ϕ_2, one component associated with each winding. It should also be noted that ϕ_1 does not link all the turns of coil 1, and ϕ_2 does not link all the turns of coil 2.

Figure 2.17b shows the region round the air gap of a rotating machine in which various components of leakage flux are indicated. Within the conductor region, the flux links only a fraction of the turns, whereas above the conductors towards the air gap, the whole of the flux links the coil side. With slots on both sides of the air gap, the resultant magnetising force can have components which are not

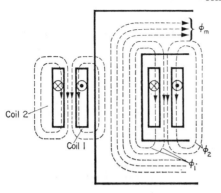

(a) Transformer leakage flux

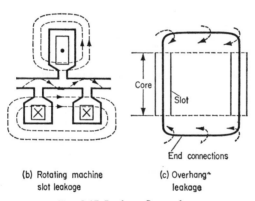

(b) Rotating machine   (c) Overhang*
    slot leakage   leakage

FIG. 2.17. Leakage flux paths.

tangential to the periphery causing flux to zigzag between the two iron
surfaces as indicated.

The slot conductors are joined at the ends by the overhang connections and there are considerable leakage fluxes associated therewith
(Fig. 2.17c). Even with mutual flux included, the overhang leakage
continues to have a separate existence, but the slot components are
superimposed on the mutual flux, modifying its distribution.

Detailed calculation of leakage flux and the associated inductance will not be attempted here but it is instructive to review the method briefly for a simple case. An open parallel slot is shown in Fig. 2.18 and the components of leakage flux which cross the slot will be calculated. The coil side consists of N turns carrying i amperes which are assumed to remain uniformly distributed over the conductor cross-section. The reluctance of the iron path is neglected, all the m.m.f.

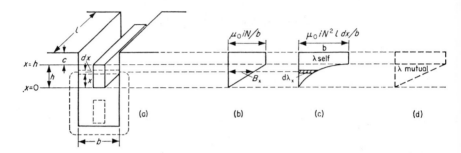

FIG. 2.18. Calculation of slot leakage inductance.

being absorbed in the air path across the slot so that $H =$ m.m.f./slot width.

If flux is proportional to current, then:

$$\mathrm{d}\phi/\mathrm{d}i = \phi/i = iN.\Lambda/i = N\Lambda.$$

If all this flux linked all the turns then the self-inductance would be $L = N\phi/i = N^2\Lambda$. This expression is not directly applicable, however, over the conductor region and it is necessary to work out the flux linkage λ by summing the incremental linkages $\mathrm{d}\lambda = N'\mathrm{d}\phi$, where N' is the reduced number of turns linked by the element $\mathrm{d}\phi$ of the total flux produced.

The flux in an element of cross-sectional area $l.\mathrm{d}x$ is first calculated. The m.m.f. enclosed by this flux element through $\mathrm{d}x$ is a fraction x/h times the total, iN, i.e. $= iN.x/h$.

The flux density at any point from $x = 0$ to $x = h$ is thus:

$$B_x = \mu_0 H_x = \mu_0 \cdot \frac{iN}{b} \frac{x}{h}.$$

B therefore increases uniformly with x from the bottom of the conductor to a maximum value when $x = h$, of $\mu_0 iN/b$ and remains so to the top of the slot (Fig. 2.18b).

Across $\mathrm{d}x$, the flux

$$\mathrm{d}\phi = B_x l \, \mathrm{d}x = \mu_0 \frac{iN}{b} \frac{x}{h} l \, \mathrm{d}x.$$

$\mathrm{d}\phi$ links only x/h of the total turns within the conductor region, i.e.

$$\mathrm{d}\lambda_x = N \cdot \mu_0 \frac{iN}{b} \frac{x^2}{h^2} l \, \mathrm{d}x.$$

$\mathrm{d}\lambda_x$ increases as x^2 from the bottom of the conductor to reach a maximum value of $\mu_0 iN^2 l \, \mathrm{d}x/b$ and remains so to the top of the slot (Fig. 2.18c).

The total flux linkage λ_c over the conductor portion is therefore:

$$\int_0^h \mu_0 \frac{iN^2}{b} \frac{l}{h^2} x^2 \, \mathrm{d}x = iN^2 \frac{\mu_0}{3} \frac{lh}{b}$$

and hence the corresponding leakage-inductance component is equal to:

$$\frac{\lambda_c}{i} = N^2 \frac{\mu_0}{3} \frac{lh}{b} = N^2 \Lambda'.$$

Note that the effective slot permeance Λ' is only one-third of the value calculated directly from the dimensions of the slot over the conductor region.

The flux which crosses above the conductor links all the coil side and the leakage inductance due to this portion can be calculated directly as $N^2 \mu_0 lc/b$.

The inductance components could have been found directly from the diagrams of Fig. 2.18b and c. The total flux is $\int B_x \mathrm{d}x$ which is the area under Fig. 2.18b. The total flux linkage is $\int \mathrm{d}\lambda_x$ which is the area of Fig. 2.18c. Remembering that the area under a parabola is one-third of the enclosing rectangle, the total flux linkage is:

$$iN^2 \frac{\mu_0 l}{b}\left(\frac{h}{3}+c\right)$$

from which the leakage inductance as before is:

$$N^2 \frac{\mu_0 l}{b}\left(\frac{h}{3}+c\right).$$

The graphical method can be used to illustrate other aspects of slot leakage calculations. For example, a bottom coil side is shown in broken lines, and in a practical case it is necessary to consider the effect of both top and bottom layers. It can be seen that the whole of the flux produced by the top layer links all the turns of the bottom layer and the corresponding mutual flux linkage diagram is shown in Fig. 2.18d. Similar diagrams could be drawn to consider the flux and flux linkage produced by current in the bottom layer alone. Self and mutual inductances are thus found.

It should be emphasised that the above example has been simplified. No account has been taken of the leakage fluxes across the top of the slot and in the overhang connections. When these are allowed for, the total leakage inductance of the coil, and hence of the whole winding, can be calculated. The winding leakage reactance x is then obtained on multiplying the result by $\omega = 2\pi f$.

CHAPTER 3

THE ELECTRICAL-CIRCUIT VIEWPOINT

3.1 SPACE PHASORS AND TIME PHASORS

A *scalar* quantity can be defined by means of its magnitude and an algebraic sign. A *vector* quantity, on the other hand, must be specified by both its magnitude and its direction in space. A vector may be represented by a line of length proportional to its magnitude, oriented at the appropriate angle to a chosen reference direction, and drawn from an origin positioned at any convenient point. The combined effect of two or more vector quantities, e.g. two forces acting on a body, can be obtained by vectorial addition illustrated in Fig. 3.1a. Using complex numbers, the addition can be performed analytically, i.e.

$$(a+jb)+(c+jd) = (a+c)+j(b+d).$$

To subtract the two vectors to get $F_1 - F_2$ say, F_2 is reversed, giving:

$$(a+jb)-(c+jd) = (a-c)+j(b-d),$$

Fig. 3.1b.

Sinusoidal Quantities

In studying machines, particularly a.c. machines, certain quantities are found to have variations which are sinusoidal in time, or sinusoidally distributed in space. When combining two or more such quantities of the same kind, it is convenient to represent them as vectors, and among many other advantages the techniques illustrated

57

in Fig. 3.1 can be used. Each line would be of fixed length equal to the r.m.s. or maximum value and set off at a particular angle to a common reference line. They are not vectors in the same sense as described earlier because they may be representing scalars such as voltage, current and flux linkage, which vary with time; or flux and m.m.f., which have an intensity varying with position round the air gap of a machine. In the latter case, the single "vector" representing the flux per pole, say, is proportional to and in space phase with the maximum value of flux

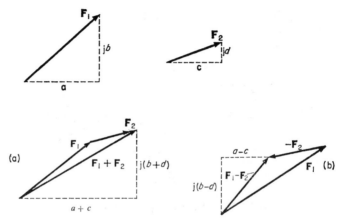

FIG. 3.1. Vector combination.

density $\hat{\mathbf{B}}$ which is a true vector. The $\mathbf{B}$ vectors decrease on either side of $\hat{\mathbf{B}}$ so that their envelope is sinusoidal as indicated in Fig. 1.2c (p. 4) and Fig. 1.6 (p. 11). Similarly, the single "vector" for m.m.f. represents, in the case of a uniform air-gap machine, an infinite number of $\mathbf{H}$ vectors decreasing along a sinusoidal envelope about either side of $\hat{\mathbf{H}}$.

Since it is permissible to use certain vector techniques only because the quantities vary sinusoidally in time or space, the terms *sinor*, *complexor* or *phasor* are sometimes used to distinguish such quantities from true vectors. Over the years the term *phasor* has become accepted for the designation of sinusoidal time variables. A logical extension of this would cover any scalar which exhibits sinusoidal changes. In machines studies, it is particularly important to emphasise the difference between time and space variations and to bring out this distinction, *time phasor* or *space phasor* will be assigned as appropriate. The term *phasor* alone, must be understood generally to refer to a time function.

Figure 3.2 shows an instance where the two kinds of phasors are closely related. From the rotating pole, a line drawn along the flux axis, i.e. at the point of maximum flux density assumed sinusoidally distributed, is of length proportional to the mutual flux per pole.

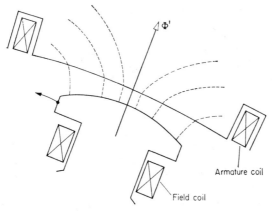

FIG. 3.2. Time phasor and space phasor of flux.

This is a space phasor and has no time variation with respect to the field coil if the field current is constant and the bore is smooth. Relative to the coil on the stator, however, the same flux is experienced as a time variation of flux linkage. When the space phasor is intended to be understood, a **bold-face** *phi* with a prime, Φ', will be used. Otherwise, if it is the time varying aspect, an unprimed Φ will represent the maximum value of flux per pole crossing the air gap to link the armature coils and induce an e.m.f. A small ϕ will be used for an instantaneous flux value.

The vectorial combination of space phasors has already been discussed in Section 1.4 (p. 16). There are further notes to make on time phasors.

Complex Number Representation

A vector **A** of modulus A at an angle α to the reference direction can be expressed in various forms. The *Cartesian* form follows from the expression $\mathbf{A} = Ae^{j\alpha} = A(\cos\alpha + j\sin\alpha) = a + jb$. The *Polar* form is A/α and is obtained from the Cartesian by the operations $A = \sqrt{(a^2 + b^2)}$ and $\alpha = \tan^{-1} b/a$. The angle α is measured anti-clockwise from the chosen reference which is the *real axis* and is usually, though not necessarily, placed horizontally. The j *axis* or

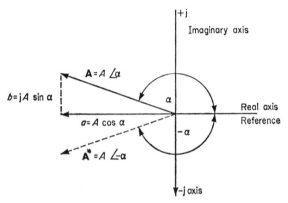

FIG. 3.3. Representation of vector by complex number.

imaginary axis is in leading quadrature with the real axis. Either or both, real and imaginary terms may be negative, depending on α. In Fig. 3.3, $\cos\alpha$ is negative and therefore a is negative; $\sin\alpha$ and b are positive. The conjugate vector $\mathbf{A}^* = A/-\alpha$ is also shown.

The operations of addition and subtraction are straightforward if the Cartesian form is used as for Fig. 3.1, but multiplication and division are more convenient with the Polar form. $A/\alpha \times B/\beta = AB/(\alpha + \beta)$ and $A/\alpha \div B/\beta = (A/B)/(\alpha - \beta)$.

Consider two voltages varying sinusoidally with time at frequency f cycles/sec and reaching their maxima at instants differing in time by one-sixth of a cycle; i.e. $\mathbf{E}_2$ lags $\mathbf{E}_1$ by a time-phase angle of $360/6$

60

degrees or $2\pi/6$ radians. These two time-varying voltages can be represented as on Fig. 3.4 by phasors, rotating at angular speed $\omega = 2\pi f$ radians/sec. The zero time reference for the phasor diagram may be arbitrary and in Fig. 3.4 is taken so that $e_1 = \sqrt{2}.E_1 \sin$ $(\omega t + \pi/4)$ and $e_2 = \sqrt{2}.E_2 \sin(\omega t - \pi/12)$; e_1 and e_2 being instantaneous values and E_1 and E_2 being r.m.s. values. The voltage values at any instant are obtained by inserting the appropriate values of t in these equations or alternatively, by rotating the phasors through an angle ωt from their position at $t = 0$, and noting their projection on the

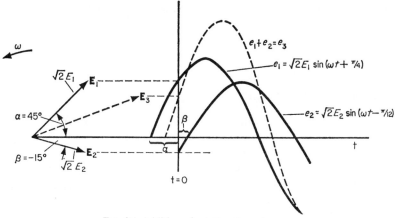

Fig. 3.4. Addition of rotating time-phasors.

vertical axis. The phasors have been drawn at time $t = 0$ but the phase difference between them, which is the important feature of their relationship, is unchanged no matter which instant of time is chosen. The actual angular position, or the instant of time, is not usually of importance except in relation to other phasors which must be drawn for the same instant. It should be clear that the time phasors must rotate at the same speed, i.e. same frequency, for a vector combination and the corresponding relative phase angles to be independent of the time chosen. Usually, one of the phasors is taken as reference and the relative phase angles of all the others are set off from this position.

61

Figure 3.4 shows that the resultant $e_1 + e_2$ obtained by adding ordinates of the sine curves at the same instants of time to give e_3, is found more conveniently by vector addition. This operation is performed analytically as follows:

$$\mathbf{E}_1 \qquad + \qquad \mathbf{E}_2 \qquad = \mathbf{E}_3$$

$$\sqrt{2}.E_1(\cos\alpha + j\sin\alpha) + \sqrt{2}.E_2(\cos\beta + j\sin\beta) =$$

$$\sqrt{2} \times [(E_1\cos\alpha + E_2\cos\beta) + j(E_1\sin\alpha + E_2\sin\beta)].$$

Note that β is a negative angle in this case.

Although maximum values of voltage have been used on the phasor diagram, it is much more convenient to use r.m.s. values to avoid unnecessary multiplication; the scale of the diagram is changed by a factor of $\sqrt{2}$ without altering the principle of representation.

Calculation of Power from Time Phasors

If two quantities varying sinusoidally with time at frequency f cycles/sec are multiplied, the result is a sine wave of double frequency, see Fig. 3.8. This different frequency would seem to preclude the calculation of power from the multiplication of two phasors $\mathbf{V}$ and $\mathbf{I}$. However, if one phasor is substituted by its conjugate the following expressions are obtained.

Referring to Fig. 3.5a:

$$\mathbf{V}^*\mathbf{I} = V\underline{/-\alpha}.I\underline{/\beta} = VI\underline{/-(\alpha-\beta)} = VI\cos(-\varphi) + jVI\sin(-\varphi)$$
$$= VI\cos\varphi - jVI\sin\varphi$$

voltamperes (VA) = active power (W) $-$ j.reactive voltamperes (VAr)

$$S \qquad = \qquad P \qquad - \qquad j \qquad Q$$

This surprisingly convenient result must be sufficient justification for the method which has been found by trial, to give the desired "components" of kVA. The Cartesian form could have been used. The product $\mathbf{VI}^*$ would have given the same numerical values but with positive sign for the reactive component. Using $\mathbf{V}^*\mathbf{I}$, a positive sign indicates leading reactive volt-amperes, (VAr).

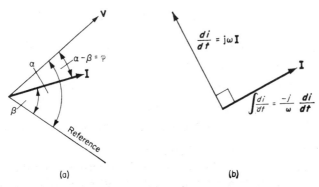

FIG. 3.5. Other operations with time phasors: (a) power,
(b) differentiation and integration.

Differentiation and Integration with Time Phasors

Differentiating a sinusoidally varying current $i = \sqrt{2}I \sin \omega t$ gives:

$$\mathrm{d}i/\mathrm{d}t = \omega.\sqrt{2}.I \cos \omega t = \omega.\sqrt{2}.I \sin(\omega t + \pi/2).$$

If i is represented by a phasor of length proportional to I, then $\mathrm{d}i/\mathrm{d}t$ can be represented by a phasor of length equal to ω times the first phasor and leading it by 90°, i.e. vectorially, $\mathbf{di/dt} = \mathrm{j}\omega\mathbf{I}$, Fig. 3.5b. This can be illustrated by the familiar example of the voltage required to sustain a sinusoidal current through a pure inductance; $v = L\,\mathrm{d}i/\mathrm{d}t$ so that vectorially $\mathbf{V} = L\mathrm{j}\omega\mathbf{I} = \mathrm{j}X.\mathbf{I}$ since $X = \omega L$.

A similar argument will show that the integration of a sine wave can be represented vectorially by a phasor lagging by 90°. Using the inductance again:

$$i = \int \frac{v}{L}\,\mathrm{d}t \quad \text{or from the voltage equation} \quad \mathbf{I} = \frac{\mathbf{V}}{\mathrm{j}\omega L} = \frac{-\mathrm{j}\mathbf{V}}{X}.$$

In a.c. power circuits, it is generally desirable to have sinusoidal waveforms, because this waveshape is reproduced by the differentiation and integration processes associated with reactive elements. Further, harmonics in the waveshape can give rise to additional power losses.

63

Application to Electrical Circuits

It will be required to find either the voltage, current or impedance knowing only two of these quantities. The "vector" expression for impedance can be found from the phasor equation or the diagram for a series circuit, Fig. 3.6a. Capacitance effects have been neglected because in general they will not be of importance for problems within the scope of this book. The symbol X will refer to inductive reactance.

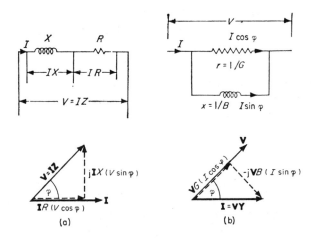

Fig. 3.6. Phasors applied to circuits: (a) series, (b) parallel.

For Fig. 3.6a the applied voltage (see p. 73), is

$$\mathbf{V} = R\mathbf{I}+j X\mathbf{I} = \mathbf{I}(R+j X) = \mathbf{IZ}. \qquad (3.1a)$$

Note: there is no significance in the order of multiplication, i.e. $j X\mathbf{I} = j\mathbf{I}X$. Both forms may be used. $\mathbf{Z} = R+j X$ is a complex number; it is neither a time phasor nor a space phasor, though it can be regarded as a vector operator. It is derived from the voltage phasor diagram, or the phasor equation on dividing throughout by $\mathbf{I}$ which is common to the three voltage components. It is convenient to use the vector notation, i.e. **bold-face** type; to distinguish it from the modulus $Z = \sqrt{(R^2+X^2)}$

The ordinary processes of complex number multiplication and division can be applied to the **V** **Z** **I** relationship. Note that addition and subtraction can only be applied to vectors or phasors of like kind.

To illustrate division, the phasor diagram of Fig. 3.6a will be considered in another way. For Fig. 3.6a the voltage is resolved into components in phase and in quadrature with the current. The in-phase component has a physical existence across R and the quadrature component exists across X. If instead the current were to be resolved along and in quadrature with **V**, there would be two components of current. These, too, can have a physical meaning if the representative circuit is arranged a different way, Fig. 3.6b. The in-phase component $I \cos \varphi$ flows through a pure resistance r, and accounts for the power consumed. The component $I \sin \varphi$ flows through a parallel pure reactance x, and accounts for the reactive volt-amperes. To find these components analytically:

$$\mathbf{I} = \frac{\mathbf{V}}{\mathbf{Z}} = \frac{\mathbf{V}}{R+jX} \cdot \frac{R-jX}{R-jX} = \mathbf{V} \cdot \frac{R}{R^2+X^2} - j\mathbf{V} \cdot \frac{X}{R^2+X^2}$$

$$= \mathbf{V}(G-jB) = \mathbf{V}\mathbf{Y}. \qquad (3.1b)$$

G is called the *conductance*; it is the in-phase current per volt so that $VG = I \cos \varphi$.

B is called the *susceptance*; it is the reactive current per volt so that $VB = I \sin \varphi$.

Y is the *admittance* and is equal to $\sqrt{(G^2+B^2)}$; it is the total current per volt.

Note that the positive sign given to an inductive reactance, $+jX$, results in a negative sign for the reactive lagging current.

A capacitive reactance has a negative sign, $X_c = 1/j\omega C = -j(1/\omega C)$.

The impedance components of the parallel circuit are given by $r = 1/G$ and $x = 1/B$. Both series and parallel circuits will be used as convenient to represent the situation indicated by a phasor diagram like Fig. 3.5a, where **V**, **I** and φ may have been obtained from input measurements to an undefined circuit; see example E.3.1 below.

Application to Three-phase Circuits

Numerical problems in this book will be confined to balanced conditions. Consequently for three-phase circuits it will only be necessary to reduce the information to phase values throughout and then treat as for a single-phase problem, noting that this will only involve one-third of the total power. Note that although the phase power pulsates, see Fig. 3.8, the three pulsating power components of a balanced three-phase system give a zero resultant when added at every instant of time throughout the cycle. The total instantaneous power is therefore the same as the average power;

$$\text{Power} = 3V_{\text{ph}}I_{\text{ph}} \cos \varphi = \sqrt{3}V_{\text{line}}I_{\text{line}} \cos \varphi = \text{VA}_{\text{total}} \cdot \cos \varphi$$

and can be measured as described in Appendix C.

Much of the complex-number manipulation associated with series, parallel, three-phase star and three-phase delta circuits required for this book, is covered by the following example. Although the solution is condensed, the student at this stage should be quite familiar with this type of problem which is inserted here for revision, and for emphasis of certain points raised in this section.

EXAMPLE E.3.1

A balanced 3-phase load consumes a total power of 100 kW at a power factor of 0·6 lagging; the line voltage being 1000 V. Determine the values of the impedance components per phase assuming:

    (a) Delta connection (i) resistance and reactance in series.
                       (ii) resistance and reactance in parallel.
    (b) Star connection (i) resistance and reactance in series.
                       (ii) resistance and reactance in parallel.

Sketch a phasor diagram for cases (a)(i) and (b)(i) showing at least one line and phase time-phasor for current and voltage.

Reference should be made to the diagrams to follow the solution.

$$\text{Line current} = \frac{\text{Power}}{\sqrt{3}V \cos \varphi} = \frac{100{,}000}{1000 \times \sqrt{3} \times 0{\cdot}6} = 96 \text{ A}.$$

Case (a)(i)

    Phase current $= 96/\sqrt{3} = 55{\cdot}5$ A

    $Z = V/I = 1000/55{\cdot}5 = 18 \, \Omega$

    $\mathbf{Z} = (V \cos \varphi)/I + \text{j}(V \sin \varphi)/I = 18(0{\cdot}6 + \text{j}0{\cdot}8) = 10{\cdot}8 + \text{j}14{\cdot}4\Omega = R + \text{j}X.$

66

Case (a)(ii)

$I \cos \varphi = 55.5 \times 0.6 = 33.3$ A

$I \sin \varphi = 55.5 \times 0.8 = 44.4$ A

$r = V/I \cos \varphi = 1000/33.3 = 30 \, \Omega$

$x = V/I \sin \varphi = 1000/44.4 = 22.5 \, \Omega$

$Y = 1/30 - j/22.5 = 0.0333 - j0.0444 \, \Omega^{-1}$.

Alternatively:

$$Y = \frac{1}{R+jX} = \frac{1}{10.8+j14.4} \cdot \frac{10.8-j14.4}{10.8-j14.4} = 0.0333 - j0.0444 \, \Omega^{-1} = G - jB.$$

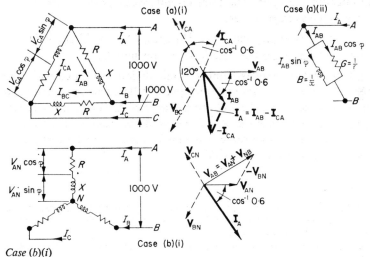

Case (a)(i)   Case (a)(ii)

Case (b)(i)

Case (b)(i)

Phase voltage $= 1000/\sqrt{3} = 577$ V

$Z = (V/I)(\cos \varphi + j \sin \varphi) = (577/96)(0.6 + j0.8) = 3.6 + j4.8 \, \Omega$

Case (b)(ii)

$I \cos \varphi = 96 \times 0.6 = 57.6$ A

$I \sin \varphi = 96 \times 0.8 = 76.8$ A

$r = 577/57.6 = 10 \, \Omega \qquad x = 577/76.8 = 7.5 \, \Omega$

$Y = 1/10 - j/7.5$; alternatively $= 1/(3.6 + j4.8) \, \Omega^{-1}$

Note, in the phasor diagrams that if V_{AB} is taken as reference $= V_{AB} + j0$ then:

$V_{BC} = V_{AB}(\cos -120° + j \sin -120°) = V_{AB}[-1/2 - j\sqrt{3}/2]$

$V_{CA} = V_{AB}(\cos 120° + j \sin 120°) = V_{AB}[-1/2 + j\sqrt{3}/2]$

For case (b) it is more convenient to take V_{AN}, the voltage to neutral, as reference.

3.2 ENERGY SOURCES AND ENERGY SINKS

An electrical machine or circuit may be either a *source* providing energy, or a *sink* absorbing energy, the rate of energy flow being measured by the product of the current in the circuit and the potential difference across the terminals. This product will be negative if current and voltage are of opposite sign. Since the choice of sign convention is arbitrary, some explanation of the implications is necessary.

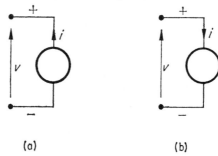

(a)   (b)

Fig. 3.7. Source and sink conventions:
(a) positive source if *v* and *i* in direction shown;
(b) positive sink if *v* and *i* in direction shown.

Figure 3.7a shows a power source; the potential rise through the element is in a direction to drive the current in the same sense. If the senses of voltage and current shown are taken to be positive, then a positive product of *v* and *i* corresponds to the device behaving as a positive source; i.e. acting as a generator. The relative senses of voltage and current as shown on Fig. 3.7a are referred to as the generator or source conventions. If in fact there occurs a reversal of *v* or *i*, the *vi* product would be negative, indicating a negative source; usually called an energy sink or a motor.

Figure 3.7b shows the motor, or sink conventions. The direction of current is in the opposite direction to the potential rise through the element. With these voltage and current senses, a positive *vi* product would mean that the device was behaving as a positive sink; i.e. acting as a motor. A reversal of either v or i would result in a negative vi product, the device being a negative sink, which is a positive source or generator.

Either of the above conventions can be adopted as convenient, but it must be remembered that the interpretation of the vi product will depend on the conventions chosen. When the functions of a series of interconnected devices are unknown, there is some advantage in assuming that they are all sinks since some of them, like machine field windings, will necessarily be energy absorbing elements. After solving

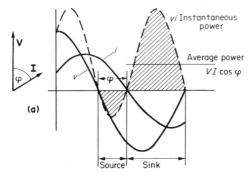

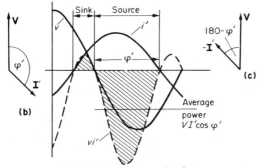

Fig. 3.8. Single-phase circuit:
(a) energy sink. Average power +ve (sink conventions);
(b) energy source. Average power −ve (sink conventions);
(c) energy source. Average power +ve (source conventions).

69

the equations, the signs of vi will then show which devices are providing and which are absorbing energy.

On alternating current, machines and reactive elements give to a circuit a characteristic which causes it to act for part of the cycle as a source and for the remaining part as a sink. The overall characteristic depends on the relative duration of these two periods. Figure 3.8a refers to a single-phase circuit which is an energy sink. Both the phasor diagram and the wave diagram are drawn assuming sink conventions, so that for most of the cycle, v and i are in the same sense corresponding to Fig. 3.7b. The time variation of the vi product shows that the average power, $VI \cos \varphi$, is positive. Figure 3.8b shows the changes which would occur, should the device change from a sink to a source. The diagrams indicate that the directions of v and i are now in opposite senses for most of the cycle so that the average power, $VI' \cos \varphi'$ is negative.

If generator conventions had been used for Fig. 3.8b, the current would have been reversed through 180° and different diagrams would then apply to the same basic circuit condition. The phasor diagram only, is shown on Fig. 3.8c, and it will be noticed that the power-factor angle φ is now less than 90°. The condition is that of an energy source *supplying* power at a leading power factor. The condition of Fig. 3.8a is that of an energy sink *accepting* power at a lagging power factor. These points will be elaborated further, with reference to Fig. 3.9 which shows two interconnected machines, one a motor and the other a generator.

On Fig. 3.9a, generator conventions have been assigned to Machine 1 and motor conventions to Machine 2. The machine e.m.f.s e_1 and e_2 are measured above the same datum. Although the phasor diagrams look similar, they must be interpreted in accordance with the different conventions chosen. Machine 1 is a generator *supplying* power at a lagging power factor, $\cos \varphi$. Machine 2 is a motor *accepting* power at a lagging power factor, $\cos \varphi$. The two power-factor angles are not usually the same, as shown here, because in general, $\mathbf{E}_1$ and $\mathbf{E}_2$ are not in phase. Their mutual phase relationship depends on the voltage absorbed in the circuit impedance as discussed later.

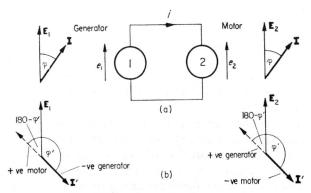

FIG. 3.9. Lagging and leading currents.
(a) Machine 1 generating, Machine 2 motoring;
(b) Machine 1 motoring, Machine 2 generating.

Should there be a change to the machine e.m.f.s such as to cause the current to lag by more than 90°, Fig. 3.9b, then Machine 1 would become a negative source and Machine 2 a negative sink. If this condition were known at the beginning, then opposite conventions would be chosen normally and the effect would be to reverse the current phasor as indicated on the figures. The power-factor angle, $180-\varphi'$, is now less than 90°. Machine 1 is a motor *accepting* power at a leading power factor and Machine 2 is a generator *providing* power at a leading power factor.

Effect of Circuit Impedance

A *resistor* is always a power sink; the voltage rise through it is always in the opposite sense to the flow of current, Fig. 3.10a. Alternatively, it can be said that the potential fall or voltage drop is in the same sense as the flow of current. The voltage and current phasors are always drawn in phase because, for a pure resistor, voltage is directly proportional to the current. If an electrical machine is regarded as a circuit element, its apparent impedance at its terminals V/I, can have a positive resistance component (motor) or a negative resistance component (generator).

71

An *inductor* is absorbing energy when current is increasing (magnetic field building up) and is providing energy when current is decreasing (magnetic field collapsing). A perfect inductor of zero resistance does not consume any power since all the energy stored in the magnetic field is recoverable and in an a.c. circuit is returned to the system. The voltage appearing across a pure inductor is $e = L.di/dt$ and if i is increasing in the positive sense shown on Fig. 3.10b, then the polarities

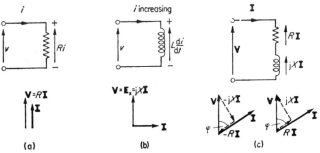

FIG. 3.10. Circuit elements.

are as shown since di/dt is positive. The potential rise through the element therefore opposes this *increase* of positive current in a similar way to the voltage rise through a resistive energy sink, which opposes the *flow* of positive current. The arrow alongside the inductor can therefore be interpreted in a way having some similarity with the interpretation of the arrow alongside the resistor. In the resistive case, the arrowhead is positive when the *current* is positive whereas for the inductor, the arrowhead is positive when the *current change*, di/dt, is positive. If the current is decreasing so that di/dt is negative, the polarities would be reversed, the inductor then giving up its stored energy. The Ri and Ldi/dt arrows will always be shown opposing the current arrow, but the actual polarities will of course reverse if i or di/dt are opposite to this arbitrarily chosen positive direction. The sign of the voltages will be taken into account automatically in any equations, in accordance with the signs of i and di/dt.

By reference to Fig. 3.5b, it will be understood that the phasor representation of the back e.m.f. Ldi/dt, appearing across the inductor, for sinusoidal current variations, will be given by $\mathbf{E}_x = j\omega L\mathbf{I} = jX\mathbf{I}$, where X is the reactance, see p. 52. This is a phasor leading the current phasor $\mathbf{I}$ by 90°. Alternatively it could be said that the current phasor lags the voltage phasor by 90°. The average power $E_x I \cos 90°$, is zero.

Figure 3.10c shows a circuit having resistance and inductance in series. It is convenient to draw arrows alongside each circuit element and these must be interpreted in the manner just described. They are both drawn in opposition to the current arrow which indicates an arbitrarily assumed positive direction of current flow. Although this means that both elements are being regarded as energy sinks, the voltage expressions, $R\mathbf{I}$ and $jX\mathbf{I}$, automatically include the appropriate allowance for any time-phase displacement between current and voltage which characterises the true nature of the circuit element. Kirchhoff's voltage law can now be applied to find the equation by summing all the voltages round the circuit to zero, taking due account of the signs indicated by the arrow directions; i.e.

$$\mathbf{V} - R\mathbf{I} - jX\mathbf{I} = 0.$$

An alternative application of the law equates the voltages between any two points of the circuit by traversing the circuit in opposite directions, giving, between top and bottom connectors in this case:

$$\mathbf{V} = R\mathbf{I} + jX\mathbf{I}. \tag{3.1}$$

Figure 3.10c shows the phasor diagrams corresponding to these two expressions which differ only in the grouping of the terms. Note that the characteristic of the circuit as a whole is that of an energy sink with a lagging power-factor angle $\varphi = \tan^{-1} X/R$. Figure 3.8a shows a similar circuit condition.

Machine Circuit Including Impedance

The simplest case to consider is the d.c. machine where only the resistive voltages and the machine e.m.f. need be included. Figure 3.11 shows two interconnected d.c. machines, one of them a generator, Machine 1, and the other a motor, Machine 2. The internal resistances of the machines are R_1 and R_2 respectively, and typical potentials, taking the bottom connector as reference potential, are indicated on the figure. Generator conventions are used for Machine 1 and motoring conventions are used for Machine 2 though it would have been quite permissible to use the opposite conventions providing that the appropriate equations were invoked in the solution. The common terminal voltage V, can be obtained from either the motor or the generator

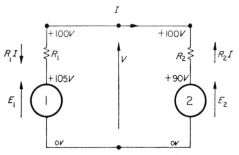

FIG. 3.11. Two-machine circuit.

having assigned the voltage arrow directions, those across the resistors opposing current flow. Applying Kirchhoff's law as when deriving eqn. (3.1):

from the motor: $V = E_2 + R_2 I$;

from the generator: $V = E_1 - R_1 I$;

in general: $V = E \pm RI$;   (3.2)

where the positive sign corresponds to a motor, using the motoring conventions and the negative sign corresponds to the generator using the generator conventions. These equations can be checked numerically using the typical potentials given on Fig. 3.11.

The internal impedance of a.c. machines includes inductive reactance and in some cases this is considerably higher than the internal resistance. To allow for this in the circuit of Fig. 3.11 and in the equations is a very simple matter. It is only necessary to replace R by $\mathbf{Z}$, where $\mathbf{Z} = R + jX$. Consequently, the general equation is:

$$\mathbf{V} = \mathbf{E} \pm R\mathbf{I} \pm jX\mathbf{I} \quad \text{or} \quad = \mathbf{E} \pm \mathbf{ZI}. \qquad (3.3)$$

74

The typical potential values shown on Fig. 3.11 could in fact correspond to a particular instant in the a.c. cycle. The indicated senses of the voltages and the current correspond to one half cycle only and further, the direction of the current may reverse either before or after the voltages reverse. The voltages themselves do not, in general, reverse together as they are not all in phase. The arrows on the diagram represent arbitrarily assumed positive directions but enable the correct circuit equations to be written down. The arrows for the impedance voltages must be in the opposite sense to the current as shown on Fig. 3.10c. Taking due account of the arrow directions, the voltage across the top and bottom conductors is therefore:

$$\mathbf{V} = \mathbf{E}_1 - \mathbf{Z}_1\mathbf{I} = \mathbf{E}_2 + \mathbf{Z}_2\mathbf{I},$$

where $\mathbf{Z}_1 = R_1 + jX_1$ and $\mathbf{Z}_2 = R_2 + jX_2$.

It will often be required to find the e.m.f. of one machine knowing all the other quantities in the above equation. If, for example, $\mathbf{E}_1$ is unknown, it can be found from:

$$\mathbf{E}_1 = \mathbf{V} + \mathbf{Z}_1\mathbf{I} = \mathbf{E}_2 + \mathbf{Z}_2\mathbf{I} + \mathbf{Z}_1\mathbf{I}. \tag{3.4}$$

At this stage, it is instructive to draw the phasor diagrams for two interconnected a.c. machines to illustrate the above equations. Figure 3.12 is drawn for the condition where No. 1 machine is generating at a lagging power-factor and No. 2 machine is motoring at a lagging power-factor. The individual diagrams for the two machines could be com-

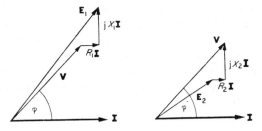

Fig. 3.12. Phasor diagrams for a.c. generator (1), and a.c. motor (2)

75

bined, but have been separated for clarity with the common I phasor and the common terminal voltage V retained on each diagram. Although I has been drawn in the horizontal position from where geometrical angles are usually measured as a reference, this choice of orientation is arbitrary, and in this case corresponds to the instant when the current is zero and about to increase positively. Any phasor, usually the terminal voltage, can be chosen for reference as convenient and can be shown for any instant, but the angular phase displacements between the phasors would of course be unchanged, see Fig. 3.4.

It should not be difficult as an exercise, to draw phasor diagrams for the case of leading power factor because this only means that the current phasor must be drawn leading instead of lagging the voltage V. It must be remembered, of course, that RI must be drawn in phase with the new position of I and jXI must lead I by 90° from its new position. A further useful exercise is to redraw the diagrams for Machine 2 generating and Machine 1 motoring. They would be the same as the ones shown except that the suffices $_1$ and $_2$ would be interchanged. As a matter of interest, it would also be possible to draw the diagrams for the original condition but with the opposite conventions chosen. For example, if motoring conventions had been chosen for the generator, then its current I' would have had to be drawn with reversed sense, I' being equal to $-I$. On applying the motoring equation to draw the phasor diagram, the difference would be that I' would have to be drawn in opposition to I and the phasors R_1I' and jX_1I' would have a negative sign, though drawn in the same position as for Fig. 3.12.

Should it be necessary to find the current knowing the magnitude and phase of both the machine e.m.f.s and also the machine impedances, eqn. (3.4) can be arranged as follows:

$$I = (E_1 - E_2)/(Z_1 + Z_2),$$

or denoting $Z_1 + Z_2$ by $Z = R + jX$

$$I = E_1/(R + jX) - E_2/(R + jX)$$

$$= (E_1/Z)\underline{\smash{/-\tan^{-1} X/R}} - (E_2/Z)\underline{\smash{/-\tan^{-1} X/R}}, \qquad (3.5)$$

which gives the current as the sum of two components. Each of these components is in the form of a complex number and the appropriate procedure for combining them must be followed. The phasor combination is shown in Fig. 3.13. Neither of the two components has a physical existence unless one of the e.m.f.s happens to be zero, in which case, the other e.m.f. will be short circuited and the current limited only by $R+jX$. It has been assumed in the equation that Machine 1 is a generator and Machine 2 is a motor but in fact Fig. 3.13 has been

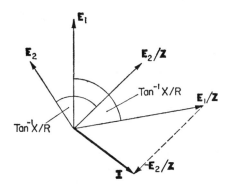

FIG. 3.13. Determination of **I**.

deliberately drawn to show what happens when this assumption does not correspond with the fact. The combination of the two components in accordance with eqn. (3.5) gives a current phasor which has a component in antiphase with both E_1 and E_2. Therefore Machine 1 must be a negative generator (motor), and Machine 2 must be a negative motor (generator). In general, Machine 1 will be a motor if $\mathbf{E}_1$ lags $\mathbf{E}_2$ and will be a generator if $\mathbf{E}_1$ leads $\mathbf{E}_2$. To find terminal voltage **V**, eqn. (3.3) must be applied.

Finally, it should be noted that the phasor diagrams are merely a pictorial representation of the phasor equations which could instead have been solved using complex numbers as expressed by the equations.

3.3 SUPERPOSITION IN ELECTRICAL AND MAGNETIC CIRCUITS

Superposition is a technique very commonly used to study complex problems in which the final effect is the result of several influences acting simultaneously. The resultant effect is assumed to be the same as would be obtained by combining the individual effects of each influence acting alone on the system. For this treatment to be valid, the system must have a linear relationship between cause and effect which is the same irrespective of the magnitude of the input quantity. For example, in the case of an electrical circuit, the ohmic values of the impedance components must be independent of the current flowing or the voltage applied. Several applications of superposition will occur in the later text and already, in Section 3.1 and Example E.3.1, sinusoidal currents and voltages have been resolved into in-phase and quadrature components to simplify calculation procedures. Another common instance of superposition is the concept of harmonics in cyclically varying quantities. The influence of harmonics in 3-phase circuits is of particular interest in electrical machines studies and a brief consideration of this topic follows.

Harmonics in Three-phase Circuits

Non-sinusoidal quantities can be broken down by Fourier analysis into a sinusoidal fundamental of the same period as the wave itself, together with harmonics having frequencies which are multiples of, and periods which are sub-multiples of the fundamental frequency and period respectively. If the average value of the original wave over a complete cycle is not zero, this average will represent the so-called d.c. component which is unidirectional and constant. Apart from tooth ripple effects, see Fig. 2.12b (p. 43), the higher the order of the harmonic, the smaller the magnitude in general, though certain generated harmonics may be suppressed by the circuit connection. In this book, the *odd* harmonics are of most frequent concern because *even* harmonics do not occur when the negative half wave follows the positive half wave exactly, apart from the sign reversal. This is the situation, for example, with the field form (Fig. 2.8) (p. 36) if produced by symmetrical north and south poles.

Figure 3.14 shows the relationships of the three main types of harmonics occurring in the 3-phase circuit. It is assumed that the

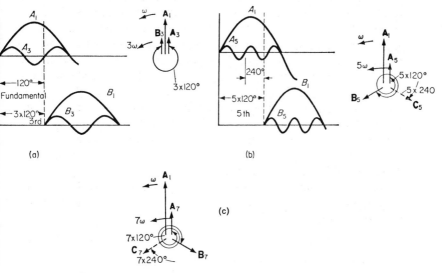

FIG. 3.14. Harmonics in 3-phase circuits; (a) 3rd harmonic. (b) 5th harmonic, (c) 7th harmonic.

waveform of each phase is identical, so that in each phase the harmonics bear the same relationship to the fundamental. It is conceivable for harmonics to arise which do not conform in this respect but such special cases will not be discussed. A further point is that each harmonic is shown starting from zero with the fundamental. This is convenient for demonstration but is not essential to the following arguments since each harmonic can be treated independently if superposition is permissible.

Due to the 120° mutual displacement of the balanced 3-phase system, the 3rd harmonics are seen to be in phase with one another. The 5th harmonics have a mutual 120° relationship but with phase sequence $A_5C_5B_5$ which is the reverse sequence to the fundamental $A_1B_1C_1$. The 7th harmonics also have the 120° relationship but the phase sequence $A_7B_7C_7$ is now the same as the fundamental. Higher order harmonics repeat this pattern, e.g. the 9th and all multiples of the 3rd are in phase. The phase sequence ABC is normally taken as the positive sequence so that the fundamental, 7th, 13th, etc., are of positive sequence; the 3rd, 9th, etc., are of zero sequence, i.e. in phase. The 5th, 11th, etc., are of negative sequence.

Consider now the effect of harmonics in star and delta connections. For the star circuit, since the line voltage is obtained from the difference of the phase voltages, in-phase components cancel, so that the line voltage is free from 3rd harmonics and those which are multiples of three; sometimes called *triplen* harmonics. All other harmonics appear, if present in the phase voltage.

It is possible to use the delta connection with all the phases short circuiting one another, because the fundamental voltages and certain harmonics sum vectorially to zero, due to the 120° displacement. Triplen harmonics, however, are additive round the mesh which offers a short circuit to them. For example, the 3rd harmonic circulating current is of magnitude $i_3 = 3e_3/3Z_3$. It follows that the harmonic impedance drop i_3Z_3 is exactly equal to e_3 the harmonic driving voltage, so that again there is no 3rd harmonic potential difference across the line terminals. Circumstances decide whether or not the circulating current is desirable and the choice of star or delta connections will be discussed again in later chapters.

Symmetrical Components for Unbalanced Polyphase Circuits

Sequence components arise not only with harmonics, but occur also with unbalanced loads. Figure 3.15 shows three sets of symmetrical 3-phase time-phasors. $A_+B_+C_+$ are of normal positive sequence, $A_-B_-C_-$ are of reverse or negative sequence, and $A_0B_0C_0$ are in-phase

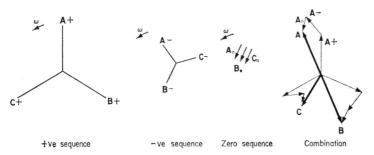

+ve sequence   −ve sequence   Zero sequence   Combination

FIG. 3.15. Symmetrical components of an unbalanced system.

or zero sequence. When these three sets are combined vectorially they give rise to a set of time phasors **A**, **B** and **C** which are unbalanced in magnitude and phase. It is not difficult to appreciate that the inverse process is possible, i.e. an unbalanced system can be replaced by a set of symmetrical, balanced components which, operating on the appropriate sequence impedances, give simple solutions in terms of the sequence components. These are readily combined for the final answer. This simple picture is merely a glimpse of the symmetrical-component transformation which is a most powerful tool with implications outside the scope of this book[2]. The idea is useful for present purposes when explaining the resolution of pulsating m.m.f.s and the operation of single-phase induction motors in Chapters 5 and 7, and when referring to sequence impedances in Chapters 4, 8 and 10.

Superposition in Non-linear Systems

Although much of electrical machine analysis is built up on the assumption that the various parameters are constant, giving rise to a linear system, in practice this is very often far from true. However this does not invalidate the analysis completely but the effects of non-linearity must be taken into account. The effects may be considerable and sometimes a solution follows by considering the problem in successive steps, each step having the same equations applied, but with different parameters substituted to correspond with the changed condition; an example occurs in connection with Fig. 6.25. In other cases, the non-linear effects can be "averaged out" and although the performance is not completely calculable by this method, certain important aspects of the behaviour can be determined with the same accuracy as by any other practicable method. Several examples of this technique will be met in the chapters on a.c. machines and the basis of the method will now be discussed.

Two familiar equations, already discussed in Section 2.4, will be examined to demonstrate and justify this extended use of superposition:

For the electrical circuit: $\quad V = IR \quad$ or $\quad I = \dfrac{1}{R}.V = GV.$

For the magnetic circuit: m.m.f. $F = \phi\mathscr{R} \quad$ or $\quad \phi = \dfrac{1}{\mathscr{R}}.F = \Lambda F.$

Consider first the electrical circuit. If several driving voltages are connected in series with a resistance, the current is given by:

$$I = G.\sum V.$$

If the conductance is unaffected by the value of current or voltage, the current could also be obtained from the following expression:

$$I = G.V_1 + G.V_2 + \ldots + G.V_n.$$

This equation, in summing the currents due to each voltage acting individually, is therefore invoking the principle of superposition.

In practice, the conductance may vary quite considerably with temperature; copper, for example, increases its resistance by about 20% from a typical ambient temperature of 20°C to a typical working temperature of 75°C. Obviously, the value of current and copper loss will affect the temperature and hence the conductance. Quite apart from this common situation, non-linear resistors are sometimes used deliberately, for special purposes. Such resistors do not necessarily depend on temperature change for their resistance variation with voltage or current. An example of such an application is discussed later in connection with Fig. 6.28.

Figure 3.16 shows an idealised current/voltage characteristic for a non-linear resistor. At low currents, the conductance, dI/dV is high (low resistance), and as the current increases there is a sharp change of slope corresponding to a higher resistance. If the resistor is supplied from two sources in series, V_1 and V_2, the value of current flowing can be read off the characteristic at a point corresponding to the resultant applied voltage. From the characteristic it can be seen that if the voltage is high enough, operation will be in the higher resistance region and superposition is no longer valid because the conductance is not constant. For example, referring to Fig. 3.16b, if V_1 and V_2 are

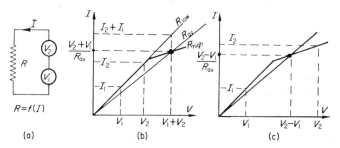

FIG. 3.16. Current through non-linear resistor.

applied in an additive sense, the component currents would be $I_1 = V_1/R_{\text{low}}$ and $I_2 = V_2/R_{\text{low}}$, but the resultant current would not be $I_1 + I_2$. The current is rather less than this. Similarly, with the voltages in opposition, Fig. 3.16c shows that the resultant current cannot be obtained from the difference of the component currents if either or both of these are beyond the low-resistance region.

The apparent resistance, in terms of the ratio between the resultant voltage applied and the current flowing, depends on the operating point and would correspond with the slope of a line drawn from the origin through this point. Such a resistance might be termed the "average" value and if it was known, i.e. if the operating current was known, then superposition could be applied since the current would be given by either $(V_1 \pm V_2)R_{\text{av}}$ or by $(V_1/R_{\text{av}}) \pm (V_2/R_{\text{av}})$.

This method of dealing with non-linearity is particularly appropriate for the ferromagnetic circuit which strictly does not have a linear region at all, though at very high and fairly low flux densities, the slope of the magnetic characteristic, $\Lambda = d\phi/dF$, is fairly constant. The magnetic analogue of the two voltage sources in series is shown on Fig. 3.17a and a typical magnetic characteristic is given on Fig. 3.17b. As for the non-linear resistor, it can be seen that the true value of the unknown quantity, flux in this case, must be read from the characteristic at a point determined by the resultant driving force, m.m.f. in this case. If the operating point is known, however, an average permeance could

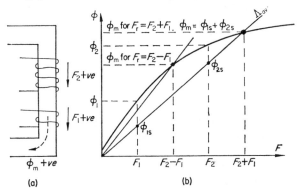

FIG. 3.17. Flux in non-linear magnetic circuit.

be used with the resultant m.m.f. to find the resultant mutual flux ϕ_m. More important, superposition could be used to find ϕ_m correctly, by combining the components ϕ_{1s} and ϕ_{2s} due to F_1 and F_2 respectively, acting alone. The simple example of Fig. 3.17 has the two coils on the same axis but on rotating machines, the component m.m.f.s are, in general, displaced in space phase. However, with sinudoidal distributions, the two m.m.f.s can be combined using vector techniques, Fig. 1.9. It is often more convenient to combine the component fluxes and if the average permeance is known, superposition can be employed as just described, to get the resultant mutual flux.

At first sight it might appear that if the operating flux is known at the beginning, then further calculation is unnecessary because the resultant m.m.f. follows from the magnetic characteristic. Usually however, it is one or other of the component m.m.f.s which is the unknown and this is then calculated using the average permeance corresponding to the resultant flux, which itself is deduced from other considerations to be discussed later. Alternatively, if the equations are in terms of inductances or reactances, the average permeance can be used to calculate saturated values of inductance as will be demonstrated in the later text. It should be emphasised, that the average permeance will only be correct for one particular value of flux, though if this value only varies over a moderate range, the errors in assuming a constant permeance will be correspondingly small. A particular case of interest occurs with a.c. magnetisation, where the flux varies cyclically, following a hysteresis loop from $+\hat{\phi}$ to $-\hat{\phi}$. Although there are appreciable permeance changes throughout the cycle, an average value corresponding to the slope of a line joining $+\hat{\phi}$ to $-\hat{\phi}$ through the origin, is used in practice almost invariably, to calculate the excitation m.m.f. with satisfactory accuracy. On a.c. machines operating at constant voltage and frequency, the peak value of flux does not vary very considerably over the whole operating range, and thus one value of permeance and its corresponding saturated reactance can be used throughout in calculations. This applies to transformers, induction machines and synchronous machines too, if power-factor changes are moderate. The cyclic changes of permeance are manifested by the appearance of harmonic components in the excitation current; c.f. Fig. 4.5.

A final comment will explain how the operating point is determined. The mutual flux ϕ_m is directly responsible for a machine induced e.m.f. E. This e.m.f. can be found independently, from the electrical circuit equation (3.3) i.e. $E = V \pm IZ$. Hence the appropriate average permeance, or its equivalent in terms of a saturated inductance value, follows, as explained in more detail in later chapters.

3.4 THE CONCEPT OF THE EQUIVALENT CIRCUIT

An electrical machine is a very complex piece of apparatus. To understand every part of it and appreciate the interaction between the various parts is a specialised job for the machine designer. Even he has to make approximations and judicious estimates to come out with a reasonable answer in a reasonable time. To do this, he may disregard factors which are known to be negligible when considering a particular problem. For example, the harmonic voltages produced in an a.c. machine contribute little or nothing to the r.m.s. value of the output voltage or to the output power, and therefore many calculations may be simplified by dealing only with the fundamental. However, if interference with communication systems, or machine vibration and noise are under consideration, this simplification is not permissible, though, in fact, the fundamental may be neglected. As a further example, the capacitance of transformer windings need not be taken into account for most calculations at power frequencies, but it becomes all important when analysing the behaviour under the impact of voltage surges. For a d.c. machine, under steady state conditions, the field winding inductance is of no interest but it is otherwise under transient conditions.

One of the first tasks when studying a new machine will be to develop an equivalent circuit. (This term must be understood as referring to an electrical equivalent circuit. There are also magnetic equivalent circuits, e.g. Figs. 2.11 and A.4.) By this is meant a circuit consisting perhaps of generated e.m.f., resistance and inductance. The terminal characteristics of this circuit will be the same as that of the machine itself under appropriate conditions. It is in fact merely a convenient and easily remembered representation of the machine equations. The circuit will be deduced by studying the reaction of the machine to electrical, magnetic or mechanical changes, in combination or alone.

One of the simplest equivalent circuits to derive is that for a battery. By loading the battery and measuring its terminal voltage, it is found within certain limits of current to be as shown in Fig. 3.18. The curve has the equation $V = E - kI$, and the battery behaves as if it were a source of constant voltage E having an internal resistance of k ohms. Whether the battery really is such a device is a question for the specialist; the above interpretation is sufficient to enable calculations of terminal voltage to be made for any value of load current. If the battery is in

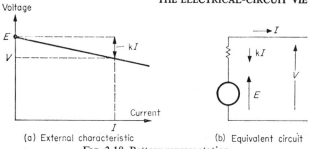

(a) External characteristic   (b) Equivalent circuit
Fig. 3.18. Battery representation.

circuit with other batteries and resistances, calculations of currents in the various parts of the circuit are readily possible from the equivalent circuit, which expresses a complex electrochemical reaction in terms of simple circuit equations.

It should be pointed out that this is not the only possible equivalent circuit; a source of constant current, E/k, shunted by a resistance of k ohms, would give the same output current and voltage to an externally connected load. However, the constant-voltage source, sometimes referred to as the Thévénin equivalent circuit, is usually the most suitable for machine studies and will be used in preference.

A d.c. machine would have the same form of equivalent circuit as the battery whether motoring or generating. An energised transformer and a synchronous machine can also be expressed in similar terms but with the addition of a series reactance (see Section 3.9).

To obtain the equivalent-circuit parameters of a machine, certain light-load tests on open-circuit and short-circuit are carried out, but it will be more convenient to discuss these under the appropriate chapter headings. However, the battery circuit can be used to illustrate the significance of these tests. When open-circuited, the current is zero and the measured terminal voltage is E. When short-circuited, the current would be I_{sc} say, but the terminal voltage V would be zero. Since $V = 0$ would also be equal to $E - kI_{sc}$ for this condition, k could be found on dividing the open-circuit voltage, E, by the short-circuit current. In practice, a battery would be damaged if short-circuited, but for a machine, the open-circuit voltage can be controlled and therefore the current can be reduced to a safe value.

3.5 PARALLEL OPERATION: *PER-UNIT* NOTATION

For many reasons, including flexibility of control, economy of operation and design limitations, it is often necessary to operate two or more machines in parallel to supply the same load. In any case, sooner or later, electrical supply equipment becomes inadequate to meet the increasing demand. Since supply voltages are normally maintained, the extra units must be connected in parallel. This raises two problems. Firstly, since one unit is connected across the other, there is a possibility of circulating currents between the two units if a voltage discrepancy exists. These currents may be small and tolerable, or excessive and dangerous. Secondly, although two connected units will share the load, it does not follow that the load sharing will be equitable. Most machines suffer a fall in terminal voltage as load is applied and since the two units are connected at the terminals, they must have a common voltage here. One machine may have to take an excessive current in order that its terminal voltage will fall to the same value as the other machine.

These two points are illustrated by Fig. 3.19a which shows the equivalent circuits of two batteries in parallel. Before the batteries are connected together, the relative polarities must of course be checked to see that the e.m.f.s are in opposition round the local circuit. If the e.m.f.s were in the same sense, they would drive a ruinous short-circuit current round the loop.

Consider first of all the no-load condition where the switch is open. $E_1 = 2 \cdot 2$ V and $E_2 = 2 \cdot 1$ V so there is a voltage difference of $0 \cdot 1$ V acting round the circuit and absorbed in the resistance of $0 \cdot 3 \, \Omega$. Thus a current of $0 \cdot 33$ A, circulating clockwise, flows out of 1 into 2. The common terminal voltage is obtained using the generator equation on 1, i.e. $V = 2 \cdot 2 - 0 \cdot 33 \times 0 \cdot 2 = 2 \cdot 133$ V or using the motor equation on 2, i.e. $V = 2 \cdot 1 + 0 \cdot 33 \times 0 \cdot 1 = 2 \cdot 133$ V.

If each battery was loaded individually, the terminal voltage would fall in accordance with $V = E - IR$. The two external characteristics

88

on this basis are plotted on Fig. 3.19b. With the batteries connected in parallel, the individual currents at any particular terminal voltage could be read off and the total I_1+I_2 could also be plotted against the same terminal voltage. This would be the total external characteristic. Now, it is possible to find how the batteries share the load at any particular

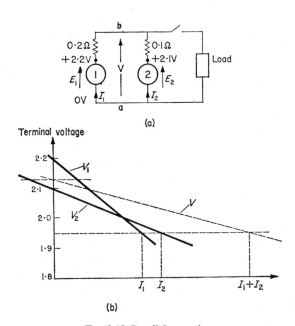

FIG. 3.19. Parallel operation.

value of total load current. In the case shown, there is a considerable difference which may or may not be undesirable depending on the individual capacity of the batteries. At zero total current, i.e. no load, battery 1 gives a positive current and battery 2 gives a numerically equal negative current, confirming the analytical solution. Battery 1 is a source being discharged and battery 2 is a sink being charged.

The reasons for the unequal load sharing are:

(a) the difference in e.m.f.s and the resulting circulating current which tends to overload one battery and unload the other,

(b) the different values of IR drops at particular values of current.

To ensure that each battery carries its proportionate share of the load it is necessary that the two e.m.f.s are equal. Further, the individual resistance drops must be equal when rated current flows in each battery. Under these conditions equal changes of battery terminal voltage are brought about by equal fractions of rated battery current.

The Per-unit Notation

This last point can more easily be appreciated by rearranging the equation $V = E - RI$. Dividing throughout by E and introducing $I_{f.l.}/I_{f.l.}$ in the last term gives:

$$\frac{V}{E} = \frac{E}{E} - \frac{I_{f.l.}R}{E} \cdot \frac{I}{I_{f.l.}}$$

$$V_{p.u.} = 1 - R_{p.u.} \cdot I_{p.u.}$$

where p.u. stands for *per unit*. In this equation, each term is expressed as a fraction of an arbitrarily chosen base quantity, the following quantities being convenient in the present case:

voltages based on the open circuit voltage giving $E = 1$ p.u. voltage;

currents based on the full-load current giving $I_{f.l} = 1$ p.u. current.

resistances based on the quotient $E/I_{f.l.}$ since $R_{p.u.} = \dfrac{R}{E/I_{f.l.}}$.

Hence, no matter what the current rating of each battery, if the open circuit voltage, and the full-load IR drop expressed as a fraction of this voltage are the same for each battery, they have the same equation for their external characteristics when plotted in terms of *per-unit* voltage against *per-unit* current. It follows that at any common terminal voltage, each battery delivers the same *per-unit* current; i.e. the current is shared in the same proportion as the individual battery ratings.

The use of *per-unit* quantities leads to economies in calculation time, the equations becoming dimensionless, a feature which also has disadvantages, since checking the equations by dimensional consistency is no longer possible. *Per-unit* parameters convey information in a way which makes certain comparisons very convenient. The value of *per-unit* resistance not only indicates the drop of terminal voltage but also the full-load efficiency since it is the fraction of generated voltage and power dissipated internally. Problems solved in the book will be worked out sometimes using numerical values and sometimes with *per-unit* quantities. It will be a good exercise for the student to recalculate the answers by whichever method has not been used. When *per-unit* values are multiplied by 100 they become *percentage* values, but this is really a wasteful operation and offers no compensating advantages over the *per-unit* system other than familiarity.

Extension of Graphical Method

The application of the graphical method for load sharing calculations can be extended in three ways:

(1) A machine does not usually have a linear characteristic and this makes an analytical solution difficult. However, since the method only involves the addition of abscissae, it is not restricted to straight line external characteristics, see Example E.6.8, p. 352.

(2) If there are three or more machines, the abscissae can still be added to give, say, $I_1 + I_2 + I_3$, and then, for any total current, the terminal voltage and the individual currents can be read off.

(3) In certain circumstances, motors have to operate in parallel mechanically, either on the same shaft as for certain rolling-mill and traction drives, or synchronised at the same speed, as when turbines are driving alternators. The same method as above can be used with speed and torque substituted for voltage and current. In this case, the situation demands that two or more units run at the same speed, and the torque and power which each supplies depends on the external mechanical characteristics.

It will be noticed in Fig. 3.19b that the battery with the more steeply sloping characteristic tends to throw off the load whilst the one with the shallow curve tends to be overloaded. For good load sharing of machinery generally, steep characteristics are preferable because in service there are changes in behaviour. If a shallow characteristic were to become even slightly flatter, it would seriously affect the load sharing, whereas a slight change to a steep curve would make little difference.

Further development of the analytical method will be deferred until the chapter on transformers.

3.6 LOSSES AND EFFICIENCY

For power apparatus, the output/input efficiency is of prime importance. Some knowledge of the power losses which occur in electrical machinery is therefore necessary. Here is not the place to discuss them in detail, but since all machines have losses, it is appropriate to explain briefly how they arise and the factors on which they depend.

Winding Copper Loss Excluding D.C. Field-circuit Loss

This is well known as I^2R, but R must be the effective resistance. The measured d.c. resistance is only the effective resistance at low frequencies. As shown in Fig. 2.5 (p. 30) a changing flux induces eddy currents in a direction to oppose the change. An alternating current establishes such a flux across the conductor section in a direction transverse to the current flow. The effects of the main and eddy currents can be considered independently if conductor resistivity is uniform throughout the copper section. Actually, the eddy currents are superimposed on the working current flowing through the conductor and result in a maldistribution of current and additional *stray* losses. This can be understood by considering the highly idealised case of Fig. 3.20 where, with current $2i$ increasing, the eddy current is additive in the top half of the conductor giving $i+i_e$ and subtractive in the bottom half giving $i-i_e$. If r is the d.c. resistance of the conductor, each half will be $2r$ since the halves are in parallel. The total loss will be:

$$(i+i_e)^2 2r + (i-i_e)^2 2r = 4(i^2r + i_e^2 r) = 4i^2 r[1 + (i_e/i)^2].$$

92

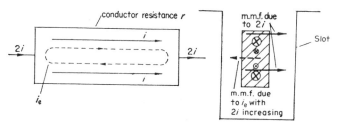

FIG. 3.20. Copper eddy-current loss; conductor in a slot.

With uniform distribution of current, the loss would just be $4i^2r$, so the eddy-current loss can in fact be computed separately and increases with frequency. The effective, increased resistance R_{ac} say, is the d.c. resistance at the working temperature, multiplied by the factor in brackets and can be measured from the power input to a machine when it is short-circuited with normal-frequency currents flowing.

Additional Electrical Losses

The alternating fluxes due to load current can cause induced currents and losses in surrounding machine parts which are made of conducting material. These add to the stray load losses and are difficult to calculate.

A further load current loss occurs when machines are provided with brushes, to make sliding contact with slip rings or commutator. The voltage drop and the associated power loss vi may be taken to vary directly with current since the contact drop is very approximately constant, of the order of 1V or less, per series brush path.

On machines provided with d.c. excited poles, there is a field I^2R loss which may sometimes be included in the fixed loss (see below).

Mechanical Losses

These consist of bearing-friction and also ventilation losses caused by the disturbance of air or other cooling medium and would have to include any power required to drive this air through the machine. Mechanical torques, including loss torques, are sometimes idealised by

93

considering them in three components giving $T_m = k_1 + k_2\omega_m + k_3\omega_m^2$. k_1 corresponds to the constant, coulomb friction load to which the bearing friction approximates. $k_2\omega_m$ is the viscous friction, a torque proportional to ω_m which arises with non-turbulent or streamline fluid flow and also characterises the torque due to eddy currents. Turbulent flow causes a torque which is approximately proportional to (speed)^2 or an even higher power of speed. The action of propellors or machine cooling fans is responsible for this torque component. The mechanical power is of course obtained on multiplying T_m by ω_m.

Fixed Losses

If the flux, frequency and speed are constant, then the iron loss, mechanical losses and excitation losses are approximately independent of load. They are sometimes grouped together as the fixed or constant loss. Equations (2.2) and (2.3) show how the power for iron losses varies with speed (frequency) and flux density.

Efficiency Calculations

Since the output is equal to the input minus the losses, the efficiency expression can be rearranged as below:

$$\eta = \frac{\text{output}}{\text{input}} = \frac{\text{input} - \text{losses}}{\text{input}} = 1 - \frac{\text{losses}}{\text{input}} = 1 - \frac{\text{losses}}{\text{output} + \text{losses}}. \quad (3\cdot6)$$

Three-figure accuracy in the final term will give four-figure accuracy in the overall expression when the efficiency exceeds 0·9 p.u. (90%). On large machines, the efficiencies may be as high as 99%, so this was a useful aid in the days before pocket calculators.

It will be noticed that the efficiency is expressed as a fraction, i.e. on a *per-unit* basis. If the losses are expressed as a fraction, say of the output, the equation becomes

$$1 - \frac{\text{losses/output}}{(\text{output} + \text{losses})/\text{output}} = 1 - \frac{\text{losses}_{\text{p.u.}}}{1 + \text{losses}_{\text{p.u}}}$$

which reduces to:

$$1/(1 + \text{losses}_{\text{p.u.}}).$$

94

Maximum Efficiency

The losses are a function of load current and can be written as $aI^2 + bI + c$. The output for a power source is $VI \cos \varphi$. The efficiency is therefore:

$$\eta = \frac{VI \cos \varphi}{VI \cos \varphi + aI^2 + bI + c}.$$

For constant V and φ,

$$\frac{\mathrm{d}\eta}{\mathrm{d}I} = \frac{(VI \cos \varphi + aI^2 + bI + c)(V \cos \varphi) - VI \cos \varphi(V \cos \varphi + 2aI + b)}{(\text{denominator})^2}.$$

Many of the terms cancel, leaving

$$\frac{\mathrm{d}\eta}{\mathrm{d}I} = \frac{(c - aI^2)}{(\text{denominator})^2} V \cos \varphi. \qquad (3.7)$$

This expression is equal to zero when the constant part of the loss c is equal to that part of the loss which varies as I^2, and this is the condition for maximum efficiency. The curve of efficiency against load current is in fact fairly flat about this point so the condition is not critical. However, if the apparatus is to operate for long periods at loads much less than the rated figure, it can be designed to balance the losses in a favourable manner to give minimum total energy loss.

3.7 MAGNETICALLY COUPLED COILS

Development of Circuit Equations

The two coils of Fig. 3.21, carrying currents i_1 and i_2 respectively, are magnetically coupled due to the flux which is common to both of them.

95

The directions of the currents shown have been chosen so that they magnetise in the same sense when they are both positive or both negative; i.e. when they both flow into or out of the dotted ends of their respective coils. The assigned positive directions of the two currents could have been, and often are, chosen to correspond with magnetisation in mutually opposing senses.

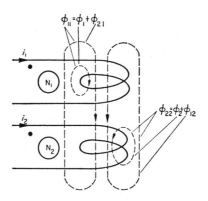

FIG. 3.21. Magnetically coupled coils.

Current i_1 acting alone, produces a self flux ϕ_{11} which is conveniently considered in two components:

ϕ_{21}, a mutual flux component linking N_1 and N_2 turns, and

ϕ_1, a leakage flux, linking N_1 turns only.

Similarly, i_2 acting alone produces a self flux $\phi_{22} = \phi_{12} + \phi_2$.

The resultant, or total mutual flux, is given by $\phi_m = \phi_{21} + \phi_{12}$.

The e.m.f. induced in N_1 if the currents and the fluxes change with time, is given by considering all the flux components linking N_1. The fluxes can then be expressed in terms of an m.m.f. and a permeance since, from eqn. (2.6):

$$\phi = F \times \Lambda = NI \times \Lambda.$$

Hence, from eqn. (1.1):

$$e_{N_1} = \frac{d}{dt} [N_1\phi_1 \qquad +N_1\phi_{21} \qquad +N_1\phi_{12} \qquad] \quad (3.8)$$

$$= \frac{d}{dt} \left[N_1.N_1i_1.\Lambda_1 + N_1.N_1i_1.\Lambda_{21}.\frac{N_2}{N_2} + N_1.N_2i_2.\Lambda_{12} \right]$$

$$= \frac{d}{dt} \left[N_1{}^2\Lambda_1.i_1 \quad +\frac{N_1}{N_2}.N_1N_2\Lambda_{21}.i_1 \quad +N_1N_2\Lambda_{12}.i_2 \right]$$

$$= \frac{d}{dt} \left[\quad l_1.\ i_1 \quad +\frac{N_1}{N_2}.\quad L_{21}.\quad i_1 + \quad L_{12}.\quad i_2 \right] \quad (3.9a)$$

In this expression, and comparing terms with those of eqn. (3.8):

$l_1 = N_1{}^2\Lambda_1$, is the leakage inductance of coil 1 with respect to coil 2.

$l_1i_1 = N_1\phi_1$, is the leakage flux linkage.

$L_{12} = N_1N_2\Lambda_{12}$ is the mutual inductance of coil 2 on coil 1.

$L_{12}i_2 = N_1\phi_{12}$, is the mutual-flux-linkage component with coil 1 due to current in coil 2.

$L_{21}i_1 = N_2\phi_{21}$, is the mutual-flux-linkage component with coil 2 due to current in coil 1.

$\dfrac{N_1}{N_2}.L_{21}i_1 = N_1\phi_{21}$ (due to the same mutual flux ϕ_{21} produced by current in coil 1), is a component of the self-flux-linkage of coil 1.

An alternative arrangement of the equations is obtained by omitting the N_2/N_2 operation on the middle term of eqn. (3.8), giving:

$$e_{N1} = \frac{d}{dt} [N_1{}^2\Lambda_1.i_1 + N_1{}^2\Lambda_{21}.i_1 + N_1N_2\Lambda_{12}.i_2]$$

$$= \frac{d}{dt} [\quad N_1{}^2(\Lambda_1+\Lambda_{21}) \quad .i_1 + \quad L_{12} \quad .i_2]$$

$$= \frac{\mathrm{d}}{\mathrm{d}t} [\quad N_1{}^2 \Lambda_{11} \cdot i_1 \quad + \quad L_{12} \cdot i_2]$$

$$= \frac{\mathrm{d}}{\mathrm{d}t} [\quad L_{11} \cdot i_1 \quad + \quad L_{12} \cdot i_2] \qquad (3.9b)$$

In this expression:

$L_{11} = N_1{}^2 \Lambda_{11}$, is the self (or total) inductance of coil 1 due to the total self-flux linkage $L_{11}i_1 = N_1\phi_{11}$ with coil 1, due to current i_1 in the coil itself.

Note 1. Similar equations could be obtained for coil 2 simply by interchanging the suffices; e.g. the self inductance of coil 2 is $L_{22} = N_2{}^2 \Lambda_{22}$, where $\Lambda_{22} = \Lambda_2 + \Lambda_{12}$.

Note 2. If the permeance to mutual flux viewed from either coil is the same, $\Lambda_{21} = \Lambda_{12}$ and so $L_{21} = L_{12} = M$ say. This is the situation for static coils. The symmetry does not hold for rotating coils when, as in some treatments of electrical machines, the rotational voltages are considered in terms of a mutual inductance; e.g. in eqn. (10.4), $X_{dq} \neq X_{qd}$.

Note 3. By comparing eqns. (3.9a) and (3.9b), it will be seen that the self inductance $L_{11} = l_1 + L_{21} \cdot N_1/N_2$ so that l_1, the leakage inductance is equal to $L_{11} - L_{21} \cdot N_1/N_2$ or, $L_{11} - M \cdot N_1/N_2$.

Note 4. The effect of coupling further coils would be to introduce further mutual terms $L_{13} \cdot i_3$, $L_{14} \cdot i_4$, etc., into eqn. (3.9b). This form of the equation is therefore very suitable when the equations of several coupled coils have to be considered; the expressions then being in terms of self and mutual inductances as used in Chapter 10.

Note 5. Equation (3.9a), in terms of leakage and mutual inductances, can be used very simply to develop the equivalent circuits of devices in which two m.m.f.s can be distinguished; e.g. primary and secondary m.m.f.s, stator and rotor m.m.f.s or field and armature m.m.f.s. Even though there are several coils on the stator say, their magnetic effects can be combined to give one resultant stator m.m.f. This transformation of the machine windings to give a simpler two-coupled-coil arrangement will be used in the major portion of the text.

Note 6. The effect of resistance is to introduce a term Ri into the equations. Using eqn. (3.9b), for example, the terminal voltage of coil 1 is:

$$v_1 = R_1 i_1 + L_{11}\, di_1/dt + M\, di_2/dt. \tag{3.10a}$$

Similarly, for coil 2:

$$v_2 = R_2 i_2 + L_{22}\, di_2/dt + M\, di_1/dt. \tag{3.10b}$$

Coefficient of Coupling k

The efficiency of flux transfer is the ratio of the mutual flux component to the self flux. By reference to the development of eqns. (3.9a) and (3.9b), the efficiency ratios for the two coils can be derived as below:

$$k_1 = \frac{\phi_{21}}{\phi_{11}} = \frac{Mi_1/N_2}{L_{11}i_1/N_1} = \frac{MN_1/N_2}{L_{11}}$$

and

$$k_2 = \frac{\phi_{12}}{\phi_{22}} = \frac{Mi_2/N_1}{L_{22}i_2/N_2} = \frac{MN_2/N_1}{L_{22}}.$$

the product $k_1 k_2 = \dfrac{M^2}{L_{11}L_{22}} = k^2$ where $k = \sqrt{\left(\dfrac{M^2}{L_{11}L_{22}}\right)}.$

k is the geometric mean of k_1 and k_2 and therefore a measure of the closeness of coupling taking the viewpoint from both coils into account. It has a maximum value of unity when the coils are perfectly coupled. On the iron-cored transformer (Chapter 4) which utilises two or more closely coupled coils, k is quite near to unity.

Effective Inductance of Two Coupled Coils Neglecting Resistance

It frequently happens under transient conditions that one of two coupled coils is closed through a low resistance and the behaviour of the circuit is then quite different from the condition when only one coil carries current. From eqns. (3.10a) and (3.10b) the voltage equations with coil 2 short-circuited are:

$$v_1 = L_{11}di_1/dt + M\,di_2/dt$$

and $v_2 = 0 = L_{22}\,di_2/dt + M\,di_1/dt$ from which: $\dfrac{di_2}{dt} = \dfrac{-M}{L_{22}}\dfrac{di_1}{dt}$.

Substituting for di_2/dt:

$$v_1 = L_{11}\,di_1/dt - \frac{M^2}{L_{22}}\,di_1/dt = L_{11}(1 - k^2)\,di_1/dt. \qquad (3.11)$$

The effective inductance viewed from coil 1 under these conditions is the self inductance L_{11} reduced as the so-called Blondel leakage factor $(1 - k^2)$. This is not the same as the leakage factor discussed on p. 34, which is equal to $1/k_1$.

If the coupling is perfect, the effective inductance is zero, the current then being limited only by the internal resistance. In the general case, $k < 1$. The effect of a closed coupled coil is to speed up the time-response of primary current i_1 as a result of the reduced inductance, and to slow up the time-response of mutual flux. Both of these effects are due to the secondary m.m.f. i_2N_2 opposing any change of mutual flux and thus reducing the corresponding primary back e.m.f. e_1. If the resistive terms had been included in the equations above, it would have increased their complexity without demonstrating so clearly the effects of the secondary-coil m.m.f. The transient response is discussed further in Section 3.10.

3.8 REPRESENTATION OF A ROTATING ELECTRICAL MACHINE BY COUPLED COILS

In this section, which may be omitted on a first reading of the book, the ideas underlying generalised analysis of electrical machines are reviewed very briefly. Though not essential for present purposes, they

are useful as a background against which the physical approach to
electrical machine theory can be considered. A more detailed discussion
of generalised circuit analysis is given in Chapter 10.

The torque of machines with rotor and stator windings is a direct
consequence of the interaction between their m.m.f.s. In a particular
case, the m.m.f. of even a simple 2-pole winding may be due to currents
in several distributed coils. Nevertheless, the same m.m.f. could be
produced by one coil if its turns were suitably distributed around the
periphery to give the same variation of field intensity. If this intensity is

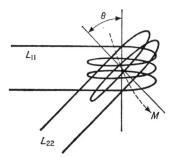

FIG. 3.22. Two-coil representation of machine.

assumed to vary sinusoidally in space, much simplification in the
analysis ensues. Further, if electrical angles are used, the analysis can
be applied directly to a 2-p pole machine providing that each pole pair
has an identical and symmetrical flux-density pattern; there being p
complete waves of flux-density variation round the periphery. For the
moment, only a 2-pole case will be considered and the stator and
rotor will each be assumed to consist of one winding section. The
mutual inductance and possibly the self inductances, will vary with the
angular position of the rotor. The coils shown in Fig. 3.22 will therefore
represent the reactions in the machine if they have the same inductance
as the machine windings themselves and the same inductance variations
with angle.

101

E.M.F. Equation in Terms of Inductances

Consider the e.m.f. equation for one coil:

$$e = \frac{d(N\phi)}{dt} = \frac{d(Li)}{dt}.$$

In the general case, both the inductance and the current may vary with time so that the expression must be completely differentiated to allow for this possibility. Hence:

$$e = L\cdot\frac{di}{dt}+i\cdot\frac{dL}{dt} = L\cdot\frac{di}{dt}+i\cdot\frac{dL}{d\theta}\cdot\frac{d\theta}{dt}.$$

The first term, for which the inductance is constant with i differentiated, is the transformer e.m.f. giving rise to electrical/electrical power conversion. The second term, the rotational e.m.f. for which the current is constant, arises if the inductance varies as a result of rotor movement causing permeance variations with time. In this case, there is mechanical/ electrical power conversion which can be related to the motional, or speed e.m.f. produced. Self inductance may not change with angular position but when there are two coils in relative motion, the circuit geometry is not constant and the mutual inductance changes through a complete positive and negative cycle as the electrical angle changes through 360°. The modification to the last term of the equation is convenient when the speed $\dot\theta = d\theta/dt$ is constant, the inductance variation being expressed in terms of angle rather than in terms of time.

Using eqn. (3.9b) and considering the two coils of Fig. 3.22:

$$e_{N1} = d(L_{11}\cdot i_1 + M\cdot i_2)/dt$$

$$= L_{11}\cdot di_1/dt + M\cdot di_2/dt + i_1\cdot dL_{11}/dt + i_2\cdot dM/dt.$$

A similar expression can be obtained for the total induced voltage e_{N2} of coil 2, by complete differentiation of $(L_{22}\cdot i_2 + M\cdot i_1)$.

The first two terms give the transformer e.m.f. used for example in eqn. (3.10a). For a rotating system, the second pair of terms is conveniently arranged as follows after introducing the speed term $\dot{\theta} = d\theta/dt$:

$$e_{N1}(motional) = (i_1\, dL_{11}/d\theta + i_2\,.\,dM/d\theta)\,.\,d\theta/dt. \qquad (3.12)$$

Torque Equation in Terms of Inductances

Neglecting self-inductances variations for the moment, the total rate of energy flow due to those e.m.f. components caused by movement is obtained from the last term of the e_{N1} and e_{N2} equations as:

$$i_1 e_{N1}(motional) + i_2 e_{N2}(motional) = i_1(i_2\, dM/dt) + i_2(i_1\, dM/dt)$$
$$= 2i_1 i_2\, dM/dt.$$

Now it can be shown that the stored energy in the mutual field of two coupled coils is given by $i_1 i_2 M$ and so with currents constant, the rate of energy flow into the mutual field is $i_1 i_2\, dM/dt$ which is exactly one half of the power supplied. The other half is converted to mechanical energy and can be re-expressed as:

$$(i_1 i_2\, dM/d\theta)\, d\theta/dt.$$

In mechanical terms the rate of energy conversion is $T_e\, d\theta/dt$. By equating the two expressions since they refer to the same energy, the torque in terms of mutual inductance changes is:

$$T_e = i_1 i_2\, dM/d\theta. \qquad (3.13)$$

The self-inductance terms in the motional e.m.f.s also give rise to a torque if L_{11} or L_{22} vary with angular position. By extending the above argument, taking into account the stored energy in the self fields, $\frac{1}{2}L_{11}i_1^2 + \frac{1}{2}L_{22}i_2^2$, it can be shown[2], that the additional torque is numerically equal to the angular rate of change of this energy with currents constant. The total torque is thus

$$T_e = \tfrac{1}{2}i_1^2\, dL_{11}/d\theta + \tfrac{1}{2}i_2^2\, dL_{22}/d\theta + i_1 i_2\, dM/d\theta. \qquad (3.14)$$

Mutual inductance always changes with θ since there is relative angular movement between the windings. Self inductance changes if the air gap is not uniform round the periphery as in the salient-pole machine, and variations of magnetic reluctance occur. Consequently any torque due to the first two terms of eqn. (3.14) is called a reluctance torque and this is discussed further in Chapters 8 and 10.

It is perhaps worth pointing out that the dynamometer instrument movement is a simple practical example of the two-coil representation. The moving coil instrument is also governed by equation (3.13) though it is preferably expressed in terms of flux linkage; i.e. since $\mathrm{d}M/\mathrm{d}\theta = \mathrm{d}(N_1\phi_{12}/i_2)/\mathrm{d}\theta$, then

$$T_{\mathrm{e}} = i_1 i_2\, \mathrm{d}M/\mathrm{d}\theta = i_1 i_2\, \mathrm{d}(N_1\phi_{12}/i_2)/\mathrm{d}\theta = i_1 N_1\, \mathrm{d}\phi_{12}/\mathrm{d}\theta.$$

The torque is thus proportional to the moving coil m.m.f. and the angular rate at which the flux produced by the other member is changing. The other member is a permanent magnet and by arranging that the coil always moves in a uniform air gap to give a uniform flux density, $\mathrm{d}\phi_{12}/\mathrm{d}\theta$ is constant and the deflecting torque is directly proportional to the coil current. A linear restraining spring then ensures the characteristic uniform scale of the moving-coil instrument.

An example of reluctance torque is provided by the moving-iron instrument movement. Here there is only one coil and only one term of eqn. (3.14) is applicable. The instrument torque is therefore $\frac{1}{2}i_1{}^2\, \mathrm{d}L_{11}/\mathrm{d}\theta$. The instrument can thus measure r.m.s. values but the scale is not uniform. By careful design of the L_{11}/θ characteristic, however, the major portion of the scale can approach linearity quite closely except at low values of θ.

Replacement of Moving Coils by Two Stationary Coils

The expressions derived for e.m.f. and torque as function of variable inductances may not seem to be very convenient as a starting point for electrical machine analysis. It is possible, however, to employ mathematical transformations which re-form the equations in terms of constant inductance coefficients which in fact correspond to the maximum or minimum values of the actual inductance occurring during a complete rotational cycle of variation. The transformed voltages and currents too, are closely related to the actual values. Providing that the speed $d\theta/dt$ is constant, the transformed differential equations have constant coefficients and can be solved by well-established methods. This method of generalised analysis permits the routine solution of steady-state, transient and certain unbalanced conditions for all types of machine, and will be explained in more detail in Chapter 10.

The physical explanation of the transformation just described can be appreciated in a general way by reference to the ideas discussed in Sections 1.3 and 1.4. It was shown that in the steady-state mechanical condition ($d\theta/dt$ constant), stator and rotor m.m.f.s must be stationary with respect to one another. Though in fact the m.m.f.s may both be moving in space, they must be in synchronism with each other at a fixed angular displacement. The magnetic effects are the same as would occur with two fixed coils, for which the inductances would be constant and the mathematical transformation confirms this. The transformed coils are not necessarily stationary in space though this is the obvious choice for the d.c. machine where the actual m.m.f. axes of the machine itself are stationary; see Fig. 1.5. The transformation is possible because in addition it is assumed that the m.m.f.s are sinusoidally distributed in space. This in turn permits the resolution of the m.m.f.s along two axes at right angles; the d- (direct) axis and the q- (quadrature) axis. This is very convenient when dealing with the salient-pole machines where there are two orthogonal axes of symmetry, one along the main-pole axis itself and the other between and in quadrature with the main poles. The different magnetic permeances along these two axes can easily be taken into account in the values of the d-coil and q-coil inductances of the transformed equations. Yet a further advantage of the method is the facility with which any number of stator and rotor coil sections can be incorporated in the analysis; each section having its own transformed d- and q-coils. Applying the standard manipulations of matrix analysis may further simplify the solution as described in reference 2.

The required assumptions inherent in the generalised approach outlined above are not really different from those idealisations common to more traditional methods. It is also true that the generalised approach may offer no particular advantages in special cases, e.g. with certain kinds of machine winding or magnetic-circuit asymmetry. It is a powerful tool, though something of a sledgehammer when employed to deal with standard, balanced steady-state conditions (see Section 10.3) and with many problems arising in design and in application engineering. A physical approach is necessary therefore both for its own sake to give a deeper understanding and also to provide a tangible background for mathematical generalisations.

Energy Balance

An alternative approach to the unified treatment of electrical machine theory is from energy considerations. Under steady-state, steady-speed conditions, the electrical energy input less the electrical loss for a motor is equated to the mechanical energy output which will include any mechanical loss. For a generator the terms electrical and mechanical are merely interchanged in the balance equation. When transient conditions are being considered there will be, in addition, stored-energy terms for the magnetic field and mechanical inertia.

The method is particularly useful for devices such as electromagnets which *only* operate in the transient state and during this time the inductance varies because of saturation and movement. The method is also useful as a check on certain fundamental equations. For example, the equations *e.m.f.* = Blv and *Force* = Bli are obtained by considering the reactions on a conductor immersed in a magnetic field having flux density B. In a practical machine the conductors are embedded in slots and it is not obvious that the same expressions hold since the flux density in the slot is much less than that in the air gap. In Section 1.2 the Blv formula was derived by considering the change of flux linkage due to movement of the coil. For a slot pitch of movement, the flux change is the same whether the conductor is in the slot or on top of the tooth so the average e.m.f. $B_{av}lv$ applies in either case.

Now consider the energy balance. If the conductor is carrying a current i, in a time dt it is associated with an amount of electrical energy $ei\,dt = Blvi\,dt$. From mechanical considerations the energy is equal to the force times distance; i.e. force . v . dt. Equating these expressions, since they refer to the same power, the force is found to be Bli. Sufficient has been said to prove that the average force is the same whether the conductors are in slots or on the surface. More refined methods would be necessary to study the instantaneous force. Actually the force is not all exerted on the conductor itself when embedded. This is fortunate because it reduces the stresses on the conductor insulation. When the conductor magnetic field is superimposed on the no load field, Fig. 2.13c, it causes the flux density on one side of the tooth tips to increase considerably. It is here that the major component of the tangential force is exerted.

The energy balance equation can be applied to the whole winding. Since in general the power per phase associated with the motional e.m.f. is $EI \cos \varphi$ where φ is the time-phase angle between the induced e.m.f. and the winding current, this can be equated to the power, in mechanical terms, required to balance the electromagnetically developed torque.

Hence:

$$EI \cos \varphi \times \text{number of phases} = T_e \omega_m.$$

The torque of the whole machine is therefore:

$$T_e = \text{number of phases} \times (EI \cos \varphi)/\omega_m. \qquad (3.15)$$

Other Methods of Analysis

Other methods of dealing with electrical machinery on a unified basis have been and are being developed. Each method has its protagonists and possibly particular advantages. No one treatment is likely to be superior in all respects though all theories contribute to the general understanding of machinery and associated circuits. A high and increasing standard of mathematical ability is called for, but this is only to be expected in view of the complex problems for which solutions can now be obtained. There is, of course, a danger in losing the physical reality behind the machine, and in this text practical

107

aspects are kept uppermost. It is still possible, even on this basis, to unify the treatment in various ways. In the next section is developed a generalised approach to the equivalent circuits of the main machine types for steady-state conditions.

3.9 EQUIVALENT CIRCUITS OF MAIN MACHINE TYPES

Section 3.8 was concerned with the application of eqn. (3.9b) for two coupled coils. This equation is in terms of self and mutual inductances and, with development, shows clearly the transformer and motional components of the induced e.m.f.s. The approach to electrical machine theory in this book is mostly associated with eqns. (3.8) and (3.9a), dealing respectively with the flux components and the leakage and mutual inductances. These factors are more readily related to the features which affect the design of the machine. For example, leakage effects are of primary importance in design and performance, particularly for transformers and induction machines. Under normal running conditions, the leakage flux components are not susceptible to saturation and therefore give rise to constant values of leakage inductances. The mutual flux components, however, are directly affected by saturation, the combination $\phi_m = \phi_{21} + \phi_{12}$ determining the operating point on the magnetic characteristic, see Fig. 3.17b.

If one coil only, of a 2-coil system is carrying current, certain voltage components arise as discussed in Section 3.7 and these are indicated on the hybrid diagram of Fig. 3.23a. Coil resistance is included and the voltage directions shown correspond to positive currents and positive rates of change of fluxes and currents. The leakage flux ϕ_1 is separated from the mutual flux since it produces a voltage only in coil 1, when flux is changing. The mutual flux ϕ_{21} however, can produce a voltage in both coils, even though coil 2 is not carrying current. If the rate of change of ϕ_{21} is the same with respect to both coils, then the voltage

induced in the coupled coil differs from that in coil 1 by the turns ratio
N_2/N_1. This is the situation in the transformer. In general, however,
the voltage in the coupled coil may be due to either a time change of
current and flux with the coils relatively at rest (transformer e.m.f.),
or due to coil 2 moving relative to coil 1 giving a motional e.m.f.,
which involves mechanical power. Both components may be present or
indeed, if the flux is steady with respect to either coil, this coil will have

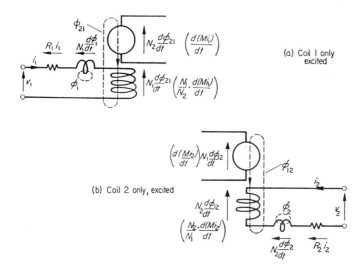

FIG. 3.23. Voltage components with one coil excited.

zero induced voltage. For present purposes, in this analysis, the
emphasis is not so much on the transformer and rotational components of the induced e.m.f., but rather on determining the appropriate
overall rate of flux change by considering the operational mode of the
machine in physical rather than mathematical terms.

If coil 2 only is excited, a similar flux pattern is set up and the voltage
components can be represented in the same way as for coil 1. Figure
3.23b shows the corresponding diagram. By combining Figs. 3.23a and

3.23b, a general diagram is obtained, Fig. 3.24, showing all the voltage components which could be produced when both coils are carrying a current. The resultant voltages due to the mutual components, e_1 and e_2, are conveniently expressed in terms of the resultant mutual flux $\phi_m = \phi_{21} + \phi_{12}$. From this general diagram, it is possible to derive the equivalent circuit of all the main machine types for the steady-state condition. Due account must be taken of each flux component and whether it has a time rate-of-change with respect to the coil being considered. The different machine types have different rates of change

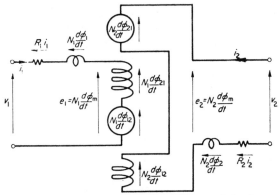

FIG. 3.24. Voltage components for two coupled coils.
General diagram.

associated with the various components. The outline of machine operational modes given in Section 1.3 should be sufficient to appreciate, in a general way, the reasons for these differences, though the understanding will be deepened after reading the appropriate chapters later on in the book.

For a.c. machines it is often convenient to express induced voltage in terms of a reactance, corresponding to the operating frequency. This reactance can then be incorporated in the circuit equation as discussed in Section 3.2. The reactance values are obtained by considering the development of eqn. (3.9a) from eqn. (3.8). Including resistance, and putting $L_{21} = L_{12} = M$, the terminal voltage for coil 1, referring to

Fig. 3.24 and assuming the coils are stationary with respect to one another is:

$$v_1 = R_1 i_1 + \frac{d}{dt}[N_1 \phi_1 + N_1(\phi_{21} + \phi_{12})], \text{ see eqn. (3.8)},$$

$$= R_1 i_1 + l_1 . \frac{di_1}{dt} + \frac{N_1}{N_2} . M . \frac{di_1}{dt} + M . \frac{di_2}{dt}, \text{ see eqn. (3.9a).}$$

$$= R_1 i_1 + l_1 . \frac{di_1}{dt} + \frac{N_1}{N_2} . M . \left[\frac{di_1}{dt} + \frac{di_2}{dt} . N_2/N_1 \right] \qquad (3.16)$$

It is worth noticing at this stage that the voltages due to mutual components can be expressed in terms of the same mutual inductance. The voltage $M.di_1/dt$ is produced in secondary coil 2, due to current i_1, but it has its direct counterpart in primary coil 1, differing only by the turns ratio N_1/N_2. Further, the expression $i_2 N_2/N_1$ has the dimensions of a current i_2' say, which, flowing through N_1 turns would produce the same m.m.f. as i_2 through N_2 turns. i_2' is the secondary current referred to the primary turns and is obtained by transforming the actual secondary current i_2, by a factor equal to the inverse of the turns ratio used to transform the secondary voltage into primary terms. It is possible also to transform or refer impedances from one coil to the other. If a secondary voltage e_2 is referred to the primary coil it becomes $e_2' = e_2 N_1/N_2$. A secondary current i_2 becomes $i_2' = i_2 N_2/N_1$.

Therefore, a secondary impedance $z_2 = e_2/i_2$ becomes:

$$\frac{e_2 N_1/N_2}{i_2 N_2/N_1} = \frac{e_2}{i_2}\left(\frac{N_1}{N_2}\right)^2 = z_2\left(\frac{N_1}{N_2}\right)^2 = z_2'.$$

This impedance transformation will be discussed in more detail in Chapter 4.

When eqn. (3.16) is concerned with sinusoidally varying currents and voltages it becomes expressed in terms of reactances and r.m.s. values (see p. 73), as below:

$$\mathbf{V}_1 = R_1 \mathbf{I}_1 + j\omega l_1 . \mathbf{I}_1 + j\omega \frac{N_1}{N_2} . M . \left(\mathbf{I}_1 + \mathbf{I}_2 \frac{N_2}{N_1}\right)$$

$$= R_1 \mathbf{I}_1 + jx_1 \mathbf{I}_1 + jX_m(\mathbf{I}_1 + \mathbf{I}_2'). \qquad (3.17)$$

x_1 is called the leakage reactance of coil 1 and is equal to

$$\omega l_1 = \omega . N_1{}^2 \Lambda_1 .$$

x_1 is also equal to $\omega(L_{11} - MN_1/N_2)$ from Note 3, p. 98.

X_m is called the magnetising reactance, and is equal to

$$\omega MN_1/N_2 = \omega . N_1{}^2 \Lambda_{21} \text{ (see p. 97).}$$

A similar equation can be written for coil 2 simply by interchanging the suffices $_1$ and $_2$ in the equations above. The magnetising reactance would then be $\omega M . N_2/N_1$ which in fact is equal to $X_m . (N_2/N_1)^2$ corresponding to the magnetising reactance in primary terms, being referred to the secondary turns.

The Transformer Equivalent Circuit

The transformer is the simplest practical illustration of a 2-coil electromagnetic device since in its basic form it consists of two adjacent co-axial coils. One coil, the primary, coil 1 say, is connected to the supply system, and the induced voltage in the secondary, coil 2, is used to supply power, usually at a different voltage, to a circuit which is electrically isolated from the primary circuit. The coils are stationary with respect to each other, being wound on a common core which is usually, though not necessarily ferromagnetic. The flux changes are therefore at the same frequency for both coils and all the voltage components shown on the general diagram of Fig. 3.24 are present. From this diagram, the primary voltage equation is:

$$v_1 = R_1 i_1 + N_1 \, d\phi_1/dt + N_1 \, d\phi_{21}/dt + N_1 \, d\phi_{12}/dt$$

or combining the mutual flux components, $\phi_{21} + \phi_{12} = \phi_m$:

$$v_1 = R_1 i_1 + N_1 \, d\phi_1/dt + N_1 . d\phi_m/dt.$$

For sinusoidal variations, the equation is rewritten as in eqn. (3.17), i.e. in terms of r.m.s. voltages, giving:

$$\mathbf{V}_1 = R_1\mathbf{I}_1 + jx_1\mathbf{I}_1 + jX_m(\mathbf{I}_1 + \mathbf{I}_2').$$

A similar equation exists for V_2 with the appropriate interchange of suffices and if this equation is now multiplied by the turns ratio N_1/N_2, all the voltage terms are referred to the primary. If in addition the secondary currents are referred to the primary, it will be found that all the impedance terms have been multiplied by the square of the turns ratio, thus referring them also to the primary winding. Taking the term $R_2 I_2$ as an example:

$$R_2 I_2 \times N_1/N_2 = R_2(N_1/N_2)^2 . I_2 N_2/N_1 = R_2'.I_2'.$$

Consequently, the secondary voltage equation referred to the primary is:

$$V_2' = R_2' I_2' + j x_2' I_2' + j X_m (I_1 + I_2'). \tag{3.18}$$

The last term, which is the resultant e.m.f. E_2 due to the mutual flux components, but referred to the primary, i.e. E_2', is in fact equal to the corresponding primary e.m.f. This is best seen by considering the instantaneous expression:

$$e_2' = e_2 \times N_1/N_2 = N_2 \, d\phi_m/dt \times N_1/N_2 = N_1 \, d\phi_m/dt = e_1.$$

Consequently, $E_2' = E_1 = j X_m (I_1 + I_2')$.

Since this last term is common to both eqns. (3.17) and (3.18), it is quite a simple matter to form a circuit having the same equations. This is shown on Fig. 3.25. It is an equivalent circuit, with the parameters all referred to the primary side though they could all have been referred to the secondary side with the appropriate transformation factors. The circuit is quite easy to remember and is useful for making calculations of performance. The practical application of the circuit will be discussed in detail in Chapter 4.

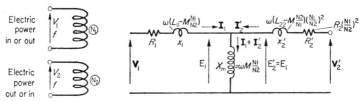

Fɪɢ. 3.25. Transformer equivalent circuit.

Note 1. The currents I_1 and I_2' are shown flowing towards X_m corresponding to magnetisation in the same sense and with both coils as energy sinks. In practice one of the coils, the secondary, is a source and the positive direction of I_2' is usually chosen in the opposite sense, indicating the through flow of energy.

Note 2. For polyphase devices, the equivalent circuit is universally taken to represent the voltage, current and power per phase. The flux components in the 2-coil representation are then due to the combined action of all phases.

The Induction Machine Equivalent Circuit

The induction machine normally has two sets of coils. One set is called the stator winding and is wound on the stationary member. The other set is called the rotor winding. The rotor is separated from the stator by a small air gap to give running clearance. One of the windings, usually the stator winding, is connected to the supply and acts as a primary to the other winding which receives power by virtue of the secondary e.m.f. induced. Each winding may have several sections or phases, and in fact the theory of the polyphase machine is rather easier to understand than that for the single-phase machine. The primary phases are disposed symmetrically around the periphery and are electrically interconnected. Due to the primary currents, a sequential system of north and south poles is established, see Fig. 1.8, which rotates with respect to the coils at synchronous speed, $n_s = f/p$ revs/sec. f is the frequency and p the number of pole pairs for which the coils are wound. The resultant primary m.m.f. could in fact be considered as produced by one specially wound coil (p. 101), rotating at synchronous speed and giving the same distribution of magnetic field intensity.

The rotor winding may be considered in the same way though in fact it may be wound for a different number of phases. When the rotor is stationary, its induced voltage is at the same frequency as the stator primary, just as for the transformer. When the rotor circuit is closed, either by a short circuit or by external impedance, the induced currents which flow, react with the primary currents and the electromagnetic force produced has a tangential component which can cause rotation in the direction of the moving field. If rotation at speed n takes place, the relative motion between the rotor coils and the moving field is changed from n_s to $n_s - n$ and is conveniently expressed as a fraction of the synchronous speed. In this form the relative motion is termed the slip s, defined as $s = (n_s - n)/n_s$. The secondary induced voltage and frequency are proportional to the relative motion and if for example the speed was to become equal to the synchronous speed, the rotor would be at rest with respect to the field and no induced voltage or current or torque would be produced; the slip would be zero.

At standstill therefore, with the rotor stationary, the induction machine behaves just like a transformer, as far as the induced voltages and currents are concerned, though the ratio of leakage to magnetising reactance is higher due to the air gap reducing the efficiency of flux interlinkage. With speed $\neq 0$, the standstill- or line-frequency values of the secondary quantities E_2, f and x_2 become sE_2, sf, and sx_2. The secondary current becomes:

$$\mathbf{I}_2 = \frac{s\mathbf{E}_2}{R_2 + \mathrm{j}sx_2} = \frac{\mathbf{E}_2}{(R_2/s) + \mathrm{j}x_2} .$$

The last expression is the same as for the transformer apart from the change to the rotor resistance term which is divided by s. Any externally connected rotor impedance would also have to be divided by s. Thus the effect of rotation viewed from the primary, from where the secondary power and reactive voltamperes are supplied, is to modify only the apparent rotor-circuit resistance. The transformer equivalent circuit applies to the induction machine too with this simple change. The secondary impedances can of course be referred to the primary on multiplying by the square of the effective primary/secondary turns ratio, giving R_2' and x_2'. This operation has been carried out on the induction-machine equivalent circuit shown on Fig. 3.26.

In mathematical terms, the simplification described has really been brought about by a transformation of variables. Although the relative motion of the windings gives rise to inductance changes with position, the equivalent circuit represents the overall effect of *all the phases* in

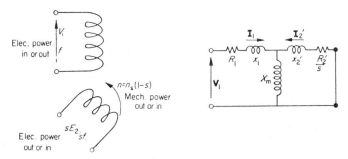

FIG. 3.26. Induction-machine equivalent circuit.

terms of constant reactances. The more detailed generalised analysis based on equations of the form (3.9b), will in fact give the same equivalent circuit if the appropriate transformations are used.[2] Furthermore, to apply this analysis, the self and mutual inductances are most conveniently derived from the equivalent-circuit parameters as measured by methods to be described in later chapters. As for the polyphase transformer, the equivalent circuit of the induction motor is in terms of voltage, current and power per phase.

The Synchronous Machine Equivalent Circuit

For the Induction mode of operation, the secondary frequency, which is a variable, is determined by the relative motion of the air-gap field and the secondary coils. The same machine can be made to operate in the synchronous mode, but this time, the secondary voltage is no longer produced by induction, but is provided by an external supply which determines the frequency of this winding voltage. In the normal, conventional operation, this supply is direct current. It has already been pointed out, that if the induction machine were to rotate at synchronous speed, there would be no induced voltage because the secondary coils would be at rest with respect to the moving field. If a d.c. current were to be applied to the secondary, however, a steady field pattern would be established which would lock in synchronism

116

with the primary field moving at the same speed, a condition indicated in Fig. 1.7; see also Fig. 1.11b. This synchronous condition can only occur at one speed corresponding to that of the rotating field established by the primary winding. Consequently, it is necessary to bring the rotor up to this speed by some other means before applying d.c. excitation. This starting problem will be dealt with in Chapter 8. The d.c. excited winding is called the field winding, suffix f. The other winding, carrying alternating currents, is usually on the stator and is sometimes called the armature winding, suffix a. The inverted arrangement, with the field on the stator, is sometimes used on small machines.

The equivalent circuit can be developed from the general diagram of Fig. 3.24. All the flux components exist as before, but on steady-state conditions there are no induced voltages in the field winding because the flux pattern is stationary with respect to the field. Assuming Coil 2 of the general figure is the field winding, the components $N_2 \, d\phi_{21}/dt$ and $N_2 \, d\phi_2/dt$ are therefore omitted. For the armature, Coil 1, conditions are as before but the mutually induced voltage $N_1 \, d\phi_{12}/dt$, brought about this time by the relative motion of the field coils with respect to the armature coils at synchronous speed, is usually designated E_f. This voltage is a function of the rotor speed, the field current and flux and also, the arrangement of the armature coils. The component $N_1 \, d\phi_{21}/dt$, due to self flux, will continue to be expressed in terms of the magnetising reactance X_m, sometimes called the armature reactance. The phasor representing the corresponding r.m.s. voltage is $jX_m\mathbf{I_a}$. The voltage due to the other component of the self flux, i.e. the leakage, will be designated $jx_{al}\mathbf{I_a}$. The total self reactance is called the synchronous reactance, $X_s = x_{al} + X_m$ and the synchronous impedance is $\mathbf{Z_s} = R_a + jX_s$. The air-gap voltage due to the resultant air-gap flux $\phi_m = \phi_{21} + \phi_{12}$ or $= \phi_a + \phi_f$ in the new terms, will be given the symbol E. ϕ_m itself can also be determined from the resultant of armature and field m.m.f.s, $\mathbf{F_r} = \mathbf{F_a} + \mathbf{F_f}$.

Figure 3.27 shows the equivalent circuit in accordance with the above derivation. Although the field winding is included, it is usually omitted since it has no induced voltage components and its effect on the armature can be expressed in terms of E_f. A motoring condition is shown,

117

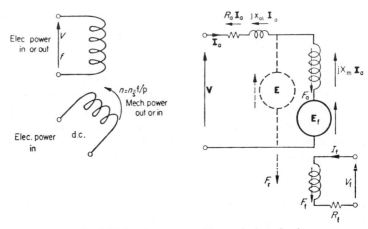

FIG. 3.27. Synchronous-machine equivalent circuit.

with the current flowing in opposition to the induced armature e.m.f. The terminal voltage can be expressed either as $V = E \pm I_a z$ where $z =$ leakage impedance $R_a + jx_{a1}$, or as $V = E_f \pm I_a Z_s$. The positive sign is for motoring and the negative sign for generating. The current would then be shown in the reversed direction on the circuit. The positive sign can be used for the generating condition too if the machine is then treated as a "negative" motor.

The D.C. Machine Equivalent Circuit

The d.c. machine is one stage further beyond the synchronous machine in that the armature as well as the field winding is connected into a d.c. circuit. The normal physical arrangement is like the inverted synchronous machine with the field on the stator. The armature, carrying alternating currents, is on the rotor. A mechanical rectifying device, the commutator, is incorporated on the rotating member, to convert the a.c. armature voltages in the coils to d.c. voltages at the pick-up terminals—i.e. the brushes—see Figs. 1.5 and 1.11a. The stationary brushes act in such a manner, in conjunction with the com-

118

mutator, that the external-circuit d.c. currents are directed through the rotating armature winding in spatially fixed paths so that the armature m.m.f. is directed along the brush axis. It is not practicable to combine the armature and field m.m.f.s vectorially, since they are far from sinusoidally distributed. Nor can the voltage components be combined as on the synchronous machine to give $\mathbf{E} = \mathbf{E_f} + \mathrm{j}X_m\mathbf{I_a}$. In fact, with the torque angle δ equal to 90° and neglecting saturation effects, the e.m.f. E would be equal to E_f, there being no total-flux change due to F_a, the armature m.m.f. The interaction of the armature m.m.f. on the flux is rather complex and discussion will be deferred till Chapter 6 where it will be found that it may be represented by a net magnetising m.m.f. F_a'—usually negative. The air-gap e.m.f. E is, as usual, due to the resultant air-gap flux ϕ_m which is the field flux ϕ_f modified in this complex fashion by F_a, the modification corresponding to the action of component ϕ_{12} in the general diagram of Fig. 3.24. The other armature self-flux component, ϕ_2 the leakage flux, has important effects on commutation and on transient performance. In so far as it has any magnetic effect on the steady-state performance, this is included in F_a' as will be described in Chapter 6. Consequently, in deducing the equivalent circuit from the general diagram, Fig. 3.24, and in accordance with the brief description of operation above, the only induced voltage component is the air-gap e.m.f. E in the armature, see Fig. 3.28. As for

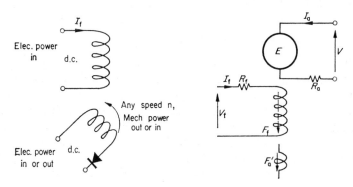

FIG. 3.28. D.C. machine equivalent circuit.

119

the d.c.-excited synchronous machine, the field winding is not usually shown, so that the d.c. machine equivalent circuit is normally represented by a simple circuit consisting of an e.m.f. E and a series armature resistance R_a.

The calculation of performance using the equivalent circuits derived in this section will be discussed under the appropriate chapter headings, Chapters 4, 6, 7 and 8. An alternative generalised derivation of equivalent circuits is given in Appendix B.

3.10 TRANSIENT BEHAVIOUR

Response for a System with a Single Time Constant

For a number of practical situations associated with the operation of electrical machinery, the change of an output quantity θ with time is governed by the following 1st order differential equation:

$$\tau \, d\theta/dt + \theta = \theta_f \tag{3.19}$$

or in operational form:

$$(\tau p + 1)\theta = \theta_f$$

where p is the operator d/dt.

Consider the transient response, i.e. where a fixed input θ_f is suddenly applied to the system when it has an initial value of θ_0.

By separating variables:

$$\int \tau \, d\theta/(\theta_f - \theta) = \int dt \tag{3.20}$$

and integrating: $t = -\tau . \log_e(\theta_f - \theta) + \text{constant}$

at $t = 0, \theta = \theta_0$

$$\therefore \text{ constant} = \tau \log_e(\theta_f - \theta_0)$$

substituting: $t = \tau \log_e(\theta_f - \theta_0)/(\theta_f - \theta)$

from which: $e^{-t/\tau} = (\theta_f - \theta)/(\theta_f - \theta_0)$

and: $\theta = \theta_f - (\theta_f - \theta_0)e^{-t/\tau} - \theta_0 + \theta_0$

$$= \theta_0 + (\theta_f - \theta_0)(1 - e^{-t/\tau}). \tag{3.21}$$

Figure 3.29 shows this solution graphically together with certain very important features. Starting from the initial value θ_0, the output variable θ approaches a final value which is equal to the suddenly applied input θ_f. If the initial rate of change were maintained, θ would reach this final value in a time τ seconds. Actually it only reaches $(1-e^{-1}) = 0\cdot632$ of the required change $(\theta_f-\theta_0) = \theta_c$. From this point it has only to change by $(1-0\cdot632)\,\theta_c$ to reach θ_f. Again, if the

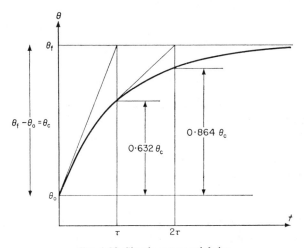

FIG. 3.29. Simple exponential rise.

rate of change at this point was maintained, θ_f would be reached in a time τ seconds, but, as before, only $0\cdot632$ of the remaining change is achieved; i.e. $0\cdot632(1-0\cdot632)\,\theta_c = 0\cdot232\,\theta_c$. Thus in a time 2τ, the total change is $0\cdot864\,\theta_c$. Similarly, in 3τ seconds the change is $0\cdot95\,\theta_c$ and in 4τ seconds $0\cdot98\,\theta_c$ at which point the accuracy of the measuring apparatus may not be sufficient to detect the remaining change very precisely. Consequently, for many practical purposes, the change may be considered complete in four times the constant τ. It can be seen that the curve is governed by τ and the base of natural logarithms. τ is quite understandably referred to as the time-constant of the system. If the differential equation of any system is rearranged so that the coefficient of its output variable is unity and it is then found to have the same form as eqn. (3.19), the solution can be written down directly as eqn. (3.21). A familiar example of this equation is provided by a circuit consisting of inductance and resistance in series, the time-constant being L/R seconds.

EXAMPLE. THERMAL TRANSIENT

If in a homogeneous body of mass M and specific heat S, there is a source of heat P watts, the temperature will rise. Part of the generated heat will be stored in the body and the remainder will be radiated to the surroundings at a rate proportional to the temperature rise θ above the ambient temperature of the surroundings, i.e. radiated heat $= k\theta$. By balancing the heat generated against the stored and radiated heat, the following equation results for a small change of temperature $d\theta$ in a time dt:

$$P \cdot dt = M \cdot S \cdot d\theta + k\theta \cdot dt \text{ watt secs (joules)}$$

which by re-arrangement comes in the same form as eqn. (3.19), i.e.

$$\frac{MS}{k} \cdot \frac{d\theta}{dt} + \theta = \frac{P}{k}.$$

The solution, for a sudden activation of the heat source, follows the exponential curve of Fig. 3.29 with a time constant $\tau = MS/k$ and a final temperature *rise* of P/k. At this point, the temperature is sufficiently high to radiate all the heat as quickly as it is being generated.

If the heat source is switched off, the cooling curve can easily be obtained from the standard form of solution, the differential equation being the same except that $P = 0$.

Using eqn. (3.21)

$$\theta = P/k + (0 - P/k)(1 - e)^{-t/\tau} = (P/k)e^{-t/\tau}.$$

Heating and cooling curves are shown in Fig. 3.30. The temperature changes of a practical machine may sometimes be approximated by

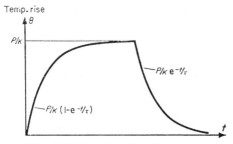

FIG. 3.30. Heating and cooling curves.

such curves. The thermal time constant may be as long as an hour or perhaps longer on the larger machines.

Graphical Solution for Single Time Constant

If the input θ_f is not constant, the solution may often be found graphically. Equation (3.20) expresses the small time interval δt as a function of the small change in output and the difference between the final value and the output. If θ_f and θ are known as functions of another common variable, the solution follows by taking a series of values $(\theta_f - \theta)$ from $\theta = 0$ to $\theta = \hat{\theta}$, calculating δt and summing δt over the whole change. In some cases, the time-constant itself varies during the transient process and this must be allowed for. Figure 3.31 illustrates

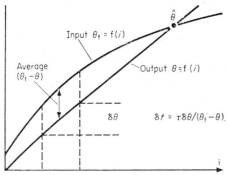

FIG. 3.31. Graphical solution for transient response.

123

the method for a typical example; the build-up of voltage on a self-excited generator, see also Example E.6.3. The value of τ in this case would be L_f/R_f and L_f varies with the saturation level as will be discussed in connection with Fig. 6.25b. Another example would be the build up of speed ω_m of the motor with a characteristic as shown in Fig. 2.16. Here the net torque $T_e - T_m$ is absorbed in accelerating the load and is equal to $J\,d\omega_m/dt$. dt can thus be expressed in terms of $d\omega_m$ and the total accelerating time found, see Fig. 6.45b.

Transient Response of Coupled Circuits with Closed Secondary Winding

Referring to Fig. 3.21 and eqns. (3.10a) and (3.10b), with coil 2 closed on itself and a constant voltage V_1 applied to coil 1, the differential equations are:

$$V_1 = R_1 i_1 + L_{11}\,di_1/dt + M\,di_2/dt$$

$$0 = R_2 i_2 + L_{22}\,di_2/dt + M\,di_1/dt.$$

These equations must be solved simultaneously, the solution being rather lengthy. The form of the solution is shown in Fig. 3.32 which is plotted in terms of coil m.m.f. The primary and secondary current both rise rapidly, the rate of rise being very approximately in accordance with the effective time constant $L_{11}(1-k^2)/R_1$ where $(1-k^2)$

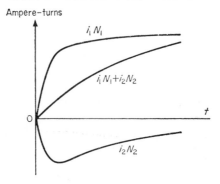

FIG. 3.32. Transient response for coupled coils.

is the Blondel leakage factor, eqn. (3.11). The current i_2 is negative, magnetising in the opposite sense to i_1 and therefore opposing the rise of mutual flux. i_2 then decays approximately in accordance with the effective time constant $L_{22}(1-k^2)/R_2$. The rapid rise of i_1 is thus no indication of the flux response which is in accordance with the net magnetising m.m.f. $(i_1N_1+i_2N_2)$, at a time constant approximately equal to $L_{11}/R_1+L_{22}/R_2$. For the case where coupling is perfect and $k = 1$, this last statement is precise and can be proved simply with a knowledge of Thévenin's theorem which is then applied to the equivalent circuit of Fig. 3.25.

If $k = 1$, there is no leakage and $l_1 = L_{11}-M.N_1/N_2$ is zero. Hence $MN_1/N_2 = L_{11}$ and the circuit becomes that of Fig. 3.33, V_2 being zero.

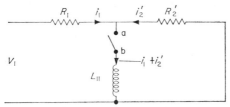

FIG. 3.33. Equivalent circuit for $k = 1$.

The voltage across ab with L_{11} open-circuited is: $\dfrac{R_2'}{R_1+R_2'}.V_1$.

The impedance between a and b with L_{11} open-circuited and V_1 short-circuited $= \dfrac{R_1R_2'}{R_1+R_2'}$.

By Thévenin's theorem, the current i_1+i_2 through L_{11} when its circuit is closed, is given by the open-circuit voltage across ab divided by the sum of the short circuit impedance and the branch impedance. Alternatively, balancing the applied voltage and the opposing circuit voltages:

$$\frac{R_2'}{R_1+R_2'}\,V_1 = \frac{R_1R_2'}{R_1+R_2'}\,(i_1+i_2')+L_{11}\,\frac{d(i_1+i_2')}{dt}\,.$$

125

Dividing by the coefficient of $(i_1 + i_2')$:

$$\frac{V_1}{R_1} = (i_1 + i_2') + \frac{L_{11}}{(R_1 R_2')/(R_1 + R_2')} \frac{\mathrm{d}(i_1 + i_2')}{\mathrm{d}t}$$

which is in the same standard form as eqn. (3.19) with the net magnetising current as the variable. The time constant is readily rearranged as:

$$\frac{L_{11}}{R_1} + \frac{L_{11}}{R_2'}.$$

Since L_{11} is also the referred secondary inductance when $L_2 = 0$, the second term is the secondary time constant L_{22}/R_2.

Switching Transients with Suddenly Applied Sinusoidal Input Voltage

For a purely resistive circuit, the current would respond immediately to follow the input voltage no matter what its waveform since $i = v/R$.

For a pure inductance, the flux variation is determined by the rate of change required to balance the applied voltage, i.e. $N \, \mathrm{d}\phi/\mathrm{d}t = v$. The initial flux is zero if the circuit is unmagnetised and it then follows a pattern dependent on the instant of switching. The current will be proportional to and in phase with the flux for an air-cored circuit.

Two extreme cases are shown in Fig. 3.34a and b. In the first case, $v = \hat{V}$, when the switch is closed and a flux wave commencing a symmetrical sequence of positive and negative half waves will give the correct values of $\mathrm{d}\phi/\mathrm{d}t$. In the second case, the switch is closed when $v = 0$, and the rate of change demanded causes a peak flux twice that of the previous case. If there is an iron core, this flux would cause excessive saturation and a very large magnetising current would be drawn from the supply as indicated.

In practice the resistance cannot be neglected and the resistive voltage drop reduces the magnitude of $\mathrm{d}\phi/\mathrm{d}t$ required. The instant of switching which would cause the current to assume its steady-state value immediately would be when $v = \hat{V} \sin \theta$ where $\theta = \tan^{-1} \omega L/R$. A current lagging the voltage by angle θ would in fact be zero when the applied voltage had this value. At any other switching instant, the flux

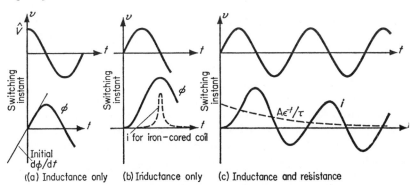

FIG. 3.34. Effect of switching instant.

and current waves would have initial asymmetry. Figure 3.29c shows a typical transient current wave for an air-cored coil. The broken line $Ae^{-t/\tau}$ where $\tau = L/R$, is a transient d.c. component superimposed on the steady-state a.c. component of r.m.s. value V/Z. The value of A is determined by the instant in the voltage cycle t', at which the switch is closed. It can be shown to have a value of $(V/Z)[-\sin(\omega t' - \theta)]$.

If a second coil with its terminals closed is coupled to the first coil when this is switched in circuit, the inrush current at the most un-favourable instant will be very much increased. This is because the inductive reactance which limits the current is reduced from ωL_{11} to $\omega L_{11}(1 - k^2)$. The exponential growth and decay components are now governed approximately by the so-called transient time constants $L_{11}(1 - k^2)/R_1$ and $L_{22}(1 - k^2)/R_2$.

In service, electrical machines are liable to be short-circuited under fault conditions. The subsequent behaviour is similar to the cases already discussed where a voltage is suddenly applied to a coil having a closed coupled circuit. In the d.c. and synchronous machines, for example, this coupled circuit would be the field winding. The driving voltage this time is generated in the machine itself and the effect of the short-circuit currents is to reduce the magnitude of the mutual flux over a transient period, the changes in the unidirectional components of

127

the induced currents being approximately exponential and governed by the transient time constants. Taking the synchronous machine as an illustration, if the terminals are suddenly short-circuited when generating a voltage E on open circuit, the a.c. component of current will decay from an initial value of $E/\omega L_{11}(1-k^2)$ to $E/\omega L_{11}$, the demagnetising action being represented by the change of inductance. The initial effective value of reactance is called the transient reactance, and the final value due to self-inductance alone is called the synchronous reactance (p. 117). In practice there are other windings even more closely coupled than the field and these give rise to sub-transient reactances and even higher initial short-circuit currents. The sudden short circuit of a machine is considered further on p. 530.

Time Response of 2nd Order System (Two time constants)

Where there are two energy stores in a system, e.g. inductance and capacitance, or inductance and mechanical inertia, the differential equation is at least of second order and the response can in fact be oscillatory, overshooting and undershooting the final steady-state value. The oscillation frequency, the overshoot and the rate at which the oscillation is damped out, depend on the system parameters. The equation for such a 2nd-order system is conveniently expressed in a standard form as below, so that like eqn. (3.19), the solution can be obtained by equating coefficients and referring to a generalised curve like Fig. 3.29. For the present case, the time axis is expressed non-dimensionally as a function of the $p^2\theta$ coefficient, i.e. $\omega_0 \times$ time. When the response is oscillatory, ω_0 is in fact the oscillation frequency in radians per second, assuming the system has no losses: i.e., the undamped natural frequency. The damping performance is determined by one of the coefficients of $p\theta$ called the damping ratio ζ, which is a dimensionless quantity. The standard form of the equation is:

$$(\omega_0^{-2} \cdot p^2 + 2\zeta\omega_0^{-1} \cdot p + 1)\theta = \theta_f. \tag{3.22}$$

By solving the quadratic characteristic equation, the two roots are:

$$\omega_0[-\zeta \pm \sqrt{(\zeta^2 - 1)}].$$

For an oscillatory response, $\zeta < 1$, the roots are usually arranged as:

$$\omega_0[-\zeta \pm j\sqrt{(1-\zeta^2)}].$$

For a step driving function θ_f, the solution in *per unit* is:

$$\frac{\theta}{\theta_f} = 1 - \frac{e^{-\zeta\omega_0 t}}{\sqrt{(1-\zeta^2)}} \{\sin[\omega_0\sqrt{(1-\zeta^2)}.t - \psi]\}$$

where ψ is $\tan^{-1}[\sqrt{(1-\zeta^2)}]/(-\zeta)$.

The solution is shown graphically on Fig. 3.35 and the effect of the damping ratio ζ on the rate of oscillation decay is clearly indicated. The polar diagram shows the relationship between the parameters on the complex plane. The real part of the root, $-\zeta\omega_0$, is the damping exponent of the quadratic solution and the imaginary part is the natural frequency of the damped oscillation, $\omega_n = \omega_0\sqrt{(1-\zeta^2)}$. The second root, i.e. $-\zeta\omega_0 - j\omega_0\sqrt{(1-\zeta^2)}$, is the conjugate of the one shown.

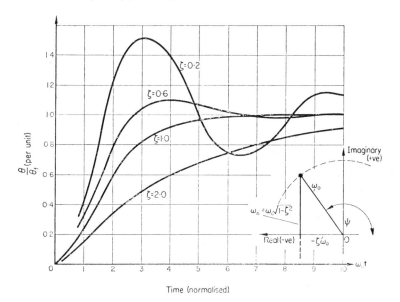

Fig. 3.35. Response of 2nd-order system to step function input.

A practical example of this equation is the series LCR circuit for which:

$$v = Ri + L\frac{di}{dt} + \int \frac{i}{C} \cdot dt.$$

Differentiating, multiplying by C and rearranging:

$$C\frac{dv}{dt} = LC\frac{d^2i}{dt^2} + CR\frac{di}{dt} + i.$$

For a suddenly applied constant voltage V the left-hand term is zero, so that ultimately the steady-state current is zero. Depending on the parameters, the response may be oscillatory however, in accordance with the equation. Meanwhile, the capacitor charges up till its voltage reaches V and the voltage response follows the pattern of Fig. 3.35. The characteristic equation has the same form as the one for current, viz:

$$(LC \cdot p^2 + CR \cdot p + 1)i = 0.$$

By equating coefficients with those of eqn. (3.22):

$$\omega_0 = \sqrt{(1/LC)} \quad \text{and} \quad \zeta = (R/2L)/[\sqrt{(1/LC)}].$$

The damping exponent $\zeta\omega_0$ is the numerator of the damping ratio, i.e. $R/2L$. It is possible to re-express the differential equation in terms of two time constants, $\tau_1 = L/R$ and $\tau_2 = CR$ as:

$$(\tau_1\tau_2 \cdot p^2 + \tau_2 \cdot p + 1)i = 0.$$

Other examples will be discussed in the text when dealing with the electromechanical equations. Very often, the response of higher-order system can be approximated by a 2nd-order quadratic if certain simplifying assumptions are made.

CHAPTER 4

TRANSFORMERS

4.1 THE TRANSFORMER ON NO LOAD

The Elementary Two-coil Transformer

The transformer is a straightforward application of Faraday's Law of Electromagnetic Induction. In the following explanation, reference will be made to Fig. 4.1 and the notation developed in Section 3.7 will be used. The basic transformer consists of two coils in close proximity. One coil of N_1 turns, say, is excited with alternating current and therefore establishes a flux ϕ_{11} which alternates with the current. The other coil is linked by most of this flux and thus has a mutually induced e.m.f. of value $e_2 = N_2 \, d\phi_{21}/dt$. This e.m.f. would drive a load current through any circuit connected to the terminals of the second coil. Energy would then be transferred through the medium of the magnetic field from coil 1 to coil 2. The transformation could be from any convenient input voltage to any convenient output voltage. This apparently simple function of the transformer makes it as vital to modern industry as the mechanical gear train which, as a "transformer" of speed and torque represents an interesting analogy.

With the coils wound on an iron core, iron losses are introduced but the value of flux per ampere is increased several hundred times because of the change of permeability from μ_0 to $\mu_0\mu_r$. The exciting or magnetising current can thus be very small. Further, the proportion of the total flux which is linked mutually by the two coils is greatly increased. Looked at another way, the leakage flux ϕ_1 which links N_1 turns alone and "leaks" between the two windings without linking the N_2 turns, is

131

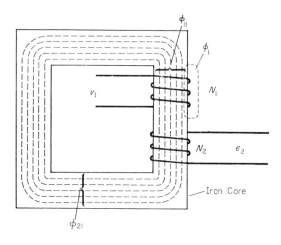

FIG. 4.1. Elementary transformer on no load.

a much smaller fraction of the total, making for more efficient energy transfer. The mutual flux would be a much larger percentage of the total than is suggested by Fig. 4.1 and the flux pattern would be rather more involved, but simplified diagrams of this kind are quite adequate for the understanding of the principles to be discussed in this chapter. The flux is changing with time, of course, and Fig. 4.1 only corresponds to one instant in the cycle.

The exciting coil which initiates the flux changes is called the primary winding, and the coupled coil which receives energy as a result, is called the secondary winding. It should not be difficult to realise that the two functions are interchangeable; if coil 2 were excited instead, a mutual e.m.f. would be induced in coil 1 which would then become the secondary winding. Since the primary winding also experiences flux changes, there is a primary e.m.f., self-induced this time. With only winding 1 excited, this primary *back* e.m.f. would be equal to $N_1 . d\phi_{11}/dt$. This is practically in direct opposition to the applied terminal voltage v_1 and when a.c. supplied, limits the current to a very much smaller value than v_1/R_1, where R_1 is the primary resistance. As

explained later, the resistance drop is normally very small so that the
back e.m.f. is virtually equal in magnitude to v_1.

If the applied voltage is alternating sinusoidally at frequency $f = \omega/2\pi$
cycles/sec, a phasor diagram can be drawn, and for an air-core transformer, neglecting resistance, would be as given in Fig. 4.2a which
should be compared with Fig. 4.2b. As shown, I_1 is the magnetising
current producing Φ_{11} which rises and falls with it in time phase. It is

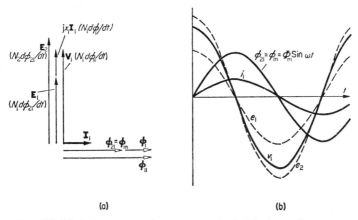

(a)   (b)

FIG. 4.2. Ideal air-core transformer on no load: (a) phasor diagram;
(b) instantaneous variations.

therefore in quadrature with V_1 and the circuit is a zero-power-factor
lagging load. On no load, therefore, with the secondary circuit open,
the primary coil is an inductor or reactor, having a voltage drop of
$N_1 d\phi_{11}/dt$ or expressed vectorially as $j\omega L_{11}I_1$ (see Section 3.9).

The two components of flux linking N_1 are shown and though I_1
is in phase with both of them, it will be found that when the secondary is
carrying current, the resultant mutual flux ϕ_m is no longer in phase
with the primary current, due to the reaction of the secondary m.m.f.
The phase of component Φ_1 is unchanged, however, since it is produced by primary current alone and is therefore in phase with it.
Further, as can be seen from Fig. 4.1, even with an iron core, the

133

reluctance offered to ϕ_1 is primarily determined by the air path it must necessarily take. Unless the iron is highly saturated, it forms a very small proportion of the total reluctance. Consequently, the leakage flux remains virtually proportional to the primary current. The e.m.f. $N_1 \, d\phi_1/dt$ may thus be treated as if it was due to an unsaturated inductance of magnitude $l_1 = N_1{}^2\Lambda_1$; see Section 3.7. The corresponding leakage reactance $2\pi f l_1 = \omega l_1 = x_1$ is a dominant factor in transformer behaviour.

This chapter is primarily concerned with the iron-core transformer. Normal operation will be at constant supply voltage and frequency, when used in the relatively low-frequency power circuits which are mostly 50 Hz or 60 Hz. The leakage component of voltage can be represented by the phasor $jx_1\mathbf{I}_1$ which therefore leads the current phasor by 90°. The second component of the primary e.m.f. $\mathbf{E}_1$, due to mutual flux $\mathbf{\Phi}_m$, leads this flux phasor by 90°. For the idealised condition of Fig. 4.2, since there is no secondary current, the mutual flux is due only to $\mathbf{I}_1$ and is in time phase with it. Consequently, $\mathbf{E}_1$ also leads the current phasor by 90° in this case and could be represented as $jX_m\mathbf{I}_1$ where $X_m = \omega M \cdot N_1/N_2$; eqn. (3.17). In the general case, though $\mathbf{E}_1$ is always in quadrature with $\mathbf{\Phi}_m$, it is not in quadrature with $\mathbf{I}_1$. It must be remembered also that when the mutual flux is in an iron core, it is not proportional to current so that the mutual inductance is not constant but depends on the operating magnetic condition of the iron.

Figure 4.2a must now be modified as in Fig. 4.3a to take account of the imperfections represented by copper and iron losses. The copper loss requires a component of voltage absorbed across the winding resistance and in phase with the current; i.e. $R_1\mathbf{I}_1$. Iron loss is a function of flux ϕ_m and therefore of E_1 so that a component of current $\mathbf{I}_p$ in phase with $\mathbf{E}_1$ would allow for this; $E_1 I_p$ being the iron loss in watts. To produce the mutual flux requires a co-phasal magnetising m.m.f. $\mathbf{I}_m N_1$ say, and therefore on no load, the primary takes a current $\mathbf{I}_1 = \mathbf{I}_0$ say, which consists of the two components $\mathbf{I}_m$ and $\mathbf{I}_p$. The no-load power factor is very low since I_p is normally much smaller than I_m. The mutual flux ϕ_m on no load is the same as ϕ_{21}, but when the secondary carries current, I_1 increases from I_0; $\boldsymbol{\Phi}_m$ is then due to the combined effect of primary and secondary ampere-turns, but is little changed from $\boldsymbol{\Phi}_{21}$ as will be seen in Section 4.2. Consequently the magnetising and power requirements of the flux are nearly constant so that $\mathbf{I}_0$ can be regarded as a substantially constant component of $\mathbf{I}_1$ at any load.

It will be noticed from the diagram that since the e.m.f.s E_1 and E_2 are due to the same flux ϕ_m, the voltage ratio E_1/E_2 is the same as the turns ratio N_1/N_2. In fact, on no load, the impedance drops $R_1 I_0$ and $x_1 I_0$ are so small that V_1 is virtually the same magnitude as E_1 and the turns ratio is equal to the no-load voltage ratio V_1/E_2. Thus any voltage V_1 can be transformed to any other voltage E_2 by suitable choice of primary and secondary turns. On load, the voltage at the secondary terminals V_2, is measurably different from E_2.

In considering Fig. 4.3, it is important to realise the relative magnitudes of the various voltage and current components, which are determined by the design of the transformer. For example, the magnetic circuit is operated in a condition where, at normal frequency, a primary induced voltage equal to V_1 can be reached without going very far

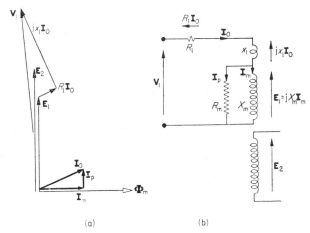

(a)   (b)

FIG 4.3. Iron-core transformer on no load: (a) phasor diagram; (b) equivalent circuit.

135

towards saturation. The high permeability which is thus maintained, means that the magnetising current I_m is very small. The power component I_p for iron losses is also kept to a low value by suitable choice of materials and lamination of the iron. The leakage reactance is adjustable at the design stage and can be kept small by arranging that the primary and secondary turns are close together. Even so, the leakage reactance is often much higher than the resistance. On Fig. 4.3, the $R_1 I_0$ and $jx_1 I_0$ phasors have been greatly exaggerated for clarity; actually, E_1 is very nearly equal to V_1. Even when the transformer is fully loaded so that the primary current increases twenty-fold or more, the primary leakage impedance drop $(R_1 + jx_1)\mathbf{I}_1$, may still be only about 3% of V_1 so that the change of E_1 and hence the change of peak mutual flux is not considerable. Consequently, the current components I_p and I_m are not appreciably affected by load changes, as already pointed out.

Figure 4.3a, deduced from physical considerations of transformer operation, represents phasor equations which could also be obtained from a suitable circuit arrangement of resistors and reactors. Figure 4.3b shows such a circuit which, considered as a power sink and applying eqn. (3.3) gives:

$$\mathbf{V}_1 = \mathbf{E}_1 + R_1 \mathbf{I}_0 + jx_1 \mathbf{I}_0,$$

where $\mathbf{I}_0 = \mathbf{I}_m + \mathbf{I}_p$. To get the correct relationship of these current components to $\mathbf{E}_1$, they must flow through a pure reactance X_m and a pure resistance R_m respectively as shown. The magnetising reactance is $X_m = E_1 / I_m$ and the relationship between E_1 and I_m, as explained in connection with Fig. 2.10, is determined by the magnetisation characteristic. Consequently, X_m is not constant but since the value of peak mutual flux and hence of E_1 are not greatly affected in normal operation as the transformer load changes, then an average slope of the curve from the origin to the operating point could be used to get a suitable value of the E_1 / I_m ratio. Section 3.3 which discusses a.c. magnetisation, should be consulted about the reasons for this procedure and its limitations. The fictitious magnetising resistance R_m is found by equating $E_1{}^2 / R_m$ to the iron loss it represents. Again, this resistance is only suitable for one voltage and frequency but in fact if frequency is constant, then the iron losses, Section 2.2, are approximately proportional to $\phi^2 \propto E_1^2$ so that R_m does not vary considerably if only voltage changes are involved.

The Transformer E.M.F. Equation

The induced e.m.f. due to mutual flux, $N\,d\phi_m/dt$, is conveniently formulated a different way to deal with the special case of sinusoidal variations. The instantaneous value of flux may then be expressed as $\phi_m = \hat{\Phi}_m \sin \omega t$ where $\hat{\Phi}_m$ is the maximum value reached in the cycle. The induced e.m.f. is thus:

$$e = N\,d\phi_m/dt = \omega N \hat{\Phi}_m \cos \omega t$$

which is another sine wave leading ϕ_m by 90°.

The peak value of this e.m.f. is $2\pi f N \hat{\Phi}_m$ since $\omega = 2\pi f$ and the r.m.s. value is:

$$E = (1/\sqrt{2}).2\pi f N \hat{\Phi}_m = 4.44 f N \hat{\Phi}_m \text{ volts.} \qquad (4.1)$$

Applying eqn. (4.1) to the two coils:

the r.m.s. primary e.m.f. due to mutual flux $= 4.44 f N_1 \hat{\Phi}_m = E_1$

the r.m.s. secondary e.m.f. due to mutual flux $= 4.44 f N_2 \hat{\Phi}_m = E_2$.

Note in each case that the *maximum* (*peak*) value of mutual flux must be used.

The time variations of flux, current and the voltages were shown on Fig. 4.2b. e_1 and e_2 are drawn in phase because they differ only by the primary/secondary turns ratio. On the transformer itself, their relationship is indicated by suitable polarity markings. Consider Fig. 4.4 which shows the two coils wound in the same direction round the core. At an instant when current i_1 is flowing into the top end of the upper coil, this gives rise to a downwards primary m.m.f. In accordance with Lenz's law, the secondary current at the same instant, i_2, would have to be flowing out of the top end of the lower coil to give an opposing m.m.f. The m.m.f.s do not quite cancel and there is a resultant downwards m.m.f. and flux. If in fact the primary current is

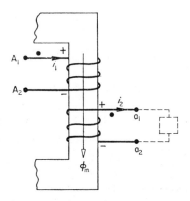

FIG. 4.4. Polarity and terminal markings.

increasing, then the induced e.m.f. must cause the instantaneous primary polarities as shown, opposing the increase of current. The secondary current must also be increasing to oppose the increasing primary m.m.f. and so the instantaneous secondary polarities must be as indicated to drive such a current through the secondary circuit. Thus the top ends of the coils have co-phasal potential variations and this is indicated by terminal markings having the same suffix; A_1 and a_1 say. Alternatively, A_1 and a_1 (or A_2 and a_2) could be marked with dots as on Fig. 3.21. Although A_1 and a_1 are in phase, corresponding to the relationship of e_1 and e_2, the phase of e_2 with respect to the secondary circuit can of course be reversed by crossing over the connections.

On circuit diagrams for the two-coil transformer, it is generally convenient to choose positive reference directions for i_1 and i_2 which correspond with those shown on Fig. 4.4. This means that positive i_2 magnetises in the opposite sense to positive i_1. It is useful, nevertheless, to have a common reference direction for m.m.f. and flux, downwards, say, on Fig. 4.4. Therefore, the secondary m.m.f. F_s, is $-i_2N_2$, whereas the primary m.m.f. F_p, is i_1N_1. Consequently the resultant m.m.f. is obtained by summing primary and secondary m.m.f.s, i.e. $F_r = F_p + F_s = i_1N_1 - i_2N_2$.

138

EXAMPLE E.4.1

A single-phase transformer core for use at 50 Hz, exhibits the following magnetic characteristics at normal frequency:

$\hat{B}$	0·8	1·0	1·1	1·2	teslas
H	128	168	209	297	r.m.s. At/m
Total core loss	1·23	1·85	2·2	2·73	watts/kg

The cross-sectional area of the core is 17500 mm², the mean length of the magnetic path is 1·6 m, and the density of the steel is 7500 kg/m³.

If a no-load voltage ratio of 6600/231 is required, choose a suitable number of primary and secondary turns so that the maximum flux density will not exceed 1·2 teslas. With these turns calculate the no-load current and power factor at normal voltage. Allow an extra 50 At for magnetising the joints; see p. 162.

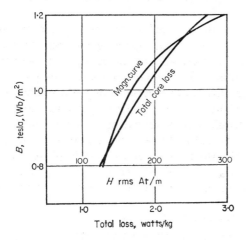

The turns must be integral numbers and finer adjustment to meet the turns ratio can be made on the winding with the larger number of turns. Therefore the low voltage turns must be chosen first.

From eqn. (4.1) $$E = 4·44\,fN(\hat{B}A)$$

where A is the cross-sectional area of the core.

Substituting: $\quad\quad 231 = 4·44 \times 50 \times N_2 \times 1·2 \times 17500 \times 10^{-6}$

from which: $\quad\quad N_2 = 49·6.$

139

The nearest higher integer is selected, to keep the flux below the limit,

i.e. $\underline{N_2 = 50}$ from which $\dot{B} = \dfrac{231}{4 \cdot 44 \times 50 \times 50 \times 17500 \times 10^{-6}}$

$$= 1 \cdot 19$$

$$N_1 = 50 \times 6600/231 = \underline{1430}.$$

From the plotted curves at $\dot{B} = 1 \cdot 19$

$H = 280$ At/m and loss $= 2 \cdot 66$ watts/kg

Total magnetising m.m.f. $= Hl_e + 50 = 280 \times 1 \cdot 6 + 50 = 498$ At.

$\therefore$ magnetising current $= 498/1430 = 0 \cdot 348$ A $= I_m$

Total core loss $= 2 \cdot 66 \times \dfrac{17500}{10^6} \times 1 \cdot 6 \times 7500 = 559$ W

$\therefore$ power component of no-load current $= 559/6600 = 0 \cdot 085$ A $= I_p$

The no-load current $= I_0 = \sqrt{(0 \cdot 085^2 + 0 \cdot 348^2)} = \underline{0 \cdot 359 \text{ A}}.$

The no-load power factor $= 0 \cdot 085/0 \cdot 359 = \underline{0 \cdot 237}.$

Waveform of No-load Current

The waveforms of Fig. 4.2b are all sinusoidal because the rate of change of a sine wave is another sine wave displaced in phase. A sinusoidal voltage applied to a constant resistance would also give a sinusoidal variation—of current in this case because i is proportional to iR. The situation is different when the circuit is a *saturable* inductance since then, L is current dependent. If i and therefore di/dt were forced to be sinusoidal, $v \rightleftharpoons L \cdot di/dt$ could not be of this form. Conversely, if the applied voltage is maintained sinusoidal, then the current must be non-sinusoidal even though $L . di/dt$ as a whole would have to be sinusoidal to balance v; the Ri voltage normally being small.

The sinusoidal flux/time wave of Fig. 4.2b is reproduced in Fig. 4.5 together with a typical static hysteresis loop for the material of the transformer core. The loop is plotted in terms of flux and current instead of B and H so that the current required to produce a particular value of flux can be read off directly. Two construction points are shown on the rising part of the characteristic and by following this same procedure round the whole loop, the current/time curve for a whole cycle can be plotted. It is far from sinusoidal and contains 3rd, 5th and higher-order harmonics which increase rapidly in magnitude if the maximum flux is taken further into saturation.

The current flows due to the combined effects of the applied voltage V_1 and the induced e.m.f. E_1, acting on the constant resistance and leakage inductance of the primary winding circuit. Either V_1 or E_1 or both, must have harmonics large enough to circulate the harmonic components of current through the harmonic impedances $R_1 + j\omega l_1$,

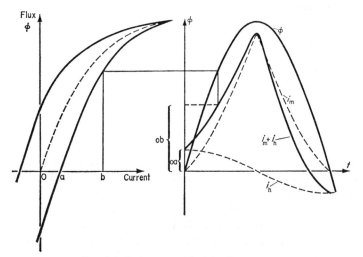

Fig. 4.5. Components of no-load current.

$R_1 + j(3\omega)l_1$, etc. Fortunately, due to other design considerations, which demand a low working flux density, quite a small departure of the flux wave from sinusoidal will give harmonic e.m.f. components of sufficient magnitude. Nevertheless, if the applied voltage is sinusoidal, the flux and e.m.f. cannot be, though the difference is slight and will be neglected subsequently unless harmonics are being discussed.

Should the primary circuit be such as to constrain the input current wave to an approximately sinusoidal shape, e.g. if there is a large series swamping resistance, the flux wave could be plotted from an assumed sinusoidal current wave employing a similar but inverse procedure to that used in Fig. 4.5. The induced e.m.f. wave shape could then be found by differentiating the flux wave, i.e. measuring the slope at various points. It would be very peaky and contain substantial 3rd, 5th and higher-order harmonics. Whatever the circuit configuration, the current, flux and e.m.f. are interdependent. The current, of whatever waveform, will produce a flux, and the resulting e.m.f. will have such a time variation that in conjunction with the applied voltage, this current will be circulated.

On Fig. 4.5 the current is seen to be slightly in advance of the flux wave. This is primarily due to hysteresis which gives rise to a sinusoidal power component of current i_h in quadrature with ϕ and hence in phase with e_1. The wave shape of i_h can be sketched in, starting from the peak value which occurs at zero flux (peak e.m.f.), and is equal to oa because if hysteresis were not present, no magnetising current would be required at this instant. The difference between i_h and the total current is i_m, the magnetising current which could instead have been constructed directly from a single magnetisation curve through the hysteresis loop. Saturation causes i_m to have appreciable harmonic components so that the value of $\mathbf{I}_m$ for the phasor diagram and the value of X_m for the equivalent circuit are not clearly defined. The r.m.s. value of i_m, or alternatively the r.m.s. value of its fundamental frequency component, could be used in calculating X_m.

A current component I_e equal to the eddy-current loss divided by E_1 must be added to I_h to give the total power component of current I_p. Hence $\mathbf{I}_0 = \mathbf{I}_m + \mathbf{I}_p = I_p - jI_m$ with $\mathbf{E}_1$ as reference phasor. If a dynamic hysteresis loop taken at the operating frequency was available, the eddy-current component would be included automatically in the construction of Fig. 4.5.

4.2 THE TRANSFORMER ON LOAD

It has been shown that a primary input voltage V_1 can be transformed to any desired open-circuit secondary voltage E_2 by a suitable choice of turns ratio. E_2 is available for circulating a load current having a magnitude and power factor determined by the secondary-circuit impedance. For the moment, a lagging power factor will be considered. The secondary current and the resulting m.m.f. $I_2 N_2$ will change the flux, reducing the peak value ϕ_m and with it E_1. Because the primary leakage impedance drop is so low, a small alteration to E_1 will cause an appreciable increase of primary current from I_0 to a new value of I_1 equal to $(V_1 - E_1)/(R_1 + jx_1)$. The extra primary current and m.m.f. virtually cancels the whole of the secondary m.m.f. This being so, the mutual flux suffers only a slight modification and requires practically the same net m.m.f. $I_0 N_1$ as on no load. The total primary m.m.f. is therefore increased by an amount $I_2 N_2$ necessary to neutralise the same magnitude of secondary m.m.f.

It is instructive at this stage to examine the redistribution of flux from the no-load condition due to the presence of secondary current. The situation will be simplified if the iron and copper losses are neglected for the moment. Consequently, the primary voltage must be balanced only by the induced e.m.f. since there is no resistance drop. This means in turn, that for a constant supply voltage, the total flux linking the primary must be unaltered. However, the primary current has increased, and with it the primary leakage flux to which it is proportional. It follows that the mutual flux has fallen and on Fig. 4.6a, this is indicated by having one less mutual flux line compared with Fig. 4.1, though the increase of ϕ_1 means that the *total* primary linkage is the same. For the secondary winding there is a further reduction in its flux linkage because the secondary leakage flux opposes the mutual flux, and as shown, there would only be two flux lines linking the secondary when ϕ_m and ϕ_2 are combined. In practice, the situation does not change so drastically, but on Fig. 4.6 all the effects of load have been exaggerated for clarity.

Using the notation developed in Section 3.7, the phasor diagram of Fig. 4.6a shows how the various flux components combine when there is a time-phase shift between their variations. The primary m.m.f. $F_p = I_1 N_1$ produces an in-phase component $\Phi_{11} = \Phi_{21} + \Phi_1$. Using the positive reference sense for I_2 as described from Fig. 4.4, opposing

143

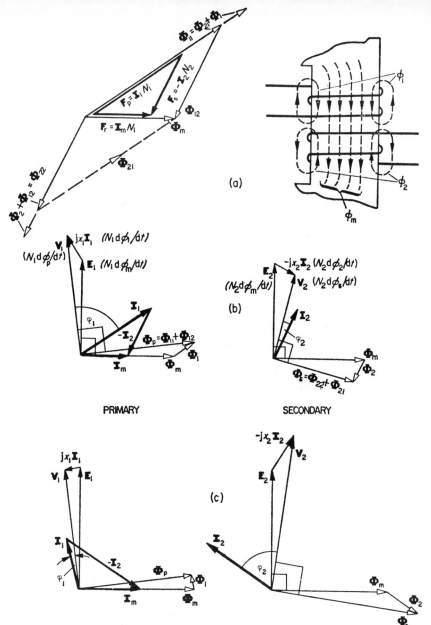

FIG. 4.6. Load conditions: (a) flux components; (b) lagging power-factor; (c) leading power-factor.

144

I_1, then the secondary m.m.f. F_s, in the same sense as the primary m.m.f. is $-I_2N_2$, so that the resultant m.m.f. is:

$$F_r = F_p + F_s = I_1N_1 - I_2N_2 = I_mN_1. \tag{4.2}$$

This produces the resultant mutual flux Φ_m in time phase with the magnetising current I_m. The mutual flux Φ_m could also be obtained by combining the individual mutual components Φ_{21} and Φ_{12} due to I_1 and I_2 respectively, if superposition is permissible; see Section 3.3.

The voltage phasor diagrams are shown in Fig. 4.6b for the above condition at lagging power factor. The primary winding is a power sink so that the equation is $V_1 = E_1 + jx_1I_1$. However, this cannot really be applied until the secondary load has been considered because this determines the phase and magnitude of I_2 and its magnetising action. It is convenient for this purpose to retain the previous phase and magnitude of Φ_m as a reference so this has been drawn the same as in Fig. 4.6a. Assuming a unity turns ratio, $E_2 = E_1$ is in leading quadrature with Φ_m. Viewed from the load, the secondary winding appears as an energy source, so with I_2 as shown, the appropriate circuit equation, (3.3), is $V_2 = E_2 - jx_2I_2$. But V_2 can also be considered in terms of the total secondary flux ϕ_s linking the coil. The two components of voltage correspond to the two components of flux ϕ_m and ϕ_2 the secondary leakage. ϕ_2 is responsible for a substantially constant leakage inductance $l_2 = N_2\phi_2/i_2 = N_2{}^2\Lambda_2$, see Section 3.7. Note that the primary and secondary fluxes and m.m.f.s are added on the phasor diagrams because their positive senses have been chosen in the same direction, though the current I_2 is taken in the opposite sense to I_1. On Fig. 4.6a, the flux ϕ_2 in the core is shown in its negative sense opposing the positive direction of ϕ_m and ϕ_1.

It is now possible to complete the primary phasor diagram using the m.m.f. balance eqn. (4.2) and the voltage balance eqn. (3.3). Since I_1 is normally the unknown, then eqn. (4.2) is rearranged as:

$$I_1N_1 = I_mN_1 + I_2N_2. \tag{4.2a}$$

In this case, $N_1 = N_2$ so it follows that $I_1 = I_m + I_2$.

The voltage drop jx_1I_1 added to E_1 gives the primary terminal voltage V_1. This could also be obtained from $N_1 . d\phi_p/dt$, where ϕ_p is the total flux linking the primary. If, as is common, V_1 and the frequency f are constant, then neglecting resistance, ϕ_p is the same as the no-load value of ϕ_{11} since ϕ_{12} is then zero. The mutual flux is *not* quite constant, as already explained, being reduced by the increased effect of ϕ_1. The secondary voltage is further reduced by the effects of ϕ_2. Thus both leakage fluxes contribute to the fall of the secondary terminal voltage. Looked at another way, the secondary leakage, by its demagnetising action on ϕ_m, as indicated on Fig. 4.6a, has caused the changes on the primary side which led to the establishment of the primary leakage flux. This interrelation of the two leakage components will be discussed again in connection with Fig. 4.35.

So far a lagging power factor has been assumed, but if the secondary load impedance has a leading power-factor characteristic, then the voltage tends to rise with load due to the secondary leakage flux having a magnetising action. The effect is shown on the phasor diagrams of Fig. 4.6c for a fairly low leading power factor. V_1 and ϕ_p are unchanged and the ratio V_2/I_2 is about the same as for Fig. 4.6b. The increase of V_2 is quite considerable, though rather exaggerated from the practical case where I_m is very much smaller relative to I_1 than shown on Fig. 4.6. Perhaps it should also be pointed out, that all the phasor diagrams are drawn for the instant where ϕ_m is zero and about to increase positively, whereas the core diagram shows a different instant in the cycle. There are also instants when the primary leakage flux is zero, when the secondary leakage flux is zero and, for very short periods, the primary and secondary leakage fluxes are in the same sense.

Equivalent Circuit

Having considered the reaction of the various flux components on the transformer load behaviour, it is a simple matter to allow for the imperfections represented by the losses. Primary and secondary resistance voltages I_1R_1 and I_2R_2 allow for the copper loss and the

146

power component I_p is added to I_m to allow for the iron loss, as on Fig. 4.3a. The complete phasor diagrams for the loaded condition are given on Fig. 4.7a. Starting from the secondary terminal voltage, current and power factor, the secondary mutual e.m.f. is given, from the source equation, by $E_2 = V_2 + R_2 I_2 + j x_2 I_2$. The mutual flux Φ_m is drawn in lagging quadrature, and this forms the link with the primary winding, for which it is sometimes advantageous to draw a separate phasor diagram as shown. Φ_m is the starting point for the primary phasor diagram, the magnetising current I_m being drawn in phase and the power component I_p drawn in leading quadrature. This is also the angular position for E_1. The m.m.f. balance equation is modified slightly by the presence of I_p. The primary current for a 1/1 turns ratio becomes

$$I_1 = I_0 + I_2. \qquad (4.2b)$$

Using the sink equation on the primary winding gives the terminal voltage as $V_1 = E_1 + R_1 I_1 + j x_1 I_1$. The primary power factor is seen to be rather more lagging than the secondary power factor due to the effects of the magnetising current and the leakage reactances. The difference is not as large as indicated on Fig. 4.7a because the various imperfections have been exaggerated for clarity.

The equivalent circuit follows simply from Fig. 4.3b by the addition of the secondary leakage impedance $R_2 + j x_2$ and is shown on Fig. 4.7b. Both the equivalent circuit and the phasor diagram are merely pictorial expressions of the appropriate transformer equations but are useful aids in remembering these equations.

Practically all transformers have a turns ratio different from unity although such an arrangement is sometimes employed for the purposes of electrically isolating one circuit from another operating at the same voltage. To explain the case where $N_1 \neq N_2$ the reaction of the secondary will be viewed from the primary winding. The reaction is experienced only in terms of the magnetising force due to the secondary ampere-turns. There is no way of detecting from the primary side whether I_2 is large and N_2 small or vice versa; it is the product of current and turns which causes the reaction. Consequently, a secondary

147

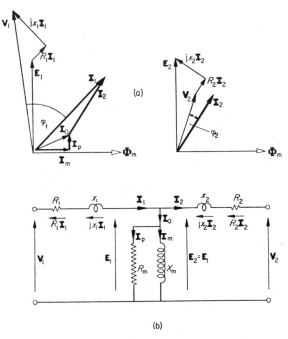

FIG. 4.7. Transformer; 1/1 turns ratio: (a) phasor diagrams;
(b) equivalent circuit.

winding can be replaced by any one of a number of different equivalent windings and load circuits which will give rise to an identical reaction on the primary. It will be convenient to change the secondary winding to an equivalent winding having the same number of turns N_1 as the primary. The phasor diagram and circuit of Fig. 4.7 will then be applicable, although changes from the actual secondary winding parameters can be expected, viz.

E_2 to E_2', I_2 to I_2', and z_2 to z_2' because N_2 changes to $N_2' = N_1$.

With N_2 changed to N_1, since the e.m.f.s are proportional to turns, $E_2' = (N_1/N_2)E_2$ which is the same as E_1.

148

For current, since the reaction m.m.f. must be unchanged, $I_2'N_2'$ $= I_2'N_1$ must be equal to I_2N_2. i.e. $I_2' = (N_2/N_1)I_2$.

For impedances, since any secondary voltage v becomes $(N_1/N_2)v$, and any secondary current i becomes $(N_2/N_1)i$, then any secondary impedance, including load impedance, must become $v'/i' = (N_1/N_2)^2 v/i$. Consequently, $R_2' = (N_1/N_2)^2 R_2$ and $x_2' = (N_1/N_2)^2 x_2$.

If the primary turns are taken as reference turns, the process is called referring to the primary side.

There are a few checks which can be made to see if the procedure outlined is valid. For example, the copper loss in the referred secondary winding must be the same as in the original secondary otherwise the primary would have to supply a different loss power. Therefore:

$$I_2'^2 R_2' \text{ must be equal to } I_2{}^2 R_2.$$

$$\left(I_2 . \frac{N_2}{N_1}\right)^2 \left(R_2 . \frac{N_1{}^2}{N_2{}^2}\right) \text{ does in fact reduce to } I_2{}^2 R_2.$$

Similarly the stored magnetic energy in the leakage field $(\frac{1}{2}Li^2)$ which is proportional to $I_2{}^2 x_2$ will be found to check as $I_2'^2 x_2'$.

The referred secondary VA is equal to:

$$E_2'I_2' = E_2(N_1/N_2) . I_2(N_2/N_1) = E_2 I_2.$$

The validity of the various transformations is thus confirmed. In fact, if the actual secondary winding was removed physically from the core and replaced by the equivalent winding and load circuit, designed to give the parameters N_1, R_2', x_2' and I_2'; measurements from the primary terminals would be unable to detect any difference in secondary m.m.f., kVA demand or copper loss, under normal power-frequency operation.

There is no point in choosing any basis other than equal turns on primary and referred secondary, but it is sometimes convenient to refer the primary to the secondary winding. In this case, if all the subscript $_1$'s are interchanged for the subscript $_2$'s, the necessary referring constants are easily found; e.g. $R_1' = R_1(N_2/N_1)^2$. It is worth noting that for a practical transformer, $R_1' \simeq R_2$, $x_1' \simeq x_2$; similarly $R_2' \simeq R_1$ and $x_2' \simeq x_1$; see Appendix E, problem 7 for Chapter 4.

149

The equivalent circuit for the general case where $N_1 \neq N_2$ is shown on Fig. 4.8. It is the same as Fig. 3.25 except that R_m has been added to the magnetising branch to allow for iron loss and an ideal lossless transformation has been included before the secondary terminals to return V_2' to V_2 and I_2' to I_2. All calculations of internal voltage and power losses are made before this ideal transformation is applied. The behaviour of a transformer as detected at both sets of terminals is the same as the behaviour detected at the corresponding terminals

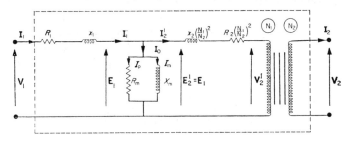

FIG. 4.8. Equivalent circuit for $N_1 \neq N_2$.

of this circuit when the appropriate parameters are inserted. The slightly different representation showing the coils N_1 and N_2 side by side with a core in between, is only used for convenience. On the transformer itself, the coils are, of course, wound round the same core.

Very little error is introduced if the magnetising branch is transferred to the primary terminals, but a few anomalies will arise. For example, the current shown flowing through the primary impedance will no longer be the whole of the primary current. The error is quite small since I_0 is usually such a small fraction of I_1. Slightly different answers may be obtained to a particular problem depending on whether or not allowance is made for this error.

With this simplified circuit, the primary and referred secondary impedances can be added to give the effective leakage impedance components:

$$R_{e_1} = R_1 + R_2(N_1/N_2)^2 \quad \text{and} \quad x_{e1} = x_1 + x_2(N_1/N_2)^2.$$

Figure 4.9 shows the final form of the equivalent circuit referred to the primary winding. If referred to the secondary winding the ideal transformation, this time from V_1 to V_1' would be on the left-hand side of the circuit with the magnetising branch across V_1'. All the impedances shown in Fig. 4.9 would have to be multiplied by $(N_2/N_1)^2$.

It should be pointed out, that the equivalent circuit as derived here, is only valid for normal operation at power frequencies; capacitance effects must be taken into account whenever the rate of change of voltage would give rise to appreciable capacitance currents, $i_c = C \, dV/dt$.

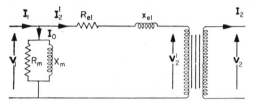

Fɪɢ. 4.9. Simplified equivalent circuit.

Under the impact of steep-fronted, high-voltage surges, the distribution of voltage across the windings is largely governed by the inter-turn and turns-to-earth capacitances. A further point is that Fig. 4.8 is not the only possible equivalent circuit even for power frequencies. An alternative, treating the transformer as a three or four-terminal network gives rise to a representation which is just as accurate and has some advantages for the circuit engineer who treats all devices as circuit elements with certain transfer properties. The circuit on this basis would have a turns ratio having a phase shift as well as a magnitude change, and the impedances would not be the same as those of the windings. The circuit would not explain the phenomena within the device like the effects of saturation, so for an understanding of internal behaviour, Fig. 4.8 is preferable.

There are two ways of looking at the equivalent circuit:

(a) viewed from the primary as a sink as in Fig. 4.8 with the referred load impedance connected across V_2', or:

(b) viewed from the referred secondary as a source of voltage $V(=V_2')$, with internal drops due to R_{el} and x_{el}. The magnetising

151

branch is sometimes omitted in this representation and so the circuit reduces to a generator producing a constant voltage E (actually equal to V_1) and having an internal impedance $R+jX$ (actually equal to $R_{el}+jx_{el}$) (Fig. 4.10).

In either case the parameters could be referred to the secondary winding and this may save calculation time, particularly when adopting the viewpoint expressed in (b) above.

The resistances and reactances can be obtained from two simple light load tests which will be described in Section 4.7.

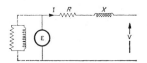

FIG. 4.10. Source equivalent circuit.

The importance of understanding this preliminary transformer theory cannot be overemphasised because it has bearings on the behaviour of all electromagnetic machines. For example, it might be appropriate at this stage to return to Section 3.9 in which the equivalent circuits of all the main machine types were derived from a consideration of the equations for two coupled coils. The appreciation of Section 3.9 should be increased as a result of the work done so far in this chapter and it will be instructive to compare the alternative treatments of the transformer. It might be helpful also to turn to Appendix B, which shows, by means of a flow diagram, the cause-and-effect sequence of the electromagnetic and circuit interactions in electrical machines.

EXAMPLE E.4.2

A 200-kVA single-phase transformer with a voltage ratio 6350/660 V has the following winding resistances and reactances:

$$R_1 = 1{\cdot}56\,\Omega \quad R_2 = 0{\cdot}016\,\Omega \quad x_1 = 4{\cdot}67\,\Omega \quad x_2 = 0{\cdot}048\,\Omega.$$

On no load the transformer takes a current of 0·96 A at a power factor of 0·263 lagging. Calculate the equivalent-circuit parameters referred to the high-voltage winding.

$$R_{eL} = R_1 + R_2(N_1/N_2)^2 = 1{\cdot}56 + 0{\cdot}016(635/66)^2 = \underline{3{\cdot}04\,\Omega}$$

$$x_{eL} = x_1 + x_2(N_1/N_2)^2 = 4{\cdot}67 + 0{\cdot}048(635/66)^2 = \underline{9{\cdot}12\,\Omega}.$$

152

For the magnetising branch, the no-load current is resolved into two components $I_p = I_0 \cos \varphi$ and $I_m = I_0 \sin \varphi$.

$$\therefore R_m = \frac{V}{I_0 \cos \varphi} = \frac{6350}{0.96 \times 0.263} = 25.2 \ k\Omega$$

$$X_m = \frac{V}{I_0 \sin \varphi} = \frac{6350}{0.96 \times \sqrt{(1 - 0.263^2)}} = 6.85 \ k\Omega.$$

Voltage Regulation

The way in which the secondary terminal voltage varies with load depends on the load current, the internal impedance and the load power factor. The change of voltage from no-load to full-load current at any particular power factor is termed the inherent regulation. It is usually expressed as a percentage or a fraction of the rated no-load terminal voltage. From the consumers' viewpoint, voltage variation due to load change is undesirable and should be kept small.

The phasor diagram for the source equivalent circuit is given in Fig. 4.11 with the impedance voltage drop phasors shown considerably enlarged relative to **E** and **V**. It is required to find the numerical difference between E, the no-load terminal voltage; and V the voltage

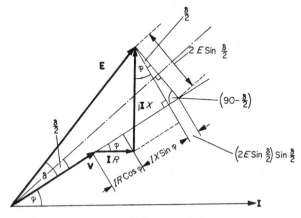

FIG. 4.11. Voltage regulation.

153

with load current I flowing at power factor cos φ. It is a little simpler here, if, unlike the general practice in this text, φ is taken positive for lagging and negative for leading power factors. There are three components of the voltage difference, with magnitudes which can be checked from the diagram giving:

$$E - V = IR \cos \varphi + IX \sin \varphi + (2E \sin \delta/2)^2/2E.$$

It can be seen that $2E \sin \delta/2$ is very nearly equal to $IX \cos \varphi - IR \sin \varphi$ and so:

$$E - V = IR \cos \varphi + IX \sin \varphi + (IX \cos \varphi - IR \sin \varphi)^2/2E. \qquad (4.3a)$$

Because IR and IX are so small relatively, the term involving sin $\delta/2$ is usually negligible except perhaps at low leading power factors. Only the first two terms will be used and expressing these in *per unit* (see p. 90), using E and $I_{f.1.}$ as voltage and current base respectively:

$$\frac{E - V}{E} = \frac{I_{f.1.}}{I_{f.1.}} \cdot \frac{IR}{E} \cos \varphi + \frac{I_{f.1.}}{I_{f.1.}} \cdot \frac{IX}{E} \sin \varphi.$$

$$\text{regulation}_{p.u.} = I_{p.u.}.R_{p.u.} \cos \varphi + I_{p.u.}.X_{p.u.} \sin \varphi. \qquad (4.3b)$$

For percentage regulation, the equation would be multiplied by 100 giving percentage impedance components, e.g.

$$\%R = R_{p.u.} \times 100 = (I_{f.1.}R/E).100.$$

Note, that although the equations are correct at any current, the *per-unit* and percentage impedances apply to full-load current, i.e. when $I_{p.u.} = I/I_{f.1.} = 1$.

Equations (4.3a) and (4.3b) apply for both lagging and leading power factors but since φ has been taken as positive for lagging power factors, the second term will be negative when the power factor is leading. Consequently, the regulation can be zero and the terminal voltage will actually rise if the power factor is low enough leading. The reason for this should be understood if reference is made to the flux diagram of Fig. 4.6c.

One of the advantages of the *per-unit* notation is demonstrated by the simplicity of eqn. (4.3b). It is also worth pointing out that

154

with this notation it is no longer necessary to refer quantities from primary to secondary side. For example, the referred secondary resistance in *per unit* is:

$$\frac{R_2' I_2'}{E_2'} = \frac{R_2(N_1/N_2)^2 . I_2(N_2/N_1)}{E_2(N_1/N_2)} = \frac{R_2 I_2}{E_2};$$

so *per-unit* values of primary and secondary impedance can be added directly.

EXAMPLE E.4.3

Using the data of E.4.2 calculate the inherent voltage regulation at unity power factor (u.p.f.), 0·8 lagging p.f. and 0·8 leading p.f. Use eqns. (4.3a) and (4.3b).

Full-load current (f.l.) $= 200 \times 10^3/6350 = 31\cdot5$ A

$I_{f.l.}.R = 31\cdot5 \times 3\cdot04 = 96$ V $I_{f.l.}.X = 31\cdot5 \times 9\cdot12 = 288$ V

Using first, eqn. (4.3a):

u.p.f. $E - V = 96 \times 1 + 288^2/(2 \times 6350) = 96 + 6\cdot4 = \underline{102\cdot4 \text{ V}}$

0·8 lag $= 96 \times 0\cdot8 + 288 \times 0\cdot6 + \dfrac{(288 \times 0\cdot8 - 96 \times 0\cdot6)^2}{2 \times 6350}$

                    $= 249\cdot7 + 2\cdot3 = \underline{252 \text{ V}}$

0·8 lead $= 96 \times 0\cdot8 + 288 \times (-0\cdot6) + \dfrac{(288 \times 0\cdot8 - 96 \times (-0\cdot6))^2}{2 \times 6350}$

                    $= -96\cdot3 + 6\cdot5 = \underline{-89\cdot8 \text{ V}}$

The corresponding terminal voltages $E - (E - V)$ are:

        in primary terms,   6248, 6908, 6440 V respectively;

        in secondary terms, 648,   633,   669 V respectively.

Note that the last term in eqn. (4.3a) is very small and only affects the answer appreciably as the power factor becomes leading. It will be neglected subsequently. Using eqn. (4.3b):

$R_{p.u.} = 96/6350 = 0\cdot0151$ $X_{p.u.} = 288/6350 = 0\cdot0453$

u.p.f. $E - V = 0\cdot0151 \times 1 = 0\cdot0511 \text{ per unit}$ $= \underline{96 \text{ V}}$

0·8 lag $= 0\cdot0151 \times 0\cdot8 + 0\cdot0453 \times 0\cdot6 = 0\cdot0392 = \underline{249 \text{ V}}$

0·8 lead $= 0\cdot0151 \times 0\cdot8 + 0\cdot0452 \times (-0\cdot6) = -0\cdot0151 = \underline{-96 \text{ V}}$

The diagram overleaf shows the equivalent circuit, currents and voltages at full load, unity power factor.

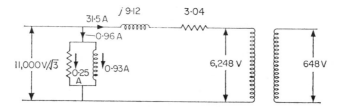

Efficiency

A detailed analysis of transformer losses would take into account the dielectric loss and stray load loss, but for present purposes they will be assumed to be included in the iron loss and the copper loss respectively. The iron losses are independent of load current if voltage and frequency are constant so they constitute the fixed loss. The copper losses are $I^2 R_e$ where R_e is the resistance, referred to either primary or secondary winding. The output is $V_2 I_2 \cos \varphi_2$ and so the efficiency follows from eqn. (3.6):

$$\eta = \text{output/input, or} = 1 - \text{losses/(output} + \text{losses)}.$$

From eqn. (3.7) the above expression will have a maximum value when the fixed loss is equal to the loss which is proportional to I^2,

$$\text{i.e. when iron loss} = I^2 . \frac{(I_{f.l.})^2}{(I_{f.l.})^2} . R_e.$$

Now $I/I_{f.l.}$ is the *per-unit* load current and $(I_{f.l.})^2 R_e$ is the full-load copper loss. Therefore the maximum efficiency occurs when the *per-unit* load current is equal to $\sqrt{}$(iron loss/full-load copper loss).

Note that the term load, refers to load current, not kW, although since the voltage is substantially constant it also means load kVA. For any particular current, the losses are virtually the same whatever the power factor. Since a machine must be designed to dissipate a certain loss, the rating basis for a.c. operation must be kVA not kW.

The absolute maximum efficiency occurs when the power factor is unity, because here, the maximum output is being obtained for any given current and loss. Figure 4.12 shows two efficiency curves, one for unity and one for 0·8 power factor corresponding to the case where the full-load copper loss is equal to the iron loss. The maxima occur at the same value of *per-unit* load and the curve is fairly flat about this region. It is possible at the design stage to adjust the iron and copper weights to alter the balance between the iron and copper losses and thus choose the load at which maximum efficiency will occur. Transformers which are directly connected to generators in

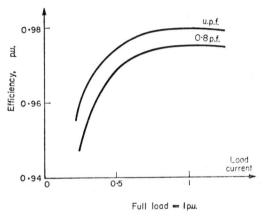

Full load = 1 p.u.

FIG. 4.12. Efficiency curves.

a power station are either out of circuit altogether when the generator is not running, or operating at full load for most of the time. Such power transformers would have their copper losses and iron losses similar at full load. On the other hand, distribution transformers which are in circuit night and day and often run on very low load, will be designed to have maximum efficiency at a figure much lower than full load, depending on the duty cycle of load variation over a 24-hour period.

All-day Efficiency

A more suitable method of assessing the efficiency of a transformer having a duty cycle is on an energy basis. The output and losses are calculated in kW hours over a 24-hour day. The all-day efficiency is then defined as:

1 − (losses in kWh)/(output in kWh + losses in kWh).

EXAMPLE E.4.4

Using the data of E.4.2 calculate the efficiency of the transformer at full load and half load when the power factor is unity and also when the power factor is 0·8. Also calculate the maximum efficiency.

The transformer is in circuit continuously. For a total of 8 hours, it delivers a load of 160 kW at 0·8 p.f.

For a total of 6 hours it delivers a load of 80 kW at u.p.f.

For the remainder of the 24-hour cycle it is on no load.

What is the all-day efficiency?

In the following solution, to avoid complications, which serve no purpose at this stage, the effects of regulation will be neglected. For example, at 0·8 lagging and 0·8 leading power-factor, the terminal voltages and hence the outputs are different at full-load current. The efficiency will be based on an output equal to the power factor, lagging or leading, multiplied by the rated kVA or a fraction of it as required. Further, the full-load copper loss will be taken as $I_{f.1}^2 . R_e$ though, in fact, the current in the primary resistance should include the small no-load current. Details concerning recommended transformer practice are included in the appropriate British Standard Specification* BS 171 covering the performance of Power Transformers.

From Example E.4.2, the iron loss = $6350 \times 0·96 \times 0·263 = 1·6$ kW

the copper loss at f.l. = $(31·5)^2 \times 3·04 = 3·02$ kW

the copper loss at half full-load = $(0·5)^2 \times 3·02 = 0·775$ kW.

∴ Total loss at full load = $1·6 + 3·02 = 4·62$ kW.

Total loss at half load = $1·6 + 0·755 = 2·36$ kW.

Unity Power-factor

F.l. output = 200 kW; Input = 204·62;

$$\eta = 1 - \frac{4·62}{204·62} = \underline{97·74\%}.$$

---

* Available from the British Standards Institution, 2 Park Street, London W1Y A44.

Half f.l. output $= 100$ kW; Input $= 102 \cdot 36$;

$$\eta = 1 - \frac{2 \cdot 36}{102 \cdot 36} = \underline{97 \cdot 7\%}.$$

0·8 *power factor*

F.l. output $= 200 \times 0 \cdot 8 = 160$ kW; Input $= 164 \cdot 62$

$$\eta = 1 - \frac{4 \cdot 62}{164 \cdot 62} = \underline{97 \cdot 19\%}.$$

Half f.l. output $= 100 \times 0 \cdot 8 = 80$ kW; Input $= 82 \cdot 36$

$$\eta = 1 - \frac{2 \cdot 36}{82 \cdot 36} = \underline{97 \cdot 14\%}.$$

The maximum efficiency occurs when $I_{\text{p.u.}} = \sqrt{\left(\frac{1 \cdot 6}{3 \cdot 02}\right)} = 0 \cdot 73.$

At unity power-factor, the absolute maximum efficiency occurs when the output is $0 \cdot 73 \times 200 = 146$ kW and

$$\eta_{\text{max}} = 1 - \frac{1 \cdot 6 + 1 \cdot 6}{146 + 3 \cdot 2} = \underline{97 \cdot 86\%}.$$

Note, in working out the efficiencies in *per unit*, as described on p. 94, the losses would first be expressed as a fraction of the rated kVA.

All-day Efficiency

At 160 kW, 0·8 p.f.   Copper loss $= 3 \cdot 02$ kW, Total loss $= 4 \cdot 62$ kW.

At 80 kW, u.p.f., copper loss $= 3 \cdot 02(80/200)^2 = 0 \cdot 48.$

Total loss $= 2 \cdot 08$ kW.

On no load: Total loss $= 1 \cdot 6$ kW.

For 8 hours, output $= 160 \times 8 = 1280$ kWh;

loss $= 4 \cdot 62 \times 8 = 37$ kWh.

For 6 hours, output $= 80 \times 6 = 480$ kWh; loss $= 2 \cdot 08 \times 6 = 12 \cdot 5$ kWh.

For 10 hours, output $= 0$ loss $= 1 \cdot 6 \times 10 = 16$ kWh.

In 24 hours, total output $= 1760$ kWh;   total loss $= 65 \cdot 5$ kWh.

All-day efficiency $= 1 - \frac{65 \cdot 5}{1760 + 65 \cdot 5} = \underline{96 \cdot 41\%}.$

Effects of Supply-voltage and Frequency Changes

Occasionally it is required to operate a transformer at a voltage and frequency different from the rated figures. It is important to realise the limitations involved in accommodating such changes.

The terminal voltage is very nearly equal to the induced e.m.f., which in turn is proportional to the flux density and frequency, i.e. $V \propto \hat{B}f$; eqn. (4.1). Thus an increase of V without a corresponding increase in f will cause the flux density to rise towards saturation, increasing the magnetising current and the generated harmonics. The same applies for a decrease of f without a corresponding decrease of V.

As far as losses are concerned, the copper losses remain the same, if changes in copper eddy-current effects are neglected. The iron losses, however, are altered. From eqn. (2.3), the eddy current loss p_e is proportional to $\hat{B}^2f^2$, or since $V \propto \hat{B}f$ it is proportional to V^2; i.e. independent of frequency since, for any particular voltage, a change of frequency is accompanied by a compensating change of flux density.

The hysteresis loss p_h, from eqn. (2.2) is proportional to

$$\hat{B}^x f \propto \hat{B}^x f^x f^{(1-x)} \propto V^x f^{(1-x)}$$

very nearly, so that it is sensitive to both voltage and frequency changes.

It follows from the above, that, if there is a departure from either rated voltage or frequency or both, the iron losses may no longer be taken as constant and may increase or decrease depending on the change. Since a machine is designed to dissipate a certain total loss without excessive temperature rise, an adjustment of the current rating may be necessary to maintain the total loss unchanged and thus give the same temperature rise approximately. The full, rated current may have to be decreased, or may in fact be increased depending on whether the iron losses have increased or decreased.

EXAMPLE E.4.5

A 3-phase, 200-kVA, 10-kV transformer gave the following readings when undergoing light-load tests at 50 Hz.

No load. Normal voltage applied; total power input = 960 W.

Short circuit. Rated current flowing with reduced voltage applied; total power input 1540 W.

What would be the permissible rating of the transformer if it had to operate at 11·5 kV, 60 Hz, 0·8 p.f.? Assume the iron losses at 50 Hz are divided equally between eddy-current and hysteresis loss, and that the Steinmetz index is 1·7.

As will be explained in more detail in Section 4·7, the no-load test gives the iron loss, and the short-circuit test at full load current gives the full-load copper loss.

Therefore, at 50 Hz, the eddy current loss p_e, and the hysteresis loss p_h are each 480 W.

$$\frac{p_{e60}}{p_{e50}} = \frac{V_{60}{}^2}{V_{50}{}^2} = \frac{(11\cdot5)^2}{10^2} = 1\cdot323 \quad \therefore p_{e60} = 1\cdot323 \times 480 = 635 \text{ W}.$$

$$\frac{p_{h60}}{p_{h50}} = \frac{(V_{60}/V_{50})^{1\cdot7}}{(f_{60}/f_{50})^{0\cdot7}} = \frac{(11\cdot5/10)^{1\cdot7}}{(60/50)^{0\cdot7}} = \frac{1\cdot269}{1\cdot135} = 1\cdot12.$$

With the calculation broken down in this way, it can readily be performed on a slide rule with a simple log-log scale if a suitable calculator is not available.

The hysteresis loss at 60 Hz = $1\cdot12 \times 480 = 536$ W.

Total iron loss at 60 Hz   = 1171 W.

For the same total loss, the copper loss must be reduced by $1171 - 960 = 211$ W.

Permissible rating at 11·5 kV, 60 Hz =

$$200 \times 11\cdot5/10 \times \sqrt{\left(\frac{1540 - 211}{1540}\right)} = \underline{214 \text{ kVA}}.$$

4.3 CONSTRUCTION AND WINDINGS

The transformer is basically a very simple device. Coils consisting of many turns are wound on a laminated iron core and insulated from the iron and from each other. The core itself forms a closed, iron magnetic circuit, thus ensuring a small magnetising current. Consequently, the windings encircle the core and the core encircles the windings. There are two main ways of achieving this in a practical transformer.

Core-type Construction (Single-phase)

The core has a simple rectangular elevation and consists of two vertical limbs around which the preformed circular coils are placed.

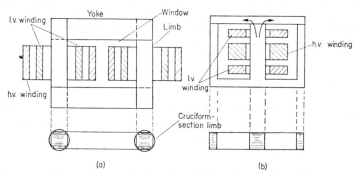

FIG. 4.13. Single-phase transformer construction: (a) core type; (b) shell type.

Top and bottom members called the yokes, connect the two limbs and are usually built up to a cross-sectional area rather greater than that of the limbs. To limit the eddy-current loss, laminations of 0·4 mm thick or less are used, successive layers being arranged to give an overlap at the corners to reduce the joint reluctance. Even so, the flux has to cross the insulation between the laminations at these places, which means that there is a virtual air gap absorbing ampere-turns over and above those required for the iron itself; see Example E.4.1. Machined butt-joints are sometimes used as indicated on Fig. 4.13b. The laminations are clamped together with more or less elaboration depending on the size of the core. If all the laminations are the same width, the cross section is often made square but this is wasteful on the larger sizes. Two or more widths are then used giving a cross section which approaches the area of the circumscribing circle and ensure that the area within the internal diameter of the coils has a more efficient iron/air ratio. Figure 4.13a shows a cruciform section requiring only two different widths. Putting the coils on two limbs instead of one, as shown in previous sketches, cuts down the average length of a turn and hence the copper weight. Each limb carries one half of the secondary and one half of the primary so that the two can be closely coupled magnetically with low leakage. The low-voltage winding is placed nearest to the core, thus separating the high-voltage winding from "earth" and reducing the amount of insulating material required.

Shell-type Construction (Single-phase)

Here, instead of the coils being divided into two groups, the core is divided. Figure 4.13b shows the more common case where only two parallel magnetic paths are used with flux encircling the single group of coils on two sides.

Modern electrical sheet steels have directional properties produced by a cold rolling process. This gives a much lower reluctance in the preferred direction. Sometimes, on the smaller sizes of transformer, the core itself is wound on a former using long strips of such steel. Two sections are formed like this and placed side by side to give an arrangement like the shell core of Fig. 4.13b with two parallel magnetic paths. The coils are then wound, turn by turn over the two adjacent half-limbs. On very small transformers and reactors the cores are first wound and then sawn across the section. The halves can be separated and then clamped together after threading the coils on the limbs. These are known as "C" cores because of their shape after sawing.

The choice of core construction depends largely on manufacturers' preference. The core type is easier to dismantle for repair and the shell type gives better support against electromagnetic forces between the current carrying conductors. These forces are of considerable magnitude under short-circuit conditions. The shell type construction is com-

163

monly used on small transformers where a square or rectangular core cross-section is suitable from economic considerations.

Core Size and Design

The kVA rating of a transformer is equal to:

$$E.I \times 10^{-3} \text{ for either winding,}$$

which at any particular frequency is proportional to:

$$\phi N.I \quad \text{or} \quad \phi.IN. \tag{4.4}$$

The flux determines the iron section and weight, so transformers are smaller for higher frequencies since the flux is less, see eqn. (4.1).

The full-load m.m.f. determines the copper section and weight. This follows because the permissible current density determines the section of each turn and by considering Fig. 4.13a, for example, it can be seen that the total copper section in the window is proportional to the number of turns and the number of amperes through each turn, i.e. to $I_1 N_1 + I_2 N_2 = 2 I_1 N_1$ very nearly.

For a particular type of transformer, the balance between iron loss and copper loss, and hence the point of maximum efficiency will be determined by the relative iron and copper weights. To maintain this balance, an increase in kVA output, for example, would require the same proportionate increase in both iron and copper. To double the output, both flux and full-load ampere-turns would have to be increased by a factor of $\sqrt{2}$. Consequently, for any particular type of transformer, the flux and the full-load m.m.f. are each proportional to $\sqrt{kVA}$ and this is a starting point for design. Once the flux is known, the voltage per turn to which it is proportional, eqn. (4.1), and the number of turns is known. Actually, the voltage per turn for 50-Hz supplies comes out from experience to be approximately equal numerically to $\sqrt{kVA}$, where maximum efficiency is required at about full load. If it is desired to reduce the iron losses, the figure would be somewhat less than this to give maximum efficiency at a fractional load.

Equation (4.4) also shows that the size and weight of a transformer increase more or less as the kVA rating which is of course to be expected.

Three-phase Transformers

For a three-phase circuit, three single-phase transformers could be used with their primaries and secondaries connected in star or delta as required. There are, however, special three-phase constructions of both shell and core types and these are much cheaper in capital cost.

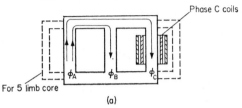

(a)

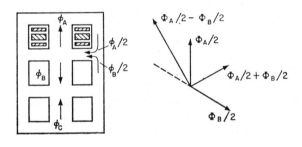

(b)

Fig. 4.14. Three-phase transformer construction:
(a) core type; (b) shell type.

Figure 4.14a shows the core type in which three equal limbs are joined by a common yoke. A primary and secondary of one phase is wound on each limb. Flux "flows" up each limb in turn and down the other two in general, so that the phase magnetic circuits are in series and therefore interdependent. The addition of two half-section outer limbs as indicated makes a five-limb core which permits the yoke section to be reduced and with it the overall height. The magnetic circuits for each phase are then virtually independent. In either case,

165

the magnetising current of the centre phase is different from that of the outer phases because the magnetic path "viewed" from the coil is not quite the same.

Figure 4.14b shows the less common shell-type construction which appears like three single-phase shell-type cores built on top of one another. Some economy in the common yoke-sections can be affected by reversing the middle coils as shown so that the flux carried corresponds to the phasor sum of the time phasors $\Phi_A/2$ and $\Phi_B/2$ instead of their difference. The phase magnetic circuits are in parallel and therefore independent, neglecting saturation effects in the common paths.

Windings

Although there are several variants in coil arrangements for special design or manufacturing reasons, basically the primary and secondary windings simply consist of a series of turns wound round the core. The conductor section may be subdivided to reduce copper eddy-current losses and the coils may be sectionalised to assist in handling, to improve cooling and to control the leakage reactance. Core-type transformers usually have cylindrical coils taking up the whole length of the limb, whereas shell-type transformers usually have a series of flat coils with the primary and secondary sections alternating in a sandwich arrangement; see Fig. 4.13. Various stages in the winding process are illustrated by Fig. 4.15.

Tappings

To permit some adjustment of the voltage or to maintain the secondary voltage against supply or load variations, tappings on the coils are brought out to terminals so that the number of turns on one winding, usually the high voltage winding, can be changed. The usual arrangement, when adjustment is required with the transformer out of circuit, is to allow for $\pm 2\frac{1}{2}\%$ to $\pm 5\%$ variation in the turns ratio. Tappings can be seen on the left-hand limb of Fig.4.15 on which both primary and secondary coils have been completed. Special arrangements must

FIG. 4.15. Three-phase, three-limb 500-kVA core-type transformer partially wound.
(By courtesy of Ferranti Ltd.)

FIG. 4.16. Single-phase auto-transformer unit for three-phase bank rated at 60 Hz, 1000 MVA, 525/241·5 kV with delta-connected tertiary rated at 75 MVA, 34·5 kV. Temporary test-caps on 525 kV and 241·5 kV bushings. Neutral terminal earthed. (By courtesy of G.E.C. Transformers Ltd.)

be made to avoid breaking the main circuit if tapping changes are to be effected with the transformer on load. One way is to divide the coil into two parallel sections with identical tappings. Each section can then be removed from the main circuit in turn while its connections are altered.

Cooling

Rotating machines create a turbulent air flow which assists in removing the heat generated internally as a result of the losses. The stationary transformer is at a disadvantage here, and many elaborate cooling arrangements have been devised to deal with the whole range of sizes.

The cooling problem can be appreciated from the following considerations. If all the linear dimensions of a transformer are increased in the same ratio, all surfaces and cross-sectional areas are increased as the square of the ratio, whereas volumes go up as the cube. Consequently, with the flux density and current density maintained, flux and m.m.f. each increase as the square giving a fourth power increase to the kVA; eqn. (4.4). Since the losses are proportional to volume with constant flux and current densities, these increase as the cube of the ratio. From these facts it follows that efficiency increases with size, because the kVA rating increases at a greater rate than the losses. However, the loss per unit surface area increases, since areas have only increased as the square of the ratio. More efficient heat transfer than can be obtained by natural radiation and convection eventually becomes imperative if the temperature rise is to be restrained.

On small transformers up to a few kVA, natural air cooling is satisfactory. For larger sizes, the transformers are usually immersed in an oil filled tank. The heat is passed to the oil which circulates round the tank by natural convection, thus carrying the heat to the tank walls whence it is dissipated. The surface area of the tank can be effectively increased by various means. A common method is to weld several vertical tubes on the sides of the tanks so that oil can circulate naturally through them; alternatively the oil is allowed to circulate through external radiators. This method can be adapted for ratings up to several thousand kVA. In the standard classification of cooling arrangements it is given the symbol "ON" for oil immersed natural cooling.

On larger sizes of transformer, means must be provided to improve the rate at which heat can be dissipated. This is done by forced cooling; blowing air over the tank or over specially designed radiators through which the oil is pumped to be cooled. For further information about the possible cooling systems the appropriate British Standard BS 171 may be consulted. Figure 4.16 shows a transformer with fan coolers. The on-load, tap-changing gear is visible on the extreme right.

4.4 TRANSFORMERS FOR THREE-PHASE CIRCUITS

A three-phase transformer is lighter, smaller, cheaper and more efficient than three single-phase units. However, it is more difficult to transport and for speedy replacement in the event of breakdown, an expensive spare must be available. Only one spare single-phase unit is necessary to give reasonable insurance against failure when a bank of three single-phase units is used. The choice of transformer arrangement is therefore governed by the relative importance of these various factors in particular cases. With regard to performance, the behaviour is the same in most respects but there are some differences which will be discussed shortly.

Standard Terminal Markings

Although there are minor variations in detail between the normal practice in different countries, essential procedures are common, in accordance with the International Electrochemical Commission (IEC) recommendations. For example, the letters U, V and W are used in some three-phase systems. Current British practice is summarised briefly below:

The high-voltage winding bears capitals A, B and C for three-phase. (A and B for two-phase; A for single-phase.)

The low-voltage winding bears small letters a, b and c.

Sometimes a third or tertiary winding is used and is labelled thus: 3A, 3B, 3C.

In addition, numbers are assigned as suffices to the letters, e.g. phase A winding has ends designated A_1, A_2. Further suffices are available for tappings and sectionalised coil arrangements. The windings are so numbered that if, on the high-voltage winding, A_1 is positive with respect to A_2 at a particular instant, then a_1 is positive with respect to a_2 and $3A_1$ is positive with respect to $3A_2$ at the same instant, see Fig. 4.4.

Alternative Three-phase Connections; Group Numbers

With three phases and at least two windings per phase, there is a large choice of possible connections. This is not only because both primary and secondary can each be connected either in star or delta, but because in addition, each phase can be reversed. When all the combinations are considered, it is found that they all fall within one of four main groups, each group being characterised by a different time-phase displacement between primary and secondary line voltages. These points are most instructively illustrated by Fig. 4.17 which shows a selection of connection diagrams extracted from BS 171 (1970). Some of these will be explained further. The later BS 171 (1978) omits the winding letters, as on the relevant IEC document (76), but the essential details of angles, group numbers and connections are the same.

In building up the time-phasor diagrams, the numbering system must be borne in mind; i.e. for any particular phase, the induced voltages rise and fall together so that A_1 and a_1 become positive at the same instant in the cycle. The time phasors used on the diagrams are the phase voltages, one end only being labelled. This can indicate the potential of that end with respect to the other. The potential variations of opposite ends are of course 180° out of phase with one another. The phase time phasors can be drawn in any angular position providing only that the B- and C-phase time phasors are displaced by 120° from

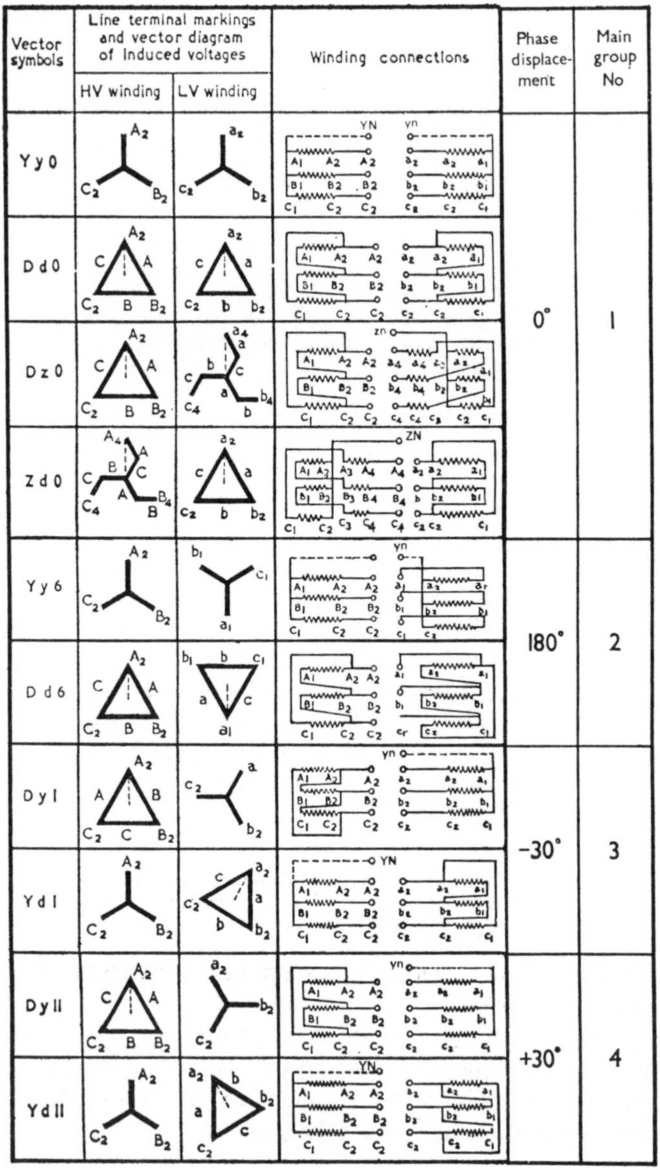

Vector symbols	Line terminal markings and vector diagram of Induced voltages		Winding connections	Phase displacement	Main group No
	HV winding	LV winding			
Y y 0	A_2 / C_2 B_2	a_2 / c_2 b_2	YN yn; A_1 A_2 A_2 a_2 a_2 a_1; B_1 B_2 B_2 b_2 b_2 b_1; C_1 C_2 C_2 c_2 c_2 c_1		
D d 0	A_2 / C_2 B B_2	a_2 / c_2 b b_2	A_1 A_2 A_2 a_2 a_2 a_1; B_1 B_2 B_2 b_2 b_2 b_1; C_1 C_2 C_2 c_2 c_2 c_1	0°	1
D z 0	A_2 / C_2 B B_2	a_4 a / b c / C_4 b_4 b	zn; A_1 A_2 A_2 a_4 a_4 a_2 a_2; B_1 B_2 B_2 b_4 b_4 b_2 b_2; C_1 C_2 C_2 c_4 c_4 c_2 c_1		
Z d 0	A_4 A / C B C / C_4 A B_4 B	a_2 / c_2 b b_2	ZN; A_1 A_3 A_4 A_4 a_2 a_2 a_1; b_1 b_3 B_4 B_4 b b_2 b_1; C_1 C_2 C_3 C_4 C_4 c_2 c_2 c_1		
Y y 6	A_2 / C_2 B_2	b_1 c_1 / a_1	yn; A_1 A_2 A_2 a_1 a_2 a_r; B_1 B_2 B_2 b_1 b_2 b_r; C_1 C_2 C_2 c_1 c_2	180°	2
D d 6	A_2 / C_2 B B_2	b_1 b c_1 / a c / a_1	A_1 A_2 A_2 a_1 a_2 a_r; B_1 B_2 B_2 b_1 b_2 b_r; C_1 C_2 C_2 c_r c_2 c_1		
D y 1	A_2 / A B / C_2 C B_2	c_2 a / b_2	yn; A_1 A_2 A_2 a_2 a_2 a_1; B_1 B_2 B_2 b_2 b_2 b_1; C_1 C_2 C_2 c_2 c_2 c_1	−30°	3
Y d 1	A_2 / C_2 B_2	c a_2 / c_2 a / b b_2	YN; A_1 A_2 A_2 a_1 a_2 a_1; B_1 B_2 B_2 b_1 b_2 b_1; C_1 C_2 C_2 c_2 c_2 c_1		
D y 11	A_2 / C A / C_2 B B_2	a_2 / c_2 b_2	yn; A_1 A_2 A_2 a_2 a_1 a_1; B_1 B_2 B_2 b_2 b_1 b_1; C_1 C_2 C_2 c_2 c_1 c_1	+30°	4
Y d 11	A_2 / C_2 B_2	a_2 b / a b_2 / c c_2	YN; A_1 A_2 A_2 a_2 a_1 a_1; B_1 B_2 B_2 b_2 b_2 b_1; C_1 C_2 C_2 c_2 c_2 c_1		

FIG. 4.17. Time-phasor diagrams for three-phase transformers.
(By courtesy of the British Standards Institution.)

the A phase. In Fig. 4.17, the line potentials of $A_2 B_2 C_2$ are maintained throughout in a fixed reference position and these ends also form the high-voltage terminals. In order to maintain this reference, the phase time phasors must be drawn in different positions for star and delta connections. Note, that although the time phasors are combined in either star or delta to line up with the physical connections, this is not

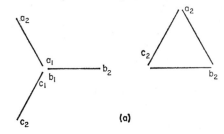

(a)

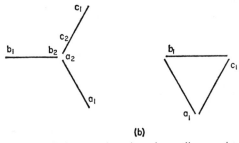

(b)

Fig. 4.18. Delta-secondary time-phasor diagrams in star and delta form: (a) group 1; a_2 connected to c_1; (b) group 2; a_1 connected to c_2.

essential. The time-phasor diagrams must not be confused with the connection diagrams, though it is sometimes useful to let them serve the double purpose. Figure 4.18 for the delta secondary shows that the individual phase time phasors when combined in a delta, indicate which phase ends must be connected to get either a 0° or 180° angle between primary and secondary line-voltages.

Group No. 1*: Phase Displacement* $= 0°$

For the Star/star (Y/y) connection, like-numbered ends have been commoned on both primary and secondary to form the neutral. The line voltages, e.g. V_{AB} and V_{ab}, are clearly in phase. For the Delta/delta (D/d) connection, the line voltage is the phase voltage, so again the phase displacement angle is zero. The zig-zag (z) connection will be discussed later.

Group No. 2*: Phase Displacement* $= 180°$

Here the phase connections to the secondary terminals are all reversed. Line voltage V_{ab}, for example, is taken from a_1 to b_1 instead of a_2 to b_2, giving a phase displacement of $180°$; see also Fig. 4.18 which shows that different phase ends are joined.

Group No. 3*: Phase Displacement* $= Minus$ 30°

The secondary and primary windings must be connected differently in both groups 3 and 4, in order to get a phase displacement different from either $0°$ or $180°$. The D/y connection is redrawn in Fig. 4.19a and shows that the secondary line voltage nearest in phase with the

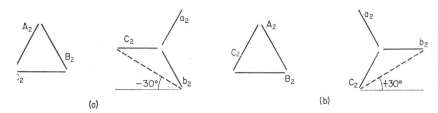

(a)   (b)

FIG. 4.19. Phase displacement angle: (a) group 3; *y* secondary; (b) group 4; *y* secondary.

172

primary line voltage has a 30° phase lag. This angle is given a negative sign ($-30°$).

Group No. 4: Phase Displacement = Plus 30°

By reconnecting the primaries as in Fig. 4.19b, the secondary line voltage is now 30° ahead ($+30°$ phase displacement).

Zig-zag Connection

It should be noted that all the possible permutations have not been demonstrated in Fig. 4.17. Where changes have been made on the secondary side, the same effect could have been obtained by a change

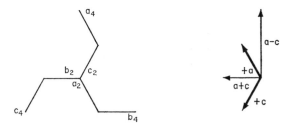

FIG. 4.20. Zig-zag connection.

on the primary side and vice versa. It is possible to obtain any desired phase shift with a polyphase transformer if the primary and/or secondary are divided into a sufficient number of sections. The zig-zag connection is one example of a sectionalised winding and its effect is to suppress third harmonics in line-to-neutral voltages, as well as in the line-to-line voltage. The half sections are connected in opposition (see Fig. 4.20), which gives the larger total fundamental voltage and causes the co-phasal third harmonics to cancel (see Section 4.5). The fundamental line-to-neutral voltage is $\sqrt{3}/2$ times the arithmetic sum of the two section voltages. This means that there must be extra turns to make up for the loss of voltage, when compared with the straight addition of the two in-phase sections. The multiplying factor is $1/0.866 = 1.15$, and this is reflected in extra copper, insulation and iron due to the extra space required. Nevertheless, the advantages of the zig-zag arrangement may offset the cost; unbalanced loads on the secondary side are distributed better on the primary side. As an extreme example of unbalanced loading, the single-phase line-neutral load can be considered. Here, although two secondary terminals do not carry current, two primary phases are involved in transformation of power. This situation arises in 3-phase, half-wave rectifier circuits because the d.c. load causes a current pulse for one-third of a cycle through each secondary terminal in turn. The current waveform on the primary, however, is much improved as a result of the interconnection on the secondary side.

Six-phase Supplies

Angular time-phase displacements corresponding to six or more phases are frequently required for rectifier equipments. Sectionalising of the secondaries is usually necessary but Fig. 4.21 shows one way of obtaining a six-phase supply directly, providing all the phase ends are available. A neutral point can be provided if the three secondaries are centre tapped.

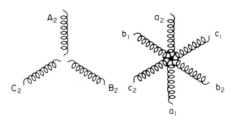

Fig. 4.21. Three/six-phase connections.

4.5 OPERATIONAL FEATURES OF VARIOUS THREE-PHASE TRANSFORMER ARRANGE-MENTS

Determination of Primary Current from known Secondary Current

Because the no-load primary current is so small, to a very close approximation the primary current on a single-phase transformer may be obtained by summing primary and secondary ampere-turns to zero, thus indicating that the net magnetising m.m.f. due to load current is zero. From eqn. (4.2b), neglecting $I_m N_1$, $I_1 \simeq I_2 N_2 / N_1 = I_2'$. The current I_0 can be superimposed vectorially for greater accuracy.

For transformers used in three-phase circuits the solution is not always so straightforward due to the electrical and magnetic inter-connection. When the magnetic circuits are independent, as on three single-phase or a shell-type three-phase unit, the m.m.f. summation can be applied directly to each phase in the simple case, and would have to include all windings on the limb. For example, with a third or tertiary winding (see Fig. 4.27), having N_3 turns carrying I_{3A}, the m.m.f. balance equation follows as below. It is simpler for the multi-winding case to take positive directions of all currents to be magnetising in the same sense. Thus:

$$\mathbf{I}_A N_1 + \mathbf{I}_a N_2 + \mathbf{I}_{3A} N_3 = 0.$$

If $N_1 = N_2 = N_3$, or if referred values of current are used, the equation is simplified and is more convenient for demonstrating the points to follow. The equation becomes:

$$\mathbf{I}_A + \mathbf{I}_a' + \mathbf{I}_{3A}' = 0.$$

Note that a phasor summation is now necessary to take account of phase differences between the currents, see Example E.4.6.

Phases B and C can be dealt with in the same way.

The three-phase, three-limb core-type transformer has a different constraint. The top and bottom ends of each limb are at a common

175

magnetic potential if the small reluctance of the yoke is neglected. Consequently, at any instant, the magnetic potential difference in ampere-turns absorbed between the top and bottom of each limb must be identical, though not necessarily zero, and may have a residual value of i_r, say, where i_r times the reference turns gives the residual m.m.f. The phase ampere-turns are now summed to i_r:

$$\mathbf{I_A} + \mathbf{I'_a} + \mathbf{I'_{3A}} = i_r.$$

By considering Fig. 4.22c it can be seen that even when i_r has a non-zero value, it does not tend to change the mutual flux since around any closed magnetic circuit including two limbs, the two residual m.m.f.s cancel.

An extreme example of unbalanced loading, viz. a single-phase line-neutral load, will be used to illustrate the above points and at the same time bring out certain distinguishing features between the magnetically dependent and the magnetically independent three-phase arrangements. A star/star connection is chosen with the primary neutral isolated. The circuit conditions are shown in Fig. 4.22a together with the current distribution which must follow.

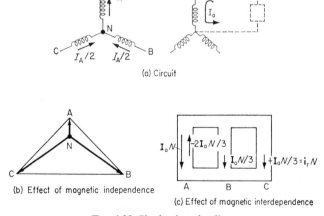

(a) Circuit

(b) Effect of magnetic independence

(c) Effect of magnetic interdependence

FIG. 4.22. Single-phase loading.

Consider first the case where the magnetic circuits are independent. The secondary m.m.f. on the loaded phase will be balanced out by an equal m.m.f. on its primary coil. The unloaded phase primaries are forced to carry one half of I_A assuming symmetry, but their secondaries are precluded by the open-circuit condition from carrying compensating current. $I_A/2$ will therefore be a magnetising current which will take cores B and C well into saturation to develop an abnormally large e.m.f. The phase voltage distribution will be considerably unbalanced, as shown in Fig. 4.22b, only a small voltage being developed across phase A. Thus, a single-phase load would cause collapse of the load voltage.

With the magnetic interdependence exhibited by the three-limb core, a different situation arises as can be seen by Fig. 4.22c. The single-phase load m.m.f. proportional to I_a is partly balanced by I_A on limb A and partly by $I_A/2$ on limbs B and C. The mathematical justification is as follows:

$$\mathbf{I_A + I_a'} = \mathbf{i_r}$$
$$\mathbf{I_B + 0} = \mathbf{i_r}$$
$$\underline{\mathbf{I_C + 0} = \mathbf{i_r}}$$
$$\mathbf{(I_A + I_B + I_C) + I_a'} = 3\mathbf{i_r}$$

and for a three-wire circuit:

$$\mathbf{I_A + I_B + I_C} = 0,$$

hence:

$$\mathbf{I_a'} = 3\mathbf{i_r}$$
$$\therefore \mathbf{i_r} = \mathbf{I_a'}/3 = \mathbf{I_B} = \mathbf{I_C} \text{ and } \mathbf{I_A} = -\tfrac{2}{3}\mathbf{I_a'}.$$

Considering any closed magnetic circuit, there is no resultant load m.m.f. so no effect on the mutual flux due to load currents. All the e.m.f.s remain balanced. This core arrangement, therefore, permits unbalanced loading with normal voltage regulation even with the primary neutral isolated.

Use of Tertiary Winding

A third or tertiary winding is frequently wound on the cores of transformers used for three-phase circuits to provide third harmonic magnetising current as will be explained shortly. It may be used for other purposes however, e.g. to provide a voltage different from either the primary or secondary to which it is magnetically coupled. The winding could then be used to supply auxiliary circuits at a suitably low voltage or to permit the interconnection of three systems entering the transformer at different voltages. The tertiary winding also assists in unbalanced loading conditions when delta connected, since it causes the load to be distributed more evenly in the primary phases. A centre-tapped secondary is required for certain rectifier circuits and this can be regarded as another example of a three-winding transformer.

Consider the same single-phase loading condition as used in the previous example, applied to a magnetically independent type of transformer but provided with a tertiary winding connected in delta. Figure 4.23 shows the solution obtained as follows:

$$\mathbf{I_A} + \mathbf{I'_{3A}} + \mathbf{I_a} = 0$$
$$\mathbf{I_B} + \mathbf{I'_{3B}} + 0 = 0$$
$$\mathbf{I_C} + \mathbf{I'_{3C}} + 0 = 0$$
$$\overline{0 + 3\mathbf{I'_{3A}} + \mathbf{I'_a} = 0} \quad \text{since } \mathbf{I_{3A}} = \mathbf{I_{3B}} = \mathbf{I_{3C}}$$
$$\text{with no external tertiary load.}$$

Therefore the tertiary current $\mathbf{I'_{3A}} = \mathbf{I'_{3B}} = \mathbf{I'_{3C}} = -\mathbf{I'_a}/3$ and substituting:
$\mathbf{I_A} = -\frac{2}{3}\mathbf{I'_a}$ and $\mathbf{I_B} = \mathbf{I_C} = +\mathbf{I'_a}/3 = -\mathbf{I_A}/2$.

Because the ampere-turns on each of the three phases can be summed to zero, unlike the case shown in Fig. 4.22b, the load current effects

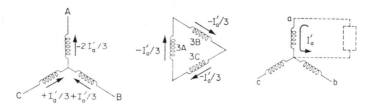

FIG. 4.23. Single-phase loading; tertiary delta-winding connected.

can be cancelled virtually, and a single-phase load can now be supplied without collapse of voltage even though the primary neutral is isolated and the cores are magnetically independent.

On fault conditions with one phase short circuited, it can be seen that the tertiary winding would carry an appreciable current. The tertiary winding must therefore be rated to carry any current which is liable to flow on unbalanced conditions of any kind, even if its function is only intended to provide a path for third harmonic magnetising current. It should be clear on reflection, that the presence of a tertiary delta and the arrangement of the magnetic circuit of a three-phase transformer will have pronounced effects on the impedance offered to zero-sequence (in-phase) currents; see Fig. 3.15.

Harmonics in Transformers for Three-phase Circuits

A star-connected winding with isolated neutral will preclude the flow of third-harmonic currents since they are in phase and would all be flowing to or from the unconnected neutral point together. Any third harmonic components in the phase voltage would appear between line and neutral but not between lines; the potential difference due to equal co-phasal voltages is zero. This applies to all triplen harmonics. For a delta winding, the third harmonic voltages are all in phase round the closed circuit and produce a circulating current (see pages 78–80). For a four-wire system, the in-phase third-harmonic currents can flow in the neutral wire circuit.

It has been explained, Fig. 4.5, that due to saturation effects, a sinusoidal flux and e.m.f. necessitate a pronounced third and less pronounced higher-order harmonic components in the magnetising current. A transformer in which both primary and secondary are star connected with isolated neutrals will present an infinite impedance to those harmonic currents which are multiples of three. Under these conditions, the third harmonics will appear in the flux wave but their relative magnitude is affected considerably by the transformer arrangement. If the magnetic circuits are independent then the magnetic reluctance to all components will be low. On the other hand, with the

179

three-limb core, the third harmonics are either upwards or down-wards in all three limbs at the same instant. They must therefore find return paths outside the iron circuit, in air (see Fig. 4.24). The reluctance is high and the third-harmonic flux components are reduced to negligible proportions. Harmonics which are not multiples of three are not co-phasal in the limbs, so these remain to distort the flux and e.m.f.

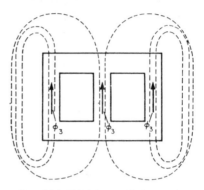

FIG. 4.24. Third-harmonic flux paths.

waveform though to a much smaller degree than when all harmonics are present.

With independent magnetic circuits the flux waveform would be particularly poor and steps must be taken to improve matters. If the neutrals must be isolated, not a usual requirement, then a tertiary delta will provide a low impedance path through which the third harmonic voltages in the induced e.m.f. wave will drive corresponding currents to rectify the deficiency in the primary no-load current. A state of balance will be reached when the harmonic voltage is just sufficient to circulate a current having the appropriate waveform, which gives rise to this voltage. Even where the neutrals are not isolated a delta winding is desirable to improve the flux waveform.

Phasor diagrams for the case where both neutrals are isolated on a Y/y, magnetically independent transformer are shown in Fig. 4.25. The fundamental phase-voltage time phasors rotate at ω radians/second

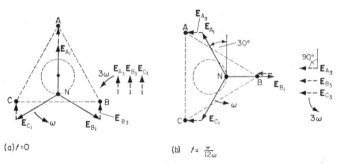

FIG. 4.25. Oscillation of neutral potential: (a) $t = 0$, (b) $t = \pi/12\omega$.

whereas the third-harmonic time phasors rotate at 3ω radians/second. Consequently, they can only be combined to get the resultant phase voltage at particular instants of time, two examples being shown. Three points are demonstrated by the diagrams:

(i) The potential of the neutral connection is not at the mean of the line potentials but oscillates about this value at third harmonic frequency with an amplitude equal to the peak harmonic voltage;

(ii) The maximum voltage across a phase winding may be as much as the arithmetic sum of fundamental and third harmonic voltages, depending on their relative phase. The insulation would thus be subjected to additional strain;

(iii) The difference between the line potentials excludes the third harmonic. Other harmonics apart from the "triplens" would appear across the line terminals.

Choice of Three-phase Connections

The star connection has advantages at the higher line voltages because the phase voltage, turns, and insulation are less than in the delta connection. At lower voltages, the availability of both phase and line voltages is useful for distribution purposes. The delta connection, apart from its beneficial effects on the waveform, behaves better on un-balanced conditions as may be gathered by considering the situation

posed in Fig. 4.22 but with a delta-connected primary winding. The
Y/d and the D/y combinations are the most commonly used, to gain the
advantages of both connections. The D/d connection has the special
advantage that a three-phase supply may be maintained with only two
phases of a 3-phase bank in circuit—the so-called open-delta or Vee
connection. Consider Fig. 4.26; since the connection of the A phase is
only practicable because the fundamental potential difference between

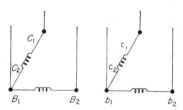

FIG. 4.26. Open-delta connection.

A_2 and A_1 is the same as that between the open ends C_1 and B_2
giving zero fundamental voltage round the mesh; the removal of
the A phase leaves the line potentials unchanged. The permissible
line current is now only equal to the phase current, so that the combination can only work at $1/\sqrt{3}$ times the normal rating.

Notes on the Rating of Transformers

The rated voltages, specified in accordance with BS 171, refer
to the line terminal voltages on no load. When the no-load voltage
ratio is expressed in terms of the phase voltages it is virtually equal
to the phase turns ratio because the $I_0 z_1$ drop is so small. The line
turns ratio has no real meaning and it is safer in calculation to reduce
all quantities to phase values and have a diagram like Fig. 4.27 in mind.
This shows a 33,000/1100/400-V D/d/y transformer with the tertiary
connected in star. When the phase voltage across, and the phase current
through each winding are known, calculations for a balanced load may
proceed as for single phase, since each phase carries one-third of the
power at the same voltage, current and power factor.

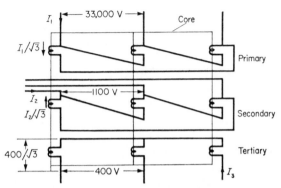

Fig. 4.27. Three-phase, three-winding transformer.

The full-load current per phase is obtained on dividing the rated kVA per phase by the no-load voltage per phase; primary or secondary as required. This is merely a question of definition and is discussed in BS 171. Note that the kVA specified is for the whole transformer, three times the kVA per phase for a three-phase transformer.

EXAMPLE E.4.6

A 3-phase, 3-winding, delta/delta/star, 33,000/1100/400-V, 200-kVA transformer has a secondary load of 150 kVA at 0·8 p.f. lagging, and a tertiary load of 50 kVA at 0·9 p.f. lagging. The magnetising current is 4% of rated load, the iron loss being 1 kW. Calculate the value of the primary current when the other two windings are delivering the above loads.

Reference should be made to Fig. 4.27 in which phase and line voltages are indicated. Working in phase values:

Referred secondary current I_2'

$$= \frac{150/3}{1 \cdot 1} (0 \cdot 8 - j0 \cdot 6) \times \frac{1100}{33,000} = 1 \cdot 21 - j0 \cdot 907.$$

Referred tertiary current I_3'

$$= \frac{50/3}{0 \cdot 231} (0 \cdot 9 - j0 \cdot 436) \times \frac{231}{33,000} = 0 \cdot 454 - j0 \cdot 22 \text{ A}.$$

Magnetising current $I_m = \dfrac{4}{100} \times \dfrac{200/3}{33} = 0 \cdot 081$ A.

Power component $I_p = \dfrac{1/3}{33} \qquad\qquad = 0 \cdot 0101$ A.

183

$\therefore$ Primary no-load current $I_0 = \underline{0 \cdot 01 - j0 \cdot 081}$ A.

Total primary current $I_1 = 1 \cdot 674 - j1 \cdot 208$ A $= I_2' + I_3' + I_0$
$$= \sqrt{[1 \cdot 674^2 + 1 \cdot 208^2]} = \underline{2 \cdot 06}\ \text{A}.$$

Power factor $= 1 \cdot 674/2 \cdot 06 = \underline{0 \cdot 811}.$

Input kVA $\quad = 3 \times 33 \times 2 \cdot 06 = \underline{204}.$

Note that the solution above neglects the small phase angles between the terminal voltages and the winding e.m.f.s. An exact solution would be more complicated.

4.6 PARALLEL OPERATION

Some general notes on parallel operation of d.c. devices have already been made in Section 3.5 and should be consulted. In the case of a.c. machinery the situation is rather more complex, although the transformer can be fairly accurately represented by a linear circuit so an analytical solution follows readily.

Before a unit can be paralleled with another unit or supply system, certain conditions *must* be fulfilled and certain conditions are desirable but not essential. For example, referring to Fig. 4.28a, the paralleling switch on the secondary side must not be closed until it is certain that the polarity of the incoming transformer is correct, i.e. a_2' must have positive and negative variation of potential at the same instant as a_2. This can easily be checked by connecting a voltmeter across the switch, which will read zero, presuming $E_A = E_B$, if the polarities are the same. The two e.m.f.s must then be in phase opposition round the local circuit. The primary winding may be connected directly because, with its secondary circuit open, it can only take the small no-load current. For the case shown, the two transformers have the same input voltage and frequency because of the common supply. If instead, each transformer primary was connected to its own generator, the input voltages and frequencies would have had to be matched as a separate adjustment.

The ideal conditions have already been indicated in Section 3.5 and can be extended to a.c. circuits as follows:

(a) The e.m.f.s should be equal in magnitude *and phase* in order to avoid circulating currents and permit good load sharing.

184

(b) The per-unit *impedances* should be equal in magnitude and have the same angle, i.e. the X/R ratios should be the same.

The implications of condition (b) are best considered in two stages and on the assumption that condition (a) has been fulfilled. Note, that with the primaries on a common supply, the e.m.f.s are bound

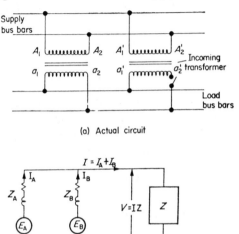

(a) Actual circuit

(b) Equivalent circuit

FIG. 4.28. Parallel operation of single-phase transformers.

to be in phase, the polarities having been checked. Referring to the equivalent circuit of Fig. 4.28b it can be seen that since E_A is being assumed equal to E_B and V is common, the internal impedance drops must be identical in phase and magnitude, though the resistive and reactive components may be different.

Consider first the case where the *per-unit* impedances are equal in magnitude, but X_A/R_A is not equal to X_B/R_B. The phasor diagram of Fig. 4.29 shows the different composition of the impedance drops, the resistive component being drawn in phase with the current. Let transformer A, say, be on full load; then $I_{Af.1}.Z_A$ must be equal to $I_B Z_B$

185

since the impedance drops are identical, with $E_A = E_B = E$. Since it has been specified that the *per-unit* impedances are the same, i.e. $I_{Af.l.}Z_A/E = I_{Bf.l.}Z_B/E$, then it follows that transformer B must be on full-load current also. The power factors are different however, and so the kW load is not shared. The phasor summation of the currents gives a slightly smaller resultant kVA capacity of the two-transformer

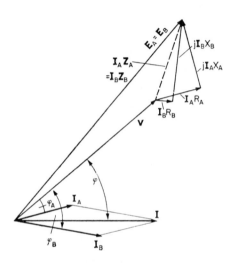

Fig. 4.29. Parallel operation, $X_A/R_A \neq X_B/R_B$.

unit than would be expected from algebraic combination of the individual ratings.

The alternative case to consider is where the X/R ratios are the same but the *per-unit* impedances are different in magnitude. From Fig. 4.29 it can be understood that the power factors will be the same but the regulation will be different. Typical regulation curves are plotted on Fig. 4.30 in terms of *per-unit* current, for a lagging power-factor and $Z_{Ap.u.} > Z_{Bp.u.}$. These external characteristics show that if transformer A is delivering 1 p.u. current, transformer B must carry more than 1 p.u. current to bring its terminal voltage down to the same

186

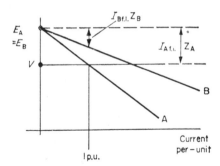

FIG. 4.30. External characteristics.

level as A. Thus, both kVA and kW loads will be out of proportion to the individual ratings.

For three-phase transformers, apart from the obvious fact that the line-voltage ratios must be the same, or nearly so, there are two further essential conditions:

(1) The transformers must have the same phase displacement between primary and secondary line voltages; i.e. they must belong to the same group number. Figure 4.31b shows the line-voltage time-vectors of two transformers, one of which has zero phase shift and the other has a phase shift of −30°. a_2 and a_2' can be connected, Fig. 4.31a, without any path being created for circulating currents, but a potential difference corresponding to the time phasors b_2b_2' and c_2c_2', comparable with full voltage, exists across the paralleling switches. If the switches were closed, a current several times greater than full load would flow and burn out the windings, even though no external load was connected.

(2) The phase sequence must be the same; faulty internal connections in the transformer tank or wrong rotation of one of the connected generators would cause an error of this kind. The phasor diagram for such a condition is shown in Fig. 4.31c. The incoming transformer has the potential of the terminal c_2' reaching a particular value in the cycle 120° later than a_2' and is followed by b_2'. Again a_2 and a_2' could be connected but the voltages b_2b_2' and c_2c_2' are sufficient to cause a ruinous circulating current if the switches are closed.

187

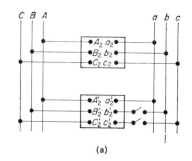

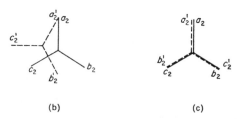

(b)   (c)

Fig. 4.31. Parallel operation of three-phase transformers: (a) circuit; (b) wrong group number; (c) wrong phase sequence.

Apart from the necessity to fulfil these additional essential conditions, the three-phase transformer can be regarded in a similar way to the single-phase transformer, calculations being carried out in either phase or *per-unit* values.

Analytical Solution for Two Transformers in Parallel

There are many ways of solving the equivalent circuit of Fig. 4.28b using one or other of the circuit theorems. For the simple case of two transformers there is little to choose between the various methods and the mesh equations are used below. The phasor diagram for the more general case of unequal e.m.f.s and per-unit impedances, but with a common primary supply, is shown on Fig. 4.32. There are two in

188

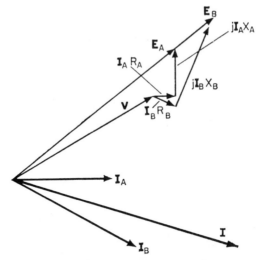

Fig. 4.32. Parallel operation; general case.

dependent mesh equations, which apply whether or not the primaries
are connected across the same supply terminals.

$$\mathbf{E_A} - \mathbf{I_A}\mathbf{Z_A} - (\mathbf{I_A} + \mathbf{I_B})\mathbf{Z} = 0, \tag{4.5}$$

$$\mathbf{E_B} - \mathbf{I_B}\mathbf{Z_B} - (\mathbf{I_A} + \mathbf{I_B})\mathbf{Z} = 0. \tag{4.6}$$

Kirchhoff's current law has been used to eliminate one of the currents,
$\mathbf{I}$, so that both $\mathbf{I_A}$ and $\mathbf{I_B}$ can be found from these equations if, as usual,
the e.m.f.s and impedances are known.

From (4.6)–(4.5)

$$\mathbf{E_A} - \mathbf{E_B} = \mathbf{I_A}\mathbf{Z_A} - \mathbf{I_B}\mathbf{Z_B}$$

and therefore:

$$\mathbf{I_B} = \frac{(\mathbf{E_B} - \mathbf{E_A}) + \mathbf{I_A}\mathbf{Z_A}}{\mathbf{Z_B}}.$$

Substituting in (4.5):

$$\mathbf{E_A} = \mathbf{I_A}(\mathbf{Z_A} + \mathbf{Z} + \mathbf{Z_A}\mathbf{Z}/\mathbf{Z_B}) + (\mathbf{E_B} - \mathbf{E_A})\mathbf{Z}/\mathbf{Z_B},$$

from which:

$$\mathbf{I_A} = \frac{\mathbf{E_A}\mathbf{Z_B} + (\mathbf{E_A} - \mathbf{E_B})\mathbf{Z}}{\mathbf{Z}(\mathbf{Z_A} + \mathbf{Z_B}) + \mathbf{Z_A}\mathbf{Z_B}} \tag{4.7}$$

189

or alternatively:

$$I_A = \frac{E_A}{Z_A + Z + ZZ_A/Z_B} + \frac{E_A - E_B}{Z_A + Z_B + Z_A Z_B/Z}. \qquad (4.8)$$

The expression for I_B can be obtained easily, by interchanging B and A in the above expressions.

It will be noticed from eqn. (4.8) that if the two e.m.f.s are not equal, the second term appears like a superimposed circulating current which has components in phase with one of the e.m.f.s, the larger one; the same current, therefore, has antiphase components with the smaller e.m.f. The circulating current will reduce the load on one transformer and increase it on the other. It will also be noticed that the superimposed circulating current is not equal to the difference of the e.m.f.s divided by the local circuit impedance except when $Z = \infty$; i.e. on open circuit, when it really is a true circulating current.

The total current is:

$$I = I_A + I_B = \frac{E_A Z_B + E_B Z_A}{Z(Z_A + Z_B) + Z_A Z_B} \qquad (4.9)$$

and the terminal voltage will be $V = IZ$ which could also be found from $E_A - I_A Z_A$ or $E_B - I_B Z_B$. An alternative expression for V follows on multiplying numerator and denominator of eqn. (4.9) by $1/Z_A Z_B$ and the result by Z to get:

$$V = IZ = \frac{E_A/Z_A + E_B/Z_B}{1/Z_A + 1/Z_B + 1/Z}.$$

The advantage of this solution is that it can be applied to any number of units in parallel simply by the addition of terms E_C/Z_C, E_D/Z_D, etc. in the numerator and terms $1/Z_C$, $1/Z_D$, etc. in the denominator. Any current can be calculated; for example, $I_A = (E_A - V)/Z_A$.

In calculations, a reference phasor must be chosen, the obvious choice being the source-equivalent-circuit e.m.f., since with primaries on a common supply, both e.m.f.s are in phase and the phasor difference terms are purely numerical values. Complex-number expressions for all quantities must be used throughout but if only the modulus of the current or terminal voltage is required, the moduli of the numerator and denominator of eqn. (4.9) may be used. The calculations can easily go awry because of the number of complex-number manipulations. A facility for changing Cartesian to Polar coordinates at the right stage in the calculation process, leads to some saving in time.

EXAMPLE E.4.7

Two transformers have the following particulars:

	Transformer A:	Transformer B:
Rated current	200 A	600 A
Per-unit resistance	0·02	0·025
Per-unit reactance	0·05	0·06
No-load e.m.f.	245 V	240 V

Calculate the terminal voltage when they are connected in parallel and supply a load impedance of $0·25 + j0·1 \ \Omega$.

Impedance in ohms = $Z_{p.u.} \times E/I_{f.l.}$.

$\mathbf{Z_A} = (0·02 + j0·05)245/200 = 0·0245 + j0·0613 \ \Omega$

$\qquad$ or in polar form = $\sqrt{[(0·0245)^2 + (0·0613)^2]}/\tan^{-1}0·0613/0·0245$

$\qquad\qquad\qquad\qquad = 0·066/\underline{68·2°}$.

$\mathbf{Z_B} = (0·025 + j0·06)240/600 = 0·01 + j0·024 \ \Omega$

$\qquad$ in polar form = $0·026/\underline{67·3°}$.

$\mathbf{Z}$, the load impedance = $0·25 + j0·1 \ \Omega$

$\qquad$ in polar form = $0·269/\underline{21·8°}$.

It will be required to find the denominator of eqn. (4.9).

$(\mathbf{Z_A + Z_B}) = 0·0345 + j0·0853$ or in polar form = $0·092/\underline{68°}$

$\mathbf{Z(Z_A + Z_B)}$ using polar multiplication = $0·269 \times 0·092/21·8° + 68°$

$\qquad\qquad\qquad\qquad\qquad = 10^{-3} \times 24·7/\underline{89·8°}$

which in cartesian form = $10^{-3} \times 24·7(\cos 89·8° + j \sin 89·8°) = 10^{-3} \times (0 + j24·7)$.

$\mathbf{Z_A Z_B}$ similarly = $10^{-3} \times 1·72/\underline{135·5°} = 10^{-3}(-1·225 + j1·201)$

$\therefore$ denominator = $\mathbf{Z(Z_A + Z_B) + Z_A Z_B}$

$\qquad\qquad\qquad = 10^{-3} \times (-1·225 + j25·9)$

$\qquad$ or in polar form = $0·0259/\underline{92·7°}$.

The numerator of eqn. (4.9) is rather easier to calculate since by taking $\mathbf{E_A}$ as reference phasor it becomes a simple numerical multiplication. $\mathbf{E_B}$, of course, is in phase with $\mathbf{E_A}$ when the transformers are paralleled on both sides.

$\mathbf{E_A Z_B} = 245(0·01 + j0·024) = 2·45 + j5·87$

$\mathbf{E_B Z_A} = 240(0·0245 + j0·0613) = \underline{5·88 + j14·7}$

$\therefore$ numerator = $\mathbf{E_A Z_B + E_B Z_A} = 8·33 + j20·57$

$\qquad\qquad$ or in polar form = $22·15/\underline{67·9°}$

$$\text{the total current } \mathbf{I} = \frac{22 \cdot 15 / 67^\circ \cdot 9}{0 \cdot 0259 / 92^\circ \cdot 7} = 855 / -24 \cdot 8^\circ.$$

Note that the phase angle is with reference to the e.m.f., not to the terminal voltage $\mathbf{V}$. The load power factor corresponds to the angle of $\mathbf{Z}$.

$$\mathbf{V} = \mathbf{IZ} = 855 / -24 \cdot 8^\circ \times 0 \cdot 269 / 21 \cdot 8^\circ = 230 / -3^\circ.$$

Although the angles have been calculated, they were not asked for and all angles could have been omitted once the numerator and denominator had been obtained.

A special case occurs when the induced e.m.f.s are equal. Manipulation of the mesh equations will show that in this case:

$$\mathbf{I_A Z_A} = \mathbf{I_B Z_B} \quad \text{and hence} \quad \mathbf{I_B} = \mathbf{I_A Z_A / Z_B} \quad \text{from which:}$$

$$\mathbf{I} = \mathbf{I_A} + \mathbf{I_B} = \mathbf{I_A}(1 + \mathbf{Z_A / Z_B}) \quad \text{and by rearrangement:}$$

$$\mathbf{I_A} = \mathbf{I} \cdot \mathbf{Z_B}/(\mathbf{Z_A} + \mathbf{Z_B}) \quad \text{and} \quad \mathbf{I_B} = \mathbf{I} \cdot \mathbf{Z_A}/(\mathbf{Z_A} + \mathbf{Z_B}). \quad (4.10)$$

Since the terminal voltage is common, the individual kVAs are found from the total kVA on multiplication by $\mathbf{Z_B}/(\mathbf{Z_A} + \mathbf{Z_B})$, or $\mathbf{Z_A}/(\mathbf{Z_A} + \mathbf{Z_B})$ as for the individual currents.

The terminal voltage need not be calculated, but if it is required, a close estimate can be made by dividing the e.m.f. into the kVA to get the current and then calculating the regulation from $E_A - V = I_A(R_A \cos \varphi_A + X_A \sin \varphi_A)$ from which V follows. A closer estimate could now be made by using this value of V with the kVA to calculate a more accurate value of current and regulation.

EXAMPLE E.4.8

Two 6600/440-V transformers have ratings of 250 kVA and 600 kVA respectively. On short-circuit test, the 250-kVA transformer requires 5% of normal voltage to circulate full-load current, the power factor being 0·23. The corresponding figures for the 600-kVA transformer are 4% and 0·16. How will they share a load of 680 kW at 0·8 p.f. lagging?

This problem will be worked out in the *per-unit* notation to demonstrate the application to apparatus of different ratings. The test data given, anticipate the work to be done almost immediately in the next section, but briefly it is meant that transformer A has a *per-unit* impedance of 0·05 and transformer B has $Z_B = 0 \cdot 04$ *per unit*. For A, $\tan^{-1} X_A/R_A = \cos^{-1} 0 \cdot 23$ and for B, $\tan^{-1} X_B/R_B = \cos^{-1} 0 \cdot 16$. Hence $Z_A = 0 \cdot 05(0 \cdot 23 + j0 \cdot 973) = 0 \cdot 0115 + j0 \cdot 0487$ *per unit*.

The polar form will be more convenient for this case where the two transformer e.m.f.s are equal; i.e. $Z_A = 0 \cdot 05 / \cos^{-1} 0 \cdot 23 = 0 \cdot 05 / 76 \cdot 7^\circ$.

It must be remembered that the *per-unit* impedance refers to the fraction of a fixed voltage absorbed at a particular value of current. *Per-unit* values of impedance are ohmic values multiplied by a scaling factor I/V. Clearly, all impedances which have to be combined must be multiplied by the same scaling factor.

Hence $\mathbf{Z}_B = 0.04/\underline{\cos^{-1}0.16} = 0.04/\underline{80.9°}$ based on the current corresponding to 600 kVA but on the smaller current corresponding to 250 kVA

$$\mathbf{Z}_B = (250/600)0.04/\underline{80.9°} = 0.0167/\underline{80.9°}.$$

In Cartesian form: $\quad \mathbf{Z}_B = 0.0167(0.16 + j0.987)$
$$= 0.00266 + j0.0165$$
$$\text{and } \mathbf{Z}_A + \mathbf{Z}_B = 0.01416 + j0.0652 = 0.0668/\underline{77.8°}.$$

The total kVA is divided in accordance with eqn. (4.10), the expression for the kVA in complex-number form being:

$$\mathbf{S} = \frac{680}{0.8} \, /\underline{-\cos^{-1}0.8} = 850/\underline{-36.9}; \text{ the power factor is lagging.}$$

Hence $\mathrm{kVA_A} = \dfrac{\mathbf{Z}_B}{\mathbf{Z}_A + \mathbf{Z}_B} \cdot \mathbf{S} = \dfrac{0.0167/\underline{80.9°}}{0.0668/\underline{77.8°}} \times 850/\underline{-36.9°}$

$$= 213/\underline{-33.8°}$$
$$= \underline{177 \text{ kW at } 0.83 \text{ p.f. lag}}$$

$\mathrm{kVA_B} = \dfrac{0.05/\underline{76.7°}}{0.0668/\underline{77.8°}} \times 850/\underline{-36.9°}$

$$= 637/\underline{-38°}$$
$$= \underline{503 \text{ kW at } 0.79 \text{ p.f. lag.}}$$

Note that the individual kW sum arithmetically to the total kW because these components of the kVA are in phase. In general, the kVAs will not sum arithmetically to the total because there is a phase difference if the ratio X_A/R_A is not equal to X_B/R_B.

The *per-unit* current on transformer A is $213/250 = 0.852$.

Therefore the regulation is approximately, from eqn. (4.3b)

$$0.852(R_{\text{p.u.}} \cos \varphi + X_{\text{p.u.}} \sin \varphi)$$

$= 0.852(0.0115 \times 0.83 + 0.0487 \times 0.556) = 0.0366$ *per-unit*, i.e. the terminal voltage is approximately $(1 - 0.0366) \times 440 = \underline{424 \text{ V}}$.

4.7 TESTING AND EFFICIENCY

It is not intended to discuss all possible tests but to select those which are relevant to the theory developed in previous pages. For details of special tests like high-voltage and impulse tests the appropriate specification may be consulted: BS 171.

Polarity Test

The recommended procedure is to connect primary and secondary windings in series in such a way that the e.m.f.s are in phase; i.e. if the polarity has been marked, A_2 should be connected to a_1; Fig. 4.33. A voltage applied across A_1 and a_2 will be opposed by the arithmetic sum of primary and secondary e.m.f.s and will therefore be divided across the two windings in proportion to the number of turns in each,

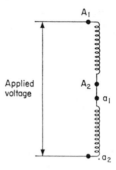

Fig. 4.33. Polarity test.

the flux being virtually the same for both windings. Should the markings be incorrect, the applied voltage will now be opposed by the arithmetic difference between the e.m.f.s. A very much larger current would flow, and it is therefore advisable to keep the applied voltage no greater than the difference between the rated voltage of the two windings. If the markings are either incorrect or not present anyway, the current cannot then exceed the no-load value.

Short-circuit Test

This test is used to determine the leakage impedance and the effective current loss $I^2 R_e$. Copper eddy-current losses within the conductors and in external conducting parts due to the load current will be present in much the same degree as at normal voltage, and will show up as an increase in resistance from the d.c. value to R_e. Figure 4.34 shows the test circuit and the effective equivalent circuit; one winding is short-circuited and a reduced voltage V_{sc} is applied to the other to cause rated current to flow. Since the windings are coupled, any current in one will be opposed magnetically by a current in the other giving the same m.m.f. The mutual flux will be very small, just sufficient to provide the secondary leakage impedance drop, so the net magnetising m.m.f. is negligible. Thus the impedance measured from the input side with the secondary short-circuited, is due almost entirely to the right-hand branch of Fig. 4.9, the current through the magnetising branch being less than 1% of the total. The ratio V_{sc}/I_{sc} may be taken as equal to z_e, the leakage impedance.

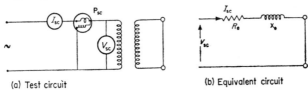

| (a) Test circuit | (b) Equivalent circuit |

FIG. 4.34. Short-circuit test.

From the power input reading P_{sc}, which is due to the loss $I_{sc}^2 R_e$, the effective resistance $R_e = P_{sc}/I_{sc}^2$ or $= z_e \cos \varphi_{sc}$.

The total leakage reactance is $x_e = \sqrt{(z_{sc}^2 - R_e^2)}$ or $z_e \sin \varphi_{sc}$. Equation (3.11) gives an alternative expression for the apparent reactance of a short-circuited transformer as $\omega L_{11}(1-k^2)$. This expression tends to the combined leakage reactance, i.e. $\omega(l_1 + l_2')$, as the coefficient of coupling k approaches unity, which is the condition on a normal iron-core transformer.

Note 1. I_{sc} need not be the rated current since the equivalent circuit is linear. It is desirable that it should be near to rated value so that the stray losses are normal. In practice, the temperature of the windings should be measured so that the loss at any temperature can be calculated by corrections to the resistance.

Note 2. The supply could be fed to either winding and for the same winding currents the power input would be the same. The measured voltage and current levels would be different, giving impedances referred to the input side. It is often convenient on the higher voltage transformers to supply the h.v. winding, thus using a smaller current. The input voltage V_{sc} which will be about 5% of the rated value may also be more suitable for test facilities.

Note 3. In laboratory experiments using small transformers, the impedances and the power consumption of the measuring instruments may have to be allowed for to get the true value of transformer terminal voltage, input current and power. The instrument positions shown usually result in minimum error for these s.c. tests at very low voltage.

Note 4. For 3-phase power measurement see Appendix C.

It is instructive to consider the flux pattern under short-circuit conditions, to understand why the secondary impedance can be reflected into the primary winding. Figure 4.35 shows two part-core diagrams to indicate the distribution of the flux. In Fig. 4.35a, the flux components ϕ_1, ϕ_m and ϕ_2 are drawn, but ϕ_2 and ϕ_m cannot exist separately in the core and they resolve into ϕ_s the net secondary flux, Fig. 4.35b. ϕ_s need only be sufficient to provide a voltage equal to

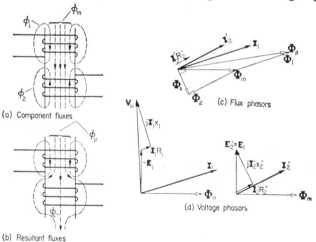

(a) Component fluxes

(b) Resultant fluxes

(c) Flux phasors

(d) Voltage phasors

FIG. 4.35. Conditions on short circuit.

$I_2 R_2$. The flux ϕ_2 in the secondary leakage paths is seen to link the primary winding giving a referred e.m.f. equal in magnitude to $I_2' x_2'$.

The flux pattern is shown for an instant when i_1 is numerically greater than i_2 which is true for most of the cycle. However, since $\mathbf{I}_1 N_1$ is not in exact antiphase with $\mathbf{I}_2 N_2$ because of the magnetising current, for short periods of time i_2 would be greater than i_1 and the leakage fluxes would then all be linking the secondary winding instead of the primary.

Figure 4.35c shows all the flux components and Fig. 4.35d shows the corresponding voltage components. It can be seen that in this case the short-circuit power-factor is very low, often less than 0·2, and special low-power-factor wattmeters must then be used to give accurate power readings.

When the transformer is loaded at normal voltage, these core diagrams still apply with the difference that the mutual flux, secondary flux and e.m.f. are much larger.

Time devoted to a thorough understanding of Fig. 4.35 is well spent, because the diagrams have a bearing on the induction machine on steady state, and on all machines in the transient state where changing conditions produce transformer e.m.f.s in coupled windings.

Open-circuit or No-load Test

The purpose here is to measure the iron losses and the components of the no-load current which in turn will give the relevant components of the equivalent circuit. One winding is open-circuited and rated voltage at rated frequency is applied to the other winding. Quite often the low-voltage winding is supplied, to reduce the test voltage required. Providing that the applied voltage per turn is normal, either winding may be used, since the flux and iron losses will then be normal; see eqn. (4.1). As in the short-circuit test, the equivalent circuit parameters will be referred to the side to which the test voltage is applied.

Figure 4.3a is the phasor diagram and Fig. 4.3b is the equivalent circuit but since $I_0 z_1$ is very small, this circuit reduces to Fig. 4.36. The actual test circuit is similar to Fig. 4.34 but with the secondary winding open and with instruments suitable for different voltage and current levels. The optimum position of the instruments to minimise the measurement errors might be different; e.g., with the current coils immediately adjacent to the transformer terminals. The power factor is again very low.

The input power P_0 will be very nearly equal to the iron loss and will be so taken, although it is realised that there is a small $I_0^2 R_1$ loss which could easily be deducted, and possibly a dielectric loss in the

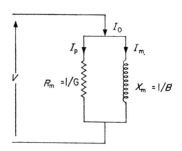

FIG. 4.36. No-load test;
equivalent circuit.

case of high-voltage transformers. The components of the no-load current, from Fig. 4.36 are $I_p = P_0/V = I_0 \cos \varphi_{oc}$ and $I_m = \sqrt{(I_0^2 - I_p^2)}$ $= I_0 \sin \varphi_{oc}$. The magnetising circuit resistance is $R_m = V/I_p$ and the corresponding conductance $G = 1/R_m$.

The magnetising reactance is $X_m = V/I_m$ and the corresponding susceptance $B = 1/X_m$.

EXAMPLE E.4.9

A 3-phase, 600-kVA, 11,000/660-V, star/delta transformer gave the following line input readings on light-load tests:

Open-circuit test	660 V	16 A	4·8 kW
Short-circuit test	500 V	30 A	8·2 kW

Calculate all the equivalent circuit parameters per phase referred to the primary, and also the currents when the secondary is delivering full-load current at 0·8 p.f. lagging.

Open-circuit test. By examining the data it can be seen that the o.c. test has been taken from the l.v. side at rated voltage, so the figures must be referred to the h.v. side to get the parameters in primary terms.

Phase turns ratio $= (11{,}000/\sqrt{3})/660 = 635/66$

o.c. power per phase $= 4800/3 = 1600$ W

o.c. power factor $\cos \varphi_{oc} = 4800/(\sqrt{3} \times 660 \times 16) = 0{\cdot}262$.

$$I_0 \text{ in primary terms } = \frac{16}{\sqrt{3}} \cdot \frac{66}{635} = \underline{0{\cdot}96 \text{ A}}, \text{ phase and line.}$$

$$I_p \text{ in primary terms } = P_0/V = 1600/6350$$
$$= \underline{0{\cdot}252 \text{ A}}, \text{ phase and line.}$$

$$I_m = \sqrt{(I_0{}^2 - I_p{}^2)} = \sqrt{[(0{\cdot}96)^2 - (0{\cdot}252)^2]}$$
$$= \underline{0{\cdot}926 \text{ A}}, \text{ phase and line.}$$

Magnetising branch reactance $X_m = V/I_m = 6350/0{\cdot}926 = \underline{6{\cdot}85\text{k}\,\Omega}.$

Magnetising branch resistance $R_m = V/I_p = 6350/0{\cdot}252 = \underline{25{\cdot}2\text{k}\,\Omega}.$

Short-circuit test. The data this time must have been taken on the h.v. side because 500 V is nearly equal to the rated l.v. terminal voltage. Working in phase values again:

$$\text{s.c. power-factor } \cos \varphi_{sc} = 8200/(\sqrt{3} \times 500 \times 30) = 0{\cdot}316$$

$$R_{e1} = P_{sc}/I^2 \qquad = (8200/3)/30^2 \qquad = \underline{3{\cdot}04\,\Omega}$$

$$z_{e1} = V/I \qquad = (500/\sqrt{3})/30 \qquad = \underline{9{\cdot}62\,\Omega}$$

$$x_{e1} = \sqrt{(z_{e1}{}^2 - R_{e1}{}^2)} \quad = \sqrt{[(9{\cdot}62)^2 - (3{\cdot}04)^2]} \quad = \underline{9{\cdot}12\,\Omega}.$$

Figure 3.6 and Example E.3.1 should be consulted for the alternative method of resolving the terminal impedances into resistive and reactive components based on the power-factor angles φ_{oc} and φ_{sc}.

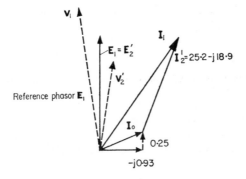

The full load-current in primary terms:

$$I_2' = \frac{600/3}{6\cdot35} = \underline{31\cdot5 \text{ A}} \text{ per phase.}$$

Neglecting the small phase angle between the terminal voltages and e.m.f.s, the phasor expressions for I_0 and I_2' are:

$$\mathbf{I_2'} = 31\cdot5(0\cdot8-j0\cdot6) = 25\cdot2-j18\cdot9 \text{ A}$$
$$\mathbf{I_0} = \qquad\qquad\quad = 0\cdot25-j0\cdot93$$
$$\mathbf{I_1} = \mathbf{I_0}+\mathbf{I_2'} \qquad = 25\cdot45-j19\cdot83 \text{ A}$$

from which $I_1 = 32\cdot1$ A, the primary power-factor being $25\cdot45/32\cdot1 = \underline{0\cdot79}$.

Note that the data per phase are the same as for the single-phase problem of E.4.2 and E.4.3, so the equivalent-circuit diagram drawn on page 156 applies to the present problem. The time-phasor diagram for one phase is shown here, E_1 being the reference phasor.

Separation of Hysteresis and Eddy-current Losses

The open-circuit curve for the transformer can be obtained by varying the input voltage at constant frequency from say 10% above rated voltage down to zero and plotting V against I_m. The applied voltage is virtually the same as the induced e.m.f. on open circuit which in turn is proportional to flux. If X_m and R_m are calculated for each reading, their variation with V, as discussed on p. 136 can be examined. I_m, which is little different from I_0 since I_p is relatively small, is the r.m.s. value of a very peaky waveform, see Fig. 4.5. It is proportional to the magnetising m.m.f. neglecting changes of waveform.

Another useful curve can be obtained if voltage and frequency can be varied together. Since V/f is virtually proportional to flux, eqn. (4.1), the flux density will then be constant. Under these conditions, the power input, which is almost entirely iron loss, can be written from eqns. (2.2) and (2.3) as:

$$P = k_e'f^2+k_h'f = \text{eddy current}+\text{hysteresis loss}$$
$$\text{and } P/f = k_e'f + k_h'$$

For a series of frequencies, with corresponding changes in input voltage, the measured input power divided by frequency and plotted against frequency will give a straight line from which the eddy-current loss and hysteresis loss can be calculated separately, Fig. 4.37.

All the above tests are applicable to single-phase or polyphase transformers, but in the latter case the line input-readings must be converted to phase values in order to work out the components of the

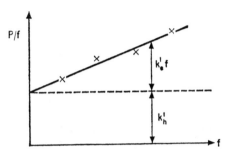

Fig. 4.37. Separation of iron losses.

equivalent circuit, all phases being assumed to be identical. A suitable measuring circuit for 3-phase devices is described in Appendix C.

Load Tests

The obvious way to carry out a load test is to provide a suitable impedance on the secondary side so that rated current will flow, at rated power-factor. However, this involves considerable expenditure of power and the provision of test facilities to dissipate the load energy. Electrical machines have the advantage that tests can be devised which permit two similar machines to be connected back to back in such a manner that one operates as a source and the other as a sink. Full current at rated voltage can circulate between the two machines, and only the internal losses need be provided from an external source.

The back-to-back test for transformers was devised by Sumpner and the circuit for two identical single-phase transformers is given in Fig. 4.38. With both primaries connected to the supply and with secondaries open, the total power supplied will be twice the no-load losses of one transformer. Between the two primaries will flow a current equal to the no-load current of one transformer. The secondary voltages are arranged to be in opposition round the local circuit and this can be checked by means of a voltmeter across the switch. The

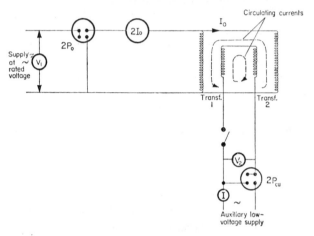

Fig. 4.38. Back-to-back load test.

reading should be zero, otherwise it will be twice the secondary voltage and one of the windings will have to be reversed. With the second source voltage V_2 at zero, the switch can now be closed and no current will flow; the two secondaries are still on no load. The second source can be obtained from a small variable-ratio transformer or a machine capable of handling the full secondary current and delivering a voltage equal to the leakage impedance drop of the two transformers in series. As V_2 is increased from zero, the voltage unbalance thus introduced will cause a secondary circulating current having in-phase components with one e.m.f. and antiphase components with the other. This current, in tending to change the mutual flux, will be opposed by a compensating current circulating between the two primaries, additive or subtractive with the no-load circulating current depending on the phase of V_2. The operating power-factor of both sources will be low and no power is being delivered to an external load. The primary supply will continue to provide the iron losses, the mutual flux being practically unchanged. The secondary supply will provide the $I^2 R_e$ losses of the two transformers, and may be adjusted to a slightly different frequency from that of the primary supply if this is more convenient.

If the secondary current is set at various values, the total loss per transformer at each value of current is one half of the sum of the wattmeter readings, say, P_I. This neglects the very slight differences in the two transformers due to their different modes of operation. The efficiency of the transformer when in service delivering this current at a power factor cos φ is:

$$1 - P_I/(VI \cos \varphi + P_I),$$

V being the terminal voltage calculated from E, minus regulation, p. 153.

Load Tests on Three-phase Transformers

The above test can be applied with suitable modification to a three-phase transformer; three-phase supplies being necessary for both primary and secondary circuits. There is an alternative method of running a dummy load-test using only one transformer, if both primary and secondary can be delta connected for the purposes of the test. The primaries are supplied at normal voltage and the secondary delta opened at one point, see Fig. 4.39. No fundamental voltage appears at these open ends though any triplen harmonics will be present since they are additive in all phases. If a single-phase low voltage is applied to the open ends, a circulating current will flow round the delta and a compensating current will flow round the primary delta.

Efficiency calculations are usually made from separate measurement of the losses, but these load tests are still necessary to enable the transformer to run for a period of a few hours at full voltage and current to see if the temperature rise is within the guaranteed figure. Heat runs can also be carried out with one transformer by short circuiting the secondary and adjusting the current to a higher-than-rated value, so that the total copper loss is equal to the rated copper

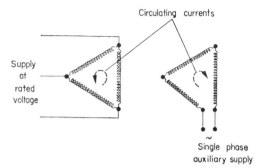

FIG. 4.39. Dummy load run on one three-phase
transformer.

loss plus the normal iron loss. Although the losses are differently distributed from normal working, the oil temperature rise will be similar to that occurring at rated kVA.

4.8 OTHER TYPES OF
TRANSFORMER

The Auto-transformer

Consider a transformer with primary and secondary connected in series aiding as for the polarity test; see Fig. 4.40. As in this test, the applied voltage divides in proportion to the number of turns. If the *total* number of turns in series is N_1 and the number of turns on the bottom section is N_2, a voltage V_2 will appear across these turns of value V_1N_2/N_1. A resistive potential divider could affect the same voltage reduction but, quite apart from its high power consumption and poor voltage regulation, it is not a reversible device; i.e. a low voltage could not be converted to a higher voltage. In the arrangement of Fig. 4.40, which is called an auto-transformer, the applied voltage could be V_2 and this would establish a flux, linking both sections and giving a potential difference between any two points on the winding proportional to the number of turns; over N_1 turns the voltage would be $V_1 = V_2N_1/N_2$.

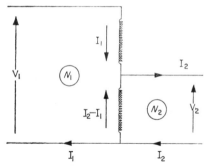

FIG. 4.40. Auto-transformer.

Consider the case where V_1 is applied. V_2 is available for supplying a load, and, neglecting losses and internal impedance drops:

$$V_1 I_1 \cos \varphi = V_2 I_2 \cos \varphi;$$

hence $I_2/I_1 = V_1/V_2 = N_1/N_2$ as for the two-winding transformers. Under these ideal conditions, the m.m.f. due to input and output currents will balance exactly, leaving the mutual flux unchanged by load currents. I_2 will therefore oppose I_1 in the common section so that less copper will be needed here compared with that required for a two-winding transformer.

The overall saving can be deduced by calculating the combined kVA rating of the two winding sections and comparing it with the two-winding transformer to give the same input/output ratios. It has already been shown that the size of a transformer is approximately proportional to kVA; see p. 164.

For the auto-transformer:

$$\text{total kVA} = (V_1 - V_2)I_1 + V_2(I_2 - I_1)$$

$$= \frac{(V_1 - V_2)}{(N_1 - N_2)}(N_1 - N_2)I_1 + \frac{V_2}{N_2}(I_2 N_2 - I_1 N_2)$$

$$= \text{Volts per turn } (I_1 N_1 - I_1 N_2 + I_2 N_2 - I_1 N_2)$$

$$= k.\text{Flux}.(2I_1 N_1 - 2I_1 N_2) \quad \text{since} \quad I_1 N_1 = I_2 N_2.$$

205

For the normal two-winding transformer:

$$\text{total kVA} = V_1 I_1 \frac{N_1}{N_1} + V_2 I_2 \frac{N_2}{N_2}$$

$$= \text{Volts per turn } (I_1 N_1 + I_2 N_2)$$

$$= k . \text{Flux} . 2 I_1 N_1 \quad \text{since } I_1 N_1 = I_2 N_2 .$$

The ratio of sizes is given approximately on dividing the first equation by the second to give

$$\frac{\text{auto kVA}}{\text{two-winding kVA}} = 1 - N_2/N_1 = 1 - V_2/V_1 .$$

The size reduction is mostly due to the saving in copper weight which is proportional to ampere-turns. The voltage per turn, and therefore the flux, is the same so the core section is not affected. However, the reduction of copper means a smaller window area and some iron saving as a result. Taking everything into account, the actual saving is only about half of that indicated, and the closer V_2 approaches V_1 the greater the saving. In practice, voltage ratios greater than about 3/1 show little economic benefit over the two-winding transformer.

The saving can be explained another way. A power $V_2 I_1$ is transmitted without transformation so that the transformer is only required for a power $V_2(I_2 - I_1)$ as against a power $V_2 I_2$ in the two-winding case. The ratio of the two values of transformed power is $1 - I_1/I_2$ $= 1 - V_2/V_1$ as before.

Disadvantages arise from the direct connection between h.v. and l.v. sides. One side is no longer isolated from the other; if an open circuit were to develop in the common section, the full h.v. would be applied to the l.v. side. This possibility must be allowed for in the insulation and protective gear arrangements.

One of the most familiar applications of the auto-transformer is the "Variac", "Regavolt" or "Regulac". These have a toroidal core and coil with a thin carbon brush, making contact with the winding in any desired position, Fig. 4.41. The output voltage can be varied between zero and usually about 20% above the input voltage.

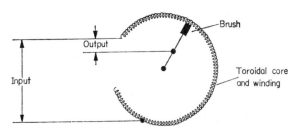

Fɪɢ. 4.41. "Variac" transformer.

Three-phase auto-transformers are used also; for example, in the interconnection of high-voltage transmission lines of various voltage levels; 132 kV, 275 kV, 400 kV, etc. Induction motors are often started through auto-transformers which reduce the starting voltage from the line value, see Section 7.6.

The Scott Three-phase/Two-phase Connection

A most useful feature of polyphase transformers is the facility with which phase conversions can be performed. A 3-phase/6-phase connection has already been mentioned, Fig. 4.21, and more elaborate 6-phase, 12-phase and even 24-phase conversions are in frequent use for rectifier transformers. A single-phase supply can be obtained from a 3-phase supply by connecting across the lines or between line and neutral, but there are special transformer conversions which balance the load a little better between the three lines. Apart from these examples, 3-phase/2-phase conversions are commonly used in electric furnace installations where it is desired to run two single-phase furnaces together and draw a balanced load from the three-phase mains. Quite aside from their usefulness in special applications, phase conversions are instructive as a study in m.m.f. relationships.

Although there are two very successful methods of performing the 3-phase/2-phase conversion, the Scott connection is the most common, the reason being apparently, that it predated the invention of the Le Blanc connection by a few months. The Le Blanc connection uses a

207

three-phase transformer with sectionalised secondaries and will not be discussed further. The Scott connection uses two single-phase transformers which may be duplicate and are electrically but not magnetically connected. A centre tap is required on one transformer winding and a tapping at 0·866 on the corresponding winding of the other transformer.

Figure 4.42 shows the connections with the BS terminal markings. Assumed positive directions of the three-phase currents are indicated.

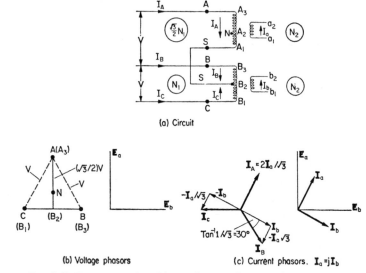

(a) Circuit

(b) Voltage phasors

(c) Current phasors. $\mathbf{I}_a = j\mathbf{I}_b$

FIG. 4.42. Scott connection. Phasor diagrams for case where $N_1 = N_2$.

Across the lines B and C is the so-called *main transformer*. The *auxiliary* or *teaser transformer* is connected between the other line terminal, A in this case, and a centre tapping on the main transformer. The voltage between A_3 and B_2 can be obtained by phasor summation of $V_{A_3B_3}$ and $V_{B_3B_2}$, B_3 being the terminal for line B. From the voltage phasors of Fig. 4.42, this is seen to have a magnitude $\sqrt{3}/2$ of the line voltage. Consequently, if the turns between A_3 and A_1 are made $(\sqrt{3}/2)N_1$, where N_1 is the number of turns on the main transformer, the voltage per turn, and therefore the flux, is the same for each transformer. With equal numbers of turns N_2 on the coupled windings, equal e.m.f.s in time-phase quadrature will be induced. A neutral point could also be provided on the three-phase side and from the geometry of the triangle, this would have to be one third of the way along the winding from A_1 as indicated.

Neglecting magnetising current, the m.m.f. difference is zero and the relationship between input and output currents follows:

$$I_A N_1 \sqrt{3}/2 - I_a N_2 = 0 \quad \text{from which} \quad I_a = \frac{\sqrt{3}}{2}\frac{N_1}{N_2}I_A. \tag{4.11}$$

For the main transformer:

$$I_B N_1/2 - I_C N_1/2 - I_b N_2 = 0 \ . \ \text{from which} \quad I_b = \frac{N_1}{N_2}\frac{(I_B - I_C)}{2} \tag{4.12}$$

and since for a three-wire system $I_C = -I_A - I_B$, substituting for I_C gives:

$$I_B N_1/2 + (I_A + I_B)N_1/2 - I_b N_2 = 0$$

from which:

$$I_B = \frac{-I_A}{2} + I_b\frac{N_2}{N_1} \quad \text{and similarly} \quad I_C = \frac{-I_A}{2} - I_b\frac{N_2}{N_1} \tag{4.13} \tag{4.14}$$

and from equation (4.11): $\qquad I_A = \frac{2}{\sqrt{3}}\frac{N_2}{N_1}I_a. \tag{4.15}$

No statement has yet been made as to whether the transformer is converting from 3-phase to 2-phase or vice versa. Either mode is possible. Equations (4.11) and (4.12) can be solved for the two-phase currents when the three-phase currents are known, and eqns. (4.13), (4.14) and (4.15) can be solved for three-phase currents when the two-phase currents are known. The equations are valid whether or not the load is balanced but in Fig. 4.42c they are applied to a two-phase balanced load, i.e. $I_a = I_b$ and lags its associated e.m.f. by the same phase angle. Internal impedance is neglected.

From the phasor diagram

$$I_B = I_C = \sqrt{\left[\left(\frac{I_a}{\sqrt{3}}\right)^2 + I_b^2\right]} = \frac{2}{\sqrt{3}} I_a \text{ since } I_b = I_a.$$

I_A is also equal to $(2/\sqrt{3})I_a$ and the angles between I_A, I_B and I_C are easily deduced to be equal from the diagram. Thus it is seen that when the two-phase load is balanced, the three-phase input is balanced and vice versa.

EXAMPLE E.4.10

A Scott-connected transformer supplies two single-phase loads at 100 V, each taking 200 kW. The load on the leading phase is at unity power-factor and that on the other phase is at 0·8 lagging power-factor. The 3-phase input line-voltage is 11,000 V. Calculate the line currents on the primary side. The magnetising current and leakage impedance may be neglected.

The solution is carried out using complex numbers but is also shown graphically on the adjoining diagram.

$$I_a = 200/0\cdot1 = 2000 \text{ A,}$$
$$I_b = 200/(0\cdot1 \times 0\cdot8) = 2500 \text{ A.}$$

Main transformer turns ratio $N_1/N_2 = 11,000/100 = 110$.

Taking E_b as reference phasor, from the diagrams, the currents can be expressed as:

$$I_a = +j2000 \text{ A,}$$
$$I_b = 2500(+0\cdot8 - j0\cdot6) = 2000 - j1500 \text{ A.}$$

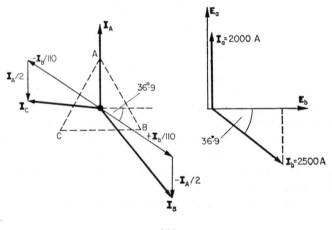

Hence, from eqns. (4.13), (4.14) and (4.15)

$$I_A = \frac{2}{\sqrt{3}} \times \frac{1}{110} \times j2000 = \underline{j21 \text{ A.}}$$

$$I_B = \frac{-j21}{2} + (2000 - j1500) \times \frac{1}{110} = \underline{18 \cdot 2 - j24 \cdot 1 \text{ A.}}$$

$$I_C = \frac{-j21}{2} - (2000 - j1500) \times \frac{1}{110} = \underline{-18 \cdot 2 + j3 \cdot 1 \text{ A.}}$$

The numerical values of the line currents are $\underline{21 \text{ A,}}$ $\underline{28 \cdot 4 \text{ A,}}$ and $\underline{20 \cdot 6 \text{ A}}$ respectively.

Instrument Transformers

Since, for a transformer the ratio between input and output voltages is approximately constant, a step-down potential transformer (P.T.) can be used for measuring purposes so that high voltages can be recorded safely with standard low-voltage instruments connected to the secondary. Such transformers are made with high quality iron operating at very low flux densities so that I_0 is very small. Careful design ensures minimum variation of voltage ratio with load and minimum phase shift between input and output voltages. P.T. secondaries are commonly designed for an output of 110 V.

A transformer winding in series with the line will produce in a closed secondary winding a current giving an m.m.f. which will balance the primary ampere-turns, i.e. it will be proportional to the primary current. This is a current transformer (C.T.), and is usually a step-up transformer in terms of primary/secondary turns ratio since the current to be measured is usually, though not necessarily, a high value. Very often, the primary is just one turn, formed by taking the line conductor through the secondary winding. The secondary currents are usually designed for either 5 A or 1 A at full rating. Again, the transformer must be carefully designed to minimise the ratio and phase-angle error. Care must be exercised to ensure that a C.T. secondary is never left open-circuited. Without opposing secondary ampere turns the primary current would grossly oversaturate the C.T. core and subject the secondary to excessive voltage. The primary current is determined by the main

211

circuit in which it is series connected, unlike the power- or voltage-transformer in which the primary current is determined by the secondary load.

Both P.T.s and C.T.s have the added advantage that the instruments are isolated from the line potential. One side of a secondary winding is usually earthed so that the instrument potential can never rise above that of the secondary voltage. Figure 4.43 shows the connections for

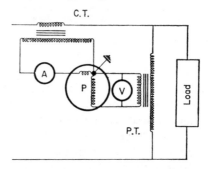

FIG. 4.43. Use of instrument transformers.

measuring single-phase voltage, current and power through instrument transformers.

High-frequency Transformers

Beyond the normal power frequency range at frequencies higher than 50–60 Hz, lamination thicknesses have to be reduced to control the iron losses to an acceptable level. If the operating frequencies extend beyond a few kilocycles per second, it becomes necessary to use iron-dust cores or the ferrites, special oxides which exhibit reasonably good magnetic properties with very small eddy-current losses.

In transformers of this kind, variable frequency is the normal operating condition and as the frequency increases, the significance of the elements in the equivalent circuit changes. At very low frequencies, for example, the magnetising reactance is small enough to be comparable with or even less than the load impedance so that this tends to be "shorted out" and the secondary terminal voltage is greatly reduced. In the intermediate frequency range, X_m is high enough for its effects to be neglected and the leakage impedance absorbs only a moderate fraction of the available voltage. With further increase of frequency, however, the leakage reactance drop tends to cause the output voltage to fall appreciably again. Capacitances between turns and between the windings and earth can no longer be ignored since these are shunted across the circuit in a distributed fashion causing behaviour similar to a leading power-factor load. The design problem here is to ensure the transfer, from primary to secondary, without excessive distortion, of an input signal having many frequency components; for example a pulse of voltage or an input having the complex frequency spectrum of a musical note. The concept of flux units in volt-seconds ($\int d\phi = (1/N)\int e \cdot dt$), is convenient for pulse-transformer design, in ascertaining the core size required (p. 164) for a given (voltage-pulse × time) product. In the high-power field too, with the advent of thyristor inverters, supply voltages may be far from sinusoidal and may appear as castellated or other compositive waveforms, see Fig. 7.31.

Impedance Matching

One further function of the transformer is worth mentioning, though usually it is only of importance in certain communication circuits. A secondary load viewed from the primary has its impedance reflected by a multiplying factor equal to (turns ratio)^2. In such circuits it is desirable from considerations of maximum power transfer to match the load impedance to the internal impedance of the source so that they are about equal. The impedance of a loud-speaker coil, for example, is only a few ohms whereas the effective impedance of the amplifier output stage may be several thousand ohms. A step-up output transformer is used to increase the apparent impedance of the load viewed from the primary. In addition, the transformer isolates the secondary load from the d.c. component of transistor current since this cannot induce a voltage in the secondary side, though it does affect the behaviour because it tends to saturate the core.

CHAPTER 5

MACHINE WINDINGS; E.M.F. AND M.M.F. DEVELOPED

5.1 GENERAL REVIEW OF WINDING ARRANGEMENTS

In an electromagnetic machine, the conversion of power takes place through the medium of the magnetic field; an e.m.f. being induced in any coils which experience a change of flux linkage. In a practical machine there are usually several coils, connected in series or in series-parallel circuits to form the winding. Some thought must be given as to the best way of placing the coils to collect their e.m.f.s in the most effective manner. The stationary windings of the transformer present no problem of this kind since they can readily be wound in close proximity on the rectilinear iron core. The situation is rather different on rotating machinery and though armature windings are a specialised study, the treatment can be simplified for present purposes. However, there are many considerations in design which make for a choice of possibilities so these aspects cannot be dealt with fully. The early sections on winding design, illustrated by simple examples, will be sufficient for straightforward problems, but the main intention is to give some practical details to support the theory in Sections 5.4, 5.5 and 5.6.

Armature-conductor Connections

For rotating machines there are many arrangements of conductors but in general they all conform in one respect. A coil is so designed that when one of its sides is under the influence of a north pole say, its other side is under the influence of a south pole for most of this time, i.e. a coil spans approximately one pole pitch; Fig. 1.2. The reason is obvious since in this way, although the e.m.f. is outwards in one conductor and inwards in the other, the "back-to-back" connection causes the e.m.f.s to be additive round the coil circuit. There are some advantages, which will be explained later, in having chorded coils in which the coil sides are not exactly one pole-pitch apart. The coils are laid in succession round the periphery of the machine and for mechanical reasons are insulated and embedded in slots which may be open, semi-closed or closed; see Fig. 5.1. New adhesives have permitted

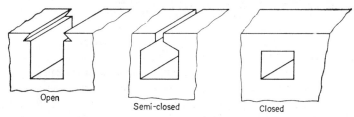

Open

Semi-closed

Closed

FIG. 5.1. Types of slot.

the direct bonding of the coils on to a smooth iron core for certain small machines. The torque pulsations due to slotting are removed but the space taken up by the coils means that the effective radial air gap may have to be longer. In addition, as may be understood by referring back to p. 107, the electromagnetic forces will be exerted on the conductors themselves, in the absence of slots. These forces can be very high. The suggestion[8] that the large air gaps of high-power turbogenerators could be utilised by having the stator and/or rotor windings in this form raises the problem of secure coil anchorage under short-circuit fault conditions.

The coils may be interconnected in a variety of ways and it is really in this that windings differ. Again there is a common feature: namely that in progressing from one coil to another in series, the e.m.f.s are additive, any time-phase displacement between the coil e.m.f.s usually being kept small by suitable choice of connection. These measures ensure that, in a winding circuit having z_s conductors in series, the total

215

e.m.f. is very nearly equal to z_s times the e.m.f. per conductor.

Single-layer and Two-layer Armature Windings

These are the two basic physical types and deal differently with the mechanical problem of arranging the coils in sequence round the periphery. Figure 5.2 shows a single-layer arrangement in which the whole of the slot is occupied by the side of one coil. This single layer usually contains many insulated turns as indicated and may also have the copper section subdivided to reduce eddy-current losses. To limit the number of awkward bends in the overhang connections, single-layer coils are placed concentrically in small groups, which are interlinked in such a way as to minimise both the space taken up outside the slot in the overhang connections, and the number of different coil

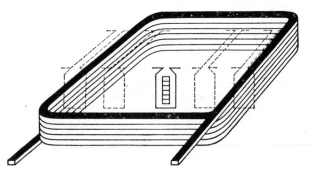

FIG. 5.2. Single-layer concentric coil.

shapes. The *mush* single-layer winding achieves this objective in a different way; see Fig. 5.16. Single-layer coils are not suitable for windings having commutators, but have the advantage that semi-closed or closed slots can easily be used; the coils being preformed at one end, pushed through the slots and connected up at the free end during the winding process. The average pitch of a coil group is made equal to a pole-pitch for practical reasons. Figure 5.3a shows a partially-wound stator using concentric coils which are all in position but

216

(a)

(b)

Fig. 5.3. Partially wound a.c.-machine stators: (a) single-layer 3·3 kV stator winding for 1100-h.p., 16-pole induction motor; (b) two-layer 11-kV stator winding for 3250-h.p., 6-pole induction motor. (By courtesy of Laurence, Scott & Electromotors Ltd.)

FIG. 5.5. Partially wound d.c. armature for double-unit, 2×4000 h.p., 2×750 V, 57/142 rev/min rolling-mill motor. (By courtesy of Parsons Peebles Motors and Generators Ltd.)

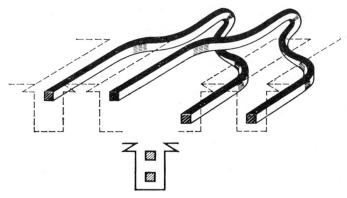

FIG. 5.4. Two-layer coils.

the free ends still have to be bent over and the joints butt-welded and insulated.

The two-layer winding is the neater arrangement and all the coils are identical, one coil-side laying in the top half of the slot and the other side in the bottom half of another slot approximately one pole-pitch away; see Figs. 5.3b and 5.4. The two-layer winding gives greater flexibility in design because of the ease with which the pitch can be chosen. Open slots are frequently used and the coils are laid in them one after the other round the periphery. While fitting the last pole-pitch of bottom layers into place, the first pole-pitch of top layers must be lifted out of the way. There are two ways of bending the front-end connections of each coil; inwards giving what is called a lap winding, or outwards to form a wave winding, see Fig. 5.6.

The Commutator

Two-layer coils are essential in practice, for windings which are connected to a commutator. The commutator itself consists of many copper segments insulated from, and running parallel to one another and clamped round the periphery of an insulated cylinder. Each segment is extended to a coil junction so there must be the same number of segments as coils. Figure 5.5 shows a d.c. lap-wound armature; the top

217

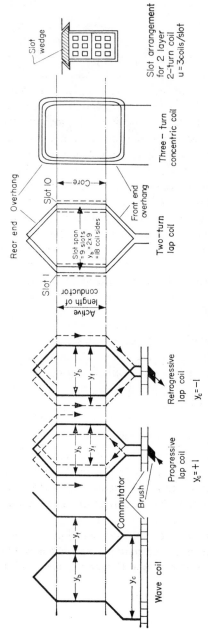

Fig. 5.6. Winding terms: y_b = back pitch or coil pitch; y_f = front pitch; y_c = coil end pitch; y_f and y_b are measured in coil sides.

half-coils in the process of being laid over the bottom half-coils which are already in the "riser" extensions from the commutator bars, prior to soldering or brazing at the coil junctions. The commutator provides access to all the coils, current being led in at one or more points through carbon brushes running on the cylindrical surface. In a similar manner, current can be taken off at one or more points. These features are demonstrated in schematic fashion by Fig. 5.14 showing the currents in a commutator winding supplied from a three-phase source. The commutator and brushes can be designed to give a long wearing life, so the situation indicated by the elementary d.c. machine of Fig. 1.5, which has effectively eight commutator segments, can be achieved in a practical manner to deal with instantaneous powers up to 30,000 kW. Speaking in general terms, the commutator and its winding provide paths for currents to flow under the influence of magnetic fields. The directions of these paths are fixed relative to the brushes. In the usual case, where the brushes are stationary, the currents flow between the brushes with a definite space orientation which is unaffected by the rotation of the commutator and winding.

Armature Winding Terms and Symbols

Reference should be made to Fig. 5.6.

Conductor. The active length of wire or strip in the slot. It may be laminated and transposed along its length to reduce eddy currents. Symbol Z will be used for the total number of conductors in the winding.

Coil. Two conductors separated by one pole-pitch or nearly so, can be connected in series to form a single-turn coil. More conductors, virtually in the same position magnetically, could be included before bringing out the coil ends; four conductors would form a two-turn coil, six conductors a three-turn coil, and so on. The symbol C will be used for the total number of coils.

Coil Side. This is one half of the coil, lying in one slot, the coil being single- or multi-turn. A commutator winding may have several coils in one slot. On Fig. 5.6 is one slot arrangement with two-turn coils and with three coil-sides per layer. The total number of coils $C = Su$ where S is the number of slots and u the number of coils per slot; 3 in this case. There are 12 conductors per slot insulated from one another, though all the conductors in one layer in the slot would be taped together to share a common insulation to "earth"; the iron core. For a single-layer winding, there is only one coil for every two slots; $u = C/S = \frac{1}{2}$, but there may be many turns in each coil and these are arranged as indicated in Fig. 5.2 to reduce eddy currents.

Pitch. There are various pitches of interest as indicated in Fig. 5.6, for which the most common terminology has been used. An additional term, *slot span*, will be used to indicate the throw of the coil in slots. This is different from y_b, the *back pitch* or *coil pitch* measured in coil sides, only when the coil is two layer and with $u > 1$. The slot span is approximately a pole-pitch, i.e. $S/2p$ measured in slots. The *coil pitch* is sometimes expressed as a fraction of the full pole-pitch (180°); or in electrical degrees.

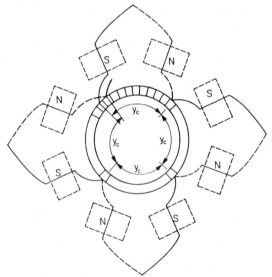

FIG. 5.7. Progressive wave winding,
$p=4$, $C=39$, $y_c=10$, $py_c=C+1$.

Coil-end Pitch or Commutator Pitch y_c. For commutator windings this is the distance measured in commutator segments between the end connections of a coil. For a simple lap winding it is ± 1, the positive sign indicating a progression to the right when tracing through the winding. A negative sign indicates a retrogression to the left. For a simple wave winding it is approximately equal to the number of commutator bars per pole pair, C/p. To be more precise, a practicable winding must follow the rule, $py_c = C \pm 1$, see Fig. 5.7. This rule is essential to ensure firstly, that after one tour of the armature periphery, the end of the last coil falls on the commutator bar next to the beginning of the first coil; and secondly, that after several (y_c) similar tours of the armature, all the commutator bars are occupied and the final coil falls naturally on the first, giving a closed winding. As before, the positive and negative signs refer to the progression of the winding, positive being to the right. The effective electrical difference between rightwards and leftwards progression is that for a particular direction of conductor current, the progression determines, for a particular brush, whether the current flows from the winding to the brush or vice versa, see Fig. 5.6; the brush polarity relative to any particular field polarity, machine function and rotation is different for the two cases.

5.2 COMMUTATOR WINDINGS

Simple Lap Windings for D.C. Armatures

Confining the discussion to d.c. machines for the moment, simplifies the treatment a little. Further, the appearance of a complete winding diagram is usually bewildering at first sight and any measures which can be taken to ease the understanding are worth adopting. For example, although it is a useful experience to trace through a winding from beginning to end, it can be confusing and strictly, is not essential for full comprehension. The group of coils in any one slot is identical with the group in any other slot, the slot pattern merely being repeated round the armature, in the normal, balanced two-layer winding. With these remarks in mind, the following examples, E.5.1 and E.5.3, have been designed with an unusually small number of slots to permit the winding to be traced through if desired, with little effort. Examples E.5.2 and E.5.4 with a few more slots but still rather impracticable, show only the minimum number of coils to indicate the whole pattern.

EXAMPLE E.5.1

Figure 5.8 shows a 4-pole simple lap winding. Note the numbering system which will be used subsequently. The top conductors take numbers progressing clockwise in sequence. The bottom conductors, if present, as on a two-layer winding, take the same number as the top conductor immediately above but with an acute accent. Further, for a two layer winding, commutator bar No. 1 (or clip No. 1) is taken to conductor 1.

The direction of generated e.m.f. in conductors under a north pole is shown into the paper, and that under a south pole is shown outwards in accordance with the clockwise rotation assumed. Commencing at commutator bar 1 and progressing round the winding in a clockwise direction, a derived diagram can be drawn, Fig. 5.8b, in which each conductor is included, together with its appropriate e.m.f. arrow. The

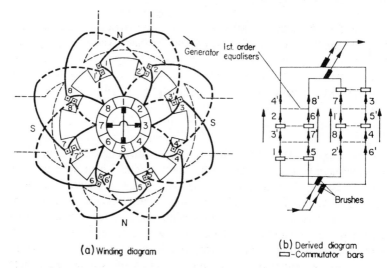

(a) Winding diagram

(b) Derived diagram
▭-Commutator bars

FIG. 5.8. Four-pole simple lap winding: $S = 8, p = 2, C = 8, u = 1, y_c = +1,$
$y_b = 2$, slot span $= 2$.

222

particular arrangement of Fig. 5.8b is deliberate, to show clearly that in one tour of the armature, with every conductor taken into account, there are four reversals of e.m.f. It should not be difficult to realise that, if there had been twice as many poles, there would have been twice as many e.m.f. reversals. Although the winding is short circuited on itself, there is no circulating current, assuming symmetry, because the resultant e.m.f. round the complete circuit is zero. However, if connection is made to the winding at four equidistant points as shown, it can be seen that four equal parallel paths are formed, from which an external circuit could be supplied. Each path consists of all the top-layer conductors under one pole, together with all the bottom-layer conductors under one neighbouring pole; i.e. in general, for a simple lap winding, there are as many parallel paths as poles; $2a = 2p$ where a is the number of pairs of parallel paths.

It should be noted that the derived diagram as shown, is only correct when the winding is in certain positions or during certain time intervals. Due to rotation, the conductors are constantly changing the d.c. circuit in which they are connected. After a movement of two slot pitches say, every conductor e.m.f. would be reversed; e.g. conductors 1, 3′, 2 and 4′ would then occupy the previous positions of 3, 5′, 4, 6′. The external circuit would still have the same polarity however, so that although alternating voltages are induced in individual conductors, the brushes pick up a unidirectional voltage; the commutator and brushes act as a mechanical rectifier.

Equalisers for Simple Lap Windings

Although the various paralleled circuits of a multipolar lap winding are nominally identical, there are slight discrepancies in the generated e.m.f.s because practical machines are not perfectly symmetrical. These e.m.f. differences will cause internal circulating currents which, from Fig. 5.8b, are seen to have a path through the brushes. Such currents, which would have a predominant unidirectional component and be superimposed on any current due to the external circuit, could cause excessive heating, sparking at the brushes and mechanical vibration.

A simple remedy is available. Consider conductors 2 and 6. Since they are exactly one pole-pair-pitch apart, S/p being an integer, they always occupy identical positions relative to any particular magnetic polarity; their e.m.f.s rise, fall and reverse in time-phase. If permanent connectors are used to join the ends of these two conductors, a short-circuited, double-pole-pitch "coil" is formed, superimposed, as it were, on the winding. With perfect magnetic symmetry, this "coil" would embrace equal and cancelling amounts of N-pole and S-pole flux in every armature position; there would be no net flux change and no net induced e.m.f. acting round the "coil" circuit. Should there be any flux unbalance, caused, for example, by unequal air gaps under the poles, there would be a difference between the conductor e.m.f.s, varying in time due to conductor movement and driving a circulating alternating current. It will be seen in Chapter 8 that when an a.c. generator is thus short circuited, the armature currents, which are at a very low internal-power-factor, produce an m.m.f. in direct opposition to the flux generating them. Any resultant flux embraced by the double-pole-pitch "coil" is therefore reduced to very small proportions, an action which is in accordance with Lenz's law. Several similar connectors are provided as indicated on Fig. 5.8b and the e.m.f. differences and d.c. circulating currents are practically eliminated. The winding must be designed so that tapping points spaced one pole-pair-pitch apart can be connected together. There are p tappings on each connector for the general case of $2p$ poles. The connections are known as first-order equalisers and may be mounted either just behind the commutator or at the rear end of the armature. A.C. commutator machines and highly rated d.c. machines are sometimes provided with second-order and third-order equalisers having a function which is rather different and beyond the scope of this book.[9]

Rules for Simple Lap Windings; $y_c = \pm 1$

Since it is essential that provision be made for equalisers, a simple rule governs the design of lap windings. There must be an integral number of slots per pole pair; S/p is an integer, which may be either odd or even. The slot span can be taken as the nearest smaller integer to $S/2p$ and the number of equalisers could be such as to provide for tappings to every alternate slot, but these are design decisions for the specialist. There are mechanical and economic limits to the number of commutator segments and the number of turns per coil must not be such as to cause the coil voltage to exceed a maximum value of 25–30 V, from commutation considerations, see p. 282. The number of coils per slot does not usually exceed five.

EXAMPLE E.5.2

Figure 5.9, constructed from the values of S, p and C given below the figure, shows the coils for one slot group only, since this pattern is merely repeated round the periphery. The full pole-pitch corresponds to $27/6 = 4\frac{1}{2}$ slots, so the slot span is taken as 4 slots, giving a short-chorded coil. The coil pitch is thus $4 \times 2 = 8$ coil sides. There is an integral number of slots per pole pair; $27/3 = 9$, so there are identical conductor positions under any particular magnetic polarity. Tappings for the first

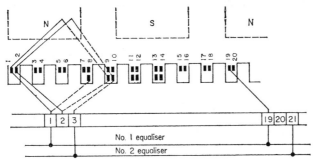

FIG. 5.9. Simple lap winding: $S = 27$, $p = 3$, $C = 54$, $u = 2$,
$y_c = +1$, $y_b = 8$, slot span $= 4$.
Equalisers No. 1 to   1. 19. 37
         No. 2 to   3. 21. 39
         . . . .
         No. 9 to 17. 35. 53

equaliser would be at conductor 1; conductor $1 + 2 \times 9 = 19$, and conductor $1 + 2 \times 9 + 2 \times 9 = 37$. The second equaliser could be connected to conductor 3, 21 and 39 and so on. This would mean one equaliser per slot, which is rather generous for a small machine. Note that the limited portion of the winding has been shown schematically, in a developed fashion, i.e. as if the periphery had been rolled out flat.

Simple Wave Windings for D.C. Armatures

EXAMPLE E.5.3

Figure 5.10, though a rather simplified illustration, will be adequate for deducing the general rules governing the design of simple wave windings. The slot span is taken as the nearest smaller integer to $S/2p = 5/4$, and is therefore made equal to one slot. The coil pitch is also 1, there being only one conductor per slot. The commutator pitch $y_c = (5 \pm 1)/2 = 3$ or 2. The retrogressive winding, $y_c = 2$ is chosen.

To introduce the general method for planning wave windings, the following table is presented and should be checked with the diagram though after practice it will

225

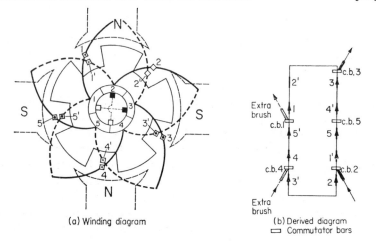

(a) Winding diagram

(b) Derived diagram
⊑ Commutator bars

FIG. 5.10. Four-pole simple wave winding: $S = 5$, $p = 2$, $C = 5$, $u = 1$,
$y_c = (5-1)/2 = 2$, $y_b = 1$, slot span $= 1$.

not be necessary in the general case to draw the diagram. The table is read left to right, along the top row and the other rows follow in sequence.

The winding has progressed through every conductor and closed on the starting point. A similar tabular method could be used for lap windings but it is hardly necessary in this case, since the conductors are taken in order and one complete tour of the armature completes the winding.

Starting at conductor 1 and drawing the derived diagram in a way similar to the case of the lap winding shows that there are only two reversals of e.m.f., i.e. two parallel circuits. This is always true whatever the number of poles, since the wave winding first picks up from every pole in succession, those conductors having a particular e.m.f. direction, before reaching the other half of the conductors with opposite e.m.f.s.

Comm. bar No.	to	Top conductor No.	to	Bottom conductor No. ($+$coil pitch $= 1$)	to	Comm. bar No. ($+y_c = 2$)
1		1		$(1+1) = 2'$		$(1+2) = 3$
3		3		$(3+1) = 4'$		$(3+2) = 5$
5		5		$(5+1) = 1'$		2
2		2		$3'$		4
4		4		$5'$		1

Placing of Brushes

The derived diagram shows that, for the particular armature position indicated, two brushes touching commutator bars 2 and 3 respectively, would pick up the maximum available voltage. However, since conductors 2 and 2′ have zero e.m.f. in this position, and conductors 1 and 3′ are nearly in the zero-flux zone, two extra brushes touching commutator bars 1 and 4 would be connected across most of this voltage and could be paralleled with the first pair. With the much larger number of conductors in the practical machine, the fact that the paralleled brushes are shorting out a few coils in the interpolar zone (1, 2′ and 2, 3′ in this case), would cause no appreciable difference in the output voltage. In practice, the brushes are placed symmetrically, as shown on the winding diagram, and make contact with conductors which are passing through the low-flux interpolar zone. It is sometimes useful for ease of accessibility, to use only two brushes on a wave wound armature, but on larger machines, since the extra brushes lower the current per brush and permit the reduction of the commutator axial length; the full number of brushes, $2p$, as on a lap winding is often used.

Rules for Simple Wave Windings

The feasibility of a wave winding depends upon whether the equation $y_c = (C \pm 1)/p = $ integer, can be satisfied; see Fig. 5.7. The total number of coils $C = Su$, and for any particular number of poles, C is determined by whether or not the winding is a possibility.

For example, if $p = 2$, then C must be odd to make y_c an integer and so both S and u must be odd. The winding may be progressive ($+1$) or retrogressive (-1)

if $p = 3$, then neither S nor u must be divisible by 3 in order that $Su \pm 1$ will be divisible by 3. There is no free choice this time between progressive and retrogressive arrangements.

If the total number of coils is not such as to permit y_c to be integral, then the ends of one coil could be cut off, insulated and the coil not used. This would be a *dummy coil* and the procedure is permissible on

227

machines of low rating to use an existing slotting number and thus avoid additional expenditure on new tools.

The coil pitch depends on the slot span which in turn is usually taken as the nearest integer lower than $S/2p$. When this is multiplied by the number of coils per slot, the coil pitch y_b is given. The front pitch need not be calculated; it follows from the values already chosen for y_c and y_b and is not of great interest for simplex windings; viz. simple lap and simple wave.

EXAMPLE E.5.4

Refer to Fig. 5.11, constructed from the values of S, p and C given below the figure.

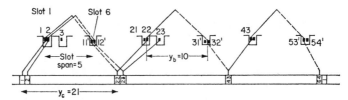

FIG. 5.11. Simple wave winding: $S = 32$, $p = 3$, $C = 64$, $u = 2$.

Slot span $S/2p = 16/3$, say 5.
Coil pitch $y_b = 5 \times 2 = 10$.
Commutator pitch $y_c = (64-1)/3 = 21$; only retrogressive possible.

Winding Table

Comm. bar No.	to	Top conductor No.	to	Bottom conductor No. (+10)	to	Comm. bar No. (+21)
1		1		11′		22
22		22		32′		43
43		43		53′		64
64		64		10′		21
21						
—						
—						
44		44		54′		1

Note. Although the winding table has been drawn up and adequate details indicated on the winding diagram, none of this is really necessary to the argument. Every slot group of coils is identical. Once the rules have been applied to find the slot span and commutator pitch, the coil laid in slots with the correct span and the coil ends aligned with the commutator segments at the appropriate pitch, the rest of the coils fall naturally into place.

Choice Between Lap and Wave Windings

The designer does not start off with the information given in the examples above. He will first decide the armature dimensions and number of poles from the required output and speed, see p. 291. From

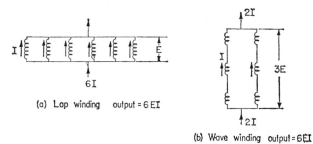

(a) Lap winding   output = 6 EI

(b) Wave winding   output = 6 EI

Fig. 5.12. Schematic diagrams of 6-pole simplex lap and simplex wave windings.

known permissible values of flux density, the e.m.f. per conductor follows and hence the number of conductors in series to give a specified e.m.f. is known. He then has to decide between a lap and wave winding and choose an appropriate number of slots and slotting arrangement. His choice of slots and commutator bars is restricted by mechanical and commutation considerations and there may be one, two, three or more permissible windings.

The schematic diagrams of Fig. 5.12 show the essential differences between simple lap and wave windings. For a given number of conductors and the same physical size and speed, the wave winding gives a higher terminal voltage and lower current than the lap winding. This arises because the wave winding has the larger number of conductors

229

in series $z_s = Z/2$; the lap winding has $z_s = Z/2p$. The wave winding is used on small machines and on low-speed large machines in order to produce a sufficiently high e.m.f. It is cheaper than a multi-turn lap winding designed for the same voltage because there are no equaliser connections. The lap winding is used on the larger machines and gives better commutation on the whole for a given rating and speed, providing single-turn coils are adequate for the voltage required, this being the case beyond a certain machine size and speed.

Multiplex Windings

If the number of commutator segments is so increased that a simple lap or wave winding only connects to alternate bars, the generated voltage would be unchanged, but another winding virtually in parallel electrically, could be interleaved between the first one and connected to the bars that had been missed. This would form a *duplex winding*. Further increase in the number of bars would permit additional folds or plexes in the winding, triplex, quadruplex, etc. A modification to the commutator pitch is obviously necessary, a duplex-lap coil being shown for demonstration in Fig. 5.13. The effect of these windings is to increase the number of parallel circuits over the simplex case; duplex giving twice as many, triplex three times as many, and so on. Multiplex windings introduce additional design flexibility and are used for various

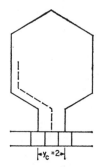

FIG. 5.13. Duplex-lap coil.

230

purposes rather too specialised to discuss here.[9] One obvious use is to bridge the gap between the number of parallel circuits given by a lap winding and the number given by a simple wave winding, when the number of poles is large.

A.C. Commutator Windings

In simple theory, these could be just the same as the d.c. commutator windings, but because commutation difficulties are more severe, special refinements are required which often involve the use of multiplex windings and of discharge windings in parallel. However, the same function of providing space oriented paths for current taken into and out of the brushes is still performed. For example, if three symmetrically spaced sets of brushes per pole pair are used, Fig. 5.14, for the two pole arrangement, shows that a delta-connected winding is formed, which, irrespective of the armature rotation, disposes the three-phase currents in the usual way, and gives them a particular orientation in space. The resultant magnetic effect of these currents will be shown later to give rise to a two-pole m.m.f. acting along an axis which, in general, is not stationary with respect to the brushes. For a

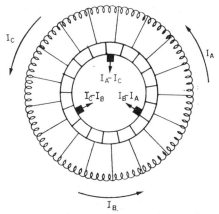

FIG. 5.14. Schematic diagram of two-pole, three-phase a.c. commutator winding.

231

four-pole m.m.f. there would have to be a different winding and another three sets of brushes in parallel with, and diametrically opposite to the first three, and the brush spacing would have to be 60° instead of 120° mechanical. The two-layer winding would be arranged with coil pitches of approximately 90 mechanical degrees instead of 180° as in the two-pole case.

5.3   A.C. OPEN-TYPE WINDINGS

Elementary Three-phase Windings

A commutator winding is closed on itself and current flowing into the winding through a brush divides into two paths which retain their spatial orientation irrespective of the angular position of the rotating winding. Even though the coils carrying this current are continually changing, the "view" of the winding from the brush remains unchanged, apart from ripple effects. For alternating current, it is not necessary to have such a closed winding and this type is only used where commutator action is required. In the open winding, the ends of each phase can be brought out to terminals and any desired interconnection can then be made externally.

The simpler forms of open windings are a good deal easier to understand because of the absence of the commutator and its special effects. For a three-phase winding, the conductors are arranged in three balanced sections. These must be identical in every way except for the mutual physical displacement which gives rise to the mutual time-phase displacement of 120° electrical, in the total generated e.m.f.s. Elementary two-pole and four-pole arrangements are shown in Fig. 5.15, there being one full-pitch coil per pole pair in each phase. A, B and C, Fig. 5.15a, and A_1, B_1 and C_1, Fig. 5.15b, are at points displaced by 120° electrical and would therefore be the corresponding starts, or the finishes, of the three phases. The two sections of each phase in Fig. 5.15b could be connected in series or in parallel. The resultant winding m.m.f. is indicated for the instant shown by the centre phasor diagram, when

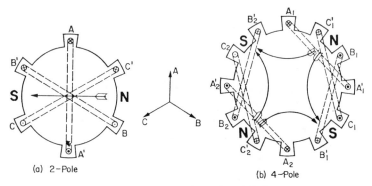

(a) 2–Pole

(b) 4–Pole

FIG. 5.15. Elementary three-phase windings.

phase A is at its maximum current and phases B and C are at one-half of their negative maxima. The two-pole and four-pole effects are clearly demonstrated.

The following additional winding terms will be used:

Slots per pole $= Q = S/2p$.

Slots per pole per phase $= q = (S/2p)/3$, for three-phase windings.

Examples of Three-phase Windings

The following five examples are all based on the same number of slots and it will be seen that the differences really stem from the way in which the slot conductors are connected together. In each case, the winding is four-pole, 720° electrical, but could easily be extended to any number of poles. The normal minimum number of slots per pole per phase is chosen, i.e. $q = 2$. The total number of slots is therefore $S = q \times 2p \times m$, where m is the number of phases, giving $S = 2 \times 4 \times 3 = 24$. For a two-layer winding the number of coils is usually the same as the number of slots, since there is no restriction on turns per coil, as on commutator windings, where commutation considerations, see Chapter 6, impose limitations. For a single-layer winding, a coil takes up two slots, so $C = S/2 = 12$ in this case. In Fig. 5.16, the slots are

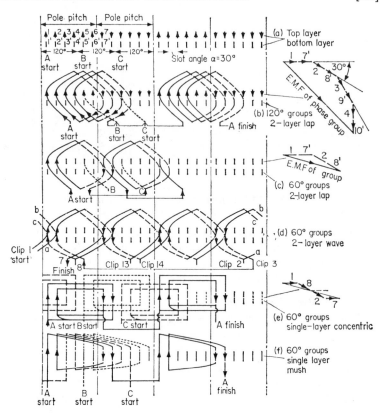

FIG. 5.16. Examples of three-phase windings: (a) slot numbering and arrangement.

divided into four equal polar zones and the beginning of each phase is readily determined; Fig. 5.16a.

Two-layer Windings. At first sight it would appear that the logical arrangement would be to take up the first four slots with phase A, the second four with phase B and the third four with phase C, and repeat the pattern on each pole pair. This would form 120° phase groups as shown in Fig. 5.16b for a two-layer lap winding. A drawing

234

of the whole winding would be rather confusing, so only phase A is shown completely and the beginnings of phases B and C are indicated. The coils are identical and the slot pattern is repeated round the whole winding.

Assuming that the flux density B is sinusoidally distributed round the periphery, the conductor e.m.f.s, which are proportional to B, will have instantaneous values determined by rotating time-phasors mutually displaced in time phase by the slot angle in electrical degrees. Taking the first slot e.m.f. as reference, the conductor e.m.f.s in the A-phase 120° group are added vectorially at the side of the winding diagram. The second 120° of the same phase will have an identical diagram. The B-phase will have a similar diagram starting from the end of the one shown, i.e. conductors 5, 11′, 6, 12′, etc., so that the resultant phasor will be displaced from that for the A-phase by 120°. The C-phase phasor will be displaced by a further 120° and composed of 9, 16′, 10, 17′, etc. It can be seen that a loss of voltage occurs due to the distribution of the winding in several slots, the vectorial sum being appreciably less than the arithmetic sum of the conductor voltages. The same effect occurs with the closed commutator winding, and with diametral brushes, which virtually divide the winding into 180° groups, the loss of voltage would be even more noticeable.

The effect can be mitigated quite simply by rearranging the coils into smaller groups, Fig. 5.16c. The coils are exactly the same as before, but now they are split into four 60° groups instead of two 120° groups for each phase. The coil-to-coil end connections are slightly different to ensure the e.m.f.s are still additive, but the phase starts are unchanged. The vectorial sum of each group voltage is now seen to be nearly equal to the arithmetic sum.

Two-layer wave windings are also used, particularly when single-turn coils or two coils per slot are adequate for the required voltage. There is then a saving in end connections and jointing, since the coil groups are interconnected naturally due to the outward spread of the coil ends. Figure 5.16d shows one complete phase of a wave winding in the same slotting arrangement as before. There are only two separate sections in each phase so only one interconnector per phase is needed instead of the three used on the 60° lap winding. This saving is more pro-nounced with a larger number of poles. The open wave winding is not restricted by the rule $py_c = C \pm 1$, because the pitches of the coil ends need not be identical. The coils are all made the same, but different end pitches can be obtained by pulling the ends apart a little, the symmetrical arrangement of the commutator connections being no longer a necessary requirement. In this particular example, if the coil-end or clip pitches are made alternately 12 and 13, a progressive winding results. After each tour of the armature, the coil ends fall naturally on to those of the next coil in sequence. However, the last coil end does not fall near the start of the first coil but this is of no consequence in an open winding. The winding table is given below and the interested

Winding Table

Clip No.	to	Slot No.	to	Slot No.	to	Clip No.	to	Slot No.	to	Slot No.	to	Clip No.
				(+6)		(+12)				(+6)		(+13)
1		1		7'		13		13		19'		2
2		2		8'		14		14		20'		connector
11		11		17'		23		23		5'		12
12		12		18'		24		24		6'		connector

reader can follow it through and prove the above statements. Clips replace commutator bars at the joints.

As for the lap windings, Phase A starts at clip 1, Phase B starts at clip 5 and Phase C starts at clip 9. The table shows the first half of Phase A and the second half of Phase B.

The choice between lap and wave is governed by many design con-siderations but it should be noted that parallel circuits do not arise in the same way as for the closed commutator winding. It is, of course, possible to connect open-winding coils in parallel groups, providing that each phase has a sufficient number of identical sections. Lap windings give a greater number of such sections than wave windings.

236

Single-layer Windings. Figure 5.16e is a *concentric winding* of a common type, but there are many other types. The phasor diagram of a phase group is similar to Fig. 5.16c but there are only two groups per phase. The vector resultant is independent of the sequence in which the individual conductors are connected within the coil group. More of the winding than in previous examples has been shown to indicate how the phases are arranged to interlace mechanically. Adjacent groups of coils, which are preformed at one end and are hairpin shaped, are pushed through semiclosed slots from opposite ends of the core. The free ends are then bent over and butt welded. The preformed ends are bent outwards from the plane of the coil so that they can cross neighbouring phase groups at a larger diameter; see Fig. 5.3a. Apart from the case where there is an odd number of pole pairs, for which a special group of coils must be formed, the diagram shows that there are as many different coil shapes are there are coils in a group; two in the present instance. One coil is over-pitched and the other under-pitched giving an average of full pitch for the group.

Figure 5.16f is a single-layer *mush winding* which is an alternative way of dealing with the problem of coil interlacing. It is suitable for small machines where the coils consist of many turns of round wire. The coils are full pitch, preformed at the ends and identical in shape. They are laid into semi-closed slots, one conductor at a time, and the coil ends form up over a portion of the periphery in a similar manner to the two-layer winding.

There are many other kinds of winding, including two-layer concentric windings and two-layer mush windings, having chorded coils of equal pitch, but space does not permit a full treatment of the subject.

Fractional-slot Windings

These are an advanced form of winding in which the number of slots per pole per phase, q, is not an integer. If the number of slots in Fig. 5.16 had been 30 say, then q on the average would have been $2\frac{1}{2}$. In practice this would mean that in one pole pair, 360° electrical, the phase groups would be alternately 3 and 2 slots giving 5 slots per pole

237

pair per phase. The top-layer sequence is shown in Fig. 5.17. The bottom-layer sequence must of course be the same, but would start from a point determined by the coil pitch. Single-layer windings can be used but they impose many more restrictions than two-layer windings. It will be noticed that the pattern is repeated every two poles in Fig. 5.17. If there had been 27 slots, q would have been $27/(3 \times 4) = 9/4$ and the

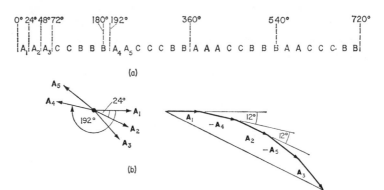

FIG. 5.17. Fractional-slot winding: (a) top-layer sequence; (b) phasor diagrams.

pattern would have required four poles to complete. The denominator determines the number of poles required for a complete pattern and the numerator determines the number of slots that each phase takes up in this distance.

This type of winding permits the use of standard slotting arrangements over a wide range of pole numbers. It is necessary for three balanced phases that the total number of slots is divisible by 3, but it is no longer necessary that it should be divisible by the number of poles as well, though there are some restrictions when the number of poles is itself a multiple of 3. Another advantage is that the phase angle between the various conductor e.m.f.s is reduced. As will be seen later, this in turn reduces the higher order harmonics in both e.m.f. and m.m.f. waveforms. For the case of 30 slots and 4 poles, the slot angle is $720/30 = 24°$. The five conductor e.m.f.s in one layer of a completed phase section are displaced in time phase as indicated in Fig. 5.17b. When they are series connected with the necessary and natural reversal of A_4 and A_5, the phase angle between the successive e.m.f. phasors is effectively 12°, since A_2 and A_3 are intermediate between A_1, A_4 and A_5.

Single- and Two-phase Windings

A single-phase winding is usually formed by taking the voltage difference between two phases of a three-phase, star-connected winding giving $\sqrt{3}$ times the voltage of one phase. Sometimes the other phase is not wound, and sometimes it is wound and kept available as a spare. A two-phase winding requires ingress to be made to the winding at points displaced in space phase by 90° electrical. The two-layer two-phase winding could use the same coils as for three-phase, but the phase groups would take up 90°. For the single layer two-phase concentric winding, some modifications would be required to the arrangements for interlinking the coil-overhang connections shown on Fig. 5.16e.

Interconnection of Phases

Star connection is usual on synchronous generators. The availability of the neutral for protective gear purposes, the dual-voltage supply and the absence of third harmonic circulating currents are the deciding factors. Induction motors use either star or delta connections, whichever is suitable to meet the design specification.

5.4 WINDING FACTORS

A winding circuit consists of many conductors in series, for which the e.m.f.s, in general, are not all in phase. This phase displacement occurs for two reasons:
   (i) distribution of the winding in several slots,
   (ii) coil pitch not equal to a pole pitch on the average; two-layer windings only.

Since the sum of these conductor e.m.f.s must be carried out vectorially, both of these effects reduce the machine voltage but this is not an overall disadvantage. The time variation of e.m.f. for a single conductor corresponds to the spatial variation of air gap flux density. This is not purely sinusoidal, being often particularly rich in lower-order harmonics. By suitable winding design, the percentage reduction of the harmonics can be made much greater than that of the fundamental. Consequently, the waveform of the circuit voltage approaches more nearly to a pure sine shape. The winding factors express mathematically the *per-unit* reduction of the fundamental and of each harmonic which takes place as a result of distribution and chording. In practice the fundamental is rarely reduced by as much as 10%, whereas the harmonics are often reduced to negligible proportions. The same winding factors apply to the space distribution of m.m.f. and so reduce the m.m.f. space harmonics.

Distribution Factor k_d

Figure 5.18 shows the components and the resultant of the e.m.f.s due to a phase group for $q = 3$ and is the same type of time-phasor diagram as on Fig. 5.16. Each component phasor is proportional to the r.m.s. value of the fundamental voltage for a slot conductor. It is displaced from the conductor voltage of a neighbouring slot by the slot angle, α electrical degrees. The distribution factor is the ratio:

$$\frac{\text{phasor sum of component e.m.f.s}}{\text{arithmetic sum of component e.m.f.s}}.$$

From the geometry of the figure:

$$k_d = \frac{2R \sin q\alpha/2}{q \times 2R \sin \alpha/2} = \frac{\sin q\alpha/2}{q \sin \alpha/2}. \qquad (5.1)$$

When α is small, the ratio approaches:

$$\frac{\text{chord}}{\text{arc}} = \frac{\sin q\alpha/2}{q\alpha/2} \qquad (5.2)$$

240

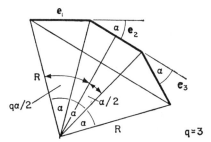

FIG. 5.18. Distribution factor k_d.

where $q\alpha$ is the phase spread in electrical *radians*. An example will illustrate the application.

EXAMPLE E.5.5

Calculate k_d for a machine with 9 slots per pole for the following cases:
  (i) One winding in all the slots.
  (ii) One winding using only the first two-thirds of the slots per pole, i.e. 120° groups.
  (iii) Three equal windings placed sequentially in 60° groups.

In each case, the slot angle $\alpha = 180°/9 = 20°$ and the values of q, the number of slots in a group are: 9, 6, and 3 respectively. Hence:

(i) $k_d = \dfrac{\sin 9 \times 20°/2}{9 \sin 20°/2} = \underline{0.640}$

 or approx. from eqn. (5.2) $k_d = \dfrac{\sin \pi/2}{\pi/2} = \underline{0.637}$.

(ii) $k_d = \dfrac{\sin 6 \times 20°/2}{6 \sin 20°/2} = \underline{0.831}$

 or approx. from eqn. (5.2) $k_d = \dfrac{\sin \pi/3}{\pi/3} = \underline{0.827}$.

(iii) $k_d = \dfrac{\sin 3 \times 20°/2}{3 \sin 20°/2} = \underline{0.960}$

 or approx. from eqn. (5.2) $k_d = \dfrac{\sin \pi/6}{\pi/6} = \underline{0.955}$.

These results illustrate one of the reasons for using three or more separate windings or phases, and confirm mathematically the conclusions deduced from comparison of Figs. 5.16b and 5.16c. Dividing the winding into six groups per pole pair instead of three, increases the voltage, and hence the power output for the same conductors, by 15%; from 0·831 to 0·96. These six groups could form a six-phase winding but in Fig. 5.16c they are connected in three groups of two sections each to form a three-phase winding. A mutual reversal between the two sections causes their voltages to add arithmetically. Example E.5.5 also demonstrates why only two thirds of the slots are used for a single-phase winding. The additional slots would only increase the output by $(9 \times 0·64)/(6 \times 0·831) = 1·15$ times for a 50% increase in copper.

Note that this slotting arrangement is not suitable for a simple two-phase winding because an exact displacement of 90° is not possible with an odd number of slots per pole. With 8 slots per pole for example, the distribution factor for two equal windings would be $(\sin 45°)/(4 \sin 11°·25) = 0·907$, which is not quite as good as for the three-phase winding with 60° groups.

Distribution Factors for the nth Harmonic, k_{dn}

Time harmonics in the e.m.f. are produced by space harmonics in the flux density wave. For example, if the flux distribution consists of a fundamental and an "in-phase" third harmonic, it will be as in Fig. 5.19. It can be imagined as being produced by three harmonic

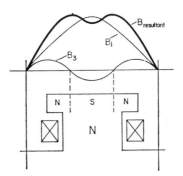

FIG. 5.19. Third-harmonic "poles".

poles superimposed on each fundamental pole as indicated. Equations (5.1) and (5.2) are based on a slot angle α electrical degrees obtained from the relationship:

$$\text{electrical degrees} = \text{mechanical degrees} \times \text{pole pairs } p.$$

Consequently, electrical angles based on the third-harmonic poles, would be three times those on the fundamental scale. Similar treatment could be accorded to higher harmonics if present, so in general, for the nth harmonic, all electrical angles are n times those on the fundamental scale. For example, the slot angle α on the fundamental scale becomes $n\alpha$ for the nth harmonic.

Thus the distribution factor for the nth harmonic becomes:

$$k_{dn} = \frac{\sin qn\alpha/2}{q \sin n\alpha/2} \quad \text{or from eqn. (5.2):} \quad k_{dn} = \frac{\sin qn\alpha/2}{qn\alpha/2}. \qquad (5.3)\,(5.4)$$

Equations (5.3) and (5.4) are the general expressions for the distribution factor, eqn. (5.4) applying when the phase spread $q\alpha$, referred to the appropriate harmonic, is less than about 20°. Under these circumstances, as demonstrated by Example E.5.5, the alternative expression for k_d gives nearly the same result as the exact expression.

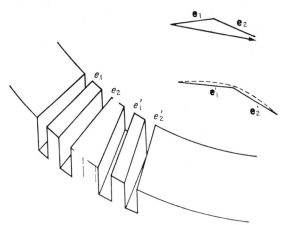

FIG. 5.20. Straight and skewed slots.

The expression (5.4) reduces continually as n, the order of the harmonic, increases so that if α can be made very small, the higher harmonics could be virtually eliminated. This can be achieved practically by skewing the slots by a full slot pitch, Fig. 5.20, and the effect over straight slots is illustrated by the next example.

EXAMPLE E.5.6

Calculate, for case (iii) in the previous example, the distribution factor for the 5th, 17th and 19th harmonics for straight slots and for slots skewed by one slot pitch, see Fig. 5.20.

$$k_{d5} = \frac{\sin 3 \times 5 \times 20°/2}{3 \sin 5 \times 20°/2} = \underline{\cdot218} \quad \text{or for skewed slots} = \frac{\sin 150°}{150 \times \pi/180} = \underline{0\cdot191}.$$

$$k_{d17} = \frac{\sin 3 \times 17 \times 10°}{3 \sin 17 \times 10°} = \underline{\cdot960} \quad \text{or for skewed slots} = \frac{\sin 150°}{510 \times \pi/180} = \underline{0\cdot056}.$$

$$k_{d19} = \frac{\sin 3 \times 19 \times 10°}{3 \sin 19 \times 10°} = \underline{\cdot960} \quad \text{or for skewed slots} = \frac{\sin 210°}{570 \times \pi/180} = \underline{-0\cdot05}.$$

17th and 19th harmonics arise with $Q = 9$ slots per pole, see Section 2.4 and Fig. 2.12b. With straight slots some harmonics are reduced but unfortunately not the pronounced tooth harmonics. With skewed slots however, there is a progressive reduction in the higher harmonics and the degree of skewing should be equivalent to a slot pitch to get full advantage from this feature. Figure 5.20 shows that with part of the conductor occupying every angular position over a slot pitch, it is equivalent to an infinite number of infinitesimally short lengths displaced by infinitesimally small angles. The winding is then said to be uniformly distributed. Skewing introduces constructional difficulties, but is used quite often on induction motors to smooth out the variations in air gap permeance. If the slots are not skewed by a full slot pitch it is necessary to multiply eqn. (5.3) by another winding factor called the skew factor. For a slot pitch of skew this brings eqn. (5.3) into the form of eqn. (5.4).

Coil Pitch Factor k_p

If a coil is chorded, i.e. under- or over-pitched, the e.m.f.s of the two coil sides are not in phase and must be added vectorially. The coils are usually, but not necessarily, short-pitched, i.e. the pitch is $x\pi$ electrical radians, being less than the full pitch by the chording angle β, taken as 30° in Fig. 5.21. x is the *per-unit* pitch. The coil pitch factor is defined as the ratio

$$k_p = \frac{\text{phasor sum of coil-side e.m.f.s}}{\text{arithmetic sum of coil-side e.m.f.s}} = \frac{2e \sin x\pi/2}{2e} = \sin x\pi/2.$$

(5.5)

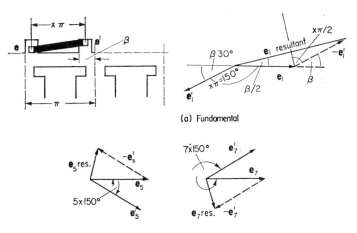

(a) Fundamental

(b) 5th and 7th harmonics

FIG. 5.21. Coil pitch factor k_p.

Whether the coil is under- or over-pitched by the electrical angle β, the pitch factor is still given by the sine of half the pitch angle.

For the nth harmonic, the pitch of the coil on the harmonic scale is $nx\pi$, so the general expression for the coil pitch factor is:

$$k_{pn} = \sin nx\pi/2.$$

(5.6)

If the coil is chorded by an angle π/n, the two nth harmonic coil-side voltages are out of phase by $n\beta = n\pi/n = \pi$ radians; i.e. they are in antiphase and cancel. The half pitch on the harmonic scale is

245

$n(\pi - \pi/n)/2 = (n-1)\pi/2$ so that $k_{pn} = 0$; the nth harmonic has been "chorded" out.

A coil pitch of 0·8–0·9 *per-unit* is usually employed on 2-layer windings and this reduces the 5th and 7th harmonics appreciably. The 3rd harmonic is cancelled between the line terminals of a three-phase winding due to the connection, see p. 80.

EXAMPLE E.5.7

A machine has 18 slots per pole and the first coil lies in slots 1 and 16. Calculate k_{p1}, k_{p5} and k_{p7}.

The slot angle $\alpha = 180°/18 = 10°$.

A full coil pitch coil would lie in slots 1 and 19 so the coil is chorded by 3 slots; i.e. $\beta = 3 \times 10° = 30°$. The pitch is $180° \times 15/18 = 150°$.

$$k_{p1} = \sin 150/2 = \underline{0·962.} \qquad k_{p5} = \sin (5 \times 150/2) = \underline{-0·259.}$$
$$k_{p7} = \sin (7 \times 150/2) = \underline{0·259.}$$

As in the case of the harmonic distribution factors, the positive or negative signs are not usually of practical significance. They are associated with the relative phase angles of the harmonics in the resultant e.m.f. wave.

For a single-layer winding, only the distribution factor is effective. Although individual coils may be short-pitched or over-pitched, the coil sides of a phase under neighbouring poles occupy the same group of slots relative to the poles; i.e. the coils are full pitched on the average. Coil pitch factors could be used in place of the distribution factor to give the same reduction factor but the procedure would be less convenient.

Winding Factors for Commutator Windings

The induced voltages in a commutator winding are alternating, and between any two points the resultant is obtained by vectorial summation. The winding factors apply as just derived but usually the slot angle is small and eqn. (5.2) = chord/arc, is used for the distribution factor. Consider, for example, two diametral tappings on a 2-pole winding, Fig. 5.22a. The vectorial summation assuming a sinusoidal distribution of flux, is proportional to the diameter, and the arithmetic summation is proportional to the half periphery; i.e. $k_d = 2r/\pi r = 2/\pi$ or using eqn. (5.2), $k_d = (\sin \pi/2)/(\pi/2) = 2/\pi$.

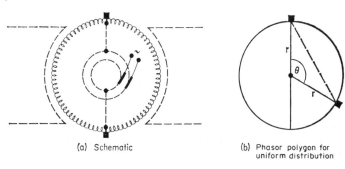

(a) Schematic   (b) Phasor polygon for
                                     uniform distribution

FIG. 5.22. Commutator winding.

In the position shown, the a.c. voltage picked up at the tappings and through them to the slip rings, if fitted, would be passing through its maximum value, the r.m.s. value of the slip ring voltage being 0·707 times this. As far as the commutator voltage is concerned, diametral brushes in quadrature with the stationary poles would pick up this maximum voltage as a substantially constant value, there being a small superimposed high-frequency pulsation or ripple, due to the commutator bars and slots, see Figs. 2.8 and 2.12.

For a d.c. machine, a sinusoidal flux waveform is of no value. In order to get the maximum output voltage, the flux density is arranged to be as high as possible over the pole pitch, bearing in mind other design considerations such as commutation performance. This leads to a field form which approaches a rectangular shape, Fig. 2.12a. A phasor summation, though feasible when the brushes are symmetrically disposed about the pole, is rather complicated in view of the large number of space harmonics, so a simpler method is used. Since only an average value, i.e. the d.c. voltage, is required, the e.m.f. calculation is based on the average flux density B_{av}, which gives the average voltage per conductor. The number of conductors in series z_s, which is equal to $Z/2$ for a wave winding and $Z/2p$ for a lap winding, is multiplied by the average voltage per conductor to find the total d.c. brush voltage. Except in special cases, the chording on d.c. machines is too small to have any noticeable effect on the output voltage.

247

The voltage at the brushes will be alternating if there is continuous relative movement between poles and brushes, as will be explained in Chapter 9. In this case the fundamental-frequency r.m.s. voltage appearing between brushes separated by an angle θ, will be equal to the arithmetic sum of the conductor r.m.s. voltages in series between the brushes multiplied by the distribution factor $(\sin \theta/2)/(\theta/2)$, θ being measured in electrical radians; Fig. 5.22b. The chording factor may be neglected if the chording angle is small.

5.5 E.M.F. PRODUCED BY AN ARMATURE WINDING

In designing a machine, the armature dimensions are decided from consideration of the output and the operating speed; see eqns. (6.1a and 6.1b). The next step in the design is to choose a winding, and this in turn is governed by the relationship between flux or flux density, speed, core length, pole pitch and the generated voltage. It is convenient to relate these parameters to one another in an expression called the e.m.f. equation. There are various ways of deriving the equation and several ways of expressing it when found. The following is an attempt to get an expression which is easy to remember, easy to develop and of general application.

It is convenient to start from the relationship:

$$e = Blv \text{ volts per conductor.}$$

The relative field/conductor velocity $v = \pi dn$, where d is the armature diameter and n the relative rotational speed in rev/sec. The average flux density is:

$$B_{av} = \frac{\text{flux per pole} \times \text{number of poles}}{\text{cylindrical area of armature surface}} = \frac{\phi.2p}{\pi dl}.$$

The number of conductors in series per circuit or per phase $= z_s$. For a d.c. machine then; with diametral brushes in the quadrature axis, i.e. with the torque angle $\delta = 90°$:

248

Total e.m.f. $= B_{av}.l.\pi dn.z_s$

$$= \frac{2p\phi}{\pi dl}.l.\pi dn.z_s = 2p\phi.n.z_s \qquad (5.7)$$

= flux cut by one conductor in one revolution
× revolutions per second × conductors in series

= flux cut per second × conductors in series,

i.e. the average e.m.f. per conductor is equal to the flux cut in one second, so eqn. (5.7) can be written down directly if this fact is remembered.

For an a.c. machine, and considering only the fundamental for the moment, the average e.m.f. per conductor from the above expression is $2p\phi n$, where ϕ is now the fundamental flux per pole. The r.m.s. value, applying the form factor for a sine wave $(\pi/2\sqrt{2})$, is therefore:

$$1.11 \times 2p\phi n \text{ per conductor.}$$

If there are z_s conductors in series per phase, the arithmetic sum of their e.m.f.s would be:

$$1.11 \times 2p\phi n.z_s.$$

But the arithmetic sum must be reduced by the winding factors as already explained, so that the final expression for the r.m.s. value of the fundamental e.m.f. of an a.c. machine is:

$$E = 1.11 \times 2p\phi nz_s.k_d k_p \qquad (5.8)$$

and since $k_d k_p$ is slightly less than unity for the polyphase machine, eqn. (5.8) is not greatly different in magnitude from eqn. (5.7) for the d.c. machine.

A further simple modification ties up eqn. (5.8) with the transformer e.m.f. equation. Since $pn = f$ and the number of conductors in series is twice the number of turns in series; $z_s = 2T_s$, then eqn. (5.8) becomes:

$$E = 1.11 \times 2\phi.f.2T_s.k_d k_p$$

$$= 4.44f T_s \phi.k_d k_p, \qquad (5.8a)$$

249

which is in the same form as eqn. (4.1) except for the effects of distribution and chording. ϕ, the flux per pole, times k_p, is in fact the maximum value of flux linked by a turn during its rotation with respect to the field. Note that the equation is still giving the *motional e.m.f.* and assumes that the flux per pole does not change with time. If it does, there is a superimposed *transformer e.m.f.* giving rise to an electrical/electrical power transfer.

For the case where a.c. voltages are produced at the brushes of a commutator winding, see Chapter 9; eqn. (5.8) gives the r.m.s. diametral voltage when z_s is taken to mean the number of conductors in series between diametral brushes and n the relative field/conductor speed. Assuming a uniformly distributed winding so that $k_d = (\sin \theta/2)/\theta/2$, eqn. (5.8) can be rewritten for $\theta = \pi$ as:

$$\frac{\pi}{2\sqrt{2}} \times 2p\phi nz_s . \frac{\sin \pi/2}{\pi/2} . k_p = \frac{2p\phi nz_s}{\sqrt{2}} . k_p, \qquad (5.9)$$

showing that the r.m.s. value of the diametral voltage, which would appear across the slip rings of Fig. 5.22a, is $1/\sqrt{2}$ times the d.c. voltage produced with quadrature brushes, stationary poles and sinusoidal flux distribution. This d.c. voltage is the peak value of the a.c. voltage occurring with relative pole/brush movement.

For any other separation, $\theta \neq \pi$, Fig. 5.22b, the a.c. voltage is reduced in proportion to the reduced number of conductors in series, but there is an improvement in the distribution factor,

i.e. $E_\theta = 1 \cdot 11 \times (2p\phi n . z_s \theta/\pi) . (\sin \theta/2)/(\theta/2)$, neglecting k_p

$\qquad = 1 \cdot 11 \times (2p\phi nz_s . 2/\pi) \times \sin \theta/2$

$\qquad = $ r.m.s. diametral voltage $\times \sin \theta/2$. $\qquad (5.10)$

This expression could have been derived directly from the geometry of Fig. 5.22b.

Harmonic Voltages

These can be calculated if the field form has been analysed into fundamental and harmonic flux densities, B_1, B_3, B_5, B_7, etc. The frequency of harmonic flux changes is increased 3, 5, 7, etc., times, but the increase of voltage which this would bring about is exactly offset by corresponding reductions of harmonic pole-pitch area and flux components, see Fig. 5.19. Consequently, the harmonic voltages can be obtained by reducing the fundamental voltage in the ratio of the new flux density and winding factors, i.e. for the nth harmonic:

$$E_n \text{ per phase} = E_1 . (B_n/B_1).(k_{dn}/k_{d_1}).(k_{pn}/k_{p_1}).$$

Alternatively: $$E_n = 1 \cdot 11 . (2p\phi_1 n z_s).(B_n/B_1).k_{dn}k_{pn} \qquad (5.11)$$

where ϕ_1 is the fundamental flux per pole.

Neglecting the relative phase angles of the component voltages, the instantaneous phase voltage can now be expressed as:

$$e = \sqrt{2}(E_1 \sin \omega t + E_3 \sin 3\omega t + E_5 \sin 5\omega t \ldots \text{etc.}).$$

The r.m.s. phase voltage would be $\sqrt{(E_1^2 + E_3^2 + E_5^2 + \ldots \text{etc.})}$, The r.m.s. line voltage for star connection would be $\sqrt{3}$ times this but excluding the third harmonic.

EXAMPLE E.5.8

The field form of a 3-phase, 50-Hz, 600-rev/min alternator has a spatial flux-density distribution given by the expression:

$$B = \sin \theta + 0 \cdot 3 \sin 3\theta + 0 \cdot 2 \sin 5\theta \text{ tesla.}$$

The machine has 180 slots, wound with two-layer 3-turn coils. Each coil spans 15 slots, the coils being connected in 60° groups. The armature diameter is 125 cm and the core length is 45 cm.

Calculate:
(a) the expression for the instantaneous e.m.f. per conductor
(b) the expression for the instantaneous e.m.f. per coil
(c) the r.m.s. phase and line voltages.

The voltage per conductor can be worked out directly using Blv or alternatively from the flux per pole using eqn. (5.8).

For this frequency and speed, since $f = pn$, the machine must have 10 poles. So:

$$\text{area of pole pitch} = \frac{125\pi}{10} \times 45 = 1765 \text{ cm}^2.$$

Fundamental flux per pole $= \phi_1 = B_{1av} \times \text{area} = 1 \times (2/\pi) \times 0 \cdot 1765 = 0 \cdot 1122 \text{ Wb.}$

(a) Fundamental voltage per conductor $= 1 \cdot 11 \times 2p\phi_1 n = \text{r.m.s. value}$
$$= 1 \cdot 11 \times 10 \times 0 \cdot 1122 \times 600/60 = 12 \cdot 5 \text{ V.}$$

251

Peak value of fundamental conductor voltage $= \sqrt{2} \times 12\cdot5 = 17\cdot7$ V.
The harmonic conductor voltages are in proportion to their flux densities:
3rd harmonic $= 0\cdot3 \times 17\cdot7 = 5\cdot31$ V.
5th harmonic $= 0\cdot2 \times 17\cdot7 = 3\cdot54$ V,

The instantaneous conductor voltage measuring t from zero voltage is:

$$17\cdot7 \sin \omega t + 5\cdot31 \sin 3 \omega t + 3\cdot54 \sin 5 \omega t \text{ V.}$$

(b) The slotting and coil span are the same as in Example E.5.7 so the same pitch factors apply. Noting that there are 6 conductors in a 3-turn coil:
the fundamental coil voltage $= 6 \times 17\cdot7 \times 0\cdot966 = 102\cdot6$ V,
3rd harmonic coil voltage $= 6 \times 5\cdot31 \times 0\cdot707 = 22\cdot6$ V,
5th harmonic coil voltage $= 6 \times 3\cdot54 \times 0\cdot259 = 5\cdot5$ V,
hence, coil voltage $= 102\cdot6 \sin \omega t + 23 \sin 3\omega t + 5\cdot5 \sin 5\omega t$ V.

Note. The expressions for resultant voltage do not include the relative phase angles. These are not usually of interest, see p. 246.

(c) The distribution factors are calculated from eqn. (5.3) for each harmonic, the slot angle α being $180°/18 = 10°$, and the number of slots per group q being $18/3 = 6$.

$$k_{d1} = 0\cdot958, \quad k_{d3} = 0\cdot645 \quad \text{and} \quad k_{d5} = 0\cdot197.$$

Total number of coils per phase $= 180/3 = 60$
Fundamental phase e.m.f. $= (102\cdot6/\sqrt{2}) \times 60 \times 0\cdot958 = 4170$ V r.m.s.
3rd harmonic phase e.m.f. $= (22\cdot6/\sqrt{2}) \times 60 \times 0\cdot645 = 627$ V r.m.s.
5th harmonic phase e.m.f. $= (5\cdot5/\sqrt{2}) \times 60 \times 0\cdot197 = 46$ V r.m.s.
These answers should be checked using eqn. (5.11).
R.m.s. phase voltage $= \sqrt{(4170^2 + 627^2 + 46^2)} = 4220$ V
R.m.s. line voltage $= \sqrt{3} \times \sqrt{(4170^2 + 46^2)} = 7220$ V

Example E.5.9

A 4-pole commutator machine has 124 lap coils each having two turns. The flux per pole is $0\cdot015$ weber.

Calculate, assuming a sinusoidal flux distribution:

(a) the d.c. voltage appearing across quadrature brushes when running at 1500 rev/min in a steady field;

(b) the r.m.s. voltage, with three sets of brushes per pole pair for 3-phase working and with relative field/conductor speed $= 1400$ rev/min.

(a) No. of conductors in series $= (124 \times 2 \times 2)/4 = 124 = z_s$
d.c. voltage $= 2p\phi n z_s = 4 \times 0\cdot015 \times (1500/60) \times 124 = 186$ V.

(b) Brush separation $\theta = 360°/3 = 120°$ electrical.
Using eqn. (5.10), r.m.s. diametral voltage $= 1\cdot11 \times 186 \times 1400/1500 \times 2/\pi = 123$ V.
Between brushes spaced at $120°$ electrical

$$\text{a.c. voltage} = 123 \sin 120°/2 = 106\cdot5 \text{ V r.m.s.}$$

5.6 M.M.F. PRODUCED BY AN ARMATURE WINDING

M.M.F. of a Distributed Winding

Any 2-pole armature with full-pitch coils, whether for a d.c. machine, a single-phase or a polyphase machine, has a current distribution such that one half of the armature carries current in the opposite direction from the other half; Fig. 5.23. The conductor currents on the

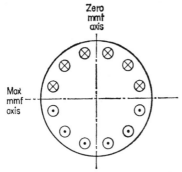

FIG. 5.23. Two-pole armature m.m.f.

polyphase machine are not all of the same magnitude at the same instants of time, but nevertheless there are only these two distinct zones. It is not difficult to imagine the armature as a solenoid with an axis along the dividing line. This axis is that of the peak m.m.f., and for equal conductor currents, the value is given by:

current per conductor × number of conductors/2;

since each turn is formed by two conductors. This is the magnetic potential difference absorbed in the magnetic circuit, analogous to the electrical potential difference absorbed in an electrical circuit. Since the normal machine is symmetrical, and one half of this m.m.f. is absorbed on either side of the armature, it is convenient to take the reference potential midway between this magnetic potential difference. Hence, since the total m.m.f. refers to two poles, the peak magnetic potential is taken as $\pm$(ampere-turns per pole); one half of the figure calculated as above. Progressing round the periphery, the m.m.f. will fall from the positive maximum say, until for an axis at 90° it is zero, since this axis is embraced by a zero resultant value of ampere-turns.

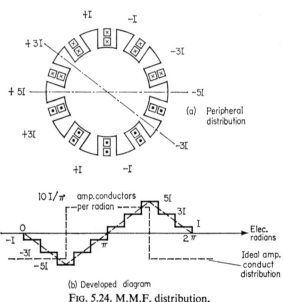

(a) Peripheral distribution

(b) Developed diagram

Fig. 5.24. M.M.F. distribution.

As a simple illustration consider Fig. 5.24a which represents diagrammatically an armature with a two-layer winding like a d.c. machine, having 20 regularly spaced conductors each carrying the same current I. On the peak m.m.f. axis, the total ampere-turns are $I.20/2 = 10I$, the m.m.f. per pole being $\pm 5I$; $+5I$ producing a north pole say, and $-5I$ producing the south pole.

Now consider another axis, displaced as shown by a slot pitch. By summing the ampere-turns as before about this new axis, the m.m.f. is: $(+16I-4I)/2 = 6I$ or $\pm 3I$ for diametral points on this axis.

Similarly for other axes, and each time the chosen axis is rotated past a slot, the m.m.f. falls by $2I$ ampere-turns per air gap. The space distribution of m.m.f. round the whole periphery is plotted on a developed diagram; Fig. 5.24b. The sharp transitions neglect the finite widths of the conductors. For a four-pole armature, the pattern would have been repeated. In general, the pattern repeats itself every pole pair.

It should be obvious that, if a larger number of slots per pole had been used, the wave would have approached a triangular shape. For a perfect triangle, the ampere conductors would have had to be distributed with a constant value per unit length of periphery, reversing sharply from positive to negative as indicated. Such an idealisation is called a uniformly distributed current sheet.

If two diametrically opposite sections of the winding were to be omitted, the m.m.f. wave would be trapezoidal, assuming uniform current distribution, Fig. 5.25. When the armature conductors are connected in phase groups, each phase develops a distribution of this kind, but with mutual, spatial phase-displacement.

It should be noted that, since the m.m.f. distribution due to a single

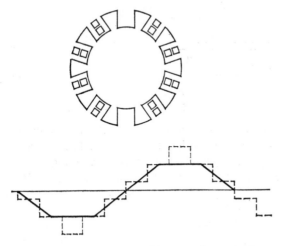

FIG. 5.25. M.M.F. wave with one section omitted.

255

full-pitch coil is a symmetrical rectangular wave, Fig. 5.24b could be obtained by summing five such waves, successively displaced in space-phase round the periphery by a slot pitch. Each of these waves could be resolved into a fundamental of full pole pitch, and a series of odd-order space-harmonics. The fundamentals, and harmonics separately could be summed vectorially in the same way as e.m.f. waves displaced in time-phase. The effects of the distribution and chording factors on the resultant m.m.f. wave would be the same, the fundamental being reduced slightly but the harmonic content being reduced considerably. This viewpoint should be clarified by studying Fig. 5.26, which shows

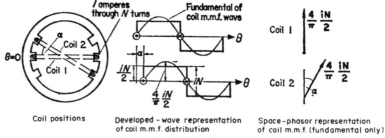

| Coil positions | Developed – wave representation of coil m.m.f. distribution | Space-phasor representation of coil m.m.f. (fundamental only) |

FIG. 5.26. M.M.F. of single coils.

the m.m.f. waves for two adjacent coils displaced by a space angle α. The rectangular wave is really an idealisation because the m.m.f. changes gradually over the width of the conductor, not sharply as shown.

Resultant M.M.F. of a Three-phase Winding

Figure 5.27a indicates the current distribution resulting from a three-phase single-layer winding when the current in phase A is maximum and the currents in phases B and C are at half their negative maxima. Two poles only are shown, but the pattern could be repeated for a larger number of poles. Each phase m.m.f. is drawn separately and, assuming uniform distribution, this results in three mutually displaced trapezia, having different magnitudes in general because

256

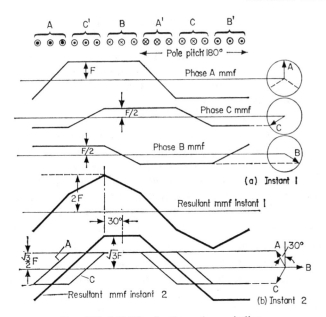

Fɪɢ. 5.27. M.M.F. of a three-phase winding.

of the mutual time-phase displacement of the currents. With r.m.s. current I, the peak value of each phase m.m.f. pulsates between $\pm$peak m.m.f. per pole per phase, i.e.

$$F = \pm \frac{\sqrt{2}.I.z_s/2}{2p}.$$

At instant 1, phase A has this value, and phases B and C have half this value, in the opposite sense to phase A. By adding the three m.m.f. space waves the resultant is seen to be nearer to a sinusoid. The 3rd order space harmonics have cancelled, a fact which can easily be checked by sketching in these components.

The condition at a second instant of time, $\omega t = 30°$ later, is shown in Fig. 5.27b. Phase B is zero and phases A and C have peak m.m.f.s equal to $\pm(\sqrt{3}/2) F$. The phase m.m.f. waves are in the same positions

257

but with different amplitudes. The resultant is now trapezoidal and of reduced magnitude.

If other instants of time where chosen, it would be found that the resultant m.m.f. varied between these two limiting shapes, repeating every 60°. More interesting is the fact that the axis of the resultant m.m.f. moves exactly 30° when the time phasors move 30° (electrical). The m.m.f. wave moves synchronously with the time phasors, passing one pole pair of the winding in the direction A, B, C in one complete cycle. For a winding with p pole pairs, one complete revolution will take p cycles. Therefore, for f cycles/sec, there will be $f/p = n_s$ revolutions per second. n_s is called the synchronous speed, and in rev/min is $N_s = 60f/p$. This rotating field is the basis for induction and synchronous machine action as already described briefly in Chapter 1 and to be described in detail in Chapters 7 and 8 respectively.

Fourier analysis of the resultant m.m.f. waves shows that only the harmonics of 5th, 7th, 11th, 13th order, etc., are present, their amplitudes decreasing inversely as the square of the order. The first three terms of the complete analytical expression are:

$$\frac{6}{\pi}.F\left[\sin(\theta-\omega t).\frac{\sin \pi/6}{\pi/6}+\frac{\sin(5\theta+\omega t)}{5}.\frac{\sin 5\pi/6}{5\pi/6}\right.$$
$$\left.-\frac{\sin(7\theta-\omega t)}{7}.\frac{\sin 7\pi/6}{7\pi/6}\cdots\cdots\right]$$

An examination of the physical interpretation of the above expression shows:

(a) The common term $6F/\pi$ is 1·5 times the fundamental component of a square wave having a value equal to $\pm$m.m.f. per phase F (see Fig. 5.26). Such a square wave would be produced if all the phase ampere conductors were concentrated in one narrow slot per pole. In practice, the conductors are distributed over the phase spread, $2\pi/6$ in this case.

(b) Since a uniform distribution has been assumed, the general nth harmonic term $(\sin n\pi/6)/(n\pi/6)$, is the appropriate harmonic distribution factor modifying each of the component terms inside the brackets.

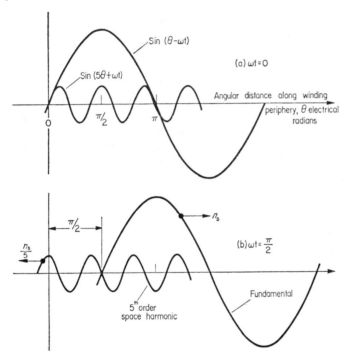

FIG. 5.28. Travelling M.M.F. wave components.

(c) The other component of the general nth harmonic term $[\sin(n\theta \pm \omega t)]/n$, is a travelling wave, having, at any angular position θ along the winding, a magnitude which varies sinusoidally at supply frequency. The effect of this variation over the whole winding periphery is to cause the m.m.f. distribution, and the flux distribution, to travel at a speed equal to the speed of the fundamental wave divided by the order of the harmonic. The direction of movement depends on the sign inside the brackets. The fundamental and the space harmonics of order 7th, 13th etc., move in the same direction, corresponding to phase sequence A B C. The 5th, 11th order, etc., with the positive sign, move in the

259

opposite direction as illustrated by Fig. 5.28, showing the fundamental and 5th harmonic space waves at two different instants.
This phenonema can be confirmed by applying the conclusions
derived from Fig. 3.14, to the m.m.f. space harmonics.
The changing shape of the resultant m.m.f. wave on Fig. 5.27, can be
accounted for by the different movement of the two groups of harmonics.
The rotational speeds of the space harmonics decrease inversely as
their order so that any e.m.f.s induced by space-harmonic flux changes
are of fundamental frequency. This fact is not surprising when it is
remembered that the fluxes are caused by currents of fundamental
frequency.

Considering only the fundamental component, this has a magnitude of
$(6F/\pi).(\sin \pi/6)/(\pi/6) = 18F/\pi^2 = 1\cdot822\ F$ which is a little less than the
average peak of the two limiting shapes of Fig. 5.27, i.e.

$$(2F+\sqrt{3}F)/2 = 1\cdot866\ F.$$

For the general case, taking into account both the distribution and
coil-pitch factors k_{dn} and k_{pn}, and substituting for F; the value of the
nth harmonic component of m.m.f. is:

$$\frac{6}{\pi} \times \frac{\sqrt{2}.Iz_s/2}{2p} \times k_d k_{pn} \times \frac{1}{n} \text{ which becomes: } 1\cdot35\frac{Iz_s}{2p}.k_{d_1}k_{p_1}$$

for the fundamental, where $Iz_s/2p$ is equal to the r.m.s. ampere conductors per phase per pole. The same fundamental of m.m.f. could be
produced by any other single-phase or polyphase winding by suitable
adjustment to the current, the turns per phase, or to both of these
quantities. On Fig. 5.27, the distribution factor k_d is virtually included
in the m.m.f. diagrams on the assumption that there is uniform distribution. For a 2-layer chorded winding, each layer would produce an
identical m.m.f. wave but because of the displaced current distribution,
the two waves would have to be drawn with a mutual phase-displacement equal to the chording angle β. k_p would then be included automatically when the two waves were added to get the total winding
m.m.f.

Time-phasor Diagram for Rotating Field of a Three-phase Winding

Consider the resultant m.m.f. space wave of Fig. 5.27a, and relate it to the position of the phase windings. It can be seen that the fundamental wave of armature flux which this m.m.f. would produce and which could be represented by a space phasor of constant magnitude Φ'_a, would have maximum linkage with the phase carrying maximum current, phase A in this case. If the diagram had been drawn for the instant when phase B was carrying maximum current and phases A and C were at half their negative maxima, the armature flux linkage would then have been maximum with phase B; similarly if phase C had been carrying maximum current. The flux is experienced by each phase as a time variation and from this point of view it must be represented by a time phasor. It will then be distinguished from the flux Φ'_a by omitting the prime, i.e. Φ_a, a distinction which has already been discussed in connection with Fig. 3.2. On the time-phasor diagram therefore, Φ_a and the current phasor I will be in alignment, since the resultant armature flux has maximum linkage with a particular phase at the same time as that phase carries maximum current. The two quantities also vary at the same frequency. Whether the two time phasors must be drawn in phase or in antiphase depends on whether the machine is motoring or generating and this distinction will be explained better in the appropriate chapters dealing with particular machines.

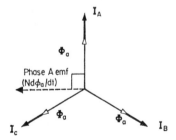

FIG. 5.29. Relationship of armature-flux time phasor to current and e.m.f. time phasors.

261

On Fig. 5.29, three flux time-phasors are shown, each being in time-phase with the individual phase currents which in fact corresponds to the motoring condition. The time phasors Φ_a, refer to the same resultant armature flux ϕ_a', due to all phases, but linking each phase in sequence, maximum linkage occurring at the same time as maximum current. The induced e.m.f. due to ϕ_a for any particular phase will be $Nd\phi_a/dt$ and the e.m.f. time-phasor will lead the flux time-phasor by 90°. The coils are effectively inductors, and if ϕ_a is taken to mean the flux crossing the air-gap, then the corresponding reactance, equal to the induced e.m.f. divided by the phase current, is the magnetising reactance of the equivalent circuit; see Section 3.9.

For phasor diagrams on a balanced three-phase machine, only one phase need be considered, but the armature flux-linkage time-phasor will be due to the resultant m.m.f. of all phases, part of it being due to self flux and the remainder due to the mutual flux produced by the other phases.

Production of Sinusoidally Distributed M.M.F. with D.C. Excitation

Figures 5.27a and 5.27b are drawn for particular values of currents in the three phases. These currents occur at particular instants in the a.c. cycle but could be supplied as constant values from a d.c. source using the circuit arrangements of Fig. 5.30. When I_{dc} is equal to $\sqrt{2}I_{ac}$, the arrangement of Fig. 5.30a will give the same m.m.f. wave as Fig. 5.27a. There is full forward current in one phase and half reverse current in the other two phases. When $I_{dc} = (\sqrt{3}/2)(\sqrt{2}I_{ac}) = 1\cdot225I_{ac}$,

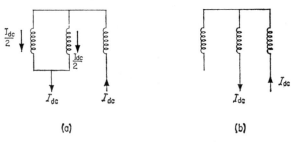

FIG. 5.30. D.C. excitation of star-connected three-phase winding: (a) three-lead connection; (b) two-lead connection.

the arrangement of Fig. 5.30b will give the same m.m.f. wave as Fig. 5.27b since one phase is carrying a forward current, a second phase carries the same current in the reverse sense, and the third phase has zero current. The d.c. current is slightly less with this arrangement but the excitation power is the same. There are corresponding arrangements for the delta connected winding, see Example E.8.8.

Thus a suitably distributed and chorded polyphase-type winding, excited from d.c., will give a very good approximation to a sine wave of m.m.f. and of flux, if the permeance is uniform in all radial directions; i.e. a cylindrical rotor running in a uniform air gap. To generate a very good sine wave voltage on a synchronous machine, such an excitation system is used, and the theory built up to explain the performance is called *cylindrical-rotor theory*. Much simplification results, because the m.m.f.s of both stator and rotor can then be combined vectorially as in Fig. 1.9, their resultant determining the gap flux ϕ_m.

Note 1. With d.c. excitation, the frequency f is zero and the m.m.f. is stationary with respect to the excitation winding.

Note 2. The ideal sinusoidal m.m.f. distribution results from a sinusoidally distributed current, i.e. a sinusoidal current sheet. The m.m.f. wave is the integral of the ampere-conductor wave as can be confirmed from Fig. 5.24b. For this figure, the ideal triangular m.m.f. wave is the integral of the idealised rectangular ampere-conductor wave.

Note 3. By switching a d.c. supply sequentially to the three phase windings, the axis of the d.c. m.m.f. can be made to rotate in angular steps at a speed corresponding to the switching frequency. With a suitable toothed rotor to produce reluctance torque, such a field could be made to drive the rotor in step with the field and permit the location of any desired angular position. Such an arrangement would be an elementary form of *Stepper Motor* (see p. 547) and in fact the system is used on *Static Variable-frequency D.C. Link Inverters* to provide a variable-speed high-power drive, see p. 430.

Resolution of Pulsating M.M.F. into Two Rotating M.M.F.s

A single-phase winding can only produce a pulsating field, varying between $\pm F$. If two sinusoidally distributed m.m.f.s each of constant magnitude $F/2$ are rotating in opposite directions, their combined effect, Fig. 5.31a, is equivalent to one pulsating field, $F \cos \omega t$, varying

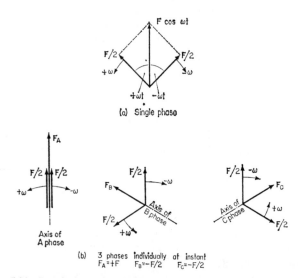

(a) Single phase

(b) 3 phases individually at instant 1
$F_A = +F$ $F_B = -F/2$ $F_C = -F/2$

FIG. 5.31. Resolution of pulsating m.m.f. into rotating components: (a) single phase; (b) three phases individually at instant 1; instantaneous m.m.f.s. $F_A = +F$, $F_B = -F/2$, $F_C = -F/2$.

between $\pm F$. Conversely then, a pulsating field can be resolved into two rotating fields of half amplitude rotating in opposite directions. This resolution is the basis of one approach to single-phase induction motor theory and other single-phase machines problems. The method can be applied to each phase of a three-phase winding to show that one set of components is in space phase combining to give a constant resultant of amplitude $3F/2$. The other set, the anti-clockwise, negative-sequence components on Fig. 5.31b, sum to zero. This explains why the resultant fundamental rotating field of a three-phase winding has an amplitude of $6F/\pi$, since the fundamental of each phase is $4F/\pi$.

Finally it should be noted that any polyphase winding with a corresponding polyphase supply will produce a rotating field. For example, a

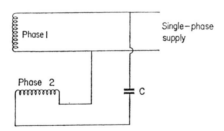

FIG. 5.32. Phase-splitting circuit.

two-phase winding is used frequently for both starting and running induction-type motors from a single-phase supply. A phase-splitting circuit is necessary to give the time-phase displacement between the currents in the two windings. Figure 5.32 shows one way in which this is achieved. By suitable choice of capacitor, the current in the associated winding leads that in the other winding by 90°.

For a 2-phase supply, it is quite easy to develop an expression for the travelling wave of m.m.f. in the same form as on page 258. Referring to Fig. 5.33 and for sinusoidally distributed m.m.f.s of peak value F per pole, the instantaneous value of resultant m.m.f. at any electrical space-angle θ is:

$$F_\theta = F.\sin \omega t.\cos \theta - F.\cos \omega t.\sin \theta = F.\sin(\omega t - \theta).$$

The expression shows that a constant-value point on the m.m.f. wave $((\omega t - \theta)$ constant), must move at synchronous speed.

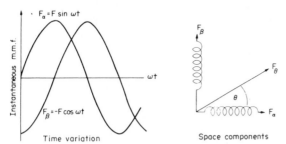

FIG. 5.33. Travelling m.m.f. wave for 2-phase winding.

CHAPTER 6

DIRECT-CURRENT COMMUTATOR MACHINES

6.1 SCHEMATIC REPRESENTATION

The first sources of electrical power were primitive batteries giving direct current so the d.c. machine was the first to be developed. Earlier chapters, 1, 2 and 3, have discussed the basic principles, electromagnetic action, the electrical circuit equation and the equivalent circuit for the d.c. machine. In Chapter 5, the operation of the commutator was explained in general terms. Figures 5.8 and 5.10 are fairly close representations of the machine showing the correct relationship between the positions of brushes, poles, commutator and coils. Other diagrams, Figs. 1.5 and 5.22, are schematic and therefore deficient in details though adequate to explain the general electromagnetic behaviour. Figure 6.1 is another representation which is convenient for many purposes. Only a few conductors in a single layer are shown, and whilst the brushes correctly indicate contact with those conductors passing through the interpolar axis where the main-pole flux density is zero, the actual brush position is normally midway under the pole due to the symmetrical bends at the ends of the armature coil, see Fig. 6.4. However, Fig. 6.1 demonstrates the important distinguishing feature of the d.c. machine, i.e. current into a brush divides in two paths, each of which maintains a fixed spatial relationship to the main poles. This in turn gives rise to an armature m.m.f. F_a, which, irrespective of any armature movement, is directed along the brush axis at a constant torque angle δ, to the main-pole axis. The relationship between the directions of F_a and the current flow through the armature from brush to brush, depends on whether the winding is progressive or retrogres-

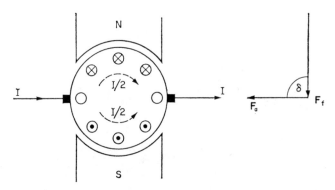

FIG. 6.1. D.C. machine schematic diagram.

sive; see Fig. 5.6. Since, neglecting the side effects of commutation, see Section 6.3, δ is constant—usually 90°—the torque is proportional only to the armature current and the field-produced flux. The machine is thus free to run at any safe speed without disturbing the electromagnetic action which produces the electromagnetic torque T_e. The value of the speed will be determined by the point at which T_e is balanced by the total mechanical torque T_m, see Fig. 2.16. These features result in simple speed and torque control, which, together with a general flexibility, make the d.c. machine ideally suitable for the large number of industrial drives which require a stepless variation of speed, with or without automatic control.

An alternative viewpoint of the machine operation considers the inherent a.c./d.c. or d.c./a.c. conversion. The commutator, in directing the flow of current in fixed paths, permits an external d.c. current even though the coil currents are alternating. The commutator performs a mechanical rectification of power from a.c. to d.c. or vice versa, quite apart from the usual mechanical/electrical power conversion process of the machine as a whole. For a.c. commutator machines, Chapter 9, this frequency conversion function is dominant.

For multipolar machines, the pattern of two poles is repeated, but the extra complication of a schematic diagram like Fig. 6.2 serves

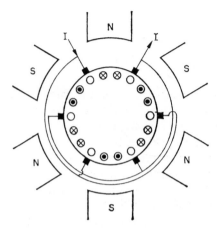

FIG. 6.2. Schematic diagram for six-pole
machine; p = three pole pairs.

little purpose except, for example, to demonstrate that any particular
conductor experiences p complete cycles of flux alternation every
revolution. Therefore the frequency of conductor currents and of
armature flux changes is given by the usual expression $f = pn$. There is
some freedom in choosing the number of poles for a d.c. machine,
the decision being partly economic though also influenced by design
limitations such as frequency and current per brush arm. The effect
of different pole numbers on the number of parallel circuits has already
been discussed in Section 5.2 with regard to the choice between lap and
wave windings.

Other schematic diagrams used in this chapter may omit the armature
conductors and poles for simplicity, but if the relationships between
main-pole flux, conductor current and motion are shown, it will be
presumed that the reader is familiar with the electromagnetic pictures
of Fig. 2.15. In particular, Fig. 2.15a shows the generating condition
and Fig. 2.15b shows the motoring condition. The student should
familiarise himself with these diagrams by redrawing them with
reversed directions of flux, current and motion, taken two at a time to
retain the same function.

268

6.2 COMMUTATION

The Commutation Process

Before discussing construction and characteristics it will be necessary to explain the nature of the main design problem associated with d.c. machines, because of its influence on the appearance and behaviour. Although the brushes perform the function of directing current into and out of the armature through a sliding contact this is by no means the end of the story. When an armature coil passes through the interpolar zone, the direction of its current must change. Whilst this is also true at unity power-factor for the coils of an a.c. machine open winding, there are two differences in the d.c. machine. Firstly, the coil current is constant while it passes under the main pole from one interpolar zone to the next, whereas in the a.c. open winding, the current rises gradually from zero to a maximum and down to zero again. The d.c. coil current has to change rapidly from full positive value $+I$, to full reversed value $-I$ in a time t_c, usually less than 1 millisecond (see oscillogram of Fig. 6.3), but dependent on the rotational speed.

The second and more important difference can be seen from Fig. 6.4. When the coil reaches the interpolar zone, it is short-circuited by the brush. Although the dynamically induced voltage is low here, the current starts at full value, and the sharp change gives a high value of self-induced voltage $L \, di/dt$, tending to retain the original current direction and prevent completion of the change. However, when the coil is eventually subjected to the influence of the next pole and circuit, its current will have to complete the transition to $-I$ suddenly, as a result of the relatively high forcing voltages which maintain the current

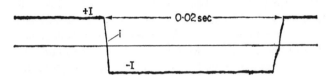

FIG. 6.3. Oscillogram of d.c. machine armature-coil current.

269

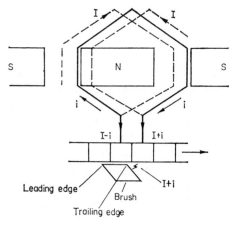

FIG. 6.4. Incomplete commutation.

in the new circuit. Referring to Fig. 6.4, the current i, in the coil which has just cleared the short circuit, will almost certainly have reversed from the positive direction shown, but has not reached $-I$, so that $I+i$ is not zero and it will be forced to flow to the brush through a path in air, constituting a spark. This is rather a simplified picture, but it is true to say that sparking is liable to occur when the reactance voltage, see below, reaches a value of a few volts. This process of current transition is called commutation and the designers' object is to achieve it without sparking, so that excessive wear and burning of the commutator and brushes may be avoided.

Reactance Voltage of Commutation

Figure 6.5 is a simplified picture of a coil in a slot just before commutation commences. The current I establishes a leakage flux ϕ_l linking the armature coils only. The other flux component, which links the field winding, can be shown[10] to be virtually constant during commutation (due to the action of the other, moving, non-commutating coils), so that it is not responsible for any inductive voltage opposing

270

current change. Some short time t_c later, when the coil current and leakage flux have reversed, ϕ_l will be linking the coil in the opposite sense. There will have been a change of $2\phi_l$ in a time t_c and therefore an induced e.m.f. of average value $N \cdot 2\phi_l/t_c$, or instantaneously, $N \, d\phi/dt$. This voltage, which opposes the change of current, can be written as $l \, di/dt$ where $l = N\phi_l/I$ is the leakage inductance. It will be noticed that there are two coil sides in each slot belonging to different coils

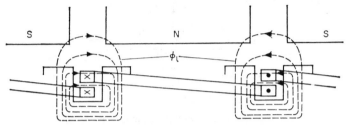

FIG. 6.5. Coil leakage flux before commutation.

though magnetising in the same sense. If these coils start commutation within the time t_c there will be mutual e.m.f.s ($M \, di/dt$), in the coils. For the simple case where the coils are full pitch, and only one coil is short-circuited at a time by the brush, the currents in top and bottom layers change simultaneously and the effective inductance of one coil $L' = l + M$, can be based on the total leakage flux produced by top and bottom layers; see Fig. 2.18 and the associated text. The induced voltage due to current change in a coil undergoing commutation is called the reactance voltage e_r, and in the general case can be expressed as:

$$e_r = l_1 \, di_1/dt + \sum_1^n M_{1n} \, di_n/dt.$$

Here, the second term takes into account all of the n coils, which commutate within the time t_c and will, in general, be out of phase; i.e. at an earlier or later stage in their current reversal than the coil under consideration. Usually the brush covers a sufficient number of commutator bars to cause coils in neighbouring slots to be commutating within this time.

Another way of considering commutation is from the energy viewpoint. The leakage field of I represents stored magnetic energy, which, within a time t_c, has to be dissipated and re-established with current flow in the opposite direction. Some of this energy may be liberated in sparks at the brushes, or may be converted harmlessly to heat or transferred electromagnetically to coupled coils.[9]

Compensation of Reactance Voltage

Since e_r opposes current change, it is important, on all but the smallest machines, to introduce a compensating voltage e_c in the coil circuit which will tend to cancel e_r. This can be achieved in two ways.

The first method is to move the brushes so that the coils commutate under the influence of a main flux having a suitable polarity. Consider a generator, Fig. 6.6. The motionally induced voltage will drive current in the direction shown, and as a coil enters the commutating zone, e_r will tend to maintain the current. By setting the brushes forward in the direction of rotation, current reversal is delayed until the coil is under the fringe field of the next main pole in sequence. This generates a voltage of reverse sense and so will oppose e_r. Such an arrangement is satisfactory for small machines but unfortunately, since the compensating voltage is determined primarily by the field excitation, it does not have the necessary functional relationship with the armature current.

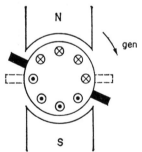

Fig. 6.6. Compensation
of reactance voltage by
brush shifting.

The average reactance voltage is directly proportional to the value of I before commutation.

The second method makes use of auxiliary poles excited from the load current and mounted in electrical space quadrature with the main poles. They are called interpoles, or more appropriately commutating poles or "compoles". Neglecting saturation, the compole flux, the voltage thereby induced, e_c, and the reactance voltage e_r, all vary in direct proportion to load current, so the compensation is automatically adjusted with load changes. The polarity of the compoles is determined

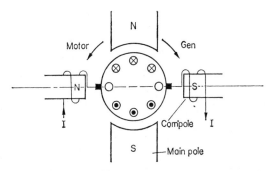

FIG. 6.7. Compensation of reactance voltage by interpoles.

from the same considerations as for the brush shift, i.e. for a generator, the polarity must be the same as the next main pole in the direction of rotation. For a motor, with the same current and main-field polarities, Fig. 6.7, the rotation would be reversed, the induced e.m.f. being in the opposite sense to the current. The north compole over the left-hand brush will, with counter-clockwise rotation, induce an e.m.f. e_c tending to drive current out of the paper, which is the sense required for the approaching bottom-half circuit. Thus the compole polarities are still correct even though the machine function is different. This is most fortunate because frequent changeovers from motoring to generating and vice versa are a common operational requirement. The student should also convince himself that a reversal of rotation for either motoring or generating functions separately, gives rise automatically to correct interpole polarity.

A further useful exercise is to check that brush shift has the correct effect only for one particular rotation and function. Only if the function is changed with the rotation will satisfactory operation be ensured. Sometimes, even though interpoles are fitted, small changes of brush position away from the quadrature axis are made, for commutation reasons or to exaggerate armature reaction effects, see Section 6.3.

Effect of Brush-contact-drops on the Nature of Coil-current Reversal

Figure 6.8 shows the local coil/brush circuit at the beginning of commutation for the simple case where the brush only short-circuits one pair of commutator bars at a time. Taking the original direction of i as positive, the circuit voltages are as shown. If i is decreasing, di/dt is negative so that $L'\,di/dt$ will therefore act in the same direction as i, tending to maintain the current direction. The compensating e.m.f. e_c opposes this reactance voltage tending to assist the current reversal. There are voltages at the brush contact, v_1 and v_2, opposing currents

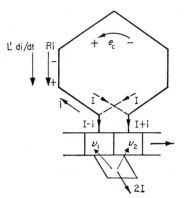

Fig. 6.8. Coil circuit during commutation.

$I-i$ and $I+i$ respectively. The local coil-circuit equation is written down readily by observing the arrow directions:

$$e_c + L'\, \mathrm{d}i/\mathrm{d}t + Ri - v_1 + v_2 = 0 \quad \text{or} \quad e_c + e_r = -Ri + (v_1 - v_2).$$

Except for small machines, the Ri voltage is negligible by comparison with the other components and v_1 will be equal to v_2 if the current density at the respective contacts is the same. In this case, e_c would have to be exactly equal to $-e_r = -L'\, \mathrm{d}i/\mathrm{d}t$, i.e. opposing the inductive voltage at every instant. If there were to be a discrepancy in these two voltages, then $v_1 - v_2$ would have to have a value to balance the equation. It will be shown, that the action of the brush drops is to reduce any discrepancy between e_c and e_r.

Since with constant commutator speed, the contact area carrying $I-i$ increases at a uniform rate and the area carrying $I+i$ decreases at a uniform rate, i itself must decrease uniformly with time if the contact current density is to be uniform; i.e. contact areas and currents carried, changing at the same rate. This condition is known as linear commutation, Fig. 6.9. The contact drops cancel out round the circuit and play no part in the current reversal. The compensating voltage cancels the reactance voltage.

Now suppose that e_c is not large enough to cancel e_r so that the current is not forced to reverse so quickly (late commutation), $I-i$ is then less than before so v_1 would fall. $I+i$ is greater than before so v_2 would rise. Consequently $v_2 - v_1$ is a voltage *rise* in the counter-clockwise direction, opposing the current flow and speeding up the reversal towards a linear commutating condition. Conversely, if e_c is too strong and the current reverses quickly (early commutation), the above arguments can be applied in reverse to show that the brush contact drops tend to slow down the current change. Summarising then, if the current departs from a linear transition there is an automatic tendency, by virtue of the brush contact voltages, to restore linearity. This brush action is sufficient on small machines to ensure sparkless commutation even without interpoles, and is referred to as *resistance commutation*. If compoles are used to produce a compensating voltage the term *voltage commutation* is used. However, the brush

275

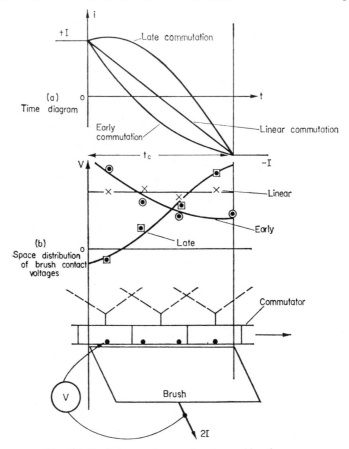

FIG. 6.9. Typical current-reversal curves and brush curves.

action is still present and is capable of developing a correcting voltage of about 2 volts without the contact densities rising so high as to cause sparking. This voltage will, as it were, absorb a corresponding discrepancy between e_c and e_r.

This last point is demonstrated by Fig. 6.10. On the assumption of linear commutation, $l\, di/dt$ and $M\, di/dt$ as functions of time can be

276

represented as rectangles since di/dt is constant. Considering the several coil sides in the top layer of a slot, commutating in sequence, and those of the bottom layer having an additional time displacement due to the chording angle β, the waveform of the reactance voltage would be of stepped shape. Figure 6.10 is a typical pattern obtained by adding a series of phase displaced rectangles due to self and mutually induced

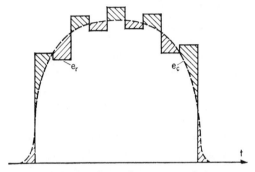

Fig. 6.10. Waveform of reactance and compensating voltages.

voltages. A practical compole field cannot produce an identical variation of flux density to generate an exact compensating voltage but could be designed to produce the smooth mean curve shown. The shaded areas represent by their ordinates the discrepancies between e_r and e_c, which must not be too large. Note that linear commutation is a simplification and Fig. 6.3 shows that in practice, there may be considerable departure from this form of current/time curve.

Brush Curves

From what has just been written, it can be seen, by reference to Fig. 6.9b, that a low-reading voltmeter measuring the contact drop at various points under the brush arc, using a carbon point touching the moving commutator surface, will indicate the nature of current reversal and show whether the compole adjustment is optimum, too weak or too strong. If optimum, the brush contact density will tend to fall a little

277

towards the brush edges, giving an average voltage drop of about 1 V per brush, for modern electrographitic brushes. If weak, the leading edge of the brush (carrying $I - i$) will have a lower voltage than the trailing edge, and may even be reversed as shown. The converse applies if the compoles are strong.

Total Ampere-turns Required on the Interpole

From Fig. 6.7 it can be seen that the armature m.m.f. is in direct opposition to the required polarity of the interpole. This is not surprising because it is the armature flux which is responsible for the reactance voltage in the first place. Having calculated the ampere-turns F_c required to magnetise the compole air gap to generate e_c, it is necessary in addition to provide ampere-turns equal to the peak armature m.m.f. per pole; viz: $F_a = (I_c Z/2)/2p$ (Section 5.6), where I_c is the armature current per conductor. This cancels the mutual field and the effects of the leakage field are cancelled, or nearly so by F_c which is usually about 20–30% of F_a.

EXAMPLE E.6.1

On a certain 8-pole d.c. machine rated at 1200-kW, 600-V, 500-rev/min, the average value of the reactance voltage per coil is 4·4 V. The machine has a single-turn lap winding with 624 conductors. The interpole length is 0·3 m, and its air gap is 8·5 mm long. The armature diameter is 1·3 m. The ratio of B_{max} to B_{av} in the commutating zone is 1·4.

Find the total number of turns required per pole on the quadrature axis, allowing 10% extra magnetising ampere-turns for the iron and slots.

Armature current per conductor $I_c = I_a/2p = \dfrac{1200 \times 10^3}{600} \times \dfrac{1}{8} = 250$ A,

Armature m.m.f./pole $= \dfrac{I_c}{2p} \cdot \dfrac{Z}{2} = \dfrac{250}{8} \times \dfrac{624}{2} = 9750$ At.

$e_r = 2Blv$

$4·4 = 2B \times 0·3 \times \pi \times 1·3 \times \dfrac{500}{60}$ from which $B = 0·216$ tesla.

This is the average value of the compole air-gap flux density. The maximum value is $1·4 \times 0·216 \doteqdot 0·302$ tesla and the magnetising m.m.f. allowing an extra 10% is:

$$H_g l_g + 10\% = \dfrac{0·302}{4\pi/10^7} \times \dfrac{8·5}{10^3} \times 1·1 = 2250 \text{ At.}$$

The total m.m.f. required on the quadrature axis is $9750 + 2250 = 12,000$ At. With full armature current through each turn this would require 6 turns. For a machine without a compensating winding, see Fig. 6.14, these turns would all be wound on the interpole. Otherwise there would usually be three on each interpole and three turns per pole on the compensating winding. If the number of turns had not been an integer the winding could have been split into parallel circuits and small adjustments made to the interpole air-gap length.

6.3 ARMATURE REACTION

Components of Armature M.M.F.

Armature reaction is the term used to describe the effects of the armature m.m.f. on the air-gap field. Neglecting commutation effects for the moment, the axis of the m.m.f. is the same as the brush axis and on Fig. 6.1 this is shown in quadrature with the main poles on what is sometimes called the geometric neutral axis, or, more generally, the quadrature axis. If the brushes are shifted clockwise by an electrical angle α, Fig. 6.11, it can be seen that the armature m.m.f. $F_a = I_c Z/4p$ can be resolved into two components as indicated, one along the quadrature axis, $(1 - 2\alpha/\pi)F_a$, and the other along the direct axis, $(2\alpha/\pi)F_a$. Because the m.m.f. is not sinusoidally distributed, these components are only approximately equal to $F_a \cos \alpha$ and $F_a \sin \alpha$ respectively. The quadrature component is little different in practice from F_a and is cross magnetising. The direct-axis component on the other hand

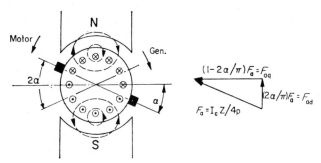

FIG. 6.11. Direct-axis (F_{ad}) and quadrature-axis (F_{aq}) components of armature m.m.f. (F_a).

279

is directly demagnetising. From this it can be deduced that forward lead, i.e. in the direction of rotation, causes direct weakening of a generator field, and conversely, backward lead strengthens the field when the armature carries current. For a motor, the opposite is true.

Although substantial brush shifts, of say 10–20°, are only made on small machines, minor adjustments are often made on large machines to trim the load performance. Even with the brushes in neutral there may be a shift of the m.m.f. axis if commutation is either early or late, in which case the current reversal does not occur at the centre of the brush arc. Consider the generator, Fig. 6.12. If the interpoles are too

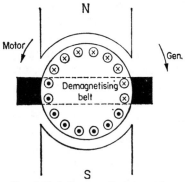

Fig. 6.12. Influence of commutation
on armature reaction.

weak, commutation is delayed so that reversal takes place nearer the trailing edge of the brush. There results a direct demagnetising belt of ampere-turns as for the forward brush lead. Figure 6.12 also shows that for a motor, commutation would have to be early, compoles too strong, for a direct demagnetising effect to arise. A direct magnetising action occurs for a generator if the compoles are strong and for a motor if the compoles are weak. Thus the commutation process may have important side effects which become more pronounced if the main field m.m.f. is reduced for low fluxes.

The quadrature-axis component of armature m.m.f. would have no effect on the total direct-axis flux if saturation were not present in the iron near the pole tips. In the air gap region, the armature winding fluxes are seen from Fig. 6.11, to strengthen the total field at one side and weaken it at the other. The strengthening effect is less pronounced than the weakening effect, due to saturation, particularly in the armature teeth. This is shown in a little more detail in Fig. 6.13, which is drawn to derive the flux density distribution on load in an approximate manner without recourse to flux plotting. The m.m.f. distributions are idealised, the armature m.m.f. being assumed triangular (Fig. 5.24), and that due to the stator coils rectangular. On no load, when neither the

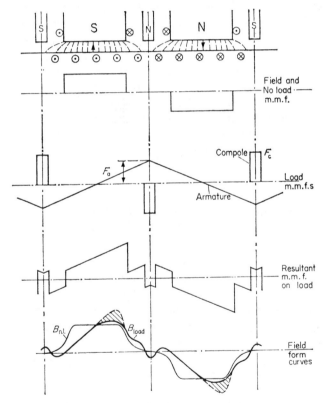

FIG. 6.13. Field form on load and no load.

armature nor interpoles carry current, the field form is a symmetrical "rectangle" with rounded corners. On load, the m.m.f.s must be summed at every point and applied to the permeance (Fig. 2.12), which under the pole face is inversely proportional to gap length, neglecting the iron parts of the circuit. The effect of the increased m.m.f. on one side of the pole, is to bring on saturation and reduce the permeance locally so that there is a smaller than proportional increase of flux density. The shaded area, which is the difference between allowing for and neglecting saturation, represents a net loss of flux even though the armature m.m.f. has no component on the direct axis.

Thus, both components of armature m.m.f. cause flux changes with load, and though the direct-axis component could be deliberately arranged to strengthen the flux by brush shifting, such a procedure is liable to upset the commutation performance and introduce instability. Generally, it may be taken that there is a net reduction of flux with load unless other steps are taken to avoid it. The actual reduction is difficult to calculate precisely since it depends on the flux level, which determines the saturation effect, and on commutation performance, which among many other factors is affected by speed of rotation.

Compensating Windings

More serious than the flux reduction on all but the smaller machines, is the local increase on overload conditions of B, and hence $2Blv$, the maximum voltage per armature coil. If the voltage between commutator segments exceeds about 30 V, there is a danger of flashover between positive and negative brush arms. An expensive, though effective, remedy is to provide another set of coils on the same magnetic axis and in series with the compoles. This compensating winding is normally embedded in pole-face slots and ideally will cancel the armature ampere-turns step by step under the main pole arc. Figure 6.14 shows a plan view of the winding together with the components of load m.m.f. The peak flux density and loss of flux on load are much reduced. On a non-compensated machine, the designer will allow extra field ampere-turns to maintain the flux on load. The allowance will be

282

10–20% of the armature m.m.f. per pole. With a compensated machine, this figure can be reduced to about 5%. Figure 6.14 shows that the total compole m.m.f. is also reduced since the compensating winding has cancelled about two-thirds of the armature m.m.f. Note that the compensating m.m.f. has been idealised, and is trapezoidal in form, as for a phase-wound armature, see Section 5.6.

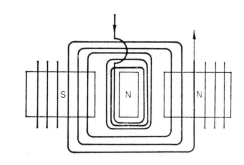

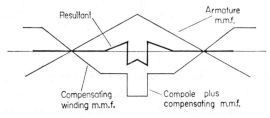

FIG. 6.14. Compensating winding.

Armature reaction effects can also be reduced by opening out the air gap at one, or usually at both ends of the pole arc. This is usually done by means of a chamfer, see Fig. 6.15, and the resulting no-load and full-load field forms are shown. The flux capacity of the machine is reduced and the field form is more nearly sinusoidal. Field and armature m.m.f.s can now be combined vectorially as an approximation, and the angular shift of the resultant m.m.f. phasor is roughly matched to the angular shift of the field form. The greater F_f compared with F_a, the less effect the load current has on the resultant flux. A longer air gap will necessitate a larger value of F_f without affecting the value of F_a which is determined by the load current. Such a machine would have a high magnetic "stiffness". In practice, F_f will tend to be designed rather larger than the value of F_a at full load, particularly on non-compensated machines. Air-gap lengths increase with the physical size

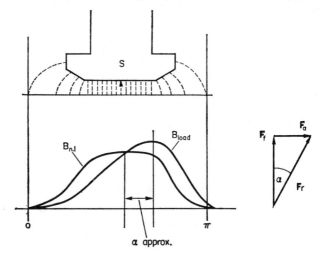

FIG. 6.15. Chamfered pole tips.

of the machine but are not usually greater than 10 mm. Interpole air gaps may be up to twice as long as the main-pole gap. In addition, there may be a rear "gap" formed by non-magnetic liners between the interpole root and the yoke. Such large total gaps improve the linearity of interpole flux with current.

Armature-excited Machines

It should not be forgotten that the torque of a machine depends on that component of armature m.m.f. which is in quadrature with the main-pole flux. The preceding discussion has tended to emphasise the undesirable effects of armature reaction. However, even the simple magnetising action is turned to good account in a certain group of armature-excited machines, among which the Metadyne, Amplidyne and the Magnicon are the most important. These three are basically the same in principle, though with differences in detail.

Consider Fig. 6.16(a); if current is led through the brushes AB, a magnetic field will be established as indicated by the N and S poles, the polarity depending on whether the armature winding is progressive

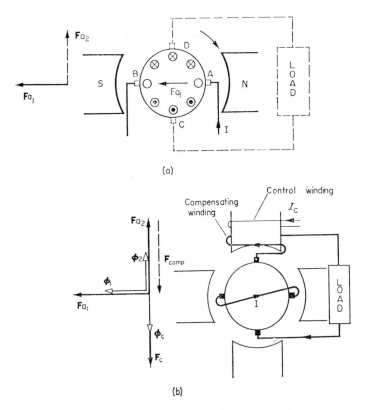

(a)

(b)

Fig. 6.16. (a) Armature excitation; (b) basic amplidyne.

285

or retrogressive. The space phasor of m.m.f. is shown as F_{a1}, its distribution being of no importance in the following explanation because all m.m.f.s are in opposition or in quadrature. Since the armature conductors are rotating in the stationary field, an e.m.f. will be generated and the maximum value would be picked up by brushes CD in quadrature with AB.

A load resistance could be connected across CD and a current would flow which could be considered as being superimposed on the existing conductor current. By considering the polarity and rotation, the direction of this generated load current would be into the paper under a north pole. Thus the second armature m.m.f. F_{a2} would arise, and because its magnetising sense would be upwards, it could be represented by a space phasor in clockwise quadrature with F_{a1}. The upwards flux due to F_{a2} would generate a voltage across brushes AB equal to KF_{a2}, where under unsaturated conditions, K would be a constant. Neglecting resistance drops, $V_{AB} = K.F_{a2} = K' \times$ load current. In this operational mode, a constant input voltage at brushes AB would be transformed to a constant current through the load connected across brushes CD. The device is then called the Metadyne Transformer. Electrical input and output powers are the same, neglecting copper losses and the only mechanical power is that required to overcome the mechanical losses. Field windings are added on the practical machine, to control the value of constant current.

The action of these types of machine in their more common Metadyne Generator mode will now be explained. The current in the horizontal brushes is generated by driving the armature with these brushes short circuited in a control flux ϕ_c, produced by a control m.m.f. F_c on poles on the vertical axis, Fig. 6.16(b). ϕ_c gives rise to F_{a1} and ϕ_1; similarly, ϕ_1 gives rise to F_{a2} and ϕ_2, which is seen to oppose F_c and ϕ_c.

Left uncompensated, F_{a2} would cancel most of F_c. In control engineering terms, it is a negative current feedback tending to suppress any increase of current, should the load circuit resistance be reduced, and permitting an increase of net vertical flux should the load current tend to fall consequent upon an increase in load circuit resistance. The

286

action is sufficient to maintain the load current approximately constant irrespective of load-resistance changes.

To produce instead, a substantially constant-voltage characteristic, the m.m.f. F_{a2} is compensated, i.e. cancelled by another winding on the vertical axis, and fed from the load current.

This device is virtually two machines in one. Current through the short-circuited brushes, which form the "load" of the first stage, acts also as the field excitation for the second stage. Mechanical input at the shaft is converted to an electrical output, modulated by the control flux. The device is a high-gain power amplifier. Figure 6.16b is only a basic diagram; other windings are usually added and the construction of the Amplidyne stator for instance is similar to an a.c. machine, thus giving a uniform air gap and a better amplifier characteristic.

6.4 STATOR WINDINGS, CONSTRUCTION AND OUTPUT EQUATION

Methods of Main-pole Excitation

To take advantage of the inherent d.c. machine flexibility of performance, the total main field m.m.f. is often made up from two or three sources, each having a particular modifying effect on the characteristics. Thus there may be several field windings. The range of total field m.m.f. required varies with the machine size from less than 1000 At to more than 10 000 At per pole. Apart from permanent magnets, which give fixed m.m.f., there are two main methods of excitation.

Separate Excitation

For this purpose, fine-wire or strap coils are usual, having perhaps several hundred turns, and carrying the appropriate current. A separate supply ensures that the excitation provided by this winding is not dependent, unless desired, upon the machine behaviour. It can be designed for any convenient voltage, usually, though not necessarily, greater than 100 V and is often similar to the machine voltage itself. It can easily be controlled by any normal method, manually through resistances, or automatically from a controlled source. Only small powers up to perhaps 3% of the machine output are required for excitation.

Self Excitation

This depends on the residual magnetism and voltage to drive a small current through a field winding in such a direction as to reinforce the residual flux and build it up. It may take two forms.

(a) *Series excitation* involves the use of a few turns per pole of large cross section through which the whole, or a large proportion of the armature current is taken. The excitation produced will be dependent on load current, a desirable feature for certain applications. Limited control can be obtained by diverting the winding with resistance or by arranging for the winding to be tapped, giving different numbers of turns. D.C. series motors are used frequently in industry. For generators however, series excitation is usually restricted to a small capacity winding, auxiliary to the main winding. Its function is to increase or decrease the total field ampere-turns as a function of load, i.e. it is cumulative or differential with the main field. For example, it could be used to offset armature reaction to maintain approximately constant flux. Alternatively, it could be arranged to give some overload protection, by causing generator voltage or motor speed to fall with increasing current.

(b) *Shunt excitation* employs a fine-wire coil of a few hundred turns shunted across the armature terminals in series with a rheostat. In this case, the excitation is dependent on the load voltage. As for series excitation, limited control can be exercised, this time by varying the series resistance.

Compound Excitation

This term refers to the arrangement whereby a relatively small capacity series winding modifies the main excitation provided by shunt-excited or separately excited coils.

288

Connection Diagrams

Figure 6.17a shows all the stator windings which have been discussed, magnetising on their appropriate axes and with the terminal markings used in BS 4999. This specification should be consulted for details concerning the correct assignment of markings. Every machine does not have all the windings shown and the quadrature windings are usually omitted from the schematic, e.g. Fig. 6.17b is a common representation of the shunt-wound generator with series modifying winding. The shunt field may be connected directly across the armature

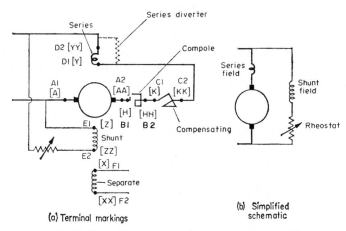

FIG. 6.17. D.C. machine stator windings and terminal markings.
(Old BS 822 markings in brackets)

terminals (short shunt) or include the series winding as well (long shunt), the only difference being a slight change to the voltage and current ratings of the two windings.

Construction

Figure 1.11a showed a simplified cross-section of a d.c. machine. Figure 6.18 is a photograph taken from the commutator end showing more clearly the details of brush assembly and mounting. It will be noticed that on each arm there are several brushes which share the current. Each brush in the tandem boxes is held against the commutator surface by its own spring. There are many other methods of supporting the brushes, see Fig. 6.19, the arrangements becoming more and more elaborate as the machine rating and speed increase. Commutation performance is highly dependent on the nature and stability of the brush/commutator contact and to maintain this satisfactorily under all operating conditions requires careful design and choice of brush grade.

The outer member of the magnetic circuit is called the yoke and is usually made in two halves bolted together. Each half consists of a thick, steel slab rolled to shape, machined, drilled and welded as required. On the smaller sizes, the yoke would be rolled and welded in one piece to form a complete cylinder. Cast steel and cast iron yokes are also used. On machines where rapid flux changes are required, the yoke is built up from thin laminations to reduce the eddy-current demagnetising effect discussed on p. 31. The reaction of these yoke eddy-currents on the field winding is similar to that which would occur if there was a closed coupled coil on the pole; see Fig. 3.32 and Reference 9.

The main poles are normally laminated using thicknesses of about 1·5 mm. This makes for simple and economic construction and reduces the iron losses in the pole face caused by local flux variations at slot ripple frequency. The compoles are often solid rectangular iron blocks on small machines but are usually laminated on larger machines and have a more involved shape. The armature core is always laminated using thicknesses of about 0·4 mm. When the frequency is low, a soft iron with good magnetisation characteristics is used. Otherwise, beyond about 40–50 Hz, a high-resistance silicon steel is used to reduce the eddy-current losses. Radial ducts are provided along the length for ventilation purposes, air being fed under the winding along axial ducts or holes in the core. A fan is often mounted on the machine shaft to increase the flow of air which may be ducted into and out of the machine for the purposes of interposing filters, coolers or external fans.

FIG. 6.18. 2400 kW, 700 V, 536 rev/min compensated d.c. generator.
(By courtesy of G.E.C. Machines Ltd.)

Fig. 6.19. Constant-tension "Tensator" brushgear: (a) split sandwich brushes and brush rocker for 6-pole 450 h.p., d.c. motor (b) 30° reaction brush and brush holder for a.c. commutator motor. (By courtesy of Laurence, Scott & Electromotors Ltd.)

Output Equation

Since $E = 2p\phi$. z_s . n volts, from eqn. (5.7), then the generated power in watts is:

$EI_a = 2p\phi$. $I_a Z/2a$. n where $2a$ is the number of parallel circuits.

$$= \underset{\text{loading}}{\underset{\text{magnetic}}{\text{Total}}} . \underset{\text{loading}}{\underset{\text{electrical}}{\text{Total}}} . n$$

or in terms of intensities:

$$= \frac{2p\phi}{\pi dl} . \frac{Z . I_a/2a}{\pi d} . \pi^2 d^2 l . n$$

$$= B_{av} . \quad ac \quad . \pi^2 d^2 l . n. \tag{6.1a}$$

The term *ac* is the symbol used for the ampere-conductors per unit length of armature periphery and is a useful design figure. It is a measure of the intensity with which the armature is loaded electrically and is limited by such considerations as insulation, heating, reactance and permissible depth of slot. B_{av} is limited by saturation and iron losses.

Thus the machine output is proportional to the armature volume, speed and to each of these design parameters. It can be seen that the armature volume (proportional to $\pi d^2 l$), is determined by the torque EI_a/n, once the permissible values of B and *ac* have been decided.

The equation with little modification is applicable to a.c. machines. The e.m.f. expression is very nearly the same, as already explained in Section 5.5. However, EI_a is no longer the generated power, except at unity power-factor. The volume of an a.c. machine is therefore determined by the kVA rating, not by the kW rating.

A slight modification to eqn. (6.1a) gives:

$$\text{generated power} = B_{av} . ac . \pi dl . v, \tag{6.1b}$$

where $v = \pi dn$, the peripheral rotor velocity; a mechanical limitation.

291

Multiplying both sides of eqn. (6.1b) by n gives:

$$\text{power} \times \text{speed} = B_{\text{av}}.ac.v^2.l. \tag{6.1c}$$

These two equations also apply to a.c. machines with the only major change that power must be replaced, in general, by volt amperes, VA. Equation (6.1c) demonstrates a common feature of most power devices in that if they are to operate up to certain design limits, the *power* × *speed* product is a constant, so that higher powers can only be obtained by operating at a correspondingly lower rotational speed. From eqn. (6.1b), since power is equal to *force* × v and since πdl is the cross-sectional area of the air gap carrying the flux, the force developed by the machine is seen to be equal to $B_{\text{av}}.ac$ per unit gap area. Proportional reductions in machine size follow from increases in B or ac. In the superconducting motor, for example, see p. 21, B can be increased well beyond the limitations normally fixed by saturation and iron losses and ultimate ratings of 200 MW have been predicted.[4] ac can be increased on some machines (see p. 529) by water cooling. Equation (6.1b) also indicates how the designer decides on the proportions between diameter and length for a given output. For the d.c. machine, the length, in conjunction with ac or B_{av}, determines the reactance voltage and bar-to-bar voltage which are limited from commutation considerations.[9] For turbo alternators, the length is restricted from mechanical considerations, see Fig. 8.3c. For hydro-electric generators, l is limited from considerations of achieving satisfactory ventilation throughout the core length, with natural fanning action. For induction machines, of moderate rating, the proportions of d and l hinge on economic considerations of manufacturing costs. Transport and weight limitations also influence the maximum possible powers, which, from eqn. (6.1b) could otherwise be increased indefinitely by increasing the diameter.

6.5   GENERATOR CHARACTERISTICS

Many characteristics of d.c. machines are deduced from tests taken when running at constant speed as a generator. The first two curves to be described come in this category.

Open-circuit Characteristic or Magnetisation Curve

Reference has already been made to this curve on page 38. It shows the relationship at a particular speed, between the generated e.m.f. $E = (2pnz_s) \cdot \phi$, eqn. (5.7), and the field current $I_f = I_f N_f / N_f$, where $I_f N_f$ is the field m.m.f. The open-circuit curve checks the excitation calculations and is vital for the determination of other machine characteristics. Figure 6.20 shows a suitable test circuit, separate

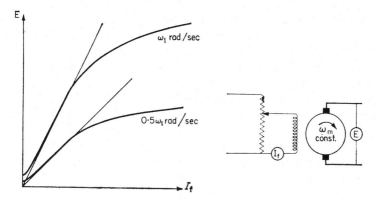

Fig. 6.20. Open-circuit (o.c.) curve and test circuit.

excitation ensuring good, independent flux control. Due to hysteresis there would be a difference between a curve taken with increasing values of field current and that taken with decreasing values. To avoid minor hysteresis loops when taking either an increasing or decreasing section of the curve, field current changes must be made smoothly in one direction.

For any particular field current and flux ϕ', the generated voltage per unit speed is constant, k'_ϕ say. With this flux, let the induced e.m.f. at speed ω_1 be E', then in SI units:

$$\text{voltage/radian per second} = \frac{E'}{\omega_1} = k'_\phi = \frac{2pn_1 z_s \phi'}{2\pi n_1} = \frac{pz_s}{\pi} \cdot \phi' \quad (6.2)$$

which is independent of speed. For any other speed ω_m with ϕ' unchanged:

$$\text{generated e.m.f.} = E = k'_\phi \omega_m \text{ volts.} \qquad (6.2a)$$

The o.c. curve at any speed can thus be deduced from the one test curve, see Fig. 6.20. Alternatively, since $k_\phi \propto \phi$, the k_ϕ/I_f relationship can be plotted after dividing E by the test speed to give k_ϕ at various values of I_f. The result will be the open-circuit curve to a different scale. k_ϕ is constant if the field current and flux are constant, and over the unsaturated region it is proportional to field current. It is a most useful parameter for calculations on variable-speed machines, which are more commonly motors.

For any particular speed, $E = k_f I_f$, where $k_f = k_\phi \omega_m / I_f$ and is constant over the unsaturated region, as shown on Fig. 6.20 by the linear portion of the curve. k_f, the voltage per field ampere, is sometimes called the transfer resistance since it has the dimensions of ohms and it transfers field current to generated voltage by multiplication. At any particular speed, k_f is constant if field current and flux are also constant. It is useful for calculations involving constant-speed machines, mostly generators. The saturated value of k_f will be designated k_{fs}.

In Section 3.8, the e.m.f. equation was derived in terms of inductances and in Section 10.1 it will be shown that for the d.c. machine, only the second term in eqn. (3.12) applies. The electrical angular velocity $d\theta/dt$ was used in this equation. For a $2p$-pole machine running at mechanical angular speed ω_m, the electrical angular velocity is $d\theta/dt = p \cdot \omega_m$. Rewriting eqn. (6.2a) in instantaneous values:

$$e = k_\phi \cdot \omega_m = i_f \cdot \frac{k_\phi}{i_f} \cdot \frac{1}{p} \frac{d\theta}{dt}.$$

This expression is of the same form as eqn. (3.12). Comparing coefficients of the second term of this equation:

$$\frac{dM}{d\theta} = \frac{k_\phi}{i_f} \cdot \frac{1}{p} = \frac{p \cdot z_s}{\pi} \cdot \frac{\phi}{i_f \cdot p} = \frac{z_s}{\pi} \cdot \frac{\phi}{i_f}.$$

It can be seen that this expression, derived from eqn. (6.2a) does have the dimensions of inductance, volts/amp/sec, radians being dimensionless.

294

The magnitude will be designated M_ϕ since it is simply related to k_ϕ, i.e. $M_\phi = k_\phi/(p \cdot i_f)$, and there will be occasion to use it again in Chapter 10.

Allowance for Armature Reaction in Calculations

From the o.c. curve, the measured terminal voltage is $V = E$, since I_a is zero. Further, the air-gap e.m.f. generated under these conditions, $E = f(\phi_m)$, is the same as $E_f = f(\phi_f)$ because in this case, the resultant gap flux is due only to the field current and ϕ' in eqn. (6.2a) is a field-produced flux. When the armature too carries current, ϕ_m is modified in a complex fashion by the action of the m.m.f.s due to armature currents, see Fig. 6.13, and a reduction in gap flux normally follows. If k_ϕ is to include this effect, the measured e.m.f. E on load should be used in the equation:

$$k_\phi = E/\omega_m \qquad (6.2b)$$

and eqn. (6.2) shows that:

$$k_\phi = (pz_s/\pi) \cdot \phi_m = f(I_f, I_a). \qquad (6.2c)$$

The relationship derived from the o.c. curve, for which ϕ_f is equal to ϕ_m, gives $k_\phi = f(I_f)$. This can be used as a near approximation to eqn. (6.2c) though for a series machine, as described below, it is possible to incorporate a fairly accurate allowance for the effect of I_a when taking the magnetising characteristic. For other d.c. machines, it will be found that less straightforward methods are required to improve the accuracy of the calculations. For certain homopolar machines, Section 1.6, as described in Reference 3, the armature reaction m.m.f. is in a different plane from the field m.m.f. and it can be virtually neutralised.

In practice, the designer and the applications engineer must make appropriate allowance for armature reaction with as much accuracy as possible but it is in general an empirical estimate based on specialised calculations and accumulated test information on similar machines. Particular test methods are illustrated in Fig. 6.21. In the normal series machine, field and armature currents are always the same. Therefore, if the magnetisation curve is taken with the armature current maintained at the same value as that in the field, and k_ϕ is found from $(V \pm I_a R)/\omega_m$, the k_ϕ/I_f characteristic would include armature reaction effects though strictly only for the particular test speed and commutating condition. Figure 6.21a shows a test circuit when it is more convenient to run the machine as a motor, and Fig. 6.21b shows a generator arrangement. Different kinds of test facilities are required and will determine the method to be used.

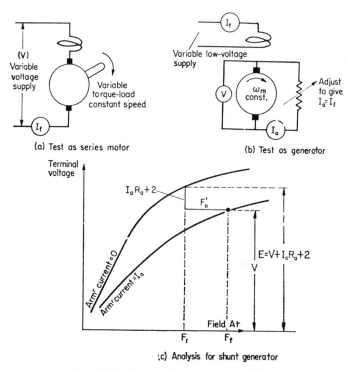

(a) Test as series motor

(b) Test as generator

(c) Analysis for shunt generator

FIG. 6.21. Allowance for armature reaction.

For shunt- or separately-excited machines, a series of test magnetisation curves with various values of armature currents would have to be taken, and analysed to find F_a', the net magnetising effect of d- and q-axis armature reaction. F_a' is neither a simple nor a unique function of armature current for reasons discussed in Section 6.3. Over a moderate range of speed and flux change however, it may be taken to depend primarily on the armature current. It is usually demagnetising as explained in Section 6.3, so that F_a' is usually negative. For any particular terminal voltage V and armature current I_a, the field excitation F_f required, can be considered as due to two components; F_r to generate the air-gap e.m.f. E, and $-F_a'$ to allow for armature reaction. F_a' for various conditions can thus be obtained by the construction shown on Fig. 6.21c which is also the basis for subsequent calculations of field excitation; see E.6.2 and E.6.9. F_r is in fact the resultant m.m.f. $F_f + F_a'$ and the required value of E is determined by the electrical circuit equation (3.2). If the brush-contact drop is allowed for, it is usually taken as constant at a value of 2 V total, i.e. for $+ve$ and $-ve$ brush-sets together.

Short-circuit Characteristic

An alternative method of allowing for armature reaction is to analyse a short-circuit test but, bearing in mind what has been said about the sensitivity of armature reaction to commutation and saturation conditions, the accuracy is only approximate. On short circuit, the machine is unsaturated and commutation performance is not the same as at normal flux and voltage. Figure 6.22 shows the test circuit and the I_{sc}/I_f curve taken from a demagnetised condition, together with the o.c. curve. For any particular value of short-circuit current I_{sc}, and the corresponding excitation F_f, only a small generated voltage equal to $I_{sc}R_a + 2$ is required. R_a is the measured armature circuit resistance and 2 V is allowed for brush drop. This internal voltage is substantially less than the voltage E_f which would be produced on open circuit with the same field current. This is because the effective or resultant m.m.f. is given by $F_r = F_f + F_a'$, and F_a' is usually negative as for Fig. 6.21c. If the construction shown, is carried out at various values of armature current, it will be found that F_a' is not directly proportional to the current because it is a function also of other operating conditions. On the synchronous machine, due to the sinusoidal distribution of m.m.f.s it will be found possible to combine field and armature m.m.f.s

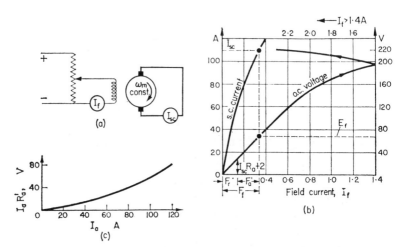

Fig. 6.22. Short-circuit (s.c.) test: (a) test circuit; (b) test curves; (c) effective resistance drop.

vectorially to obtain the resultant F_r. The total armature m.m.f. F_a will then be used, the extent of magnetising or demagnetising action being dependent on the power factor.

For the d.c. machine, a convenient approach for demonstration purposes is to consider armature reaction in terms of the voltage loss it causes. Since F_f on o.c. would generate E_f, it would appear that this voltage is being absorbed internally in circulating I_{sc}. The effective internal resistance is thus $R'_a = E_f/I_{sc}$ ohms, greater than R_a due to armature reaction. The value of this internal drop $E_f = I_{sc}R'_a$ can be read off at various currents and plotted as shown on Fig. 6.22c. It is not a straight line but a mean slope could be found to derive a value for the effective resistance. A similar approach, when adopted for synchronous machines (cf. p. 481), can be made to yield quite accurate results.

Load or External Characteristic of Separately Excited Generator

With constant speed and a constant value of separate excitation, the terminal voltage falls as load current increases, the construction of

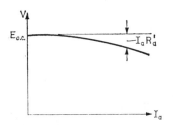

FIG. 6.23. External characteristic, separately excited generator.

the external characteristic, Fig. 6.23, being obvious. E_{oc} is the voltage produced on no load by the particular value of field current, but as load is applied, the internally generated voltage E falls, normally, due to armature reaction. This voltage plotted against load would give the so-called internal characteristic but it is of more interest to include the additional voltage drop due to internal resistance which then gives the external characteristic of terminal voltage against current as shown.

With a series winding connected in circuit, the characteristic is modified, a cumulative series winding raising the voltage with increasing load and the differential series winding depressing the voltage, Fig. 6.24. Compounding can be adjusted by means of a series-field diverter. If the full-load voltage is thereby made the same as the no-load voltage, this is known as a level-compound characteristic, though the curve is not actually flat because armature reaction demagnetising effects are not exactly linear with current. In calculations involving multiple excitation it may be convenient to express the data in ampere-turns, because the different windings will in general have different numbers of

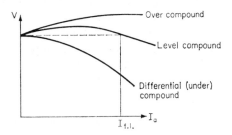

FIG. 6.24. Effect of series winding.

299

turns. It may also be necessary to use tabular methods in the calculation of the non-linear d.c.-machine characteristics and some of the worked examples in this chapter will use such methods.

EXAMPLE E.6.2

The open-circuit (o.c.) and short-circuit (s.c.) test data of a d.c. generator rated at 200 V, 100 A and 1000 rev/min are given by the curves of Fig. 6.22b. The armature is wave wound with 708 conductors.

Deduce

(a) the demagnetising m.m.f. due to full-load armature reaction expressed as a percentage of full-load armature m.m.f.;

(b) the no-load voltage when the separately-excited field current is adjusted to give 200 V at full load;

(c) the new field current and the no-load voltage when a cumulative series winding having 25 turns per pole is connected in circuit and the voltage again adjusted to 200 V on full load;

(d) the value of the series diverter to give a level compound curve at 200 V.

Armature circuit resistance 0·14 Ω. Series field resistance 0·01 Ω. Separately excited field resistance 100 Ω, 2500 turns per pole. Brush drop 2 V. The machine has 4 poles.

(a) Internal drop at full load $= 100 \times 0.14 + 2 = 16$ V

Full-load armature m.m.f. $F_a = \dfrac{100/2}{4} \times \dfrac{708}{2} = 4420$ At/pole.

Total excitation for 100 A on s.c. $= 0.28$ A $= F_f/2500$

Excitation for 16 V from o.c. curve $= 0.08$ A $= F_r/2500$

∴ Armature demagnetising m.m.f. in
terms of field turns $= 0.2$ A $= -F_a'/2500$.

As a percentage, $F_a'/F_a = 100 \times 0.2 \times 2500/4420 = 11.3\%$

(b) E at full load $= 200 + 16 = 216$ V

Required ampere-turns F_r from o.c. curve $= 2.1 \times 2500 = 5250$

$$-F_a' = 0.2 \times 2500 = \underline{\;\;500\;}$$

Since F_a' is negative, total excitation required $F_f = 5750$.

The total field current required is thus $\underline{2.3\text{ A}}$ giving $\underline{220\text{ V}}$ on no load, from the o.c. curve.

(c) Neglecting the small, additional series-field resistance drop the total excitation must still be: 5750 At

the series winding contributes 25×100 $= 2500$ At

the separate winding must provide: 3250 At $= \underline{1.3\text{ A}}$.

The no-load voltage with 1·3 A (series current $= 0$) is $\underline{192\text{ V}}$.

(d) For 200 V on no-load, the separate excitation would have to be 1·5A or
$1 \cdot 5 \times 2500$ = 3750 At
∴. the series excitation must be reduced to   2000 At
giving as before a total at full load   5750 At

The series current is therefore 2000/25 = 80 A.

The required diverter resistance is $\dfrac{80}{20} \cdot 0 \cdot 01 = 0 \cdot 04 \ \Omega$.

The Process of Self Excitation

Figure 6.25a shows the circuit, the field winding being connected in
series with the armature and a variable resistance. For a series machine,
this variable resistance would represent the load, but for a shunt
machine it would represent the field regulator and the circuit would
only correspond to the no-load condition, the load being connected
across the armature terminals as indicated. The shunt generator will
build up its voltage without the load circuit being closed, whereas the
series generator of its nature requires rated load current for rated
terminal voltage. In either case, the connection of the field winding
must be such that the small initial current, which the residual voltage
drives through the circuit resistance, is in a direction to increase the
flux and cause the voltage to build up. It is clear that unless residual
magnetism is present, the machine will not build up at all. In this case,
it will be necessary to apply separate excitation temporarily, to establish
the desired residual polarity.

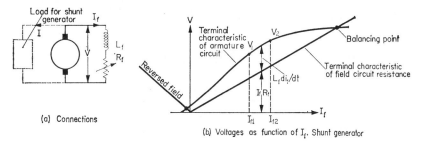

(a) Connections

(b) Voltages as function of I_f. Shunt generator

FIG. 6.25. Self excitation.

301

The full-line circuit can be regarded as two parallel branches, the voltage across each of which is a function of the current flowing through it. For the armature it is a complex non-linear function expressed by the magnetisation curve reduced by the internal voltage drop. For the field-circuit resistance, it is a linear function of current for any particular value of this total resistance R_f. Since the two branches are connected, their terminal voltages must be identical and so the current will eventually settle down to a steady value corresponding to the point where the two terminal characteristics intersect. The intersection determines the final terminal voltage which is reached after the transient build-up process is completed. If the winding connections are reversed so that the field current opposes the residual flux, the field circuit resistance line will have a negative slope and the intersection will now be lower than the residual voltage, Fig. 6.25b. When this connection is employed deliberately to hold the voltage to low value, it is called a suicide connection for obvious reasons. The reader should be able to appreciate that, in a similar manner, a reversal of rotation would prevent the voltage from building up. Consequently, if a generator is to self excite with reversed rotation, the field connections to the armature must be reversed.

More specifically, consider the shunt machine. The current I_f and the associated armature-circuit voltage drop are relatively small so the terminal voltage is only slightly smaller than the open-circuit voltage. On Fig. 6.25b the o.c. curve is used. Commencing from the residual voltage, and until the terminal voltage builds up to the balancing point, there is a difference between V and $I_f R_f$ at any particular value of I_f. This difference is absorbed in the field circuit inductance as $L_f \, di_f/dt$ and will exist as long as the field current is changing. The actual time to change from say V_1 to V_2 can be obtained as outlined briefly on p. 123.

The mean value of this voltage difference is, from Fig. 6.25b:

$$L_f \frac{di_f}{dt} \simeq L_f \frac{\Delta I_f}{\Delta t} \simeq L_f \frac{(I_{f2} - I_{f1})}{\Delta t};$$

from which:

$$\Delta t = L_f (I_{f2} - I_{f1})/(L_f \, di_f/dt).$$

302

By taking several such increments of voltage from the residual value up to the balancing point and measuring the mean value of $L_f \, di_f/dt$ for each interval, the values of Δt can be found and summed to give the total time of build up. Allowance can also be made for the changing values of inductance caused by saturation. Reference to Section 3.3 will show why the average permeance will correspond to the average slope of the o.c. curve over any particular interval. The unsaturated field inductance corresponds to the slope of the o.c. curve over the initial linear region, see Example E.6.3. The reduction of slope, for any

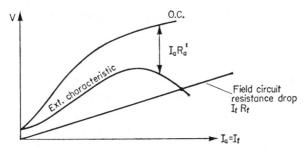

FIG. 6.26. Series generator.

particular interval, gives the appropriate reduction factor to be applied to the unsaturated inductance, for that interval. The transient response is very slow, because the field circuit has to wait for its supply voltage to build up. In control engineering terms, this is a system with positive feedback in which slow response is inherent.

In considering the series generator, the internal drop due to the much larger field current flowing through the armature cannot be ignored, so the external characteristic must first be obtained by deducting the value $I_a R_a'$ from the open-circuit curve. The operating point for a given external resistance is obtained by drawing a straight line of appropriate slope, Fig. 6.26. As will shortly be explained, there are stability difficulties over the rising part of the characteristic and operation is on the descending portion. The straightforward series generator has limited application but the function is useful for electrical braking of series motors. Here, the circuit conditions can be so changed that the motor becomes a generator, converting its stored mechanical energy into electrical energy in the process of stopping; see p. 345 *et seq.* Series generators can also be used satisfactorily as boosters in series with a distribution line. Increase of load current causes the machine to generate additional voltage to compensate for the line *IR* drop.

Critical Resistance and Critical Speed

Consider the shunt generator again. If the field circuit resistance is increased from the value shown in Fig. 6.25b, the slope of the line will increase, reducing the value of V at the intersection. Eventually the line will coincide with the straight part of the magnetisation characteristic and under these conditions the circuit will be in indifferent equilibrium since the equation $V = I_f R_f$ is satisfied at all voltages along the region of coincidence. Rheostatic control becomes very sensitive and the voltage usually collapses. When the voltage is being *increased* from the residual figure, by reducing the field-circuit resistance from its maximum

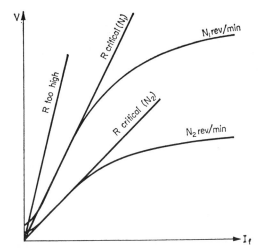

FIG. 6.27. Critical field-circuit resistance and critical speed.

value, there is little change in voltage until the resistance line coincides with this linear region, i.e. when it reaches the *critical resistance*, whereupon the voltage tends to rise out of control to a point just beyond the linear portion. This region is unstable for both increasing and decreasing values. It should be noted that failure of a shunt generator to build up its voltage is often due to a high resistance contact in the field circuit, which prevents the resistance from being reduced below the critical value.

If the speed of a machine were to be reduced, the critical resistance would go down *pro rata* since all the e.m.f. ordinates and hence the slope of the linear portion are proportional to speed; Fig. 6.27. If, for a given field-circuit resistance, the speed is brought up from zero with the field connected, it is clear that there will be a *critical speed* below which the voltage will not build up. Actually, due to hysteresis effects on the magnetisation curve, these critical values are not precisely defined and are different for increasing than for decreasing changes.

Thus, the range of control on a self-excited machine is limited, though it can be improved by deliberately introducing curvature in one or other of the characteristics. By restricting the magnetic circuit at some point, local saturation tends to "bend" the magnetisation curve after a shortened linear portion. Alternatively, non-linear resistors in the field circuit can cause a more clearly defined inter-section, see Fig. 6.28.

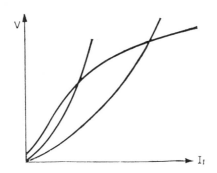

Fig. 6.28. Use of non-linear field-circuit resistors.

External Characteristic of a Shunt Generator

Under steady-state conditions the terminal voltage V is also equal to $I_f R_f$, Fig. 6.25. Choosing a particular value of I_f will therefore give V and also the internal e.m.f. E_{oc}, Fig. 6.29. The difference between these two voltages must be absorbed internally as $I_a R_a'$, so by finding a matching ordinate on the R_a' characteristic, Fig. 6.22c, the value of I_a is determined. Repeating the procedure for various values of I_f will enable V to be plotted against I_a or against the line current, $I = I_a - I_f$. It will be noticed that the difference between the two voltages has a

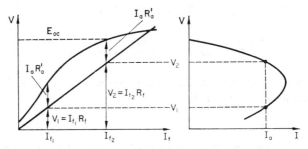

FIG. 6.29. External characteristic of shunt generator.

maximum value and there are duplicate values at lower currents, for a given field-circuit resistance. It follows therefore, that the external characteristic must double back on itself. By comparison with the separately-excited generator, the fall of terminal voltage is much greater, being exaggerated by the reduction of the field current as the voltage falls.

To calculate the terminal voltage of a shunt generator with some series excitation is rather more difficult. Figure 6.17b gives the circuit. As before, I_f will give $V = I_f R_f$, but the two equations for the open-circuit voltage are now:

$$E_{oc} = V + I_a R_a' \quad \text{and} \quad E_{oc} = f(I_f N_{sh} + I_a N_{se}).$$

If R_a' can be taken as constant, a graphical solution follows on the lines of Fig. 6.30. Otherwise, for each value of I_f, the value of I_a must be

306

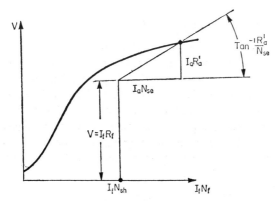

FIG. 6.30. Cumulatively compound generator; construction
for armature current, given field current.

found by trial and error to satisfy the two equations given, and also the
I_a/R_a' relationship of Fig. 6.22c.

EXAMPLE E.6.3

The open-circuit curve for the machine of Example E.6.2 taken from normal
residual voltage, is shown on p. 308.

If the series winding is not in circuit, find:

(a) the no-load voltage with the main field connected in shunt directly across the
armature circuit;

(b) the full-load voltage under these conditions with an armature current of 100 A;

(c) the time for the field current to change from 0·4 A to 0·6 A when building up
with conditions as in (a);

(d) the critical circuit resistance;

(e) the critical speed with no external field circuit resistance.

(a) The construction is shown on diagram. The resistance line corresponding to
100 Ω drawn through any convenient point on the line, say 100 V and 1 A, cuts the
magnetisation curve at 220 V.

(b) The difference between the o.c. curve and the terminal voltage absorbed across
the field, $V = I_f R_f$, is due to the internal drop $I_a R_a + 2$ V and the armature reaction
F_a'. At 100 A these are respectively 16 V and -500 At (from E.6.2), and form the
two sides of a right-angled triangle as shown in Fig. 6.21c. The triangle must be
fitted between the two curves as indicated. It can be seen that when the terminal
voltage is 192 V, this circulates a field current, giving an m.m.f. which, after adding

307

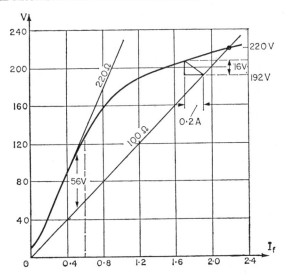

F_a' at 100 A, (F_a' is negative), is just sufficient to generate the terminal voltage, plus the internal voltage drop, at the same value of current. The method is quite accurate providing only that the correct value of F_a' is known. The approximate method of Fig. 6.29, employing the R_a' concept, was used only for demonstration purposes.

The present method could be extended to take a series winding into account, the additional ampere-turns being combined with F_a'. This is left as an interesting exercise for the student.

Note that the voltage regulation from 220 V to 192 V is greater than when the machine was separately excited in E.6.2. The reason for this is the additional reduction due to the lower field voltage as already explained on p. 306.

(c) The field inductance $L_f = N_f \phi / I_f$ must first be calculated, and since the response is required over a portion of the curve which is practically of the same slope as the air-gap line, the ratio ϕ / I_f is constant and may be taken at any convenient point, say 100 V where the field current is 0·46 A.

The flux per pole to generate 100 V from eqn. (5.7) is:

$$\phi = \frac{E}{2pnz_s} = \frac{100}{4 \times 1000/60 \times 708/2} = 0.004\ 24 \text{ Wb,}$$

leakage being neglected, although this is not difficult to take into account by increasing the flux if the leakage factor is known.

$$L_f \text{ for 4 coils in series} = 4 \times 2500 \times \frac{0.004\ 24}{0.46} = 92 \text{ henrys.}$$

308

From the figure, the average value of $L_t \, di_t/dt$ over the interval is 56 V.

$$\therefore \Delta t = \frac{L_t \Delta I_t}{L_t \, di_t/dt} = \frac{92 \times (0 \cdot 6 - 0 \cdot 4)}{56} = 0 \cdot 33 \text{ sec.}$$

To calculate the total time of build up, several such intervals from $I_t = 0$ to $I_t = 2 \cdot 2$ A would have to be taken and the incremental times summed. Over the saturated region average inductances would have to be used, these corresponding to the average slope of the magnetisation curve over the interval. An alternative treatment is to use the relationship:

$$L_t \, di_t/dt = N_t \, d\phi/dt \simeq N_t \Delta\phi/\Delta t$$

though there would be no difference in the final answer.

(d) The construction is shown. The critical resistance at 1000 rev/min is $220/1 = \underline{220\,\Omega}$.

(e) The critical resistance must now be $100\,\Omega$ and for this to be tangential to the air-gap line, the latter must be reduced in the ratio $100/220$ by a corresponding reduction of speed, i.e. critical speed for $100\,\Omega = 1000 \times 100/220 = \underline{455 \text{ rev/min.}}$

Below this speed the generator will not build up its voltage.

6.6 EQUIVALENT CIRCUIT AND TRANSFER FUNCTIONS

Limitations of Equivalent Circuit

Figure 6.31a is the usual equivalent circuit representation for a generator, and Fig. 6.31b is for a motor, typical potential values being inserted, assuming a common terminal voltage $V = 100$ V. The two circuits are identical apart from the current reversal. The current direction depends only on the relative magnitudes of V and E. If E is greater than V, power will flow into the external circuit; the machine is a source. If E is less than V, power will flow from the system into the machine, which is then a sink. Although the equivalent circuit resembles that for a battery, Fig. 3.18b, the e.m.f. is not a simple constant and is here expressed in the most likely form suitable for the particular function. E is due to the combined effects of total field and armature ampere-turns and so will vary in a complex manner as load increases, rising even if a strong enough cumulative series winding is provided. The accuracy of this simple equivalent circuit therefore depends on having a correct knowledge of E, which is not really a problem for either the battery or transformer equivalent circuits. Figure 3.28 showed one way of including armature reaction effects in terms of F_a', a method which, in the earlier sections of this chapter, has been elaborated and the difficulties explained. The simple expressions given for E in Fig. 6.31, are very convenient for calculations, but their limitations must be borne in mind. If k_f and k_ϕ are derived only from the o.c.

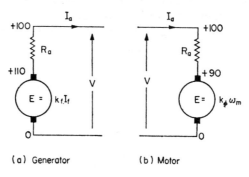

FIG. 6.31. Equivalent circuits.

curve, calculation errors *could* be small but might be as much as 10% or even more in certain cases. To facilitate the solution of numerical problems, it will be assumed unless otherwise stated, that the machine is fully compensated or that allowance for armature reaction is included in the magnetisation data. If brush contact drop is allowed for, it can be assumed constant at 2 V total though it will often be neglected, as in Fig. 6.31.

There are other aspects of d.c. machine calculations for which the equivalent circuit alone is inadequate. Some of the graphical and tabular methods necessary to deal with machine non-linearities and multiple excitation have already been introduced, and more will follow. It has been seen that non-linearity is essential for controlling and limiting the voltage of a self-excited machine and in many other problems it cannot be neglected.

310

Transfer Function of a Generator

A d.c. generator may be an element in a control system. A small, field-input power is received and variation of this, by virtue of its effect on the flux, controls the rate of energy conversion from mechanical input to electrical output; i.e. the generator is a power amplifier with a power gain of 50 perhaps, or more if specially designed. To represent the behaviour under both steady state and transient conditions, it is convenient to have an expression relating the output voltage to the input voltage. This is called the transfer function of voltage and is the ratio output/input.

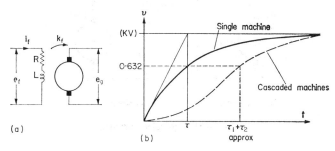

FIG. 6.32. Generator transfer function and transient response.

For a generator running at constant speed with field circuit inductance and resistance L and R respectively, the output and input voltage equations from Fig. 6.32a are:

$e_g = k_f i_f$ on open circuit.

$e_f = R i_f + L \, di_f/dt = (R + Lp)i_f$, where p is the operator d/dt. Hence the transfer function is equal to:

$$\frac{e_g}{e_f} = \frac{k_f/R}{1 + \tau p} = \frac{\text{steady-state voltage gain } K}{1 + \tau p}$$

where $\tau = L/R$ and $K = k_f/R = k_f i_f/R i_f =$ (output voltage)/(input voltage) when the steady state is reached.

311

The differential equation connecting output and input voltage can be written down from the transfer function by cross-multiplication; i.e.

$$(1 + \tau p)\, e_g = K e_f$$

which is a standard form having a simple exponential solution when a sudden change of input voltage e_f is applied. For example, if the field circuit is suddenly closed on to a constant voltage V, the behaviour of the output voltage with time is deduced from eqn. (3.21) and is shown graphically in Fig. 6.32b. This is known as the transient response, the time constant in this case being $\tau = L/R$ seconds.

The time delay in the response is due to the time taken to store the field magnetic energy $\frac{1}{2} L i^2$. On a machine which is specifically designed as a power amplifier, this stored energy is kept as small as possible by having a small air gap. This improves the response. To improve the voltage and power gain the machine is run at as high a speed as possible to give a large voltage per field ampere.

Transfer Function for Cascaded Machines

For further increases in gain, two machines could be cascaded as shown in Fig. 6.33.

The overall transfer function is:

$$\frac{e_3}{e_1} = \frac{e_2}{e_1} \cdot \frac{e_3}{e_2} = \frac{\check{K}_1}{1 + \tau_1 p}\,\frac{K_2}{1 + \tau_2 p} = \frac{K_1 K_2}{(1 + \tau_1 p)(1 + \tau_2 p)}.$$

The block diagrams drawn for each stage are of assistance in making up the overall transfer function.

The differential equation for the whole system can be obtained as before, but the solution is not quite as simple. The transient response is indicated in Fig. 6.32b. The output voltage reaches 0·632 of its final value in a time slightly greater than the sum of the individual time constants, $\tau_1 + \tau_2$. The voltage gain is the product of the individual gains if the internal resistance is neglected, but the time constants are only added, or nearly so. Cascading is an efficient method of improving

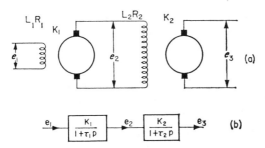

FIG. 6.33. Cascaded machines: (a) circuit; (b) block diagram.

the performance as an amplifier and explains the use of machines like the amplidyne, where this is achieved with one armature and field system.

Transfer Function of a Motor

The transfer function of a motor depends on the mode of operation and will be derived for the case where the flux is constant so that $k_\phi = E/\omega_m$ is constant. k_ϕ is also equal to the torque per ampere T_e/I_a, as will be shown in the next section when deriving eqn. (6.4); i.e. $T_e = k_\phi . I_a$. When writing down the electrical equation, the armature-circuit inductance will be neglected since this normally involves only a relatively small amount of stored energy, though this may be important in some circumstances, see Fig. 6.52 and Appendix E, Problem (9) for Chapter 6. The motor will be assumed to drive a mechanical load corresponding to a torque T_m. The total coupled inertia is J, the inertia torque therefore being $J . d\omega_m/dt$. Any torque due to mechanical resilience will be neglected, which is equivalent to assuming a rigid shaft and coupling.

The electrical circuit equation for instantaneous values is:

$$V = Ri_a + e = Ri_a + k_\phi . \omega_m.$$

Similarly, by balancing the developed torque against the absorbed torque, the mechanical torque equation is $T_e = T_m + J \, d\omega_m/dt = k_\phi i_a$

313

from which $i_a = (T_m + J \, d\omega_m/dt)/k_\phi$, and substituting for i_a in the voltage equation:

$$V = \frac{RT_m}{k_\phi} + \frac{RJ}{k_\phi} \cdot \frac{d\omega_m}{dt} + k_\phi \omega_m$$

which by rearrangement becomes:

$$\frac{V}{k_\phi} - \frac{RT_m}{k_\phi^2} = \frac{JR}{k_\phi^2} \cdot \frac{d\omega_m}{dt} + \omega_m = (1 + \tau_m p)\omega_m, \quad \text{where} \quad \tau_m = \frac{JR}{k_\phi^2}.$$

The equation is of a standard form again, eqn. (3.19), and will give a simple exponential change of speed for either a sudden change of load torque with voltage constant, or a sudden change of voltage with load torque constant. τ_m is an electromechanical time constant, being a function of the circuit resistance and the electromechanical conversion coefficient k_ϕ. In the general case, it is also a function of the mechanical load characteristic. For the case where the load torque is negligible by comparison with the inertia torque, the equation gives the transfer function between speed and voltage. Putting $T_m = 0$:

$$\omega_m/V = 1/k_\phi(1 + \tau_m p).$$

Further consideration will be given to Electromechanical Transients in the next section and in Appendix B the use of transfer functions will be extended to illustrate their application to simple feed-back control circuits.

Frequency Response

As more and more machines and circuit elements are included in a control system, the differential equation for the whole system becomes more and more complicated. Although the transient response is one of the most important requirements, it becomes easier to solve the equations by considering the response of the system not to a suddenly applied input, but to a sinusoidally varying input of constant peak value but of varying frequency. Since any particular system will have a unique transient response and a unique frequency response, there

314

must be a mathematical connection between the two. The frequency response is easier to calculate and when measured it conveys more information about the potentialities of the system.

By definition, the frequency response is the phasor ratio of output to input as a function of ω, the angular frequency. It can be obtained from the transfer function by substituting $j\omega$ for p. This can be seen by considering as an example the equation relating field current to field voltage on a generator. From Fig. 6.32:

$$e_f = (R+Lp)i_f$$

for instantaneous values, and it is known that:

$$\mathbf{E}_f = (R+j\omega L)\mathbf{I}_f$$

for sinusoidally varying values of frequency $f = \omega/2\pi$.

Multiplying by k_f: $k_f\mathbf{E}_f = (R+j\omega L)k_f\mathbf{I}_f = (R+j\omega L)\mathbf{E}_g$

from which:

$$\frac{\mathbf{E}_g}{\mathbf{E}_f} = \frac{k_f/R}{1+j\omega\tau},$$

which is the frequency response, $j\omega$ replacing p in e_g/e_f.

The magnitude ratio is:

$$\frac{E_g}{E_f} = \frac{k_f/R}{\surd(1+\omega^2\tau^2)} = \frac{\text{steady-state voltage gain}}{\surd(1+\omega^2\tau^2)}.$$

The phase angle is: $-\tan^{-1}\omega\tau = -\tan^{-1}\omega L/R$.

$\mathbf{E}_g$ is a phasor lagging the phasor $\mathbf{E}_f$ by this angle. In this particular case as frequency increases, $\mathbf{E}_g$ falls and it lags behind $\mathbf{E}_f$ by an increasing phase angle.

6.7 MOTOR CHARACTERISTICS: SPEED AND TORQUE CONTROL: GENERAL MACHINE EQUATIONS.

For each generator field-arrangement there is a corresponding one for the motor. There is no need for self-excitation of course, because in the motoring mode, power in electrical form is already available.

However, if field control is not required, a permanent-magnet field system would give independence from a separate field supply, see Appendix A. The shunt motor has its field across the supply and therefore experiences any voltage fluctuations to which the armature is subject. A series motor has an excitation varying directly as load current. A separately excited motor could refer to one in which the field was fed from an independent source so that it was shielded from armature supply-voltage variations. With constant supply voltage assumed, the shunt motor would behave in the same way as the separately excited motor. Separate excitation is used when special field-circuit control is required.

One of the useful features of rotating electrical machinery is the ease with which the transition from a motoring to a generating mode can be made. The transition is particularly easy to control for the d.c. machine and this accounts for its popularity in many applications. The function is determined, as already explained, by the magnitude of the machine e.m.f. and this in turn is affected by changes to either speed, flux or to both quantities. In this section, both operational modes will be discussed.

Derivation and Application of D.C.-machine Control Equations

If eqn. (3.2) is multiplied by armature current, a power balance equation is formed like eqn. (3.15). Thus:

$$\underset{\substack{\text{Electrical terminal}\\\text{power}}}{V.I_a} = \underset{\text{Mechanical power}}{E.I_a} \pm \underset{\text{Armature copper loss}}{RI_a.I_a.} \qquad (6.3)$$

Equating the mechanical power $E.I_a$ to the torque times speed product $T_e . \omega_m$ gives:

$$T_e = (E/\omega_m).I_a = k_\phi . I_a. \qquad (6.4)$$

It is seen that k_ϕ has the alternative dimensions of e.m.f. per radian/sec or torque per ampere in newton-metres/ampere. It has the nature of an electromechanical conversion coefficient and as seen already on p. 295 it is directly proportional to the air-gap flux ϕ_m. Equation (6.4) can be rearranged to correspond with eqn. (1.2a) if, for E, is substituted the e.m.f. equation (5.7) modified to allow for *any* torque angle δ.

316

For $\delta = 0$, $E = 0$ and for any other angle, assuming a sinusoidal flux distribution, $E = 2p\phi.n.z_s \sin \delta$. Therefore:

$$T_e = \frac{2p\phi.n.z_s.\sin \delta}{2\pi n}.I_a = \frac{pz_s}{\pi}.\phi.I_a.\sin \delta. \tag{6.4a}$$

Note that the constant of proportionality referred to in Section 1.4 is pz_s/π and the angle δ must correspond to that between the two chosen m.m.f. axes. If $\phi_f(F_f)$, and $I_a(F_a)$, are chosen, then the angle must be that of the brush axis δ_{fa}, which is nearly always set at 90°. However, in eqn. (6.4), $E = f(\phi_m)$ is used and E can be calculated very conveniently from the circuit equation. If E is to be involved in the derivation of the equation it would appear that the angle should be δ_{ra}. At this point the practical realities must be emphasised. The flux distribution for a normal d.c. machine is far from sinusoidal and in any case, saturation effects are considerable and preclude a too simple interpretation of eqn. (1.2a). Equation (6.4) on the other hand is derived from a linear circuit equation (which could include the brush contact loss for greater accuracy), so that it provides a more reliable basis. From eqn. (6.4), $\sin \delta$ must therefore be unity ($\delta = 90°$) and k_ϕ must correspond to the resultant gap flux generating E, in accordance with eqn. (6.2b), but note the comments on p. 310 discussing the accuracy of Fig. 6.31.

If it is desired to work in the units sometimes favoured by mechanical engineers, then by equating EI_a to the mechanical power in watts:

$$E.I_a = 746 \times hp = 746 \times 2\pi \times NT_e'/33\,000,$$

from which $T_e' = 7.04\,(E/N).I_a$ lbf ft,   (6.4b)

where $E/N = k_N$ is the generated voltage per rev/min and is proportional to k_ϕ. It will be noticed that the torque per ampere in lbf ft is not the same as the voltage per rev/min, the constant 7.04 being involved because of the change of unit system.

Not all the electromagnetic torque T_e developed by the motor is available at the coupling because there is a torque loss T_{loss}. The loss torques are often similar in nature to the actual mechanical load torque, see p. 93, and it is sometimes convenient to group all the mechanical torques together as T_m, which is equal to T_e for steady-speed running. During the speed transients; see p. 347, there is an additional mechanical torque $J \cdot d\omega_m/dt$ and during current changes there will be an inductive voltage $L \cdot di/dt$, but for the moment, steady-state operation only is being considered. Figure 6.34 shows the power-flow diagram for this condition and it illustrates both the equation for an electrical sink,

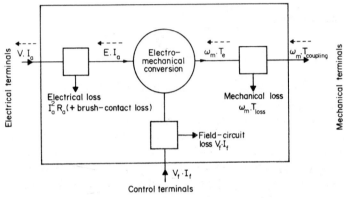

FIG. 6.34. Power flow for motor (←-- generator).

$V = E + RI_a$, and the equation for a mechanical source, $T_e = T_m = T_{coupling} + T_{loss}$, analogous to $E = V + RI$ for an electrical source. $T_{coupling}$ is the mechanical "terminal" torque. For generating operation, the power flow is reversed so that both electrical and mechanical equations take a negative sign. A control input is also shown, which for the d.c. machine would be the field circuit. It involves a power loss $V_f I_f$ under all operating conditions. Figure 6.34 should be compared with the diagram on the front cover of the book.

It is convenient at this point to list all the equations derived which are relevant to this section:
Electrical circuit equation:
$$V = E \pm RI_a. \tag{3.2}$$
Power balance equation:
$$VI_a = EI_a \pm RI_a^2. \tag{6.3}$$

Magnetisation characteristic:
$$k_\phi = f(I_f, I_a) = (pz_s/\pi).\phi_m \simeq f(I_f) \qquad (6.2c)$$

E.m.f. equation:
$$\omega_m = E/k_\phi = (V \pm RI_a)/k_\phi \qquad (6.2b)$$

Electromechanical relationship:
$$T_e = k_\phi.I_a. \qquad (6.4)$$

Series-machine equation:
$$I_a = I_f.$$

Mechanical system:
$$T_e = T_{coupling} \pm T_{loss} = T_m. \qquad (6.5)$$

Mechanical load characteristic, p. 93:
$$T_m = f(\omega_m) \simeq k_1 + k_2\omega_m + k_3\omega_m{}^2. \qquad (6.6)$$

One of the attributes of an engineer might well be said to be his ability to derive, interpret and apply the equations of a system. The above list provides a ready-made exercise of this kind and a few suggestions will now be made for the reader to follow through and verify by manipulating the equations. It will be noted that they are all linear apart from the magnetisation and load characteristics. These two are therefore used either at the beginning or at the end of the calculation; e.g. having found k_ϕ, I_f is read off from the magnetisation curve, or, given I_f, the value of k_ϕ is first read off from the curve to use in the solution of the system equations. Similarly, if the mechanical torque is specified, the speed follows from the load characteristic; or given ω_m, the corresponding value of T_m can be read off the curve.

Consider first the main factors affecting the machine speed. E is approximately equal to V normally, because the RI_a drop is usually small. Consequently, from eqn. (6.2b), speed is directly proportional to voltage very nearly and is inversely proportional to flux, which means that speed is approximately inversely proportional to field current. For a series motor, this means that the speed falls appreciably as *armature* current rises; see p. 331. Let us suppose that the actual field current is required for a given voltage, speed, resistance and mechanical load. Then by manipulating eqns. (3.2), (6.2b) and (6.4) above:
$$k_\phi = [V \pm \sqrt{(V^2 - 4R.\omega_m.T_m)]/2\omega_m}$$
and the curve of eqn. (6.2c) then gives I_f and the field-circuit voltage

and/or resistance to give this field current. The larger value of k_ϕ is the one required as will be understood later when examining Fig. 6.36a. If instead, k_ϕ had been given, then the required voltage follows from:

$$V = R \cdot T_m/k_\phi + k_\phi \cdot \omega_m.$$

If the electromagnetic torque is unknown then:

$$T_e = k_\phi(V - k_\phi \omega_m)/R,$$

and if the required total series resistance for a particular condition is wanted:

$$R = k_\phi(V - k_\phi \omega_m)/T_m,$$

assuming that voltage, field current, magnetisation and load characteristics are known.

The armature current which will be drawn from the supply to drive a particular mechanical load, with V and R known, follows from eqns. (6.3) and (6.4) as:

$$I_a = [V \pm \sqrt{(V^2 - 4 \cdot R \cdot \omega_m \cdot T_m)}]/2R,$$

the smaller value being the one required.

Finally, the equations can be used to show the value of the apparent resistance of the machine viewed from the electrical terminals as:

$$\frac{V}{I_a} = \frac{k_\phi^2 \cdot \omega_m}{T_m} + R.$$

This interesting equation shows that the mechanical load appears from the electrical terminals like a resistance, equal to $k_\phi^2 \cdot \omega_m/T_m$ and if *either* the speed or the torque reverses, the apparent resistance is negative so that the machine must be behaving as a "negative" motor; i.e. a generator. This kind of mechanical-load representation is particularly useful for the induction machine as will be seen in Chapter 7.

When solving numerical problems in this section, these derived equations will not necessarily be used, since it may be more instructive to proceed step by step. Further, not all the possibilities have been exhausted; the most important relation, for speed versus torque, has yet to be discussed. Before a more detailed treatment of this curve is given, a general appraisal of d.c. motor control features will now be made by using the equations in *per-unit* form (see p. 90). In general, it is necessary to define the 1 *per-unit* reference or base values for V, I_a, $\phi(k_\phi)$, T_e, *power* (P), ω_m and *impedance* (R), of which three may be chosen arbitrarily and the others derived therefrom. Taking rated values V_R, I_{aR} and $k_{\phi R}$ as 1 *per-unit*:

$$V \text{ per unit } (V_{\text{p.u.}}) = V/V_R; \quad I_{a\text{ p.u.}} = I_a/I_{aR}; \quad \phi_{\text{p.u.}} = k_\phi/k_{\phi R}$$

$$R_{\text{base}} = \frac{V_{\text{base}}}{I_{\text{base}}} = \frac{V_R}{I_{aR}}, \quad \therefore R_{\text{p.u.}} = \frac{R}{V_R/I_{aR}} = \frac{R \cdot I_{aR}}{V_R}$$

$$\omega_{\text{base}} = V_{\text{base}}/k_{\phi\text{base}} = V_R/k_{\phi R} \quad \text{and using eqns. (3.2) and (6.2b):}$$

$$\omega_{m\text{ p.u.}} = \frac{[V - R \cdot I_a(I_{aR}/I_{aR})]/k_\phi,}{V_R/k_{\phi R}}$$

which, by rearrangement, using the expressions derived above, becomes:

$$\omega_{m\text{ p.u.}} = \frac{V_{\text{p.u.}} - R_{\text{p.u.}} \cdot I_{a\text{ p.u.}} \cdot}{k_{\phi\text{ p.u.}}}$$

From this expression:

$$I_{a\text{ p.u.}} = (V_{\text{p.u.}} - k_{\phi\text{ p.u.}} \cdot \omega_{m\text{ p.u.}})/R_{\text{p.u.}} \cdot$$

$$\text{Torque}_{\text{base}} = \frac{\text{Power}_{\text{base}}}{\text{Speed}_{\text{base}}} = \frac{V_R \cdot I_{aR}}{V_R/k_{\phi R}} = k_{\phi R} \cdot I_{aR}$$

from which:

$$T_{e\text{ p.u.}} = (k_\phi \cdot I_a)/(k_{\phi R} \cdot I_{aR}) = k_{\phi\text{ p.u.}} \cdot I_{a\text{ p.u.}} \cdot$$

It can be seen that the *per-unit* expressions for ω_m, T_e and I_a are identical in form with the normal expressions. This is generally true in the *per-unit* system providing that suitable base quantities are chosen. In the present case, 1 *per-unit* speed must correspond to 1 *per-unit* voltage; i.e. it must be $V_R/k_{\phi R}$ if the terminal voltage is used, or $(V_R - I_{aR} \cdot R)/k_{\phi R}$ if using rated e.m.f. as the base. Constants of proportionality in the original expressions do not appear in the *per-unit* expressions. Since a *per-unit* quantity is really a fractional ratio of two values of the same parameter, any constant would be common and would cancel.

The expression for speed can be rearranged in terms of torque instead of armature current by substituting T_e/k_ϕ for I_a. Hence:

$$\omega_m = \frac{V - RT_e/k_\phi}{k_\phi} = \frac{k_\phi V - RT_e}{k_\phi^2}. \tag{6.7}$$

This is a general speed/torque equation for all d.c. machines and applies when used either with *per-unit* quantities or with the actual values, as explained above. It will first be used to demonstrate the equations of the separately excited motor for which k_ϕ can be controlled independently of the other terms. The *per-unit* suffices will be omitted from the equations for simplicity. The speed expression in the following table is easily obtained from eqn. (6.7), by inserting the appropriate *per-unit* value of the fixed quantity or quantities given in the first column. The starting-torque expression, where applicable, is obtained from the starting current V/R multiplied by k_ϕ.

Condition	Speed ω_m	Starting torque
Rated $V = V_R$. ($V = 1$ per unit)	$(k_\phi - RT_e)/k_\phi^2$	k_ϕ/R
Rated V and k_ϕ. (V and $k_\phi = 1$)	$1 - RT_e$	$1/R$
Rated T_e and k_ϕ. (T_e and $k_\phi = 1$)	$V - R$	
Zero torque	V/k_ϕ	
Rated V, T_e and k_ϕ. (rated speed)	$1 - R = \omega_R$	

On Fig. 6.35 are drawn speed/torque curves at rated voltage for various conditions. Only two points are needed to draw the straight lines: the speed at zero torque, $\omega_{n.l.}$ and the torque at zero speed, the starting torque. Note that the speed at rated torque is negative in one case though normally the speed regulation is quite small, since without external resistance in the armature circuit, $R = R_a$ and has a relatively low value. The full-load resistance drop is perhaps 3% of normal voltage (0·03 p.u.) on a large machine though increasing as a *per-unit* value as the machine size decreases.

322

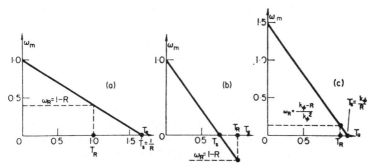

FIG. 6.35. Speed/torque curves in *per unit* for separately excited d.c. motor, $V = 1$ p.u.: (a) $R = 0.6$ p.u., $k_\phi = 1$ p.u.; (b) $R = 1.33$ p.u., $k_\phi = 1$ p.u.; (c) $R = 0.6$ p.u., $k_\phi = 0.67$ p.u.

Further insight is obtained into the characteristics of the d.c. machine if the speed/torque characteristic of the mechanical load is incorporated into eqn. (6.7). Two idealised ω_m/T_m relationships will be considered; firstly the coulomb-friction load for which the torque is independent of speed and secondly, the viscous-friction load for which torque is directly proportional to speed. For the first case, substituting $T_e = T_m = 1$ *per unit* in eqn. (6.7) gives

$$\omega_m = (k_\phi V - R)/k_\phi{}^2.$$

For the second case, $T_m = k_2\omega_m$ and presuming that rated torque occurs at rated speed, then $T_R = k_2\omega_R$ and $\omega_R = 1 - R_a$ *per unit*, from the previous table. Hence: $T_e = T_m = k_2\omega_m/k_2\omega_R = \omega_m/(1 - R_a)$ *per unit*.

Substituting in eqn. (6.7) gives:

$$\omega_m = \frac{k_\phi V - R\omega_m/(1 - R_a)}{k_\phi{}^2} \quad \text{from which} \quad \omega_m = \frac{k_\phi V}{k_\phi{}^2 + R/(1 - R_a)}.$$

On Fig. 6.36 these two expressions for speed are plotted against k_ϕ for various values of R and for rated voltage $V = 1$ *per unit*, maintained. Note that for the constant-torque load, k_ϕ must first be increased to a high enough value to produce this torque before the machine will start; e.g., when $R = 0.8$ *per unit*, the current at zero speed is $1/0.8$

$= 1.25$ *per unit* and a k_ϕ of 0·8 is necessary to produce 1 *per-unit* torque. For the viscous-friction load, however, no torque is required at zero speed, so the ω_m/k_ϕ curve starts from zero speed and zero flux.

For high values of R, the speed increase is approximately linear with flux and for the constant-torque load, simple relationships exist. When the value of R is less than 0·5 per unit, increase of field current gives an increase of speed initially but it reaches a maximum. Thereafter the speed decreases as the field current is increased. This second region of field control is by far the most commonly used since *per-unit* resistances are less than 0·1 except on machines smaller than about $\frac{1}{2}$ kW. More important, if the armature current is calculated, it will have very high values on the early part of the characteristic. With $R = 0.1$ *per unit*, for example, the starting current would be 10 *per unit*. For even smaller values of R, the maximum speed too would become excessive at normal voltage. For constant-voltage systems, therefore, a series resistance is temporarily inserted in the armature circuit to limit the current at starting, and the field current is set, initially, at or near the maximum value and never reduced below a certain minimum, to prevent over-speeding, see p. 346.

By differentiating the expressions for speed, it will be found for the constant-torque load that $d\omega_m/dk_\phi$ is always positive up to $k_\phi = 1$ per unit, providing that R is greater than 0·5 per unit. Conditions are different for the viscous-friction load unless $R = R_a$. Otherwise, for ω_{max} to occur at $k_\phi = 1$, R must be $1 - R_a$. With these conditions fulfilled, a simple speed control is possible because there is a continuous increase of speed with field current, up to the rated value. The armature-circuit resistance would have to be supplemented to the appropriate value and the circuit losses would be proportionately large. However, if the power itself is measured in tens of watts, or less, then the extra loss is not significant. Small d.c. *servo-motors* operate on such field-control schemes with the field current provided by an electronic amplifier. Even in the higher power applications up to several hundred kilowatts, field control of speed in this manner is sometimes desirable. The armature-circuit losses would be intolerable in this case, however, and so the supply generator is incorporated in a high-gain control

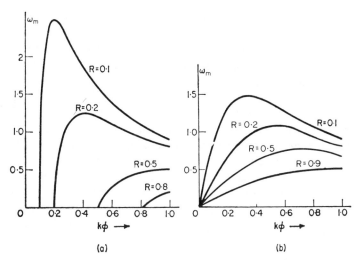

FIG. 6.36. Speed/flux (k_ϕ) curves in *per unit.* $V = 1$ p.u. constant.
$R_a = 0\cdot1$ p.u. $T_m = 1$ p.u. at rated speed. $T_e = 1$ p.u. at rated flux and current.
(a) T_m constant; $\omega_{max} = 1/4R$ (generally, $\omega_{max} = V^2/4RT_m$);
(b) T_m proportional to speed: $\omega_{max} = V.\sqrt{[(1 - R_a)/4R]}$.

scheme which maintains the system-current constant against any load
change or condition. Such a scheme is mathematically explicable in
terms of an "infinite" circuit resistance since the current is fixed, and
changes to the apparent "resistance" presented by the motor at its
terminals are insignificant by comparison with infinity! Constant-
current systems of this kind are sometimes used on specialised rolling-
mill and marine drives.

Speed/Torque Characteristics

The relationship between speed and torque corresponds to the
relationship between voltage and current for a generator, i.e. it is the
motor external characteristic. The speed at which a motor will run
depends on the balancing point of the electromagnetically developed
torque T_e and the torque T_m arising mechanically, see Fig. 2.16. Both
torques are functions of speed though they may be approximately

constant, as in the case of the shunt motor and the coulomb-friction load; or they may vary considerably with speed, as for the series motor and for the fan-type load. There is some analogy here with the generator external characteristic and its load electrical characteristic, which are both functions of current. As an example, the output voltage of a self-excited generator is decided by the intersection of the field-circuit resistance and the armature-circuit terminal characteristics, which are both functions of field current, see Fig. 6.25.

The present discussion is concerned primarily with the speed/torque characteristic of the machine itself; i.e. ω_m/T_e. The mechanical speed/torque characteristic can then be superimposed on the same graph as on Fig. 2.16, to obtain the balancing speed. The speed/current and speed/torque equations have already been used and are reproduced in the most convenient forms below, as derived from eqns. (3.2), (6.2b) and (6.4).

$$\omega_m = \frac{V}{k_\phi} - \frac{R}{k_\phi} . I_a = \frac{V}{k_\phi} - \frac{R}{k_\phi^2} . T_e . \tag{6.7a}$$

R could include external resistance but if this is zero, there is only the machine armature-circuit resistance R_a, which in the descriptive matter following will be taken to include armature, interpole and compensating windings and the brush drop for simplicity. The series winding if present will be considered separately as R_f, which symbol will also be used for the shunt-excited or separately excited field resistance.

The final term in eqn. (6.7a), the influence of the load, is normally only a second order effect if $R = R_a$ which is small. The first term V/k_ϕ immediately shows the general effects of voltage and flux on speed as already discussed. To draw the speed/torque or speed/current curves merely requires the substitution of various values of torque (or current) and a knowledge of all the other parameters on the right-hand side of the equation will permit the corresponding speed values to be worked out. If field current is being varied, then the values of k_ϕ must first be obtained from the k_ϕ/I_f relationship of eqn. (6.2c). This is particularly relevant to the series machine as will soon be explained. Perhaps it should be emphasised at this stage, that in the solution of the equations on pp. 318 and 319 for any condition, k_ϕ is related independently to both the magnetisation curve and to the rest of the system, but nevertheless it must satisfy both of these relationships simultaneously for steady-state conditions. Similar considerations apply to the torque $T_e = T_m$, which is related to both the electromagnetic and the mechanical systems.

Shunt-excited and Separately Excited Motors are characterised by having control of k_ϕ which is approximately independent of supply voltage or the machine load. For any constant value of k_ϕ, eqn. (6.7a) and Fig. 6.35, show that the speed/torque curve is a straight line starting from V/k_ϕ at zero torque and falling by an amount proportional to the resistance drop and hence to the torque. If the field current I_f and therefore k_ϕ are reduced, the no-load speed is increased in inverse proportion and the slope of the characteristic becomes steeper. In practice, the lower field m.m.f. permits the armature reaction to exert a proportionately larger weakening effect, reducing k_ϕ still further and possibly causing the speed to rise with load, the first term of eqn. (6.7a) being dominant. This is dangerous, see later under *Compound Motor*, p. 334. Figure 6.37 shows characteristics based on eqn. (6.7a) and the above reasoning. They are bounded by the limits of maximum torque, maximum armature

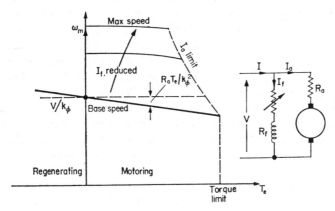

FIG. 6.37. Speed/torque for shunt-excited or separately excited motor.

current and maximum speed. Note that increase of field current is limited from considerations of winding heating and saturation, so that there is a lower limit of speed too, for this method of speed control. However, the simplicity of speed control above this base speed is one of the main virtues of the d.c. motor, and even though the torque per ampere is reduced as field is weakened, this is quite acceptable for many practical applications. With properly compensated machines, the ratio maximum-speed/base-speed can be as great as 5/1.

If the machine were somehow to gain speed so that ω_m became higher than V/k_ϕ, the developed torque would reverse; i.e. a generating condition. This situation, in which the net mechanical torque reverses, can

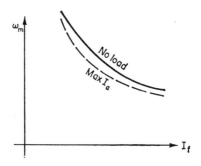

6.38. Speed/field-current for shunt motor.

arise when the mechanical system has an active component due to gravitational forces, e.g. an electrically driven vehicle moving down a gradient, or due to another coupled machine which can supply mechanical power. The d.c. machine is then said to be regenerating and the characteristic is continued into the negative-torque region as shown. The friction-loss torque component still opposes motion of course.

Figure 6.38 shows how speed varies with field current for machines of moderate size and above, where R_a *per unit* is less than about 0·1. The curve shape could be inferred from Fig. 6.36a with small values of R_a but in any case it follows from the dominant first term of eqn.

(6.7a), armature current not having a very large effect on the characteristic. The speed rises as field current is reduced and the reduction of I_f must be limited to avoid overspeeding. The external field-circuit resistance required for any particular speed and field supply-voltage V_f, is given by (V_f/I_f) minus the field winding resistance.

EXAMPLE E.6.4

A 25-hp, 500-rev/min, d.c. shunt-wound motor operates from a constant supply voltage of 500 V. The full-load armature current is 42 A. The field resistance is 500 Ω, the armature resistance is 0.6 Ω and the brush drop may be neglected.

Calculate:

(a) the field current required to operate at full-load torque when running at 500 rev/min. What would be the no-load speed with this field current?

(b) Calculate the speed, with this field current, at which the machine must be driven in order to regenerate with full-load armature current.

(c) Calculate the extra field-circuit resistance required to run at 600 rev/min; (i) on no load, (ii) at full-load torque.

(d) Calculate the external armature circuit resistance required to operate at 300 rev/min with full-load torque.

The magnetic characteristic was taken at 400 rev/min.

The problem will be solved in mechanical engineers' units so that $E = k_N N$ and $T_e = 7.04\,k_N\,I_a$ (p. 317). N is the speed in rev/min and k_N is the generated e.m.f. per rev/min. From the data given below, k_N is calculated immediately and the curve k_N/I_f plotted.

Field current	0·4	0·6	0·8	1·0	1·2	A
Generated e.m.f.	236	300	356	400	432	V
k_N = e.m.f./400	0·59	0·75	0·89	1·0	1·08	V per rev/min.

(a) E at full load $= V - I_a R_a = 500 - 42 \times 0.6 = 474.8$ V

$\therefore k_N$ required $= 474.8/500 = 0.9496$.

From the k_N/I_f curve this requires 0·9 A and the field circuit resistance must be 500/0·9 $= 555$ Ω, i.e. an external 55 Ω.

On no load, $E = 500$ V neglecting the small $R_a I_{n.1.}$ drop.

$\therefore N = E/k_N = 500/0.9496 = \underline{527\ \text{rev/min.}}$

(b) when regenerating $E = V + I_a R_a = 525.2$ V

$\therefore N = 525.2/0.9496 = \underline{554\ \text{rev/min.}}$

(c) (i) On no load, $E = 500$ V $\therefore k_N$ required $= 500/600 = 0.833$.

From k_N/I_f curve this requires 0·7 A as the field current.

The extra field circuit resistance is $500/0.7 - 555 = \underline{159\ \Omega.}$

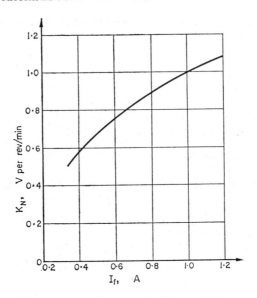

(ii) Full-load electromagnetic torque $T_e = 7.04 \times 0.9496 \times 42$
$$= 280 \text{ lbf ft.}$$

Useful torque at full load, from the rating particulars given:

$$T_{\text{coupling}} = 25 \times \frac{33,000}{2\pi \times 500} = 262.5 \text{ lbf ft,}$$

i.e. there is a mechanical loss torque of 17·5 lbf ft due to the iron and mechanical losses. Assuming this varies as speed, the total torque developed at 600 rev/min must be $262.5 + 17.5 \times 600/500 = 283.5$ lbf ft.

(This is only a nominal allowance for the change of loss torque and no great error would follow from assuming it to be constant.)

Hence, $T_e = 7.04 \, k_N I_a = 283.5$, from which $k_N = 40.2/I_a$.

Further, $E = 500 - 0.6 \, I_a = k_N N = (40.2 \, I_a) \times 600$, from which a quadratic equation in I_a can be obtained; i.e. $I_a{}^2 - 833 \, I_a + 40,200 = 0$—see alternative development of equation on p. 320.

The lower and only practicable value of I_a is found to be 51·5 A and this gives $k_N = 40.2/51.5 = 0.78$, and $E = 500 - 0.6 \times 51.5 = 468.8$ V.

I_f from the curve is 0·625 A, and the extra field circuit resistance is $500/0.625 - 555 = \underline{245 \, \Omega.}$

330

Note that k_N is very nearly $0.9496 \times 500/600 = 0.79$. This neglects the small change in E due to the additional $I_a R_a$ drop necessary to maintain the torque at full load with reduced flux.

(d) The developed torque must be $262.5 + 17.5 \times 300/500 = 273$ lbf ft.

With the flux maintained as is usual for speed reductions, the current becomes $42 \times 273/280 = 41$ A.

The generated e.m.f. will be $k_N N = 0.9496 \times 300 = 284$ V.

The total resistance drop in the armature circuit must be 216 V.

$\therefore$ the external resistance must be $216/41 - 0.6 = \underline{4.66\ \Omega}$.

Note: It would be a useful exercise for the student to recalculate this problem in SI units and check the answers, using the equations developed from those on pp. 318 *et seq.*

The Series Motor is shown on Fig. 6.39. With modern control techniques it is possible to make the field current a function of any desired parameter of the system. The series motor is a much older and simpler arrangement whereby the field current is made equal to the armature current or any desired fraction of it by using field diverters. Neglecting saturation, k_ϕ is proportional to I_a so that $T_e = (kI_a)I_a = kI_a^2$, where k is a constant. Approximately then, $\omega_m = V/k_\phi = V/kI_a$, so that speed is inversely proportional to current. In terms of torque, since it follows from the first equation that $I_a = \sqrt{(T_e/k)}$:

$$\omega_m \simeq V/k\sqrt{(T_e/k)} = V/\sqrt{(kT_e)}. \tag{6.8}$$

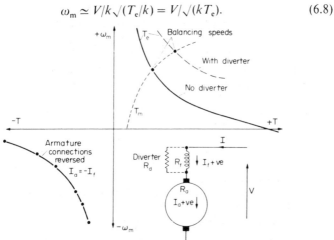

FIG. 6.39. Speed/torque for series motor.

331

Thus, neglecting resistance and saturation, the speed is less than inversely proportional to torque, a quadruple increase in torque being necessary to halve the speed. However, the characteristic is still steep enough to limit the motor input considerably, by comparison with the shunt motor, should there be a torque overload. Further, since flux increases with load, a given torque demand can be met by a smaller increase of current.

As the load and current decrease, k_ϕ falls and the speed could rise to dangerously high values. It is essential that the torque losses of the motor and the minimum load of any coupled machinery are sufficient to keep the no-load speed within safe mechanical limits.

Saturation cannot be neglected in practice and though it is not difficult to take into account, a tabular method or its equivalent becomes necessary. Since $\omega_m = E/k_\phi$, $T_e = k_\phi I_a$ and E, I_a and k_ϕ are all functions of field current, ω_m and T_e can be calculated at several values of I_f and the speed/torque curve plotted. Referring to the equations on pp. 318 and 319 and with the simple series connection:

$$k_\phi = f(I_f) \text{ from the magnetisation curve.}$$
$$E = V - I_f(R_a + R_f); \qquad \text{and } I_a = \pm I_f.$$

Note that the $\pm$ sign indicates the two possibilities of field/armature relative connection, giving rise to either positive or negative torque and mirror-image speed/torque curves, as shown on Fig. 6.39.

With a series field diverter R_d, the resistance becomes $R_a + R_f R_d/(R_f + R_d)$ and the armature current as a function of field current becomes $I_a = I_f(R_f + R_d)/R_d$. With these new values substituted where appropriate, the calculation is the same as before. Field diverter control is sometimes used and the effect is to lift the speed ordinates as shown in Fig. 6.39. Armature diverters are also used sometimes, see below.

EXAMPLE E.6.5

The following input data were taken on a series motor tested as indicated in Fig. 6.21a when running at a constant speed of 300 rev/min.

Field current I_f	=	20	30	40	50	A
Terminal voltage V	=	162	215	250	274	V

The resistance R_a is $0.3\ \Omega$ and the series-field resistance $R_f = 0.1\ \Omega$. Neglecting brush drop, determine the speed/torque curves when connected to a 250 V supply:

(a) without any external resistance;

(b) with an external series resistance of $2\,\Omega$;

(c) with a series-winding diverter of $0\cdot1\,\Omega$;

(d) as for (b) but with an armature diverter of $10\,\Omega$.

The test data must first be analysed to get k_ϕ as a function of I_f. Since ω on test $= 2\pi \times 300/600 = 10\pi$ radians/sec, then $k_\phi = E/10\pi$, where $E = V - I_f(R_a + R_f)$ for the test condition.

Field current $I_f =$ 20   30   40   50   A

$E =$ Test voltage $-0\cdot4\,I_f =$ 154   203   234   254   V

$k_\phi = E/10\pi =$ 4·9   6·45   7·45   8·1

It is now necessary to obtain expressions for E and I_a as functions of I_f, and then at each value of I_f, $T_e = k_\phi I_a$ and $\omega_m = E/k_\phi$.

(a) $I_a = I_f$; $E =$ Terminal voltage $- I_a R = 250 - 0\cdot4\,I_f$.

(b) $I_a = I_f$; $E = 250 - I_f(0\cdot4 + 2) = 250 - 2\cdot4\,I_f$.

(c) $I_a = \dfrac{0\cdot1 + 0\cdot1}{0\cdot1}\,I_f = 2\,I_f$; $E = 250 - I_a\left(0\cdot3 + \dfrac{0\cdot1 \times 0\cdot1}{0\cdot1 + 0\cdot1}\right)$

$$= 250 - 2\,I_f(0\cdot35) = 250 - 0\cdot7\,I_f.$$

(d) Referring to the adjacent figure:

$$I_a = I_f - I_d = I_f - \frac{250 - 2\cdot1\,I_f}{10}, \qquad \therefore\ I_a = 1\cdot21\,I_f - 25.$$

$E = 250 - 2\cdot1\,I_f - 0\cdot3\,I_a.$ Substituting for I_a; $E = 257\cdot5 - 2\cdot46\,I_f.$

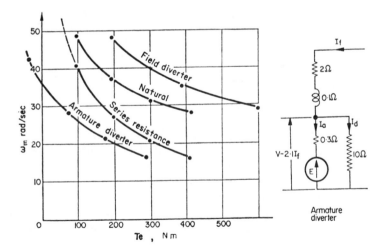

Using the above expressions, the following tabular method is used to calculate points on the speed/torque curves, which are plotted on the adjoining figure.

(a)

k_ϕ		4·9	6·45	7·45	8·1	
$I_a = I_f$		20	30	40	50	A
$E = 250 - 0.4\,I_f$	= 242	238	234	230		V
$T_e = k_\phi I_a$	= 98	193	298	405		Nm
$\omega_m = E/k_\phi$	= 49·3	37·8	31·4	28·4		rad/sec

(b)

$I_a = I_f$; torque					
figures as for (a)	98	193	298	405	Nm
$E = 250 - 2.4\,I_f$ = 202	178	154	130		V
$\omega_m = E/k_\phi$ = 41·2	27·6	20·7	16·1		rad/sec

(c)

$I_a = 2\,I_f$	= 40	60	80		A
$E = 250 - 0.7\,I_f$	= 236	229	222		V
$T_e = k_\phi \cdot I_a$	= 196	386	596		Nm
$\omega_m = E/k_\phi$	= 48·3	35·3	29·8		rad/sec

(d)

$I_a = 1.21\,I_f - 25$	= -0·8	11·3	23·4	35·5	A
$E = 257.5 - 2.46\,I_f$ = 208·3	183·7	159·1	134·5		V
$T_e = k_\phi \cdot I_a$	= -3·9	73	174	287	Nm
$\omega_m = E/k_\phi$	= 42·6	28·6	21·3	16·7	rad/sec

Note that circuit connection (d) gives a finite no-load speed at zero torque.

The Compound-wound Motor, see p. 288, has a more complex functional relationship for the total field excitation. The series winding could be differential so that the flux was weakened with load, causing the speed to rise. However, the differential connection is rarely used because it is unsafe, particularly with low values of main-field current. Any increase of load leads to an increase of speed and probably a further increase of load, which is an unstable condition. Even without a series winding, the speed tends to rise with load on weak field conditions due to the more powerful influence of armature reaction. It is usual on shunt motors to provide a few turns of cumulative series-stability-winding to counteract this tendency. A strong cumulative series winding is sometimes used to cause an appreciable drop of speed with load, perhaps 20%, to enable a coupled flywheel to give up some of its stored energy to a recurrent peak load and thus shield the electrical supply system from the surge. Calculations of compound-motor characteristics are not difficult but it may be convenient to express k_ϕ as a function of ampere-turns. For any particular main-field current, various values of armature current will yield corresponding values of total field m.m.f., k_ϕ, E, T_e and ω_m.

EXAMPLE E.6.6

The machine of E.6.4 has a cumulative series-winding added, its strength at full-load current being equivalent to 0·2 amperes through the shunt turns. By how much will this reduce the full-load speed if the field current is maintained at 0·9 A? What will be the new torque and output horsepower? The additional resistance due to the series turns can be neglected.

Effective excitation in shunt terms = $0·9 + 0·2$ A = $1·1$ A.

This gives a k_N of 1·04 volts per rev/min.

Hence $N = E/k_N = 474·8/1·04 = 455$ rev/min.

Developed torque $T_e = 7·04 \times 1·04 \times 42 = 307$ lbf ft.

Useful torque $= 307 - 17·5 \times 455/500 \quad = \underline{291 \text{ lbf ft.}}$

Output horsepower $= \dfrac{2\pi \times 455 \times 291}{33,000} \quad = \underline{25·2 \text{ hp.}}$

N.B. As for Example E.6.5, a check using SI units would be appropriate.

Ward–Leonard Speed and Torque Control

If a variable voltage is provided, either from a d.c. generator or a controlled rectifier, a most flexible control circuit becomes available, and in its original form it was called the Ward–Leonard system; Fig. 6.40. The direction and speed of rotation, the direction and magnitude of power flow are all decided by the relatively small powers in the field circuits. The two machine e.m.f.s E_1 and E_2 differ only sufficiently to maintain the IR and brush drop in the local circuit. A small change to either E_1 or E_2 can change the current considerably, or interchange the functions of the two machines. The motor can be made to convert stored mechanical energy into electrical energy, regenerating thus right down to zero speed simply by maintaining the other machine e.m.f. lower than that of the regenerating "motor" as the speed falls. At zero speed, the other machine e.m.f. will have become negative and further

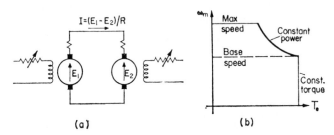

(a)   (b)

FIG. 6.40. Ward Leonard control.

increases in negative voltage will cause the motor to run up to speed with reverse rotation. In view of the fact that armature currents can change so widely for small e.m.f. changes, it is not surprising that d.c. machines are subject to heavy peak currents in service. This is particularly true of machines in steelworks,[11] where rapid control of the various processes is required and current overloads are typical of normal operation.

The envelope of motor characteristics is shown in Fig. 6.40b, see also Fig. 6.37. Speed control is by armature voltage variation up to base speed with the flux maintained at its rated value, thus giving the maximum torque per ampere. Motor-field weakening is employed beyond base speed up to the maximum permitted by mechanical or commutation considerations. Although the reduction in k_ϕ causes the torque to fall in this region, with any fixed value of I_a, the rise of speed means that the power itself is approximately constant as is

336

confirmed by the fact that beyond base speed, V has been maintained at its rated value and I_a too has been taken at a fixed value.

The envelope is only shown for positive torque and positive speed, i.e. forward motor-rotation. The speed and torque can both be reversed to give similar diagrams in each of the four quadrants. The electromagnetic pictures of Fig. 6.41 show the four different modes of operation. Although the diagrams indicate that the field polarity is maintained and the armature current is reversed to get negative torque, the same

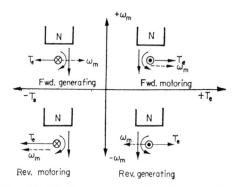

FIG. 6.41. The four quadrants of machine operation.

effect is obtained by reversing the field polarity and maintaining the armature-current direction. Field reversal is necessary with some forms of rectifier control. These four quadrants of operation are feasible for any a.c. or d.c. rotating machine but the d.c. machine is much freer to transfer its operation between quadrants and run satisfactorily at any point within the envelope.

One of the most economical variable-speed drives is provided by the thyristor/d.c. motor combination,[12] in which a transformer and bank of thyristors replace the Ward–Leonard motor-generator (M.G.) set; i.e., the Ward–Leonard d.c. generator normally driven by an a.c. motor. When d.c. motor operation only is required, this controlled-rectifier system is relatively simple, giving voltage variation by gate firing-control which, in conjunction with the anode/cathode polarity, determines when a thyristor becomes conducting. The provision of regeneration facilities poses a difficulty, however, since current can only flow in one direction through a rectifier. To get reverse *power* flow, it is necessary to ensure that the polarity of the voltage at the terminals of the thyristor circuit is reversed, by delaying the gate firing-control. The d.c. machine too must have a polarity reversal with respect to the thyristor terminals of the previous connection. This can be achieved by a switch reversal of its connections or alternatively by reversing its field after de-energising the thyristor gates to inhibit conduction and avoid an otherwise excessive current surge. Figure 6.42 shows the

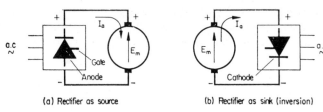

(a) Rectifier as source   (b) Rectifier as sink (inversion)

FIG. 6.42. Motoring and generating with thyristor converter.

different circuit conditions for motoring and regenerating, the transformer/thyristor circuit being represented by one controlled-rectifier unit. Note that the positive machine terminal is connected to the cathode in Fig. 6.42a but is connected to the anode in Fig. 6.42b. Since the positive terminal is now at the anode, the required reversal of both the thyristor/machine connections and the thyristor terminal polarity has been achieved. These two figures correspond respectively to operation in the upper right-hand and the upper left-hand quadrants of Fig. 6.41. For operation in the remaining two quadrants, the machine and rectifier polarities would have to be reversed from those shown on Fig. 6.42, though the connections and the current directions would, of course, be unchanged. Thus if two thyristor banks are provided and both are connected across the one machine in the opposing manner indicated, 4-quadrant operation becomes possible without a power-switching changeover. One group of thyristors is fired only for one direction of machine current and the other group is fired only when it is

338

desired to have a current flowing in the opposite direction through the machine. Interlocking protection is necessary to prevent the two rectifier units acting as short circuits on one another. This anti-parallel connection is used quite often even on large power units, though it is relatively complicated and rather expensive due to the duplication of thyristor equipment.

The circuit voltages are not steady d.c. quantities but consist of portions of sine waves which the rectifier unit selects by its switching action from the a.c. supply. Figure 6.43a shows a typical waveform obtained from a group of diodes giving m voltage pulses per fundamental cycle. The mean no-load voltage E_{do}, which would be read by a d.c. voltmeter (neglecting the 1–2 V diode voltage-drop), is readily shown by integration, to have a value $E_{do} = (m/\pi) \sin \pi/m \times E_{peak}$.

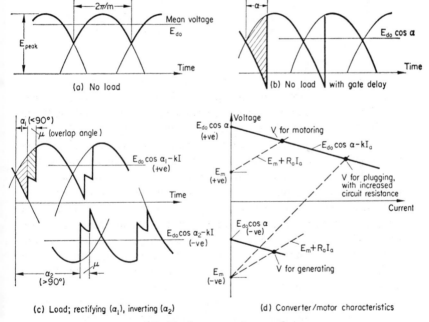

(a) No load

(b) No load with gate delay

(c) Load; rectifying (α_1), inverting (α_2)

(d) Converter/motor characteristics

FIG. 6.43. Thyristor-converter waveforms.

339

With the diodes replaced by controlled rectifiers like thyristors, the normal changeover or firing point can be delayed by an angle α say and for a highly inductive d.c. load, the mean output voltage is then reduced to $E_{do} \cos \alpha$; see Fig. 6.43b. If α is delayed beyond $90°$, the mean voltage becomes negative, permitting reverse power flow since the product, $E_{do} \cos \alpha . I$, will have a negative value. When the thyristor circuit is loaded, there is a further drop in voltage due to the a.c. supply inductance which delays the transfer of current from one thyristor to the next. During this overlap angle μ, two thyristor circuits share the current at their mean voltage as shown in Fig. 6.43c and the loss of voltage is proportional to the current value. References 21 and 33 discuss these various circuit conditions and mathematical relationships.

Consequently, for a controlled rectifier system, the d.c. component of the terminal voltage is given by $V = E_{do} \cos \alpha - kI_a$, whereas from the machine viewpoint, the terminal voltage is $V = E_m + R_a I_a$, using motoring conventions. These two straight-line relationships are shown on Fig. 6.43d for the following three conditions ($I_a +$ ve throughout).

Machine motoring: $E_m . I_a +$ ve, $E_{do} \cos \alpha . I_a +$ ve (rectification).

Machine generating: $E_m . I_a -$ ve, $E_{do} \cos \alpha . I_a -$ ve (inversion).

Machine plugging: $E_m . I_a -$ ve, $E_{do} \cos \alpha . I_a +$ ve (rectification).

Note: in the plugging mode (or reverse-current-braking mode), which will be discussed on pp. 345 and 346, the machine and the supply-voltages are in the same sense around the circuit and both feed electrical power into this local electrical circuit, where it is dissipated.

Typical waveforms for the voltage due to the thyristor circuit are shown on Fig. 6.43c. Note, however, that due to the special uni-directional current characteristics of rectifying devices, the current through a connected d.c. machine may become zero during part of the cycle if the voltage across thyristor units falls below the necessary voltage drop to maintain conduction. During this discontinuous current condition, for which overlap no longer takes place, the terminal voltage rises to the value of the machine e.m.f., the current being zero.

EXAMPLE E.6.7

A d.c. shunt motor has a rated voltage of 500 V and when providing rated torque at this voltage and at rated flux, it takes an armature current of 25 A and runs at 1000 rev/min. The armature resistance is 1 Ω. The machine is to be supplied from a 3-phase thyristor bridge energised at a line voltage of 440 V. At a d.c. output current of 25 A, the voltage regulation of the thyristor circuit is 20 V. Find:

(a) the value of the firing angle α and the rated flux factor k_ϕ, for operation at rated voltage and torque;

(b) the value of the firing angle α for motor operation at half speed with rated flux, assuming that the torque is proportional to speed;

(c) the value of α for machine operation as a generator, at rated terminal voltage and current and with k_ϕ unchanged but with speed increased;

(d) the value of series resistance in the armature circuit which is necessary to limit the current to 50% overload at standstill with α as in (a);

(e) the *per-unit* flux to limit the current to 100% overload when the machine is plugging at full speed with α as in (a) and external resistance as in (d).

The d.c. circuit may be assumed highly inductive so that there is no discontinuous-current condition. This means that the solution for one or two unknowns may be obtained from the following equations:

$$V = E_{do} \cos \alpha - kI_a = k_\phi \omega_m + RI_a.$$

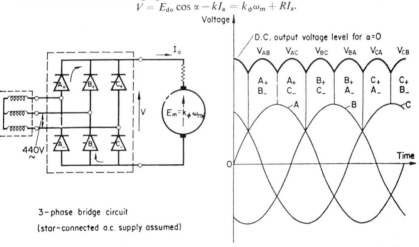

3-phase bridge circuit

(star-connected a.c. supply assumed)

The attached figure shows the arrangement of the popular 3-phase bridge circuit. The a.c. phase-voltage waveforms, the d.c. output voltage waveform for zero firing delay, and the rectifier units which conduct during particular intervals are also shown. The rectifier units automatically switch off in sequence as a result of a reverse potential appearing across them; e.g. during the interval when A_+ and B_- are conducting as shown, the anodes of B_+ and C_+ are negative with respect to the anode of A_+, which is virtually at the potential of the positive d.c. line. This switching effect results in the d.c. terminal voltage being equal to the maximum a.c. line voltage available at any particular instant and the ripple voltage corresponds to $m = 6$ pulses per fundamental cycle. Consequently:

$$E_{do} = (6/\pi) \sin (\pi/6) \times \sqrt{2} \times 440 = 0{\cdot}957 \times 621 = 593 \text{ V.}$$

The terminal voltage at 25 A is $V = 593 - 20 = 573$ V with $\alpha = 0$.

In general $V = 593 \cos \alpha - (20/25) \cdot I_a$, using source conventions.

(a) $V = 500 = 593 \cos \alpha - 0{\cdot}8 \times 25$ giving $\alpha = 29°$.

$\quad k_\phi = E_m/\omega_m = (500 - 25 \times 1)/(1000 \times 2\pi/60) = 4{\cdot}54 \text{ } N \text{ m/A.}$

(b) If T_e is proportional to ω_m, then I_a must be 12·5 Å.

$\quad E_m = k_\phi \cdot \omega_m = 4{\cdot}54 \times (2\pi/60) \times 500 = 237{\cdot}5$ V

$\quad\quad V = 593 \cos \alpha - 0{\cdot}8 \times 12{\cdot}5 = 237{\cdot}5 + 12{\cdot}5 \times 1 = 250. \quad \therefore \alpha = 64°.$

(c) $593 \cos \alpha - 0{\cdot}8 \times 25 = -500$ V ($-$ve source) $\therefore \cos \alpha = -480/593; \alpha = 136°.$

$\quad E_m + 25 \times 1 = -500$ V ($-$ve sink) $\quad \therefore E_m = -525$

$\quad \therefore$ speed $= (-525/4{\cdot}54) \times 60/2\pi = -1155$ rev/min.

Note that the same condition could occur with positive speed, providing there was a reversal of either the flux or the machine terminal connections.

(d) $593 \cos 29° - 0{\cdot}8 \times 37{\cdot}5 = 0 + I_a (1 + R_{ext})$

$\quad\quad 518 \quad - 30 \quad = 37{\cdot}5 (1 + R_{ext}) \quad \therefore R_{ext} = 12 \text{ } \Omega.$

(e) $\quad\quad 518 \quad - 0{\cdot}8 \times 50 = k_\phi \times (2\pi/60) \times (-1000) + 50 \times 13$

$\quad \therefore k_\phi = -172/-104{\cdot}7 = 1{\cdot}64 \quad$ and $\quad \phi = 1{\cdot}64/4{\cdot}54 = 0{\cdot}36 \text{ per unit.}$

It would be instructive for the student to draw the voltage diagram corresponding to Fig. 6.43d for all the circuit conditions above. Note that it is not usually necessary to incorporate additional resistance in the thyristor/motor circuit because $E_{do} \cos \alpha$ is a controllable quantity. Normally, plugging only occurs during transients.

Reversal of D.C.-Machine Rotation

Once the interpole- and compensating-winding polarities are set correctly for any particular function, they are automatically correct for motoring or generating with either direction of rotation. From the previous discussion, this can be regarded as most fortunate because otherwise it would detract from the flexibility which permits operation in any of the four quadrants and with Ward–Leonard control, repeated and rapid reversals are a common requirement. A further fortunate occurrence is that a series winding, if cumulative as a motor and therefore affording some overload protection, becomes differential as a generator and again is in a safe sense. However, if rotation is reversed, this happy state of affairs is upset, and to maintain a cumulative motor or differential generator, the series winding connections must be reversed relative to the armature connections. To achieve the reversal in practice it may be convenient to bring out the armature terminals to the reversing switch, or alternatively to reverse both shunt and series connections together. To reverse a series motor, again field or armature must be reversed, see Fig. 6.39. It is a useful exercise for the student to check all these statements with connection diagrams and in conjunction with the electromagnetic pictures shown on the four quadrant diagrams of Fig. 6.41.

Maximum Torque

Many d.c. motors, perhaps most of them in fact, operate under conditions where heavy torque-overloads are part of the normal duty. The maximum permissible torque is limited primarily by the overload commutation performance. However, on non-compensated machines particularly, operating at reduced flux, there may be an absolute limit to the torque. The loss of flux due to armature reaction at high currents becomes excessive so that all torques are less than otherwise expected. Eventually the flux falls at a greater rate than the armature current increases, so that the product of flux and current, which is proportional to torque, passes through a maximum value.

Starting: Armature Resistance Variations: Braking

To bring a motor from rest up to speed from a constant supply voltage is a special case of speed control in which the armature-circuit resistance is varied. At standstill the e.m.f. is zero so the armature resistance alone limits the starting current and torque. With a value of say 0·05 p.u., the starting current at full voltage would be 1/0·05 = 20 p.u. Extra resistance is therefore required, and is connected in series with the armature.

Referring first to the shunt motor, eqn. (6.7a) shows that the downwards slope of the speed/torque curve is directly proportional to the armature resistance with constant flux and voltage. By increasing the

343

resistance, the characteristic will cut the zero-speed axis to give a lower, though adequate starting torque and a reasonable starting current. Usually a figure of 50–100% more than the full-load value is chosen. Providing this torque is greater than the stiction torque of the motor and load, the speed will rise along the curve R_1, Fig. 6.44b. At some point, the resistance is changed over quickly to give the curve R_2, such that the current again rises to the starting value, and the

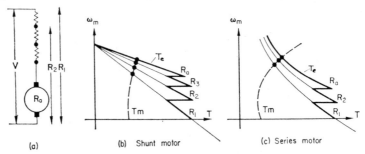

FIG. 6.44. Starting and resistance control.

speed continues to build up along this new characteristic. The process is continued through perhaps 2 to 10 steps until the balancing speed on the natural characteristic is reached.

For a series motor, the effects of the additional circuit resistance can easily be calculated using the tabular method and a typical set of curves is shown on Fig. 6.44c. Rather fewer steps of starting resistance are necessary due to the shape of the characteristics. As already explained, a given overload torque on a series motor is obtained with less overload current than on a shunt motor. This is a useful feature for loads where a high stiction torque is present, e.g. traction applications.

In the case of both the series and the shunt motor, it can be seen that series resistance could be used for speed control, as illustrated by the intersections of the typical T_m curves with the machine characteristics. Limited speed control by this means is sometimes employed but it is wasteful in power. Further, as resistance is increased, the greater sensitivity of speed to load changes is a drawback and eventually it becomes impossible to hold the speed steady at all.

Another point to notice is that the characteristics run into the negative-speed, positive-torque quadrant at high values of torque. This generating condition can arise if there is an active component in the load torque, as in a hoist. If this is sufficient to stop the machine and reverse its rotation, electrical braking then occurs, the generating torque opposing the load torque, see Fig. 2.16. A similar braking condition arises when running in the forward direction if the armature

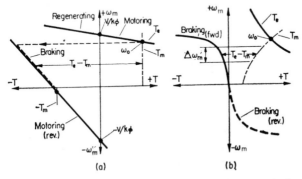

Fig. 6.45. Electrical braking: (a) reverse-current; separately excited machine;
(b) rheostatic (dynamic); series machine.

connections are reversed, tending to reverse the rotation. The speed/torque relationship for this condition is obtained simply, by substituting $-V$ for $+V$ in eqn. (6.7a). The curve is a reflection of the characteristic for $+V$, but when braking, extra armature-circuit resistance is inserted to limit the reverse current to a reasonable value. Figure 6.45a shows the path of operation following such a change to reverse-current braking from normal motoring operation against a constant load torque T_m; the speed being virtually constant during changeover since the mechanical system is relatively slow. To avoid running up in the reverse rotation, the armature circuit would have to be broken when the motor reached zero speed. This braking method is used for rapid stopping and since the machine e.m.f. and the supply voltage are in the same sense until the speed reverses, electrical power is being supplied from both sources and is dissipated in the local circuit resistance. The operation is sometimes referred to as plugging. The series machine can also be operated in this mode and the armature connections are brought out separately so that when they are reversed, the field current flows in the original direction as for the shunt machine. A reversal of torque always requires a reversal of field or armature current.

Another method of braking, for which the supply may be completely disconnected, involves dissipating the stored kinetic energy in a resistor connected across the machine terminals, through which the machine e.m.f. drives a current as the speed falls. This is called rheostatic- or dynamic-braking and the speed/torque characteristic is obtained by putting $V = 0$ in eqn. (6.7a). If the shunt field remains connected to the supply, k_ϕ could be maintained constant and the characteristic would then be a straight line passing through the origin. If, as with a series field, the machine is braking as a self-excited generator, then k_ϕ falls as the speed, voltage and current fall, so that a tabular method of calculating the speed/torque curve is necessary, as in Example E.6.5. Figure 6.45b shows a typical characteristic for both rotations. The transition through a rotation reversal could only occur if the load had an active component like a hoist. Unlike the friction loss-torque component, this would not reverse with rotation but would pull round the machine rotor against its developed electromagnetic torque until the speed was constrained at the intersection of the T_e and T_m characteristics, in a manner similar to that indicated on Fig. 2.16.

Starters for D.C. Motors

Figure 6.46 shows a simplified diagram of a shunt-motor starter for manual control, and includes the usual two protective devices. When the starter handle is moved from the "off" position, connection is made to the starting resistors causing the armature to be energised through the overload coil; the field being energised through the "no volt" coil and any external regulating resistance. The "no volt" magnet is sufficiently strong to hold the starting handle against the force of the return spring, provided the field current is not lower than the designed minimum. Should the supply fail, the handle will be released, so ensuring that full voltage is not applied directly to the armature when the supply is restored. This device also affords some protection

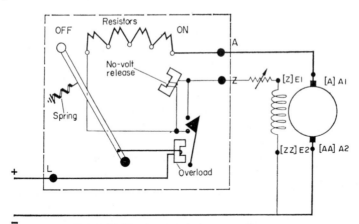

FIG. 6.46. Shunt-motor starter connections.

against field failure and the consequent rise of speed. For motors with a wide range of field control, the "no-volt" coil is excited directly across the supply. The second protective device is the overload relay carrying the line current and arranged to short out the "no-volt" coil if the load becomes excessive. This causes the starter to be tripped as soon as the "no-volt" magnet becomes weak enough.

A series-motor starter is similar, but there are fewer steps and the "no-volt" coil is connected across the supply. The overload protection must then be arranged to open circuit the "no-volt" coil. Automatic starting for either motor can be arranged for push-button control. This initiates a sequence of switching operations, shorting out the starting resistors in turn at pre-arranged time intervals, or in accordance with specified speed or current values.

Electromechanical Transients

The equations for the electrical and mechanical systems associated with a motor can be interlinked by balancing the electromagnetic torque T_e against the total mechanical torque. This has already been done for the steady state in eqn. (6.5) (p. 319) and for the transient state when deriving the motor transfer function in Section 6.6. In this

347

last case, if the armature inductance had been included, it would have given rise to a quadratic characteristic equation; see p. 128, eqn. (3.22). Under certain conditions, see Appendix E, problem 9, for Chapter 6, this means that the transient responses of current, torque and speed can be oscillatory. The effect of inductance is often quite small and will be neglected here for simplicity. If the rotational inertia of the mechanical system is J kg m^2, then:

$$T_e = T_m \pm J . d\omega_m/dt. \qquad (6.9)$$

The negative sign would apply for the generating condition.

Note that if it is desired to convert this to a *per-unit* equation, each term would be divided by T_{base} (usually rated torque $T_{eR} = k_{\phi R} . I_{aR}$). Further, since the actual speed $\omega_m = \omega_{m\,per\,unit} \times \omega_{base}$ ($\omega_{base} = V_R/k_{\phi R}$, see p. 321), then the coefficient of $d\omega_{m\,p.u.}/dt$ would become:

$$\frac{J . \omega_{base}}{T_{base}} = \frac{J . (\omega_{base})^2}{T_{base} . \omega_{base}} = \frac{J . (\omega_{base})^2.}{Power_{base}}$$

This is the inertia expression which must be used in *per-unit* dynamic equations. It is exactly twice the stored-energy constant H, see p. 509.

In a related manner, the per-unit inductance for such equations is obtained from the voltage expression divided by V_{base}, i.e.

$$\frac{L . di/dt}{V_{base}} \times \frac{I_{base}}{I_{base}} = \frac{L}{Z_{base}} \times \frac{d(i_{per\,unit})}{dt}$$

The expression as a whole is dimensionless, though, per-unit L (and J) have the dimensions of time.

Three applications of eqn. (6.9) will now be considered. Firstly, refer to Fig. 6.45a, for which the mechanical torque T_m and the flux factor k_ϕ are both constant. By substituting the value of T_e, as derived from eqn. (6.7a) and given on p. 320 but with voltage negative, the symbol V itself representing a + ve number, eqn. (6.9) after some little rearrangement becomes:

$$\frac{JR}{k_\phi^2} . \frac{d\omega_m}{dt} + \omega_m = -\frac{V}{k_\phi} - \frac{RT_m}{k_\phi^2}.$$

This is the same equation as derived previously in Section 6.6, except for the negative voltage term. The electromechanical time constant $\tau_m = JR/k_\phi^2$, is the same, and since the equation is of standard form, see eqn. (3.19), the solution, for a sudden change to the value of the

348

right-hand-side of the equation from an initial speed ω_0, can be written down immediately by reference to eqn. (3.21); i.e.:

$$\omega_m = \omega_0 + \left[-\frac{V}{k_\phi} - \frac{RT_m}{k_\phi^2} - \omega_0 \right](1 - e^{-t/\tau_m}).$$

The time taken for the motor to come to rest is obtained by substituting $\omega_m = 0$ and the other known quantities to find t. The same equation applies for acceleration, with the V sign positive, and for dynamic braking, with V equal to zero. R is the total armature-circuit resistance.

For the second example, refer to Fig. 6.45b. Here, both T_e and T_m may not have simple equations as functions of speed but the graphical solution described with reference to Fig. 3.31 is applicable. The speed change is considered in several discrete intervals, $\Delta\omega_1$, $\Delta\omega_2$ etc., from ω_0 down to zero. The time taken to change the speed over any interval $\Delta\omega_m$, is obtained by finding the mean decelerating torque $T_e - T_m$ during the interval. This average torque, from eqn. (6.9), is equal to $J\Delta\omega_m/\Delta t$ and hence Δt can be found for each interval and the incremental times summed to give the total time for ω_0 to zero speed. A similar method could be applied to Fig. 2.16 to find the accelerating time.

The third example concerns a pulse load which will be idealised as a peak torque T_m suddenly applied to the motor shaft for a time t_p, the motor having been running previously at a torque T_0; Fig. 6.47a. During this time t_p, energy is supplied from the motor and from the stored mechanical energy $\frac{1}{2}J\omega_m^2$ in the rotating system, which may have been augmented by a coupled flywheel. To relieve the motor and the

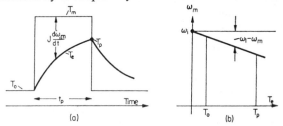

FIG. 6.47. Pulse loading: (a) load requirements; (b) motor characteristic.

electrical supply from the surge demand, the motor characteristic can be modified to have a steep speed regulation so that the speed falls substantially during the pulse and permits more energy to be extracted from the inertia. The motor characteristic is shown on Fig. 6.47b as a straight line, falling in proportion to the speed difference $\omega_i - \omega_m$. Let $(\omega_i - \omega_m)/\omega_i = s$, then:

$$T_e = ks \quad \text{and} \quad J \cdot \frac{d\omega_m}{dt} = J \cdot \frac{d\omega_i(1-s)}{dt} = -\omega_i J \cdot \frac{ds}{dt}.$$

Hence, from eqn. (6.9):

$$ks = T_m - \omega_i J \cdot ds/dt.$$

Rearranging:

$$\frac{J\omega_i}{k} \cdot \frac{ds}{dt} + s = \frac{T_m}{k}.$$

which from eqns. (3.19) and (3.21), has the standard solution:

$$s = s_0 + (T_m/k - s_0) \cdot (1 - e^{-t/\tau_m}),$$

where $s_0 = T_0/k$ and the electromechanical time constant $\tau_m = J\omega_i/k$. This equation in s can be solved for any one unknown: for example, the inertia required for a given pulse and maximum motor torque $T_p = k \cdot s_p$, or the permissible duration of a pulse with a given inertia and maximum motor torque, or the required speed regulation to limit the motor torque with a given pulse and inertia. See Appendix E, problem 16 for Chapter 7 and Ref. 33 for further work on Electrical Drives problems.

6.8 PARALLEL OPERATION

Much of what has been written in Section 3.5 is directly applicable to d.c. machines. An analytical solution is possible if the equivalent circuit is taken to be a constant e.m.f. in series with a constant internal resistance. The eqns. (4.7)–(4.10) developed for the transformer are applicable, but all the impedances are purely resistive.

In general, a graphical solution is necessary to deal with the non-linear external characteristics. The closer these are matched, the better will the load be shared, and with d.c. machines the characteristics can be adjusted in many ways; e.g. by varying the strength of the series winding, if present; by brush shifting to introduce direct-axis armature reaction, or by adjusting the shunt field-circuit resistance setting to lower or raise the whole characteristic. Figure 6.48 shows another

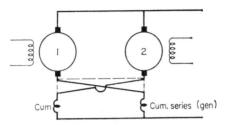

Fig. 6.48. Load sharing connections.

method which is very good for load sharing under transient conditions. With the series windings cross-connected and arranged to be cumulative generating, differential motoring, any load discrepancy will cause automatic correcting tendencies by virtue of the series ampere-turns. For example, if No. 1 machine takes an excessive generating current, No. 2 flux is strengthened and No. 1 flux is weakened. A less efficient method, but which avoids the cross connections, is to parallel the two series windings with a low resistance "equalizer bar" as indicated. This keeps the series currents the same, even though the armature currents are out of balance.

When motors are in "parallel" mechanically, i.e. driving the same shaft, they can be electrically connected in series, if this is more convenient from the viewpoint of voltage and current ratings. However, if the mechanical connection is not rigid, e.g. if driven through a belt, it is possible at starting for one machine to "get away" and leave the other stationary. The machine which starts first develops a back e.m.f. and hence tends to absorb all the supply voltage. This condition must be avoided by enforced voltage sharing. Series motors operate in parallel very satisfactorily because of their steep characteristics, see Section 3.5.

EXAMPLE E.6.8

The external characteristics of two 220-V, 110-kW differentially compounded d.c. generators are given by the following readings:

Output Current	0	200	400	500	700	900	A
Machine 1 Terminal voltage	229·5	226·5	222·5	220	213	205·5	V
Machine 2 Terminal voltage	224	223	221	220	217·5	214	V

351

How will they share a total load of 1500 A and what will be the terminal voltage at this load?

The curves are plotted on the adjoining figure and the construction to find the individual currents and terminal voltage at 1500 A total, follows that of Fig. 3.19b. It will be noticed that the characteristics are not well matched except at full load. On the overload, Machine 2 takes 850 A while Machine 1 only takes 650 A. Further,

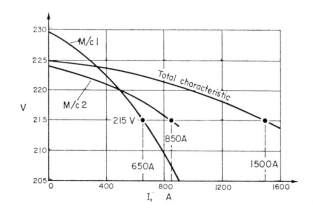

when the net load is zero, Machine 1 generates about 300 A which causes Machine 2 to motor. This unsatisfactory behaviour could be improved by two modifications:

(a) divert the series winding on Machine 1 so that its characteristic only falls by about the same amount as that of Machine 2,

(b) reduce the shunt field excitation of Machine 1 so that its no-load voltage is similar to that of Machine 2.

With these changes, the two curves would be quite close to one another and the machines would share the load quite well. Note that modification (a) could be achieved by "rocking" the brushes of Machine 1 forward, see Fig. 6.11, providing that the commutation performance was not thereby impaired.

6.9 TESTING AND EFFICIENCY

Open-circuit tests and short-circuit tests to obtain the magnetisation curve and the equivalent circuit parameters have already been described in Section 6.5. Tests on short circuit are also carried out to check the commutation and temperature performance when it is inconvenient to carry out a load test at normal voltage and current. Such a pro-

352

cedure provides, of course, only an estimate of performance and the designer must interpret the results with care. It is not proposed to go into the subject of commutation tests, which are rather specialised; they are required for the purpose of finding the optimum adjustment of interpole air gap, brushgear arrangement and angular position. Brush curves have been mentioned in Section 6.2 and other commutation tests involve the weakening and strengthening of interpole flux to determine the "black band", or limits of sparkless commutation at various armature currents.[9, 13] Subsequent discussion is confined to the measurement of losses and the calculation of efficiency.

Efficiency Calculations (*Refer also to Section* 3.6)

The individual losses for the d.c. machine are:

Armature circuit copper loss $I_a^2 R_m$, where R_m includes all the series connected windings;

Field-circuit copper loss $V_f I_f$. This will include the regulating resistance loss. It will form part of the armature circuit loss for the series machine;

Brush contact loss; usually taken as $2I_a$ watts, the brush material itself having negligible resistance by comparison with that of the sliding contact;

Iron loss at working flux and speed;

Friction and windage loss at working speed;

Stray load loss, $(I_a^2 R_{ac} - I_a^2 R_{dc})$, see p. 92.

For a generator, with output VI, $\eta = \dfrac{\text{output}}{\text{input}} = 1 - \dfrac{\text{losses}}{VI + \text{losses}}$.

For a shunt or series motor, $\eta = \dfrac{\text{output}}{\text{input}} = \dfrac{\text{losses}}{\text{line input } VI_{\text{line}}}$.

or $\eta = 1 - \dfrac{\text{losses}}{746 \times \text{hp} + \text{losses}}$.

The rating of the machine corresponds to the useful output, electrical or mechanical, depending on whether the machine is a generator or motor.

353

Measurement of Losses

The d.c. resistances can be measured by any technique suited to the ohmic value, the armature-circuit resistances being very low except on small machines. In the case of the armature itself, current may be passed through the stationary armature winding with all the brushes down and potential measurements taken between adjacent brush arms; on the commutator bars nearest to the centre of the brush contact. The armature resistance thus measured is equal to the resistance of one coil, multiplied by the number of coils in series and divided by the number of parallel circuits; $2p$ for a lap winding and 2 for a wave winding. For the purposes of declaring efficiency in accordance with the appropriate British Standard Specification, all resistances must be corrected to a temperature of 75°C or 115°C, depending on the insulation. There are several relevant specifications, for example BS 4727 dealing with terms used in Electrical Machines and BS 4999 dealing with General Requirements for Rotating Electrical Machines. These specifications cover matters such as definitions, temperature rises, tests, rating particulars, enclosures, etc.

To measure the mechanical and iron losses, there are many methods available. One method requires the measurement of the input power with the machine running on no load, at constant speed but with varying voltage and flux. For a motor, the field would have to be weakened to maintain the speed with decreasing voltage. The electrical input could then be corrected for known losses at the brush contact $2I_a$, and the small copper loss $I_a^2 R_a$. The remainder of the input would be the sum of mechanical loss, which is constant because the speed is constant, and the iron loss varying approximately as B^2 and hence as the (applied voltage)^2. At zero voltage, the iron loss is zero, and whilst the motor could not be run at this condition, the curve, Fig. 6.49, could be extrapolated as shown, a process made more accurate if the net input is plotted against V^2, since the points then lie on a line which is very nearly straight. The iron losses at normal e.m.f. and the friction and windage loss at this particular speed can be read off from the curve as indicated on the figure.

For a generator, which would have to be driven by a motor for which the losses were known, the total motor input would be the

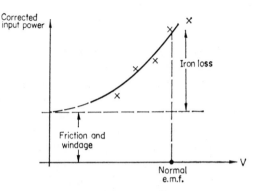

FIG. 6.49. Separation of iron and mechanical losses.

generator mechanical and iron losses plus the losses of the calibrated motor itself. The curve of net input to the generator shaft would be of the same shape as that of Fig. 6.49.

Load Tests

The back-to-back test for d.c. machines was devised in its original form by Hopkinson and is similar in principle to the Sumpner test for transformers. The main circuit connections for one form of the test, with losses supplied electrically, are shown in Fig. 6.50. The machines are mechanically coupled, and having started up the set with one machine only connected to the supply, the terminal voltage of the other machine is adjusted to that of the supply, so permitting the circuit breaker between the machines to be closed. No circulating current will flow unless one or other of the machine e.m.f.s is changed. If E_2 is increased for example, machine 2 will generate, circulating a current through machine 1 as a motor, of value determined by the difference between E_2 and E_1 and the resistance in the local circuit between the machines. The mechanical power from machine 1 will be converted

355

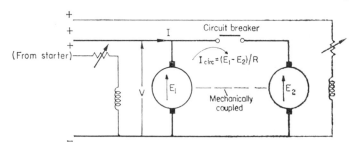

FIG. 6.50. Back-to-back test on shunt machines.

to electrical output from machine 2, apart from the losses in machine 2. Any fall in speed can be corrected by weakening the field and e.m.f. of machine 1 which will cause further increases of current. By a series of such adjustments, the speed and the circulating current can be set to any desired values; or, by making E_1 greater than E_2, the functions of the two machines can be interchanged. In either case, the input VI is the total loss apart from the field circuit losses. If the machines are nominally identical, a close estimate of the efficiency can be obtained since the actual stray loss on load is included.

For efficiency calculations, more instruments should be provided to measure the field currents and circulating current. From measured values of armature circuit resistances, the individual armature current losses, $I_a^2 R_a$ and brush contact losses can be calculated. Deducting the sum of these losses from VI gives the total iron-, friction-, windage- and stray-load-loss for the two machines. Assuming these are equally divided, the efficiency of each machine at the particular values of armature current can be calculated.

Even when the machines are not identical, this test connection is useful for load tests such as temperature runs and commutation checks. There are various other forms of the circuit in which the losses are supplied either mechanically by a separate driving motor, or electrically by means of a low-voltage machine inserted in series with the circulating current to provide the resistance drop and loss. The circuit can also be adapted for testing series motors.

Load tests, in which the machine output is dissipated directly as heat, can be carried out on a generator by loading it on to resistors. For a single motor, a brake pulley and friction band can be mounted, with arrangements to measure the tension in the two sides of the band so that the torque and mechanical power output can be calculated.

Another method of carrying out a load test is possible if the motor, or generator, can be coupled to a d.c. dynamometer. This is a d.c. machine with its stator mounted in bearings so that its torque reaction can be transmitted to "ground" through a pressure sensor. Figure 6.51a shows how the spring-balance reading multiplied by the torque-

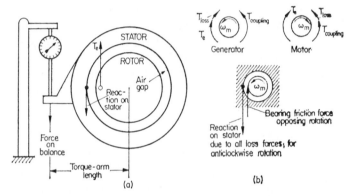

FIG. 6.51. Dynamometer machine for loading and torque measurement: (a) air-gap forces; (b) loss forces.

arm length will measure the net stator-reaction-torque. This in turn is made up of two components, the one due to the electromagnetic torque T_e exerted in the air gap, and the loss torque T_{loss} due to the friction, windage and iron losses which always oppose rotation. If, with the air-gap force directions shown, the dynamometer is generating, then rotation must be anticlockwise, the driving torque opposing T_e. Consequently, the total force on the balance will correspond to $T_e + T_{loss}$. But this is the necessary driving torque to sustain such a condition as shown by the shaft torque-diagram above Fig. 6.51b so that the balance reading corresponds to the torque at the coupling.

357

If on the other hand the dynamometer was motoring, the rotation would be in the direction of T_e, clockwise, so T_{loss} would be reversed giving $T_e - T_{loss}$ as the recorded torque. This is, of course, the net useful torque for the motor as exerted at the coupling, see eqn. (6.5) and Fig. 6.34. Consequently the product of the dynamometer torque-reading and the rotational speed will always give the dynamometer coupling power; the mechanical input to a coupled generator say, or the mechanical output from a coupled motor, i.e. the "mechanical terminal" power. Hence the efficiency of the coupled machine and the dynamometer too, can be determined if the electrical powers are also measured. The dynamometer can be used just as a generator on resistance load or preferably, connected to a d.c. supply of voltage V. It can then run either as a generator or a motor to control both the sense and magnitude of the power flow through the coupled machine by simple adjustments to the dynamometer field and e.m.f. E. The dynamometer armature current can thus be made positive or negative since $I = (V - E)/R$; see Fig. 6.40.

EXAMPLE E.6.9

The d.c. machine of E.6.2 is run on no load as a shunt motor at 200 V, 1000 rev/min, and takes a line current of 7 A. Calculate the efficiency as a shunt generator with full-load armature current of 100 A and as a shunt motor with the same armature current. Use the data and results of E.6.2 where appropriate.

When motoring on no load at 200 V, the internal drops could be neglected, the brush drop in practice being less than 2 V. However, it will be allowed for here and so the generated e.m.f. is:
$$E = 200 - 7 \times 0.14 - 2 = 197 \text{ V}.$$
With negligible armature reaction, I_f from o.c. curve = 1.42 A.
Hence,   no-load field loss   = 200×1.42 = 284 W
              no-load $I_a{}^2 R_a$ loss = $(7 - 1.42)^2 \times 0.14 =$ 4 W
              no-load brush loss = $(7 - 1.42) \times 2$ = <u>11 W</u>
                                                299 W
∴ mechanical and iron loss = $200 \times 7 - 299 = 1101$ W.
   Motoring at full load, $E = 200 - 100 \times 0.14 - 2 =$ 184 V
      from the o.c. curve $F_r = 1.16 \times 2,500$ = 2900 At
                $F_a{}'$ = <u> 500</u> At
                F_t = 3400 At
giving a shunt field current of 3400/2500 = 1.36 A.

Efficiency		Generator	Motor	
Armature copper loss	$= 100^2 \times 0{\cdot}14 =$	1400	1400	W
Brush loss	$= 100 \times 2 =$	200	200	W
Field loss	$= 200 \times I_f =$	460	272	W
Iron and mechanical loss	$=$	1101	1101	W
Total loss		3·16 kW	2·97 kW.	

Generator output $= V(I_a - I_t) = 200(100-2{\cdot}3) = 19{\cdot}5$ kW.

Input $= 22{\cdot}66$.

Efficiency $= 1 - 3{\cdot}16/22{\cdot}66 = \underline{86{\cdot}1\%}$ neglecting stray loss.

Motor input $= 200(100+1{\cdot}36) = 20{\cdot}3$ kW.

Efficiency $= 1 - 2{\cdot}97/20{\cdot}3 = \underline{85{\cdot}3\%}$ neglecting stray loss.

6.10 APPLICATIONS OF D.C. MACHINES AND OTHER MODES OF OPERATION

Generators and Motors of Normal Construction

With the advent of controlled rectifiers of various kinds, the importance of d.c. generators has subsided somewhat since it is cheaper to convert a.c. to d.c. with a transformer and rectifier rather than with an a.c. motor and d.c. generator. The rectifier performance is inferior in terms of power factor, simplicity of control, regenerative and reversing ability and harmonic generation, but scores on cost (in most cases), space, noise and maintenance; the latter considerations weighing more heavily in a modern age. D.C. generators will continue to be made for their virtues but will lose certain traditional fields of application. They are invaluable for use as dynamometers to control and measure the output (or input), when coupled mechanically to engines and motors of various kinds including the "rolling roads" used for vehicle testing. Generators are also useful for imparting special characteristics to a load in a simple manner, e.g. automatic current limitation for d.c. welding generators utilising series windings and exaggerating armature reaction demagnetising effects.

As far as motors are concerned, there are continual attempts to obtain infinitely variable-speed motors which can be directly connected to the a.c. mains and thus dispense with the intermediate a.c./d.c. conversion apparatus. Some success has been achieved with these various methods but there are always disadvantages and it seems likely that the demand for d.c. motors will continue for many years to come. The availability of compact, variable-voltage thyristor d.c. supplies, having built-in provision for closed-loop speed control, has greatly improved the economic viability of the rectifier/d.c.-motor drive. Shunt-, compound- and separately-excited motors are used for drives which have to operate over a wide range of constant speeds and in automatic control systems. Series motors are used when heavy starting torques and overloads are characteristic of the load, as on crane drives and traction. The series characteristic automatically limits the speed and power demand and gives a high torque per ampere on overload.

A very popular application for d.c. motors is on battery-powered vehicles.[14] The problem here is the size and weight of the battery which limits the operating range and it is desirable therefore that any control-circuit losses should be kept to a minimum. The armature voltage must be reduced for starting and speed control, and if this is done by series resistance, there will be appreciable power losses. A few discrete speeds can be obtained economically by switching the battery cells into series, series–parallel or parallel groups, but intermediate speeds still pose problems. Considerable savings in losses are effected by employing switching control which continually interrupts the battery circuit several times per second, reducing the average applied voltage in direct proportion to the ON/OFF (mark/space) ratio. The motor inductance smooths out the current waveform and Fig. 6.52a shows how the circuit can be arranged with a by-pass diode to maintain the motor current flow during the OFF period. This "chopping" action is usually performed by a series thyristor or transistor and to vary the average voltage, the durations of the ON or OFF periods, or the chopping frequency itself, are all possible variables. This system has also been applied to railway traction.

A simple analysis of performance is possible if saturation is allowed for by using a mean value of volts/field-amp k_{fs} (p. 294). With this approximation, the equations derived in Section 3.10 can be used. For

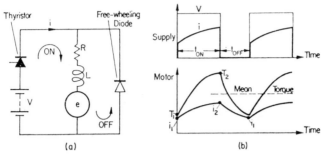

FIG. 6.52. ON/OFF control for d.c. series motor: (a) circuit; (b) performance.

the series motor, the field current is the same as the armature current, so the e.m.f. $= k_\phi \omega_m = k_{fs} i$, if ω_m is constant. Although the torque pulsates in an ON/OFF system, the mechanical inertia is sufficient to damp out an oscillation of this frequency so the speed ω_m can be taken as constant with little error. Hence, from Fig. 6.52a, the voltage equation in the ON condition is:

$$V = Ri + L.di/dt + e \quad \text{and} \quad e = k_{fs} i.$$

Therefore: $V = (R + k_{fs})i + L.di/dt$

or: $V/(R + k_{fs}) = i + \tau.di/dt,$

where $\tau = L/(R + k_{fs})$ is the time-constant of current response.

Referring to eqns. (3.19) and (3.21) with $\theta_f = V/(R + k_{fs})$ (ON) and $\theta_f = 0$ (OFF), the maximum current when the oscillation has settled down to "steady" limits is:

$$i_2 = i_1 + \left(\frac{V}{R + k_{fs}} - i_1 \right) (1 - e^{-t_{ON}/\tau}).$$

The minimum current is:

$$i_1 = i_2 + (0 - i_2)(1 - e^{-t_{OFF}/\tau}).$$

By solving these simultaneous equations for i_1 and i_2, knowing all the other quantities, the maximum and minimum torques can be found from $T_e = k_\phi i = k_{fs} i^2 / \omega_m$. Hence $T_1 = k_{fs} i_1^2 / \omega_m$ and $T_2 = k_{fs} i_2^2 / \omega_m$. From Fig. 6.52b it can be seen that the mean torque is slightly more than the arithmetic mean of T_1 and T_2 but approximately, the total motor power developed is $\omega_m T_{mean} = k_{fs}(i_1^2 + i_2^2)/2$ watts; see Appendix E, problem

20 for Chapter 6. Regeneration can be achieved by interchanging the positions of thyristor and diode, "charging up" L during the ON period and discharging L through the supply in the OFF period.[33]

In concluding this sub-section, it may be said that the normal construction d.c. motor can be used for almost any application, though it has a lower limit on maximum power and maximum speed than the synchronous and induction motors.

Use of Thyristors to Supplement or Supplant the Commutator Function

Most of the operational modes of the d.c. machine have been discussed in this chapter. However, there are some recent developments which should be mentioned for completeness.

The undoubted advantages of d.c. machines have been obtained with the penalty of a sliding-contact arrangement, the commutator and brushgear, which is more complicated than the arrangements required for synchronous and induction machines. Further, for motors, the necessary provision of a d.c. supply might be regarded as a disadvantage. Even though years of development have made the d.c. machine reliable and relatively trouble free, the possibility of removing its inherent limitations whilst retaining its special features, has always been attractive. Now that small, static, controlled rectifiers like the thyristor are available, the question might well be asked whether such an ideal is within sight of achievement.

The most obvious way of dispensing with a generator commutator is to use a controlled-rectifier supply for a d.c. source. The advantages are obvious and the limitations have been pointed out. It is fortunate that in many cases, standard d.c. motors, with little or no modification, seem to be able to run satisfactorily from such supplies, notwithstanding their high harmonic content. As a result, the demand for d.c. motors has risen appreciably, since this combination of transformer, thyristors and motor, forms what is perhaps the most economic general-purpose variable-speed drive. For the higher ratings and for high performance, laminated yokes become necessary, see p. 290. But quite apart from this well-tried method of using rectifier supplies in place of generators, there have been other developments concerned with motor-commutator replacement.

In the first of these to be explained,[15] an otherwise normal d.c. motor is provided with two commutators, each of which has only a few segments round the periphery, compared with a normal machine which, on the very large sizes, may have up to a thousand or more. Adjacent tappings from the armature winding are taken alternately to the two commutators so that only half of the segments ever carry current. Figure 6.53 shows the arrangement schematically for just a portion of the machine but including brushes of both polarities. The path of current through the coils is determined by which thyristors are switched into the conducting state and whether one brush, or both paralleled

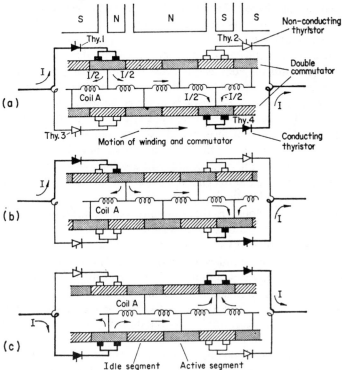

FIG. 6.53. Thyristor-assisted commutation: (a) coil A before commutation; (b) coil A during commutation; (c) coil A after commutation.

363

brushes are connected to an active segment. Figure 6.53a shows a condition where each of the brushes at a conducting set carries current. Figure 6.53b for a later instant, shows that one brush has stopped carrying current because it is on an idle segment. In this same position, one brush on each of the non-conducting sets, has made contact with a coil and so, as soon as the associated thyristors are switched on, they will be ready to take over the current feed to the machine. In this position also, it is arranged that coil A is moving through a suitably strong interpole field which induces a large commutating voltage sufficient to reverse-bias thyristors 1 and 4 with a negative voltage, and thus switch them off. Thyristors 2 and 3 are then switched on so that they take over the current feed. Figure 6.53c shows the transition complete; coil A has been commutated.

There are several possible advantages with this scheme. Firstly, the commutator may be cheaper because there are only a few segments and further, there are less restrictions on the number of turns per coil, from reactance-voltage considerations. A second point is that the interpole flux adjustment is not so critical as it is on highly rated machines. The interpole field must be strong enough to induce adequate commutating voltage for the thyristors but it does not matter if this is greater than is strictly necessary. (An auxiliary commutating voltage must be introduced in the circuit by external means to supplement the interpole induced voltage at very low speeds.) A third, perhaps the most important point, is that commutation of coils, which involves coil switching, does not have to take place while a brush is breaking contact with a segment as on a normal machine. Consequently, there is not the same chance of sparking occurring. The disadvantage of the scheme is in having a double commutator with 50% idle segments which do not get any electrical wear. Whether, in selected high-speed applications, this possibility of easing the commutation problem is worth adopting these measures and providing the necessary control equipment, is a decision which must await the test of time.

The second development[16] consists in completely replacing the commutator by power transistors or thyristors. Consider Fig. 6.54. For the normal d.c. machine, field and armature m.m.f.s are in space quadrature as shown by Fig. 6.54a. Suppose now that the armature is provided with slip rings instead of a commutator and at the instant shown in Fig. 6.54b the polarities are as indicated. No torque will be produced, as can be seen by examining the m.m.f. space phasors. However, if the armature is somehow turned through a small angle, δ changes from 180° so that $F_a F_f \sin \delta$ (eqn. 1.2a) has a non-zero value and there is a clock-

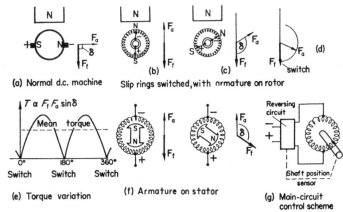

(a) Normal d.c. machine   Slip rings switched, with armature on rotor

(e) Torque variation   (f) Armature on stator   (g) Main-circuit control scheme

FIG. 6.54. Commutator replacement by static switching circuit.

wise torque; Fig. 6.54c. If the movement continues, this torque will persist until $\delta = 0$. If now an external switch changes over the slip-ring polarity, the cycle will repeat itself so that the machine develops a unidirectional torque even though it pulsates about a mean value and falls to zero. For a sinusoidal distribution of the fields the torque variation would be as shown in Fig. 6.54e and it can be seen that the mean torque is considerably less than would be obtained on a normal d.c. machine with $\delta = 90°$. However, if the number of slip rings and tapping points was doubled and switching took place every 90° instead of every 180°, there would be a marked reduction in the torque pulsation and there would be no torque zero to give rise to starting problems.[17] The advantage of these schemes is that it is no longer necessary to put the high-power member, the armature, on the rotor, since switching would be equally effective if the armature was on the stator and the field was on the rotor. Such an arrangement is shown schematically on Fig. 6.54f. The armature power no longer has to be transmitted through a sliding contact. The instant of switching is determined by the rotor field position relative to the stator. A shaft-position sensor is therefore required and this is used to switch over the thyristors in the main d.c. reversing circuit which controls the polarity of the armature-winding section being switched, see Fig. 6.54g.

Many forms of brushless d.c. motor working on the general principles just described are now available in the smaller sizes, and usually for special applications. Parallel developments of variable-frequency systems using power transistors or thyristors are removing the disadvantages of a.c. machines with regard to their inherent tendency to run at constant, or nearly constant speed. These developments include the substitution of the commutator on a.c. commutator machines by a static switching arrangement.[4]

Motors with Permanent-magnet Excitation, see also Appendix A

With the improvements in permanent-magnet materials, these motors are becoming more popular. Their efficiency is relatively high since there is no field power-loss, the penalty being the absence of field control. These motors are especially competitive in the low-speed, high-torque range where the machine may be physically quite large even though the power is small; see eqn. (6.1a) and associated text. Armature voltage and speed control is effected by power-electronic circuits.

Permanent-magnet construction is also used on low and very-low power ratings. These may be *low-inertia motors* having requirements which might mean several hundred stop/start cycles per second as on computer tape and printer drives. Low inertia is sometimes achieved by employing thin, ironless armatures of disc or cup shape, rotating in a double-sided air gap between fixed outer and inner iron circuits. The disc-type armature can have either wound coils or printed-circuit windings, etched on the two sides of a copper-covered insulated disc. Brushes then bear directly on the "windings". Brushless designs are also used, in general accord with the principle illustrated in Fig. 6.54. The switching periods, however, are usually 60°, Fig. A.6, or even 30° (3-phase or 6-phase windings), to reduce the torque ripple. Much valuable information on these topics is contained in recent IEE conferences on Small and Special Electrical Machines.[18]

CHAPTER 7

INDUCTION MACHINES

7.1 BASIC THEORY AND CONSTRUCTION

The Rotating Field

In Section 1.3, induction-motor action was explained briefly by reference to the travelling field produced by polyphase currents flowing in a polyphase winding. In Section 5.6, a more detailed explanation of the three-phase winding and its m.m.f. was given. Since the proper understanding of this phenomenon is so important for the appreciation of the remaining chapters, a brief recapitulation will be made.

Figure 7.1 shows separately the three phase-windings of a two-pole machine, together with the assumed positive directions of currents and of m.m.f. space phasors which are taken to represent a sinusoidal distribution. As the currents alternate, one of them, together with its m.m.f., will, in general, be of opposite sign to the other two. In Fig. 7.2 the situation is considered at the same two instants as for Fig. 5.27, the time-phasor diagram being deliberately drawn with a different orientation from the space phasors in order to avoid confusion between

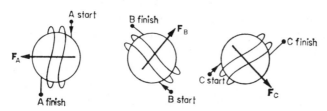

FIG. 7.1. Individual m.m.f.s of three phases.

367

them. The time phasors are of constant length, rotating at constant speed, whereas the individual space phasors vary in length with time, but maintain their orientation along the axes of their respective phase windings. The resultant armature m.m.f. $\mathbf{F_a}$ is line with the axis of phase A and its m.m.f. $\mathbf{F_A}$, at the instant when I_A is maximum, Fig. 7.2a.

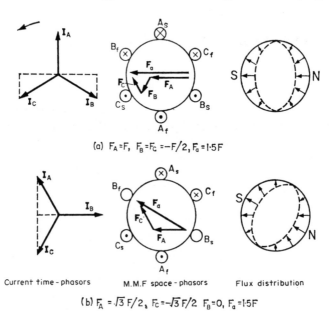

(a) $F_A = F$, $F_B = F_C = -F/2$, $F_a = 1\cdot5F$

(b) $F_A = \sqrt{3}\,F/2$, $F_C = -\sqrt{3}\,F/2$ $F_B = 0$, $F_a = 1\cdot5F$

Current time - phasors M.M.F space - phasors Flux distribution

FIG. 7.2. Rotating field.

At certain other instants, $\mathbf{F_a}$ is in line with $\mathbf{F_B}$ when I_B is maximum, and is in line with $\mathbf{F_C}$ when I_C is maximum. Because of the sinusoidal distribution assumed, $\mathbf{F_a}$ can be obtained by vectorial combination of $\mathbf{F_A}$, $\mathbf{F_B}$ and $\mathbf{F_C}$, and at any instant would be found to have a constant value $1\cdot5F$ where F is the maximum value of the phase m.m.f. With a practicable distribution of the winding, as explained in Section 5.6, the fundamental of the resultant m.m.f. space wave is slightly more than this at $(6/\pi)Fk_dk_p$, which is $1\cdot5 \times$ fundamental component of F.

368

At a later instant, corresponding to a 30° movement of the time phasors, Fig. 7.2b, $\mathbf{F_B}$ becomes zero, but $\mathbf{F_a}$ is unchanged in magnitude though it has moved through 30° also. The fact that this is clockwise, opposite in direction to the time phasors, is only due to the way the phases are placed round the periphery. If the start of phase B as indicated by $\mathbf{B_s}$ had been placed 120° counterclockwise from $\mathbf{A_s}$, and phase C advanced to take up the old position of phase B, the space phasor $\mathbf{F_a}$ would have rotated counterclockwise. In fact, reversal of the field rotation is accomplished quite easily by interchanging any two phase-supply-connections; all phase windings are identical, and the lettering merely indicates the required supply phase-sequence for a particular rotation. Interchanging two supply leads means that the windings will reach their maximum m.m.f.s in a different sequence.

The speed of $\mathbf{F_a}$ with respect to the coils is seen to be synchronous with the time phasors and of value $n_s = f$ rev/sec, f being the frequency of the coil currents in cycles/sec. A simple 4-pole winding would have coils spanning only half the mechanical angle of the 2-pole winding. The speed of the field relative to any point on the winding would now be halved, $\mathbf{F_a}$ still moving past one pole-pair of coils in one cycle. In general, for p pole-pairs, the synchronous speed $n_s = f/p$ rev/sec.

In an induction machine, the winding which initiates the rotating field, the primary winding, is usually, though not necessarily, placed in the stator and the term *stator winding* is often used synonymously with *primary winding*. The stator, as indicated by Fig. 1.11c is made up of thin slotted laminations mounted coaxially and clamped together. Reference should also be made to Fig. 5.3. The rotor is made up similarly but mounted on a shaft in bearings, two examples being given on Figs. 7.6 and 7.8. Apart from the small slot openings, the radial air gap is uniform and only large enough (usually <2·5 mm, depending on the rating), to give safe mechanical clearance under the action of the gravitational, magnetic and rotational forces present over the operational speed range. Neglecting the m.m.f. of the iron, the flux distribution produced by F_a acting on the air gap would give a sine wave of flux density as indicated in Fig. 7.2, and this would move like a travelling wave round the periphery at synchronous speed. Even when saturation distorts the flux wave, it still travels at synchronous speed of course.

Conditions with Rotor-winding Stationary and Open-circuited

Another 3-phase winding could be placed in rotor slots. Every coil of this winding would experience flux changes with the primary excited. All the primary coils contribute to the flux, their combined effect being conveniently treated in terms of the resultant m.m.f. F_a. As the flux axis rotates, each rotor phase senses a flux linkage varying between positive and negative maxima at supply frequency. The three rotor phases are in a very similar situation to the secondary windings of a 3-phase, 3-limb transformer. The phase magnetic circuits are interdependent and the induced phase voltages rise and fall in sequence. There are differences in physical construction and coil arrangement for the purposes of producing a rotating field. Nevertheless, if a voltage V_1 per phase is applied to the primary winding, an induced voltage will appear at the open ends of the secondary which, neglecting the small internal voltage drops, will be of magnitude E_2, equal to V_1 multiplied by the effective turns ratio per phase. In subsequent work, the primary, unless otherwise stated will be assumed to be the stator winding. The

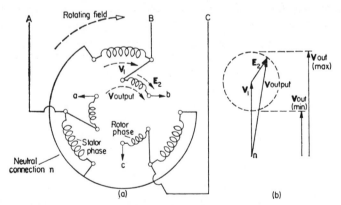

Fig. 7.3. Three-phase induction regulator: (a) physical arrangement; (b) time-phasor diagram for one phase.

370

inverted arrangement, i.e. with the primary on the rotor, is sometimes used.

By moving the rotor in the direction of field rotation to a new position, the phase of the rotor e.m.f. is delayed though its magnitude is unchanged; the travelling flux wave reaches a particular rotor phase at a later instant. This useful phase-shifting feature can also be employed to produce a variable voltage from a constant voltage supply. In the 3-phase *Induction Regulator*, Fig. 7.3, the primary and secondary of each phase, are conductively connected as in the Auto transformer, so that the vectorial combination of V_1 and E_2 will vary in magnitude as the relative phase of E_2 is changed. Although the induction machine can have these special transformer features with limited rotor movement, the performance as a transformer is poor. Even a small air gap requires many magnetising ampere-turns, giving a relatively high I_0 and a low, primary power-factor. However, transformer theory is applicable and when modified slightly will be used to explain the behaviour even when the rotor is not stationary.

Rotor Stationary; Winding Closed

With the rotor held in position and its circuit closed, supply frequency currents will flow in the secondary phases, the m.m.f.s opposing the primary m.m.f.s as in the transformer. In addition however, due to the positioning of the windings and the time displacement between the phase currents, the resultant secondary field will rotate in step with the primary field. As indicated in Fig. 7.4, there will be established a

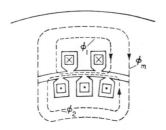

FIG. 7.4. Mutual and leakage fluxes.

371

secondary leakage flux, a modified mutual flux and an increased primary leakage flux, all rotating at synchronous speed. The leakage fluxes and the corresponding reactances will be higher than for the transformer since primary and secondary are separated by an air gap.

Secondary Winding Closed; Rotor Released

Consider one conductor of the rotor winding and its reaction to the rotating flux wave, Fig. 7.5. The direction of induced current can be determined by the *whiplash rule* as already explained in connection

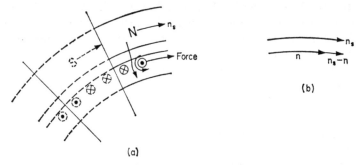

(a)

(b)

Fig. 7.5. Induction motor action.

with Fig. 2.15 and demonstrated for this particular case by Fig. 2.15d. The force on the conductor is seen to be clockwise, so it will be urged in the direction of the field motion. Most of the other rotor conductors, though in different positions with respect to the field, will also experience a clockwise force. The remainder will be carrying current in the wrong direction for this, if the rotor power-factor is not unity, a condition indicated in Fig. 7.5 and discussed further in connection with Fig. 7.15. If the resultant torque developed is greater than the mechanical resisting torques, the rotor will accelerate to some speed n revs/sec, where the developed torque and the opposing mechanical torque are balanced. The field/conductor relative motion will now be reduced to $n_s - n$, as indicated by the velocity vectors of Fig. 7.5b. As far as the rotor coils are concerned they experience a slower rate of flux change which lowers

372

the induced e.m.f., current and frequency. Note that the torque resulting from the induced current reduces the relative motion which is in fact responsible for the current and torque in the first place.

Since the rotor e.m.f. and frequency are proportional to the relative motion, the actual values at speed n become:

$$\frac{n_s - n}{n_s} \times \text{standstill e.m.f.} \quad \text{and} \quad \frac{n_s - n}{n_s} \times \text{supply frequency.}$$

The coefficient, $(n_s - n)/n_s = s$, is called the fractional or *per-unit* slip, being proportional to the speed at which the rotor is slipping back relative to the rotating field. It is absolutely essential to understand how the expression arises, because the slip is continually appearing as a term in the equations of the induction-type machine. For example, by rearranging the expression, the actual rotor speed is given as $n = n_s(1 - s)$ and the relative field/conductor (slip-) speed as $n_s - n = sn_s$.

In practice, the operating speed is very close to synchronism, so that s is quite small even at full load. On the large, efficient machines it may be less than 0·01 or 1%. If the slip were to become zero, there would be no relative motion to produce rotor e.m.f., current and torque. Therefore, unless the current is supplied conductively from an external source, induction motor operation at synchronous speed ($s = 0$) is not possible. In fact the current would have to be of zero frequency, a condition which is sustained on the normal synchronous machine by employing d.c. excitation. Note that at standstill when n is zero, the slip is unity.

The rotor currents produce their own rotating field, so with the frequency changed to sf, the rotor field-speed relative to the rotor coils is also changed, to $sf/p = sn_s$. However the rotor itself is running at speed n with respect to space; therefore, combining the velocity vectors:

speed of rotor field in space = speed of field relative to rotor

+ speed of rotor relative to space

$$= sn_s + n_s(1 - s) = n_s.$$

This means that no matter what the value of slip, even if negative with $n > n_s$, the rotor field in space is automatically adjusted to synchronism with the stator field. The fields are stationary with respect to one another; an essential condition for the production of a uni-directional torque as discussed in Section 1.3. From this point of view then, there is no bar to the running of the induction machine at any speed, though in practice controlled speed-variation introduces complications.

Alternative Rotor Arrangements

So far, only the phase-wound rotor has been discussed and Fig. 7.6 is a photograph of such a rotor showing brass alloy rings mounted on, but insulated from the shaft. Carbon or copper–carbon brushes take

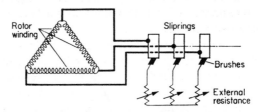

FIG. 7.7. Phase-wound rotor circuit connections.

current into and out of the rotating secondary winding through these slip rings with which they make sliding contact. Figure 7.7 is the connection diagram, the winding being either in star or delta depending on design considerations. The normal operation is with the rings short-circuited, but for starting and speed control, the slip rings permit the insertion of additional impedance or a voltage source, in the rotor circuit.

This arrangement is by no means the only one nor even the most popular. Providing there are paths for induced currents to flow, and the induced e.m.f.s do not cancel round the local circuits, reference to Fig. 7.5a will confirm that the number of phases is immaterial. Under the north poles, most of the conductor currents are in one direction and under the south poles they are in the opposite direction. If all the conductors are simply shorted together at the ends of the slots, paths

Fig. 7.6. Phase-wound, slip-ring rotor for 1100-h.p. 10-pole induction motor. (By courtesy of G.E.C. Machines Ltd.)

FIG. 7.8. Squirrel-cage rotor for 1000-h.p. 4-pole induction motor. (By courtesy of G.E.C. Machines Ltd.)

will be provided for the conductor current streams to coalesce. This is the most common arrangement and is called the *squirrel-cage rotor*, Fig. 7.8. The cage may be built of copper bars brazed on to brass or copper end-rings, or the rotor conductors may be die-cast in aluminium to form the cage and end rings in one operation. The squirrel-cage construction is simple, cheap and robust. A further advantage becomes evident when the stator is provided with coils and connections which permit the number of poles, and therefore the synchronous speed, to be changed. In this case, although the sequence of conductor current reversals round the periphery is altered, the end rings still provide free paths for the currents to flow; the cage winding adapts itself readily to a different number of poles. A phase-wound rotor would have to be reconnected to deal with this situation because the coils impose constraints on the current paths.

One of the disadvantages of the simple cage rotor is its fixed characteristic; no external rotor circuit impedance can be added, to reduce the starting current for example, which might be six times rated value, with full voltage applied. The limitations can be overcome largely, by designing the slot bars with special shapes so that eddy-current effects, see Section 3.6, are pronounced and cause a high effective resistance at starting when the secondary frequency is high, and a low resistance at normal speed when the slip frequency is very low. Such a resistance variation is beneficial for starting characteristics as will be explained in Section 7.3. The use of simple rectangular bars, if deep enough to enhance the eddy-current effect, will result in considerable improvements over round or square bars.

Occasionally, to impart special characteristics, two or even three concentric sets of bars are used in the rotor. On occasions, a solid-iron cylinder or shell is used as a rotor without any additional conductors, the currents circulating within the iron mass causing torque and rotation. The general performance of such a solid-rotor machine is poor but adequate for certain starting duties and might serve for other special cases.

Summarising these remarks on construction; the primary winding is of conventional type such as those illustrated in Fig. 5.16, and designed

375

for voltages up to and beyond 11 kV. Windings which can use semi-closed slots are sometimes preferred since, with a small air gap, open slots cause the tooth harmonics, Fig. 2.12b, to have a relatively large magnitude. The rotor winding may take various forms, providing current paths are available to accommodate the induced e.m.f. pattern. On the high-power, high-speed induction motors, for mechanical reasons the rotor iron may be of solid steel instead of laminations. Maximum motor ratings may exceed 20,000 kW.

7.2 EQUIVALENT CIRCUIT

The Phasor Diagram

It has been pointed out that the induction machine when stationary, behaves in a similar manner to, though less efficiently than a transformer. It has primary and secondary windings, producing mutual and leakage fluxes which in turn give rise to induced e.m.f.s in the windings. At standstill, whether the secondary winding is closed or open circuited, the phasor diagram is the same as for the transformer, though the leakage reactance voltages and the current I_0 are relatively higher. Further, the primary still senses the secondary reaction in terms of its m.m.f., the resultant m.m.f. wave rotating at synchronous speed in space, irrespective of the rotor motion. It is not possible to detect from the primary terminals whether the rotor is squirrel-cage or phase-wound; whether it has the same number of turns per phase or even the same number of phases. It is often convenient, therefore, to express the secondary m.m.f. in terms of the primary turns and phases, and to use referred values of voltage, current and impedance. The referring coefficients are virtually the same functions of turns ratio as for the transformer with the slight difference that primary and secondary windings may not be distributed and chorded in the same way as one another. Since the winding factors k_d and k_p reduce the e.m.f.s and m.m.f.s, the effective turns per phase must be used in calculating the ratio; i.e. the actual turns for both primary and secondary must be multiplied by the appropriate values of $k_d k_p$. The rotor quantities are referred to the primary in the phasor diagram of Fig. 7.9, and in the following discussion, but when dealing with rotor circuit alone, it may be more convenient to use the actual rotor parameters. It will be remembered for example, that the secondary copper loss $I_2'^2 R_2'$ is the same as $I_2^2 R_2$, this being one of the conditions to be fulfilled to make the referring process valid.

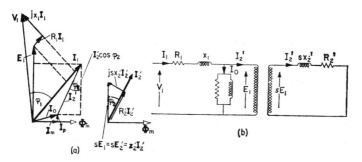

FIG. 7.9. Phasor diagram and equivalent circuit per phase at slip s.

When the machine is rotating, the field/rotor relative motion is no longer at the synchronous speed n_s but is equal to $(n_s - n)$, and all rates of flux-change affecting the rotor are modified in proportion to the fractional slip $s = (n_s - n)/n_s$. Consequently the standstill e.m.f. $E_2' = E_1$, becomes sE_1; the rotor-circuit frequency becomes $f_2 = sf$, and the rotor leakage reactance changes from the standstill value x_2' to sx_2' because of the frequency change. The whole of the rotor induced voltage sE_1 due to ϕ_m, is absorbed in the referred secondary impedance $R_2' + jsx_2'$ plus any external impedance. Although the stator phasor diagram is for frequency f and the rotor phasor diagram is for frequency sf, it must be remembered that the stator and rotor m.m.f.s are always in synchronism. Consequently, with the mutual flux Φ_m as a common reference phasor, the correct angle at which I_2' is added to I_0 to get the primary current I_1, can be obtained as for the transformer. The validity of this construction can be confirmed by studying Fig. 7.15.

377

Development of the Equivalent Circuit

It has already been explained, when developing the circuit of Fig. 3.26 that at standstill with the secondary short-circuited, the equivalent circuit per phase of an induction machine is drawn as for the transformer in the same condition. The primary input per phase is $V_1 I_1 \cos \varphi_1$ and after deducting the primary loss components $I_1(I_1 R_1)$ and $E_1 I_p$ as indicated on Fig. 7.9a, there remains a power $P_g = E_1 I_2' \cos \varphi_2$ which is transferred magnetically across the air gap. At standstill with $s = 1$, the whole of this power is absorbed in the rotor electrical circuit. However, with rotation the rotor parameters change and it appears from Fig. 7.9b that an amount of power $(1-s)E_1 I_2' \cos \varphi_2$ has been lost.

Consider the expression for the referred rotor current:

$$\mathbf{I}_2' = s\mathbf{E}_1/(R_2' + \mathrm{j}sx_2') = s\mathbf{E}_1/\mathbf{z}_2'.$$

The same value of current and power factor can be obtained if the expression is divided throughout by s to give:

$$\mathbf{I}_2' = \mathbf{E}_1/(R_2'/s + \mathrm{j}x_2') = \mathbf{E}_1/\mathbf{Z}_2'.$$

The effect of rotation is now manifested only as an apparent change to the rotor-circuit resistance. This is not just a mathematical trick; if the rotor was stationary and its resistance was supplemented to say $R_2/0.05$, the magnitude of the rotor current, m.m.f. and reaction on the stator would be the same, neglecting slotting effects and extra iron losses, as if the machine were running at slip $s = 0.05$ with only R_2 in circuit. The input measured at the stator terminals would also correspond to the running condition apart from the presence of line-frequency rotor iron losses. An electrical analogue of the machine could be made up of resistors and reactors on the basis of the equivalent circuit and rapid estimation of performance would then be possible by varying the value of the unit representing the apparent rotor resistance and measuring the appropriate voltage, current and power. Reference back to p. 114 might now be appropriate.

An alternative way to look at the effects caused by rotation, which has already been discussed on page 14, is to consider the reduction of

378

the referred rotor e.m.f. from E_1 to sE_1. The difference $(1-s)E_1$, which is proportional to speed, could be regarded as a back e.m.f. arising with rotation as on the d.c. motor. Consider the revised expression for I_2'. This suggests that a modification to the equivalent circuit is required which will increase the rotor impedance from $R_2'+jsx_2$ to $(R_2'/s)+jx_2'$, and increase the voltage absorbed from sE_1 to E_1. The change is shown in Fig. 7.10 together with the phasor diagram. The additional voltage $(1-s)E_1$ is absorbed across the extra impedance:

$$(1-s)R_2'/s+j(1-s)x_2'$$

having the same power factor angle, $\tan^{-1} R_2'/sx_2'$ as the actual rotor impedance. Figures 7.9b and 7.10a therefore give the same rotor current

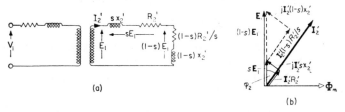

(a)

(b)

FIG. 7.10. Electrical representation of mechanical load.

and power factor, but Fig. 7.9b is a true electrical circuit in which the power dissipated is the true electrical loss $I_2'^2R_2'$ whereas Fig. 7.10a has an extra "loss", part of the total power transferred across the air gap and of value $(1-s)I_2'^2R_2'/s$. This is the amount of power missing from Fig. 7.9b, viz. $(1-s)E_1I_2' \cos \varphi_2$. It is absorbed in the only other sink of energy available, i.e. in turning the rotor against the mechanical resisting torques. The mechanical power is therefore represented by a variable resistance loss $(1-s)I_2'^2R_2'/s$, the extra reactance being necessary to satisfy the circuit relationships between voltage and current. Reference back to p. 320 will show that the mechanical power of a d.c. machine was also represented as a resistance loss.

Figure 7.11 shows one phase of the complete equivalent circuit, differing from the transformer only in the relative magnitudes of the

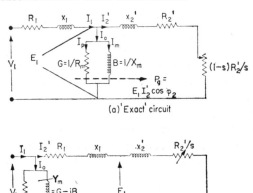

(a) 'Exact' circuit

(b) Approximate circuit

FIG. 7.11. Complete equivalent circuits (per phase).

parameters and in having a purely resistive load which is a function of slip. Separating the secondary resistive components emphasises the difference between copper loss and mechanical output, but they can be combined to give a total R_2'/s as in the simplified circuit of Fig. 7.11b. Here the magnetising branch has been transferred to the terminals. In the transformer case, this led to very slight errors, but for the induction machine, the voltage drop due to I_0 flowing through the primary leakage impedance is appreciable. Its exclusion from Fig. 7.11b can give rise to errors of 10% or more in some cases. Such errors are too large for the machine designer, but will be tolerated here because the approximate circuit enables a variety of performance characteristics to be estimated simply, and with reasonable accuracy.

There are other approximations which may be more accurate for certain purposes, as when R_m, and/or R_1 if $\ll x_1$, are omitted from Fig. 7.11a; or may be very much less accurate, as when the primary impedance is omitted, this giving very large errors at high values of slip. The expressions which will be derived can easily be adapted to this later approximation by omitting all terms in R_1 and x_1.

380

Note that as already explained for the transformer, p. 136, the magnetising circuit parameters may be assumed to have constant values if voltage and frequency are constant. R_1 and R_2 change with temperature of course and the leakage reactances fall due to saturation when the currents are very large, e.g. during starting, pp. 422 and 439. Since in normal operation, all three phases are excited and interact to produce a common stator or rotor flux as shown by Fig. 7.2, the equivalent-circuit parameters per phase must correspond to this magnetic condition. Discussion on the significance of this situation as it affects the inductive elements of the circuit, will be deferred until Chapter 10, p. 587.

7.3 ANALYSIS OF MACHINE EQUATIONS: SPEED/TORQUE CURVES

Power Distribution

The air-gap power per phase $P_g = E_1 I_2' \cos \varphi_2$, transferred magnetically from stator to rotor, has been considered in two components:

(i) $I_2'^2 R_2'$ is converted to electrical power and is associated with a voltage sE_1. It is dissipated as a heat loss.

(ii) $(1-s)I_2'^2 R_2'/s$ is converted to mechanical power and is associated with a voltage $(1-s)E_1$.

Referring to Fig. 7.11a, a power balance equation can be written. Rotor input/phase $E_1 I_2' \cos \varphi_2 = P_g = I_2'^2 R_2'/s$ or:

$$E_1 I_2' \cos \varphi_2 = E_1 I_2' \frac{R_2'/s}{Z_2'} = I_2'^2 R_2' + (1-s)I_2'^2 R_2'/s$$

$$= I_2'^2 R_2'/s = \text{rotor Cu loss} + \text{mech. output } P_m$$

$$= P_g = sP_g + (1-s)P_g. \qquad (7.1)$$

This equation brings out the distinguishing feature of dual power conversion and determines how the power is divided for any value of slip, positive or negative. For a motor, P_m will be taken as positive. It includes the torque loss which is virtually due to friction and windage only, since the rotor iron loss is negligible at normal slip frequency. At large slips, the effect of rotor iron loss must be allowed for in practice. For a generator, P_m would be negative.

Most of the theory for any polyphase induction machine in the steady state can be deduced from this simple but important equation. This chapter is devoted mainly to the 3-phase machine, and eqn. (7.1) will therefore only represent one-third of the total power since it is expressed in phase values. The equation is true for both the exact and approximate equivalent circuits because it deals only with the secondary power. The equation must be modified for the single-phase induction motor as will be seen in Section 7.7.

Torque

The total torque developed electromagnetically can be obtained from either eqns. (3.15) or (7.1) and is given by:

$$T_e = 3\frac{P_m}{\omega_m} = \frac{3P_g(1-s)}{2\pi n_s(1-s)} = \frac{3P_g}{2\pi n_s} \text{ Nm.} \qquad (7.2)$$

For any particular frequency and synchronous speed the torque is directly proportional to the rotor input $3P_g$ since both P_m and ω_m vary as $(1-s)$. Note that P_g itself is a function of speed. The torque is sometimes expressed as $3P_g$ synchronous watts and from eqn. (7.2) this is equal to the torque in newton metres multiplied by the synchronous speed $\omega_s = 2\pi n_s$. A torque of one synchronous watt would develop one watt of power if acting at synchronous speed.

The torque in lbf-ft, see eqn. (6.4b), p. 317, would be $7\cdot04 \times 3P_g/N_s$ where N_s is the synchronous speed in rev/min.

An expression for torque could have been obtained in terms of flux, current and conductors by summing the contributions of each rotor conductor taking due account of the conductor positions and flux density variations round the periphery. The procedure is not as straightforward as that used for deriving eqn. (7.2) but in fact the expression can be obtained from (7.2) if the e.m.f. equation (5.8) is substituted for E_1, using the resultant flux ϕ_m due to stator and rotor m.m.f.s combined. By reference also to the relationship $P_g = E_1 I_2' \cos \varphi_2$;

$$T_e = \frac{3E_1 I_2' \cos \varphi_2}{\omega_s} = \frac{3(1 \cdot 11 \times 2p\phi_m n_s z_s \cdot k_d k_p) I_2' \cos \varphi_2}{2\pi n_s}$$

$$= 1 \cdot 11 k_d k_p \times \frac{p \times 3z_s}{\pi} \times \phi_m I_2' \cos \varphi_2. \qquad (7.3)$$

Note that the synchronous speed must be used in the e.m.f. equation because this is the speed of the flux wave relative to the stator conductors, and the stator e.m.f. has been used. The expression is not quite the same as that derived for the d.c. machine in eqn. (6.4a), though the torque per ampere at u.p.f. is not very different. $1 \cdot 11 k_d k_p$ is very nearly unity, and $3z_s$ is the total number of conductors which are series connected though in three separate groups or phases. The term $\cos \varphi_2$, will be shown later, when discussing Fig. 7.15, to have the same value as the sine of the load angle δ. For the d.c. machine, $\sin \delta$ is usually unity. Equation (7.3) corresponds to eqn. (1.2c) with ϕ_m replacing F_r and I_2' replacing F_a. The constant of proportionality is seen to be $1 \cdot 11 k_d k_p \times p \times 3z_s/\pi$, which is not greatly different numerically from that for the d.c. machine.

To write eqn. (7.2) in terms of the equivalent circuit parameters it is simplest to calculate P_g from $I_2'^2 R_2'/s$. From the approximate circuit since:

$$I_2' = \frac{V_1}{(R_1 + R_2'/s) + j(x_1 + x_2')} \qquad (7.4)$$

then:

$$T_e = \frac{3}{2\pi n_s} \times \frac{V_1^2}{(R_1 + R_2'/s)^2 + (x_1 + x_2')^2} \times \frac{R_2'}{s}. \qquad (7.5)$$

Equation (7.5) can now be examined to find the effect of changes to one or other of the various parameters involved. It can already be seen from eqn. (7.3) that the torque is very nearly proportional to $I_2' \cos \varphi_2$ because the peak mutual-flux is approximately constant as in

the transformer. Increase of speed from standstill causes the secondary
e.m.f. sE_1 and the current I_2' to fall, but the rotor power-factor,
$\cos \varphi_2 = R_2'/sZ_2'$ improves, so that initially there is an increase of
torque with slip until the fall of current is greater than the rise of power
factor. However, analysis of eqn. (7.5) is more convenient than that of
eqn. (7.3).

Torque/Slip Curve at Constant Input Voltage V_1

Though each parameter in eqn. (7.5) is a potential variable and will at
least be considered briefly as such, the effects of variations in s alone
are the most informative and will be studied in some detail. The
equation can be re-formed as:

$$T_e = \frac{k_t R_2'}{s}\left(\frac{1}{(R_1 + R_2'/s)^2 + x_{e1}^2}\right) \tag{7.6}$$

where $k_t = 3V_1^2/\omega_s$ and $x_{e1} = x_1 + x_2'$.

Maximum Torque. This is also referred to as the *stalling torque* or
breakdown torque and can be obtained by differentiation. Rearranging
eqn. (7.6):

$$T_e = k_t R_2'/[sR_1^2 + 2R_1 R_2' + (R_2'^2/s) + sx_{e1}^2]$$

$$dT_e/ds = -k_t R_2'[R_1^2 - (R_2'/s)^2 + x_{e1}^2]/(\text{denominator})^2$$

from which $dT_e/ds = 0$ when $R_2'/s = \pm\sqrt{[R_1^2 + (x_1 + x_2')^2]}$.

When this relationship is satisfied, the slope of the torque/slip curve
is zero and the torque therefore has its mathematical maximum or
minimum value. It is interesting to note that this expression could have
been written down directly by applying the general theorem for maximum power transfer in such a series circuit; i.e. the load resistance must
be equal to the impedance modulus of the remaining circuit components.
In this case, the maximum power $I_2'^2 R_2'/s$ is required, so R_2'/s is the
appropriate load resistance.

It follows that the slip $\hat{s}$ at which maximum torque occurs is:

$$\hat{s} = \pm R_2'/\sqrt{[R_1^2 + (x_1 + x_2')^2]} \tag{7.7}$$

and the corresponding $\varphi_2 = \tan^{-1} x_2'/(R_2'/s) \simeq 45°$, since $x_1 \simeq x_2'$ and R_1
is small.

For the negative value of slip the speed is supersynchronous, $n > n_s$, a generating condition as will be explained later.

Substituting $R_2'/\hat{s}$ in eqn. (7.6):

$$T_e = k_t[\pm\sqrt{(R_1{}^2 + x_{e1}{}^2)}]/\{[(R_1 \pm \sqrt{(R_1{}^2 + x_{e1}{}^2)})]^2 + x_{e1}{}^2\}$$

which when multiplied out becomes:

$$k_t[\pm\sqrt{(R_1{}^2 + x_{e1}{}^2)}]/[2(R_1{}^2 + x_{e1}{}^2) \pm 2R_1\sqrt{(R_1{}^2 + x_{e1}{}^2)}]$$

$$= \frac{k_t}{2[\pm\sqrt{(R_1{}^2 + x_{e1}{}^2)} + R_1]}. \tag{7.8}$$

Equation (7.8) shows that the maximum torque is independent of the R_2' value but the value of slip $\hat{s}$ at which this maximum occurs is directly proportional to R_2' from eqn. (7.7). For negative slips, the maximum negative torque is greater in magnitude than for positive slips, but occurs at the same numerical value of slip. In calculations it is sufficient to remember eqn. (7.7), find the value of $\hat{s}$ and then substitute it in eqn. (7.5) to get the maximum torque.

Starting Torque. This is obtained by substituting $s = 1$ in eqn. (7.5) or (7.6) giving $T_s = k_t R_2'/[(R_1 + R_2')^2 + (x_1 + x_2')^2]$. This *is* subject to variation if R_2' changes so that with a phase-wound rotor, the starting torque can be increased up to the maximum figure merely by inserting an external resistance to bring R_2' up to $\sqrt{[R_1{}^2 + (x_1 + x_2')^2]}$ (substitute $\hat{s} = 1$ in eqn. (7.7)). This value is much higher than the natural rotor resistance. For maximum torque at any other value of slip s', the total rotor resistance per phase must be adjusted to $s'.\sqrt{[R_1{}^2 + (x_1 + x_2')^2]}$.

From eqn. (7.6) it is a straightforward matter to plot the variations of torque as slip takes all positive and negative values, see Fig. 7.12.

When $s = 0$, the rotor must be running at the same speed as the field so there is no relative motion, rotor e.m.f. or current; $T_e = 0$.

When $s = 1$ the starting torque is developed; perhaps 20–30% of the maximum torque for the simple cage-rotor machine.

When $s = \hat{s}$, the maximum torque obtains as already described.

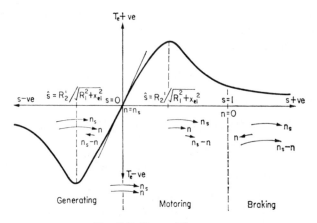

FIG. 7.12. Torque/slip curve.

When s is very small, all denominator terms are swamped by $s(R_2'^2/s^2)$ so the torque is approximately $k_t s/R_2'$. Near $s = 0$ then, the curve is asymptotic to a straight line and departs from this to reach the maximum value. Thereafter the curve falls to approach the hyperbolic asymptote $k_t R_2'/s(R_1^2 + x_{e1}^2)$ as s becomes very large. For negative values of slip, the curve is similar but with a higher value of maximum torque. Since speed is proportional to $(1-s)$, the same curve can be used to show the torque/speed relationship by reversing the abscissae scale as shown.

Motoring Operation $0 < s < 1$: $P_m = (1-s).I_2^2 R_2/s$ positive.

The various regions of the torque/slip curve will now be examined, but by far the most important and frequently used, is that between standstill and synchronous speed. Figure 7.13 shows a typical speed/torque characteristic replotted from Fig. 7.12 on axes which line up with the forward-motoring quadrant, see Fig. 6.41. The thin-line portion at the lower end of the curve is one practical feature caused by slotting effects and shown here to demonstrate some stability considerations. Figure 7.13 also shows some typical load characteristics, the normal one being T_{m1}. At standstill, $s = 1$, the accelerating torque $T_e - T_{m1}$ is positive, causing the speed to rise at an increasing rate until maximum torque is reached, and thereafter the acceleration will fall to zero at the balancing speed, point A. Though sE_2 is quite small here,

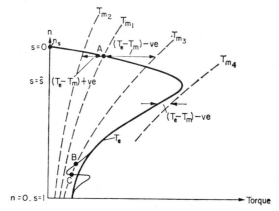

FIG. 7.13. Motor speed/torque characteristic.

it is sufficient to produce the rotor current and hence the torque necessary to drive the load. Full-load operation is usually designed to occur at about half the maximum torque so there is about 100% overload capacity.

Stability. It is not sufficient for stable operation that the characteristics intersect. In practice they change with time slightly; for example, friction coefficients vary with temperature and T_m would then change. A change of any kind will cause a momentary unbalance of torque $T_e - T_m$, which if positive will cause acceleration and if negative, deceleration. The condition for stability (mathematically, $d\omega_m/dT_e$ − ve, or, if + ve, $d\omega_m/dT_m < d\omega_m/dT_e$), is that the unbalance should become less as speed changes under its influence. If it were to become greater, the speed would continue to change at an increasing rate; an unstable condition.

387

On this basis, a change to T_{m_2} from point A would give rise to a positive acceleration but with a decreasing value of $T_e - T_{m_2}$. If the change were towards T_{m_3} there would be a negative but decreasing acceleration. Point A is stable. However, if an overload T_{m_4} were to persist long enough, an unstable condition would arise with $T_e - T_{m_4}$ becoming numerically greater once the stalling torque had been passed. Unless the overload were removed quickly, the machine would stop, otherwise, depending on the drop of speed, the operating point would fall on say B or C. B is not a stable operating point since if the load were to fall a little more, $T_e - T_m$ would be positive and become larger as speed built up again.

Point C is stable because a tendency for load to reduce a little further would be met by a decreasing positive value of $T_e - T_m$; conversely if the load were to rise a little. This stable operation at low speeds is called crawling and is liable to occur on an incorrectly designed cage-rotor machine at a speed which is about a seventh of normal, the seventh harmonic in the m.m.f. space wave being the most prominent with forward sequence; see Figs. 3.14 and 5.28. In addition to this low-speed *induction* torque, certain combinations of rotor and stator slots can react to interfere with the speed/torque curve as suggested on page 44, and cause low-speed *synchronous* locking torques. Very often the slots are skewed on cage rotors to give a torque curve nearer to the smooth ideal case.

Stability thus depends on the way in which the curves intersect. Whilst it is possible to have a T_m characteristic which intersects the T_e characteristic satisfactorily between $s = 1$ and $s = \hat{s}$, operation in this region would be at high current and low power factor. For most practical purposes then, the drooping portion of the curve between $s = 0$ and $s = \hat{s}$, is used and is sometimes referred to as the stable region since it would be difficult to arrange a T_m characteristic which did not intersect here in a stable manner. Thus, the normal induction motor characteristic is very like that of a shunt motor with constant voltage and flux, a situation which virtually exists in this case also.

Stable behaviour is also affected by changes of T_e and this is considered on p. 402, when discussing the effect of system voltage changes.

EXAMPLE E.7.1

A 500-V, 3-phase, 50 Hz, 8-pole, star-connected induction motor has the following equivalent-circuit parameters: $R_1 = 0.13\,\Omega$, $R_2 = 0.32\,\Omega$, $x_1 = 0.6\,\Omega$, $x_2 = 1.48\,\Omega$; magnetising branch admittance $\mathbf{Y_m} = 0.004 - j0.05\,\Omega^{-1}$ referred to primary side. The full-load slip is 5%. Determine the full-load electromagnet torque, stator input current and power factor using both approximate and "exact" equivalent circuits. The effective stator/rotor turns ratio per phase is $1/1.57$. Neglect mechanical loss.

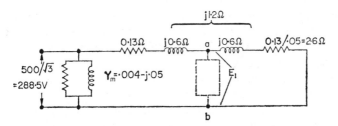

$$R_2' = 0.32 \times (1/1.57)^2 = 0.13\,\Omega. \quad x_2' = 1.48 \times (1/1.57)^2 = 0.6\,\Omega.$$

The equivalent circuit for one phase is shown for a slip of 0.05.

Using the approximate circuit:

$$\mathbf{I_2}' = \frac{\mathbf{V_1}}{\mathbf{Z}} \quad = \frac{288.5}{(0.13 + 2.6) + j1.2} = \quad 88.8 - j39 \quad = 97\,\text{A}$$

$$\mathbf{I_p} = \mathbf{V_1}G \quad = 288.5 \times 0.004 \quad = \quad 1.15 - j0 \quad = 1.15\,\text{A}$$

$$\mathbf{I_m} = -j\mathbf{V_1}B = 288.5 \times -j0.05 \quad = \quad \underline{\quad -j14.4} = 14.4\,\text{A}$$

$$\mathbf{I_1} = 89.95 - j53.4 = \underline{105\,\text{A}.}$$

Input power factor $= 89.95/105 = \underline{0.856}.$

$$\text{Torque} = \frac{3P_g}{\omega_s} = \frac{3I_2'^2 R_2'/s}{2\pi f/p} = \frac{3 \times 97^2 \times 2.6}{2\pi \times 50/4} = \underline{935\,\text{Nm}.}$$

Using the "exact" circuit, the admittance between a and b is:

$$\mathbf{Y_{ab}} = 0.004 - j0.05 + \frac{1}{2.6 + j0.6} = 0.369 - j0.1345\,\Omega^{-1}.$$

The corresponding impedance, $1/\mathbf{Y_{ab}} = \mathbf{Z_{ab}} = 2.4 + j0.872 \ = 2.55\,\Omega.$
Adding $R_1 + jx_1$, total impedance $\quad = \mathbf{Z_{in}} = 2.53 + j1.472 = 2.92\,\Omega.$

$$\text{Total input current} = \frac{288.5}{2.53 + j1.472} \quad = 85.5 - j49.6 \ = \underline{98.5\,\text{A}.}$$

Input power factor $= 85.5/98.5 = \underline{0.858.}$

To find the e.m.f. E_1, observe that the same current I_1 flows through z_1 and Z_{ab} in

389

series. The voltage across each impedance is therefore proportional to the impedance modulus, i.e.:

$$E_1 = V_1(Z_{ab}/Z_{1n}) = 288 \cdot 5(2 \cdot 55/2 \cdot 92) = 252 \text{ V},$$

hence $I_2'^2 = \dfrac{252^2}{(2 \cdot 6)^2 + (0 \cdot 6)^2} = 8880$ and $I_2' = \sqrt{8880} = 94 \cdot 3$ A.

Torque $= 935 \times (94 \cdot 3/97)^2 = 886$ Nm.

$$I_p = E_1 G = 252 \times 0 \cdot 004 = 1 \cdot 01 \text{ A}$$
$$I_m = E_1 B = 252 \times 0 \cdot 05 = 12 \cdot 6 \text{ A}.$$

The figures calculated for the two circuits should be compared.

Note, that if this problem is worked out in *per-unit* values, it must be remembered that the choice of the three base quantities is arbitrary, though rated terminal voltage and synchronous speed are usually taken as 1 *per unit*. For current, the rated value can be used if it is known, though it is more usual to know the output power from which 1 *per-unit* current could be defined as rated power per phase/rated voltage per phase. Rated current would then be greater than 1 *per-unit* and is easily shown to be: 1/(effy × power factor) at rated load. Similarly, rated torque $= 1/(1 - s_{rated})$, *per-unit*. 1 *per-unit* impedance will be V_{base}/I_{base} Ω. Applying this method to the exact-circuit results of the above example gives:

1 *per-unit* power $\quad = 886 \times 0 \cdot 95 \times 2\pi \times 50/4 = 66$ kW $\; = \omega_{base} \cdot T_{base}.$

$\qquad\qquad\qquad\qquad\qquad\qquad\qquad\qquad\qquad\qquad\qquad\qquad$ (22kW/phase)

1 *per-unit* current $\quad = 22000/(500/\sqrt{3}) \qquad = 76$ A. $\quad \therefore I_1$ rated $= 1 \cdot 29$ p.u.

1 *per-unit* impedance $= (500/\sqrt{3})/76 \qquad\quad = 3 \cdot 78$ Ω

1 *per-unit* torque $\quad = 66000/(2\pi \times 50/4) \quad = 841$ Nm $\therefore T_e$ rated $= 1 \cdot 05$ p.u.

Magnetising reactance $X_m = (1/0 \cdot 05)/3 \cdot 78 \qquad\qquad\qquad\qquad = 5 \cdot 3$ p.u.

This last figure shows that the magnetising current is approximately 1/5·3 p.u., which is only about 15% of the rated I_1. This figure is in fact rather smaller than occurs in practice normally, where figures up to half rated-current or even more are not unusual, depending upon the machine design.

Generating Operating; s negative: $P_m = (1-s) \cdot I_2{}^2 R_2/s$ *negative*

The conditions when s is a positive fraction and when s is a negative fraction are illustrated by Figs. 7.14a and 7.14b respectively. The direction of the induced rotor-conductor e.m.f. is found by considering the flux/conductor relative motion $(n_s - n)$. Figure 7.14a is similar to Fig. 7.5 and is shown here for direct comparison with the case when the conductors are moving faster than the field, with $(n_s - n)$ in the opposite sense; Fig. 7.14b. This figure shows the conductors, as it were, pushing the lines of force emanating from the poles so that under a north pole the field established by an induced current would be clockwise round the

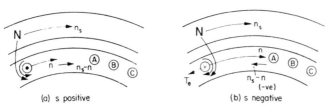

FIG. 7.14. Motoring and generating action.

conductor. With the induced e.m.f. determining the current direction, the torque is seen to oppose the rotor motion. To maintain such a condition, mechanical torque must be exerted, driving the conductor against T_e and doing mechanical work in the process. The only outlet for this applied power is in a conversion to electrical power so the machine must be generating.

Although the rotor motion is faster than n_s, its m.m.f. is still moving forward *in space* at synchronous speed. The reversed field/conductor relative motion causes currents in particular rotor conductors or phases, to rise and fall in opposite sequence, CBA instead of ABC. The rotor m.m.f. wave thus moves backwards relative to the conductor motion giving a net speed in space of:

$$+n+(n_s-n) = +n_s.$$

For a real physical understanding of the operation it will be necessary to examine the space relationship between mutual flux and rotor m.m.f. with some care. An easy solution which avoids this discussion will be given shortly, but meanwhile Fig. 7.15 has been prepared to justify the step which will then be taken. Consider first the motoring condition. In Fig. 7.15a, which shows both rotor and stator conductors, the rotor m.m.f. wave I_2N_2 is indicated on one full-pitch coil representing the centre conductors of a phase group. The stator compensating m.m.f. $I_2'N_1$ is similarly represented. The resultant m.m.f. $I_1N_1-I_2N_2$ establishes the mutual air-gap flux, its axis being indicated by the position of the North pole rotating in space at synchronous speed ω_s radians/sec. The m.m.f. phasor diagram, in which I_pN_1 is omitted for simplicity, explains the phase difference between the mutual flux and

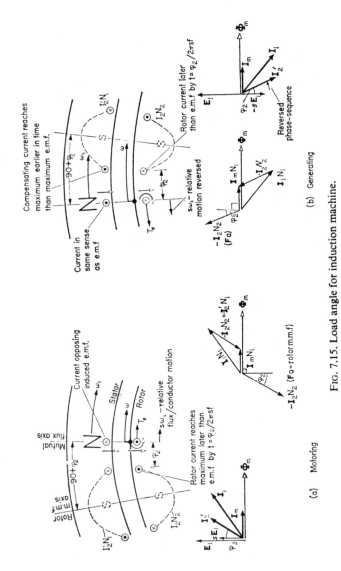

Fig. 7.15. Load angle for induction machine.

the rotor m.m.f. Maximum rotor e.m.f. per phase $s\hat{E}_2$, occurs when the centre conductors of a phase are cutting maximum flux density, under the North pole. If $\cos \varphi_2$ is unity, maximum currents in this rotor phase occur at the same time, so that the rotor m.m.f. would be in space quadrature with the mutual flux. Consider now the more general case where there is a time-phase angular delay φ_2, on the rotor slip-frequency scale, between the rotor current and e.m.f. By the time the phase current has reached its maximum value, the North pole would have passed the centre conductors of a rotor phase by a *mechanical* angular distance:

$\theta = $ time $\times$ relative mechanical angular velocity of flux wave

$$= (\varphi_2/2\pi sf) \times s\omega_s = (\varphi_2/2\pi sf) \times s(2\pi f/p) = \varphi_2/p$$

so θ is φ_2 converted to mechanical degrees; i.e. the time-phase angular delay between rotor current and e.m.f. corresponds to a space-phase angular delay of the same magnitude. Defining the load angle δ as that between mutual flux $\mathbf{\Phi}_m$ and rotor m.m.f. $-\mathbf{I}_2 N_2$, i.e. between $\mathbf{F}_r$ and $\mathbf{F}_a$, using the notation of eqn. (1.2c), it is seen that $\delta = \pi/2 + \varphi_2$ and $\sin \delta$ has the same numerical value as $\cos \varphi_2$, a fact which has been noted without explanation when discussing eqn. (7.3).

Figure 7.15b is the corresponding generator diagram in which a lagging rotor power-factor again causes the rotor m.m.f. to be delayed in space phase by φ_2 in the opposite direction to the flux/rotor-conductor *relative* motion as before. However, the stator reaction to the rotor m.m.f. in terms of $I_2' N_1$ is *advanced* in space phase by φ_2 in the direction of the flux/*stator*-conductor relative motion. The position of the rotor m.m.f. phasor, $-\mathbf{I}_2 N_2$, is therefore deduced to be *leading* $\mathbf{\Phi}_m$ by $\pi/2 + \varphi_2$. This load-angle expression is thus the same for motoring and generating. For leading rotor power-factors, which can be brought about by injecting a rotor voltage from an external source, φ_2 becomes negative in the load-angle expressions. Note that the direction of stator current is the same as that of the induced voltage, confirming generator operation.

The effects of these factors on the equivalent circuit and phasor diagram are taken into account simply and automatically if negative values of slip are inserted where appropriate. For convenience in the algebraic expressions, s itself will be taken as a positive number so that the apparent rotor resistance becomes $R_2'/(-s)$, and the voltages $- sx_2'I_2'$ and $- sE_1$ indicate phasors in antiphase with their position for normal motoring operation with s positive. Since the flux, both mutual and leakage, is cutting the rotor conductors in the opposite sense, this reversal of e.m.f. components and the phase sequence of rotor currents, is only to be expected.

If the machine is still considered as a power sink, i.e. using motor conventions, but with a negative value of s, eqn. (7.4) gives:

$$I_2' = \frac{V_1}{(R_1 - R_2'/s) + j(x_1 + x_2')}.$$

If s is large enough, the effective resistance becomes negative and:

$$I_2' = \frac{V_1}{-A + jB} = V_1 \left\{ \frac{-A}{A^2 + B^2} - \frac{jB}{A^2 + B^2} \right\} = V_1(-a - jb).$$

This is a current having antiphase components with V_1, i.e. the machine is a "negative" motor, which is a generator. The primary current I_1 is obtained by adding I_0 vectorially as for the motor, see Fig. 7.16.

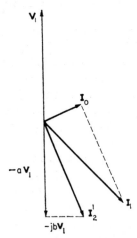

FIG. 7.16. Effect of negative resistance.

Generator Phasor Diagram. The diagram is constructed most conveniently by treating the machine as a motor, i.e. using the same
equations as before. The difference from the motor diagram then
comes out naturally as a result of the negative slip value and the effect
of this on the relevant voltage components and their related currents.
To convert to a diagram with generator conventions requires a relative
reversal of either voltage or current. On Fig. 7.17 $-V_1$ has been indi-

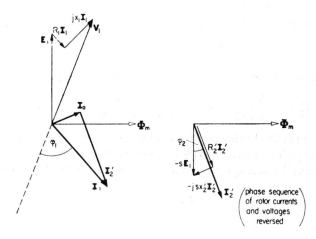

FIG. 7.17. "Negative"-motor (generator) phasor diagrams for
"exact" circuit.

cated and it can be seen that a leading power-factor is a feature of
induction-generator operation.

Generator Equivalent Circuit. This is the same as Fig. 7.11 in either the
exact or approximate versions. The slip is negative however, so that $-s$
must be substituted for s, if s itself is still to symbolise a positive
number. The effect is to change the apparent rotor circuit resistance to
$-R_2'/s$ and the resistance representing the mechanical power to
$-(1+s)R_2'/s$. The air-gap power $P_g = -I_2'^2 R_2'/s$ is negative, signifying

395

a power source and reversed power flow from the motoring condition. The mechanical power, still using motor conventions is:

$$P_m = P_g - I_2'^2 R_2' = -I_2'^2 R_2'/s - I_2'^2 R_2' = -(1+s)I_2'^2 R_2'/s$$

and is also negative. Mechanical power must be supplied of value $3P_m +$ mechanical loss. As for the motor, the electromagnetic torque can be obtained either from $T_e = 3P_m/\omega_m$, or more conveniently from $T_e = 3P_g/\omega_s$, being a negative quantity in either case, since P_m and P_g are negative. Finally, the slip can be calculated as for the motor from:

$$s = I_2'^2 R_2'/P_g = \text{Rotor-circuit copper loss/power across the air gap.}$$

EXAMPLE E.7.2

Using the approximate circuit of E.7.1 calculate the output of the machine when driven at a speed of 780 rev/min.

Synchronous speed $= 60f/p = 60 \times 50/4 = 750$ rev/min

$\therefore$ slip $= (750-780)/780 \quad = -0.04$.

Apparent rotor circuit resistance $= 0.13/-0.04 = -3.25\ \Omega$.

$$I_2' = \frac{288.5}{(0.13-3.25)+j1.2} = \frac{288.5}{-3.12+j1.2} = -80.5 - j31.$$

$$\begin{array}{ll} I_0 \text{ from approximate circuit of E.7.1} = & 1.15 - j14.4 \\ \hline I_1 & = -79.35 - j45.4 = 91.5 \text{ A.} \end{array}$$

Output kVA $= \sqrt{3} \times 500 \times 91.5 = 79.3$ kVA.

Power factor $= -79.35/91.5 \quad = \underline{0.865 \text{ leading}}$.

Source of Magnetising Current; Self Excitation. It will be noticed from Fig. 7.17 that the reactive component of I_1, which includes the current I_m necessary to excite the machine and produce Φ_m, can only be supplied via the line terminals since the power source is in the form of mechanical energy. To operate as an induction generator then, it is necessary to have a system connected across the line which can provide this lagging current, or what amounts to the same thing, can accept a leading current. (See Fig. 3.9 and the associated text.) This is not usually a problem, because induction generators, which have a limited application, are normally connected in parallel with synchronous machinery for which the reactive volt-amperes are readily controllable.

However, it is possible to employ self-excitation using a capacitor of suitable value to provide the necessary current/voltage relationship. To simplify the picture, consider the no-load condition and neglect the small slip and the stator leakage-impedance drop. With zero slip, the speed of the generator n gives the frequency directly as $f = pn$. In Fig. 7.18 the capacitor current I_c is the same as the machine current, which for this condition is the magnetising current I_m. The relationships

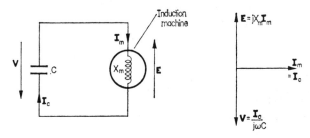

FIG. 7.18. Circuit for self excitation. No-load condition.

are similar to those occurring in a resonant LC circuit, as shown by the phasor diagram. The induction machine has the characteristic of a pure reactance such that the induced e.m.f. E, which with the assumptions stated is also the terminal voltage, is given by $\mathbf{E} = jX_m\mathbf{I}_m$. X_m itself is a non-linear function of the current magnitude and a direct function of frequency. The relationship is best expressed by the magnetisation curve for various frequencies corresponding to various rotational speeds; as shown in Fig. 7.19.

The voltage across the capacitor is given by $\mathbf{V} = -j\mathbf{I}_c/2\pi f\,C$, being directly proportional to the current but increasing as the frequency decreases. Both voltages can therefore be plotted against current at various frequencies and since the circuit constrains them to be numerically equal, then:

$$I_c/2\pi f C = f \times F(I_m) \quad \text{where} \quad I_c = I_m.$$

The graphical solution of this equation is shown on Fig. 7.19, stable operating points occurring where the characteristics at a particular frequency intersect. In a similar manner to the self-excited d.c. generator, there is a critical value of capacitance below which the induction generator will not build up at a particular speed. There is also a critical speed for a particular capacitance. In practice, the effect of leakage impedance is to cause another resonant condition at a higher speed.

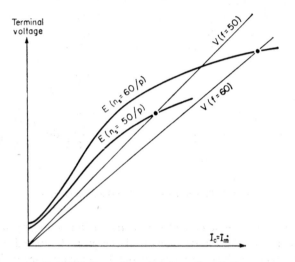

Fig. 7.19. Operating points for self excitation.

The induction generator is useful when the prime mover does not run at constant speed; for example, in certain hydro-electric power stations having a variable low-head water supply. Where the generator and turbine run submerged in a river to utilise tidal power, induction generators might be considered. The utilisation of wind power is another suitable application for this type of generator since the stator frequency depends on that of the paralleled synchronous machines, not on the rotor speed. Sometimes an induction motor drives a load with an active component due to gravity, such that the speed can rise above synchronous value. In this case regeneration will take place automatically and exert a braking action, see Fig. 7.21.

Reverse-current Braking, $s > 1$: $P_m = (1-s).I_2{}^2 R_2/s$ negative

Figure 7.20a indicates the electromagnetic conditions when $s > 1$, i.e. with the rotor running in the opposite direction to the rotating field. Although the flux/conductor relative motion is the same as when motoring, the torque now opposes motion and mechanical energy must therefore be supplied as well as electrical energy from the line. Figure 7.20b shows that I_1 still has a component in phase with V_1, the line current being high and the power factor low, if the rotor

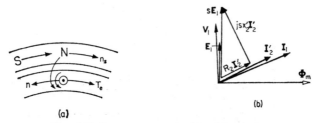

(a)

(b)

FIG. 7.20. Reverse-current braking (plugging).

resistance is not supplemented externally. In this plugging mode, the whole of the electrical and mechanical energy is dissipated within the machine circuits as a heat loss as for the d.c. machine (see p. 345). The equivalent circuit is the same as Fig. 7.11 but the slip values are always greater than unity. The increased rotor frequency, which is now higher than the line frequency, means that there are additional rotor losses in both copper and iron. In practice, therefore, the torque values are somewhat higher than predicted from this circuit.

This mode of operation is very useful when it is required to stop an induction motor quickly. In order to achieve the relative reversal between n and n_s, it is necessary to interchange two stator leads since initially the machine rotation is maintained. Mechanical work is provided by the stored energy of the rotating parts as the machine slows down. Electrical energy must still be provided by the line supply.

Figure 7.21 which is the four-quadrant diagram corresponding to Fig. 6.41 for the d.c. machine, shows the induction machine character-

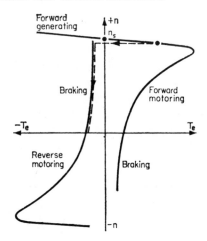

FIG. 7.21. Operation in the four quadrants.

istics over the range of slip values as discussed in the last few pages. They are drawn for both directions of the rotating field and show, as an example, the path followed on reversing the field from normal speed to give reverse-current braking. It can be seen that when the speed falls to zero, the supply must be disconnected unless it is required that the machine should run up to speed in the opposite rotation.

Speed/Torque Curves with Changes of Rotor-circuit Resistance

With all other parameters constant, eqn. (7.7) shows that variations of R_2' change the slip at which maximum torque occurs, but from eqn. (7.8) the value of the maximum torque is the same. The form of eqn. (7.5) shows that any particular value of torque requires a unique value of R_2'/s so that to maintain this torque a change to R_2' necessitates a corresponding change to the slip. Alternatively, as an approximation at low slips, since $T_e \simeq k_1 s/R_2$, an estimate of the torque can be made at particular slips and different values of rotor-circuit resistance. Hence, either precisely or approximately, the shape of the speed/torque

400

curves with varying R_2 can be obtained without detailed calculation. Figure 7.22 shows the results. From the intersections with the typical mechanical characteristic it is seen that speed control is possible by varying R_2, a method analogous to the insertion of armature circuit resistance in the d.c. motor, and suffering from the same disadvantages. The speed regulation becomes large as R_2 is increased so that the speed is very sensitive to load changes. The method is unsatisfactory for large speed changes and uneconomical in power also, for continuous low-speed running. Its main use, as for the d.c. motor, is in building up the speed from zero to normal, i.e. starting. The curves show that with

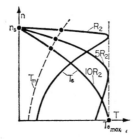

FIG. 7.22. Rotor resistance variation.

R_2 increased tenfold, maximum torque will be obtained in this particular case when $s = 1$ which is almost certainly sufficient to start any load which the motor is capable of driving. As speed picks up, rotor circuit resistance is cut out and the operation is along a different characteristic until the next resistance change. Eventually all the external resistance is cut out by short-circuiting the slip rings and the motor then runs with its natural rotor resistance only. External resistance is also inserted when operating in the reverse-current braking mode. This reduces the current, improves the power factor and increases the braking torque. Rotor resistance control is of course confined to phase-wound rotors with slip rings. For starting, however, centrifugally operated switches have been used on a rotor without slip rings to change the internal rotor-circuit resistance during acceleration.

Speed/Torque Curves with Varying Voltage

With V_1 as the only variable, eqn. (7.5) shows that all torque abscissae at particular values of speed are reduced as the square of the voltage, Fig. 7.23, voltage *increases* being precluded by saturation effects. This sensitivity of torque to voltage change is sometimes troublesome, particularly at starting. On squirrel-cage motors, the voltage is often reduced at starting to limit what is virtually the short-circuit current, and many schemes have been devised to achieve this

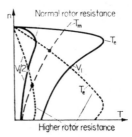

FIG. 7.23. Voltage variation.

without suffering such a heavy penalty in terms of reduced starting torque. The ratio between starting torque T_s and full-load torque $T_{f.l.}$ is usefully expressed in terms of the currents. Since the referred rotor current is approximately equal to the input current, and at starting, $s = 1$:

$$\frac{T_s}{T_{f.l.}} = \frac{I_s^2 R_2/1}{I_{f.l.}^2 R_2/s_{f.l.}} \text{ from which } T_s = \left[\frac{I_s}{I_{f.l.}}\right]^2 . s_{f.l.} . T_{f.l.} \qquad (7.9)$$

For squirrel-cage rotors in which this ratio of currents may be 6/1 or more, and the full load slip say 3%, the starting torque at full voltage, from eqn. (7.9), would be about the same as the full load torque.

A further feature of voltage variation which must be considered, concerns the ability of a motor to recover speed, following a momentary reduction in its terminal voltage. Such reductions occur due to the voltage drop in the supply system reactance if there is a fault condition, or even due to the large, low power-factor currents taken by neighbouring motors during their starting periods. The fall of voltage reduces the torque available for acceleration and increases the current which in turn could cause a further fall of voltage. The situation could become unstable with all the connected motors stalling to standstill.

Voltage variation can be used to give a relatively inefficient, though simple method of speed control which is suitable when, as on fan-type loads, the torque required falls considerably with speed. A high-resistance rotor is necessary in practice, to give a reasonably wide speed-range. The general effect can be understood by considering the intersection points of the typical T_m characteristic shown on Fig. 7.23.

Effect of Variations in R_1, x_1, f and x_2

It is a simple matter from eqn. (7.5) to find the effect of changing any of the parameters by substituting the new values. Occasionally on small motors, R_1 is supplemented externally to limit the starting current, but the method is not very effective for speed control. An increased R_1 leads to an increased $I_1 z_1$ drop and hence a fall in E_1, flux and torque; the speed falls but the reduced value of T_e may be insufficient to hold the load. On the other hand, when R_2 is changed it directly affects the rotor current and the flux is virtually unaltered. From eqn. (7.5) it can be seen that with small values of slip, the term R_2'/s has much more pronounced effects on the result than R_1.

x_1 is occasionally supplemented externally by means of inductors to limit the starting current. The starting power-factor becomes still lower though there is less power wastage in the starter than when using resistance.

x_2 is not normally used as a controlling parameter.

The frequency can be changed to give speed control and this affects the value of n_s, the reactances and the magnetising current. In practice, the voltage must also be changed as described below.

Effects of Varying V_1 and f together

This kind of variation can occur for example, if the supply generator is subject to speed changes, either due to momentary overloads or because of a noticeable speed regulation. Both output voltage and frequency will vary as the speed if automatic correction is absent. On some large marine drives, the propellor motors are induction type and are speed-controlled from such a local supply which is provided by synchronous generators coupled to variable-speed turbines.

From the e.m.f. eqn. (5.8a), it can be understood that a change of frequency will give rise to a change of flux level unless the induced e.m.f. is changed in the same ratio. An imbalance in this ratio will cause either an excessive flux and saturation, or diminished flux and reduced torque per ampere; see eqn. (7.3). Constant flux per pole is approximated if the terminal volts per cycle V/f is maintained constant since $V_1 \simeq E_1$. However an idealised variable-frequency control scheme might aim to ensure that any particular torque should be obtained at the same flux (E_1/f), as when operating at normal voltage and frequency. For any specified flux level, eqn. (7.3) shows that the torque is then proportional to:

$$I_2' \cos \varphi_2 \propto \frac{sE_1}{z_2'} \cdot \frac{R_2'}{z_2'} \propto \frac{\omega_s - \omega_m}{\omega_s} \cdot E_1 \cdot \frac{R_2'}{(z_2')^2} \cdot$$

With this criterion of controlling flux, (E_1/ω_s) to a constant value for a particular torque, the above expression shows that the speed difference $\omega_s - \omega_m$ is also fixed for the same torque, since $\omega_s - \omega_m$ in turn determines the slip frequency f_2, the rotor e.m.f., impedance and current, the magnetising current and the primary current, which are also unchanged. From this it follows, as shown on Fig. 7.24, that as the frequency is reduced from a value corresponding to ω_{s1} say, the speed/torque curves are of the same shape, but displaced vertically. The required supply frequency f_s for any particular speed n follows from the expression for the synchronous speed at that frequency; i.e.

$$n_s = n + sn_s \quad \text{or} \quad pn_s = pn + p \cdot (f_2/f_s) \cdot n_s, \quad \text{i.e.} \ f_s = pn + f_2.$$

f_2 will be known for the particular value of torque required. The required terminal voltage V_1 can be obtained from the equivalent circuit; see Appendix E, Problem (4) for Chapter 7, for a numerical

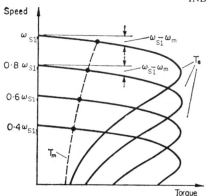

FIG. 7.24. Variable-frequency control with constant flux per pole.

example. The speed can be increased beyond the value set by rated voltage and flux by increasing frequency alone. The flux and torque capacity will fall as for the comparable field-weakened condition described on p. 336 for the d.c. machine.

Calculations at different voltages and frequencies can be carried out using the expressions so far developed after correcting *all* the frequency-sensitive parameters, the value of the slip s being based on the new supply frequency. In the equivalent circuit of Fig. 7.11a, all the reactances are changed directly as the frequency. X_m will have to be corrected for changes of flux level also unless E_1/f is maintained constant. For this condition, if R_m is unchanged, the iron loss it represents (E_1^2/R_m) would vary as f^2 which is only approximately true. The value of R_m should be derived from the actual iron loss but in fact it becomes less significant in the equivalent circuit as X_m falls with frequency. The departure from the simple relationships of Fig. 7.24 which occur if V_1/f is maintained constant, instead of fixing E_1/f, is shown in the next example, part (c).

EXAMPLE E.7.3

On a certain induction motor the power factor at starting with the slip rings short circuited is 0·242. It is required to estimate some salient features of the performance with changed parameters and for this purpose the magnetising branch of the equivalent circuit may be neglected and the stator and referred rotor resistances may be assumed identical. Compare the normal values of slip at maximum torque, the maximum torque itself and the starting torque, with those occurring as a result of the following changes:

(a) Normal voltage and frequency, R_1 doubled.

(b) Normal voltage and frequency, R_2 doubled.

(c) Circuit unchanged, voltage and frequency both reduced by 30%.

Hence sketch the speed/torque characteristics.

At starting, the slip is unity and the impedance is

$$(R_1 + R_2') + j(x_1 + x_2') = R_{e1} + jx_{e1} = z_{e1},$$

The power factor $\cos \varphi$, is therefore:

$$R_e/z_e \text{ and } \sqrt{(1 - \cos^2 \varphi)} = x_e/z_e$$

Substituting s.c. test data:

$$0.242 = 2R_1/z_e \text{ and } \sqrt{[1 - (0.242)]^2} = 0.968 = 2x_1/z_e$$

since $R_1 = R_2'$ and $x_1 = x_2'$.

Consequently $x_1/R_1 = 0.968/0.242 = 4$.

Let $R = R_1 = R_2'$, then $x_1 = x_2' = 4R$ and $x_1 + x_2' = 8R$.

Normal conditions will be taken as reference; all torques being expressed as a fraction of normal maximum torque $\hat{T}_n$ occurring at slip $\hat{s}_n$. Using eqn. (7.7) the slips at maximum torque are:

$$\hat{s}_n = R_2'/\sqrt{(R_1^2 + x_{e1}^2)} = R/\sqrt{[R^2 + (8R)^2]} \quad = \underline{0.124}$$
$$1/\hat{s}_n = 8.05$$

$$\hat{s}_a = \qquad\qquad R/\sqrt{[(2R)^2 + (8R)^2]} \quad = \underline{0.122}$$
$$1/\hat{s}_a = 8.2$$

$$\hat{s}_b = \qquad\qquad 2R/\sqrt{[R^2 + (8R)^2]} \quad = \underline{0.248}$$
$$1/\hat{s}_b = 4.02$$

$$\hat{s}_c = \qquad\qquad R/\sqrt{[R^2 + (0.7 \times 8R)^2]} = \underline{0.175}$$
$$1/\hat{s}_c = 5.72.$$

Note that in case (c) the reactance is reduced because of the frequency change.

To get the torques in terms of $\hat{T}_n$, a general expression for the torque ratio can be written from eqn. (7.5) as:

$$\frac{T}{\hat{T}_n} = \frac{(3/\omega_s)(V^2/z^2)(R_2'/s)}{(3/\omega_n)(V_n^2/\hat{z}_n^2)(R/\hat{s}_n)}.$$

Certain terms will cancel in particular cases, if there is no change. Dealing first with the starting torques, T_n, T_a, T_b and T_c,

$$\frac{T_n}{\hat{T}_n} = \frac{\hat{z}_n^2}{z_n^2} \cdot \frac{\hat{s}_n}{1} \qquad \text{voltage, rotor resistance and frequency being unchanged. Therefore,}$$

$$\frac{T_n}{\hat{T}_n} = \frac{(R+8\cdot05R)^2+(8R)^2}{(R+1R)^2+(8R)^2} \times \frac{0\cdot124}{1} = \frac{18\cdot1R^2}{68R^2} = \underline{0\cdot266}$$

$$\frac{T_a}{\hat{T}_n} = \frac{\hat{z}_n{}^2}{z_a{}^2}\cdot\frac{\hat{s}_n}{s_a} = \frac{145\cdot8R^2}{(2R+R)^2+(8R)^2}\cdot\frac{0\cdot124}{1} = \underline{0\cdot248}$$

$$\frac{T_b}{\hat{T}_n} = \frac{\hat{z}_n{}^2}{z_b{}^2}\cdot\frac{\hat{s}_n}{s_b}\cdot\frac{R_2'}{R} = \frac{145\cdot8R^2}{(R+2R)^2+(8R)^2}\cdot\frac{0\cdot124}{1}\cdot\frac{2R}{R} = \underline{0\cdot496}$$

$$\frac{T_c}{\hat{T}_n} = \frac{\omega_n}{\omega_s}\cdot\frac{V^2}{V_n{}^2}\cdot\frac{\hat{z}_n{}^2}{z_c{}^2}\cdot\frac{\hat{s}_n}{s_c} = \frac{1}{0\cdot7}\times\frac{0\cdot7^2}{1^2}\times\frac{145\cdot8R^2}{(R+R)^2+(0\cdot7\times8R)^2}\cdot\frac{0\cdot124}{1} = \underline{0\cdot357}$$

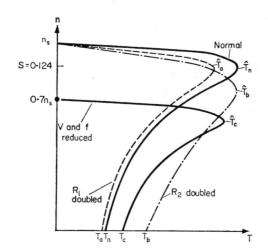

The maximum torques will be found in the same way apart from the changes to the value of slip and apparent rotor resistance,

$$\frac{\hat{T}_a}{\hat{T}_n} = \frac{145\cdot8R^2}{(2R+8\cdot2R)^2+(8R)^2}\cdot\frac{0\cdot124}{0\cdot122} = \underline{0\cdot882}$$

$$\frac{\hat{T}_b}{\hat{T}_n} = \frac{145\cdot8R^2}{(R+4\cdot02\times2R)^2+(8R)^2}\cdot\frac{0\cdot124}{0\cdot248}\cdot\frac{2R}{R} = \underline{1\cdot0}$$

$$\frac{\hat{T}_c}{\hat{T}_n} = \frac{1}{0\cdot7}\cdot\frac{(0\cdot7)^2}{1^2}\cdot\frac{145\cdot8R^2}{(R+5\cdot72R)^2+(0\cdot7\times8R)^2}\cdot\frac{0\cdot124}{0\cdot175} = \underline{0\cdot942}$$

7.4 TESTING AND EFFICIENCY

Ratio Test

On a transformer, the turns ratio is very nearly equal to the phase voltage ratio on no load because the $I_0 z_1$ voltage drop is so small and therefore:

$$N_1/N_2 = E_1/E_2 \simeq V_1/E_2.$$

With an induction machine, I_0 is much larger and $I_0 z_1$ cannot be neglected. On a phase-wound rotor where it is possible to measure the secondary voltage, a close estimate of the turns ratio can be made if primary and secondary are each supplied in turn.

Let $r_1 = V_1/E_2 = |\, E_1 + I_{01} z_1\,|/E_2$ be the measured phase-voltage ratio when the stator is supplied and the secondary voltage at the slip rings is measured on open circuit.

Let $r_2 = V_2/E_1 = |\, E_2 + I_{02} z_2\,|/E_1$ be the ratio when the slip rings are supplied and the stator winding is open circuited. Then:

$$\frac{r_1}{r_2} = \frac{|\, \mathbf{E}_1 + \mathbf{I}_{01}\mathbf{z}_1\,|}{E_2} \cdot \frac{\mathbf{E}_1}{\mathbf{E}_1} \times \frac{E_1}{|\, \mathbf{E}_2 + \mathbf{I}_{02}\mathbf{z}_2\,|} \cdot \frac{E_2}{E_2}$$

$$= \left|\, 1 + \frac{\mathbf{I}_{01}\mathbf{z}_1}{\mathbf{E}_1}\,\right| \frac{\mathbf{E}_1}{\mathbf{E}_2} \times \frac{1}{\left|\, 1 + \dfrac{\mathbf{I}_{02}\mathbf{z}_2}{\mathbf{E}_2}\,\right|} \frac{E_1}{E_2} \simeq \left(\frac{E_1}{E_2}\right)^2,$$

i.e. $E_1/E_2 = N_1/N_2$ is very nearly equal to $\sqrt{(r_1/r_2)}$ since the *per-unit* impedances for primary and secondary windings are about the same and the terms within the modulus brackets cancel.

Short-circuit (Locked Rotor) Test

This is very similar to the s.c. test for the 3-phase transformer, except that the rotor must be locked in position otherwise it will rotate. Within a rotor slot-pitch of movement, there will be a slight variation of the readings and the rotor should be held in a position which gives the mean of the maximum and minimum current values. A low voltage of normal frequency is applied to the primary windings, V_{sc}, and the slip rings, if fitted, are connected through a low-resistance short circuit.

Neglecting the magnetising branch of the equivalent circuit, the power input will be $3 \times (I_{phase})^2 R_{e1}$, where $R_{e1} = R_1 + R_2'$ includes the effect of stray load-losses at full frequency, an overestimate in the case of the rotor winding where the frequency may be less than 2 Hz under normal running conditions. For a phase-wound rotor, the a.c. resistance thus measured can be divided into the components R_1 and R_2' in the ratio of the d.c. resistances measured normally, but using the referred value of R_2'. For squirrel-cage rotors, this means of segregation is not available, though special tests could be taken to get a nearer estimate of R_2' than that obtained merely by deducting the d.c. value of R_1 from R_{e1},

The effect of the magnetising branch on the short-circuit impedance, V_{sc}/I_{sc}, is usually small enough to be neglected, with little error in the final calculation of performance. Consequently, the leakage reactance $x_1 + x_2'$ as for the transformer is $\sqrt{[(V_{sc}/I_{sc})^2 - R_{e1}^2]}$. x_1 is usually taken to be equal to x_2', unless the ratio of x_1/x_2' is known from design data.

No-load Test

The object of this test is similar to that for the transformer, but the method is different in application. It is not sufficient to open circuit the slip rings, because at standstill the rotor would be magnetised at full frequency, whereas under normal conditions at slip frequency, the rotor iron loss is negligible. For a squirrel-cage rotor such a test would be impossible anyway. Reference to the equivalent circuit will show that at synchronous speed when $s = 0$, the rotor branch is non-conducting. The real fact is that there is no relative motion between the field and the rotor conductors and therefore no induced e.m.f. or current. If the rotor were to be running at synchronous speed then the equivalent circuit would have the right-hand (rotor) branch of Fig. 7.11a omitted and the power input to the stator at normal voltage and frequency would only be the stator iron loss plus the small $I_0^2 R_1$ copper loss. No torque could be developed and no mechanical power produced, not even to supply the mechanical loss. Such a test is not possible, unless an external driving motor is coupled to provide the mechanical loss-torque. However, when the induction machine alone is running "light", i.e. supplying only its own friction and windage losses, the slip will be very small indeed to produce the necessary induced e.m.f., current and torque.

The friction and windage loss can be separated from the stator iron loss, by a similar method to that used for the d.c. motor; see Fig. 6.49 and the associated text. The frequency is maintained, but the voltage is reduced from say 10% overvoltage down to perhaps 20% of normal voltage, below which the machine is liable to stall, see Fig. 7.23. The speed will only fall slightly during this test, so that the mechanical loss is substantially constant. Input readings of power and current are plotted against input voltage, Fig. 7.25. Extrapolating the power curve to zero voltage will give the friction and windage loss at the y-axis intercept. Normal iron loss and no-load current can be read off the curve at full voltage. As in the case of the d.c. machine, the extrapolation is assisted if power is plotted against (voltage)^2. For greater accuracy a correction for the stator copper loss should be made by deducting the appropriate values of $I_1{}^2 R_1$, before plotting the power

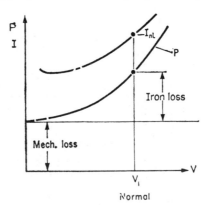

FIG. 7.25. No-load losses.

curve. Eventually it is necessary to convert to phase values, but the power for three phases and the line current can be used on Fig. 7.25, providing that phase values are extracted for calculations of the equivalent circuit.

From the equivalent circuit, $E_1 I_p$ is the iron loss per phase and the test value can be found from the construction of Fig. 7.25. The $I_0 z_1$ drop is no longer negligible as in the case of the transformer no-load test, but a close estimate of E_1 at normal voltage is obtained from the *algebraic* difference, $V_1 - I_{nl} x_1$, the no-load power factor being very low; see Fig. 4.3a. Hence the value of I_p follows and the magnetising resistance is $R_m = E_1/I_p$. The low no-load power factor also means that I_m is very nearly equal to $I_{nl} \sin \varphi_{nl}$, being little different in practice from I_{nl} itself. The magnetising reactance is $X_m = E_1/I_m$. The current which would be taken at synchronous speed is therefore $\mathbf{I_0} = I_p - jI_m$, with the $\mathbf{E_1}$ phasor as reference. For certain purposes, e.g. in constructing the circle diagram (Section 7.5), it may be sufficiently accurate to use $\mathbf{I_{nl}}$ as an approximation to $\mathbf{I_0}$. It should be noted that the magnetisation curve can also be obtained from the no-load test if for each reading, E_1 and I_m are calculated and the E_1/I_m relationship plotted. The variation of X_m with e.m.f. (E_1) or magnetising current (I_m) is also of interest and can be plotted from the same data.

Back-to-back Test

This is not a very common test on induction machines but it will be described briefly with reference to Fig. 7.26. The two induction machines are mechanically coupled and supplied electrically from two different sources. The motoring machine may be connected to the line, but the generating machine must be supplied from a source of variable frequency. In order that it may run as an induction generator, the speed, which in terms of the motor parameters is $(1 - s_m)f_m/p_m$, must be higher than the synchronous speed f_g/p_g. Since the speed in generator terms is $(1 + s_g)f_g/p_g$, where s_g here is a positive number, the two expressions can be equated giving:

$$(1 - s_m)f_m/p_m = (1 + s_g)f_g/p_g.$$

411

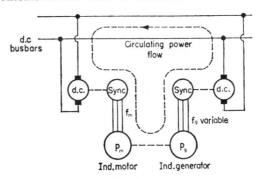

FIG. 7.26. Back-to-back test.

If, as is usual in this test, the machines are identical and $p_m = p_g$, then $f_g = f_m(1 - s_m)/(1 + s_g)$. The load flow is controlled by varying f_g until the motor is operating at full-load slip. The motor output, less the losses, is fed back to the system through the variable-frequency machine and a d.c./a.c. conversion. One possible circuit is shown in which the supply is d.c. and forms a link between the two a.c. systems. There is an alternative test circuit in which the two machines can be connected to the same supply frequency. In this case, the speed of the generator is increased above the synchronous speed by using a special mechanical-differential-coupling.

Reverse-rotation Test

This special test is used to determine the rotor stray load-loss by direct measurement. The major component of this loss is due to the stator m.m.f. harmonics. These rotate at their own synchronous speeds; see p. 259, and induce high-frequency rotor currents at the appropriate slip relative to the particular stator-harmonic field. In the first part of the test, the total stator copper loss and iron loss are measured directly, by removing the rotor and circulating full-load current through the stator windings. The flux is small as a result of removing the rotor so that only a low voltage is required and the iron loss, too, is small. The rotor is replaced and the assumption is now made that the rotor stray load-loss occurs only with rotation so that it must be provided by a mechanical input power. The rotor is driven against the stator rotating field at synchronous speed, i.e. in reverse rotation, and the measured mechanical input will therefore be equal to the friction and windage, plus the rotor stray loss, plus any term due to $P_m = I_2{}^2 R_2 (1-s)/s$. With $s = 2$, this term is equal to $-I_2{}^2 R_2/2$ which is just one half of the rotor current loss at $sf = 2f$. The other half is supplied from the air-gap power $P_g = I_2{}^2 R_2/s = I_2{}^2 R_2/2$, per phase. The difference between the stator input and the mechanical input to the rotor is now formed, giving

$$[\text{stator Cu loss} + \text{Fe loss} + 3(I_2{}^2 R_2/2)]$$
$$- [\text{stray h.f. loss} + 3(I_2{}^2 R_2/2) + (f+w)] = \text{measured difference}$$

The rotor copper-loss terms due to the fundamental-frequency stator field cancel and since all the other terms will be known apart from the stray loss, this can now be found from the expression.

The theory behind this test involves several assumptions which affect the validity of the method in certain cases. The matter is discussed in detail in Reference 19.

Efficiency

The earlier tests described above give the parameters of the equivalent circuit which can be used in either the "exact" or approximate versions. The full-load slip $s_{f.l.}$ at rated power is obtained from:

$$\text{rated hp} \times 746 = (1 - s_{f.l.}) . 3P_g - \text{mechanical loss}$$

or rated $\text{hp} \times 746 + \text{mech. loss} = (1 - s_{f.l.}).3I_2'^2 R_2'/s_{f.l.}$

$$= \frac{(1 - s_{f.l.}).3V_1{}^2 R_2'/s_{f.l.}}{(R_1 + R_2'/s_{f.l.})^2 + (x_1 + x_2')^2}$$

from the approximate circuit. The equation is a quadratic in $s_{f.l.}$, two answers being expected from the shape of the torque/slip curve. The smallest value corresponds to the stable operating point.

413

Knowing $s_{\text{f.l.}}$, I_2' and I_1 can be calculated from the equivalent circuit and the total losses are:

mechanical loss + iron loss + $I_2'^2 R_2' + I_1^2 R_1$

and efficiency $\eta = \dfrac{\text{output}}{\text{input}} = 1 - \dfrac{\text{total losses}}{\text{hp} \times 746 + \text{total losses}}$.

The use of the approximate circuit gives rise to a few anomalies in the calculations and slight errors in the answers.

EXAMPLE E.7.4

A 1330-h.p., 6600-V, 50-Hz, 8-pole, 3-phase star-connected induction motor, when supplied at normal frequency and varying voltage, gave the following readings when running on no load:

Line voltage	6600	6000	5000	4000	3000	V
Line current	40					A
Total input power	45	40·2	31·5	26·7	21	kW

With rotor locked and short circuited, the line input readings were:

$$1400 \text{ V}, \qquad 80 \text{ A}, \qquad 50 \text{ kW}.$$

If stator and referred rotor impedances are assumed to be identical, calculate the efficiency.

By plotting the no-load input against V or V^2 it will be found that the intercept on the zero voltage axis is 15 kW which is the mechanical loss. The remainder of the input at full voltage is 30 kW; equal to the iron loss plus the stator copper loss at 40 A. It will be necessary to find the stator resistance before the iron loss can be calculated. The short-circuit test data can be analysed as for the transformer, p. 195, to find the leakage impedance. An alternative method based on the in-phase and quadrature current components, $I_{\text{sc}} \cos \varphi$ and $I_{\text{sc}} \sin \varphi$ follows as below:

$$I_{\text{sc}} \cos \varphi = \frac{50}{\sqrt{3} \times 1\cdot4} = 20\cdot6 \text{ A} \quad \text{and} \quad I_{\text{sc}} \sin \varphi = \sqrt{[80^2 - (20\cdot6)^2]} = 77\cdot2 \text{ A}.$$

At full voltage, $\mathbf{I}_{1\text{sc}} = (20\cdot6 - j77\cdot2)(6\cdot6/1\cdot4) = 97\cdot2 - j364 \text{ A}.$

The leakage impedance is virtually equal to $\dfrac{\mathbf{V}_1}{\mathbf{I}_{1\text{sc}}} = \dfrac{6600/\sqrt{3}}{97\cdot2 - j364} = 2\cdot62 + j9\cdot8 \ \Omega.$

Dividing this equally between stator and rotor:

$$x_1 = x_2' = 4\cdot9 \ \Omega \qquad R_1 = R_2' = 1\cdot31 \ \Omega.$$

It is now possible to calculate the magnetising impedance after making allowance for the no-load stator copper loss, $3 \cdot I_{n1}^2 \cdot R_1 = 3 \cdot 40^2 \cdot 1\cdot31 = 6\cdot3 \text{ kW}.$

The iron loss at full voltage is therefore $30 - 6\cdot3 = 23\cdot7 \text{ kW}.$

414

The primary e.m.f. at full voltage on no load is:

$$E_1 \simeq V_1 - I_{n1}x_1 = 6600/\sqrt{(3)} - 40 \cdot 4 \cdot 9 = 3614 \text{ V}.$$

No-load power factor $\cos \varphi_{n1} = \dfrac{45,000}{\sqrt{3} \cdot 6600 \cdot 40} = 0 \cdot 0985.$

$$\sin \varphi_{n1} = \sqrt{(1 - (0 \cdot 0985)^2)} = 0 \cdot 995.$$

Hence:

$$I_m = I_{n1} \sin \varphi_{n1} = 40 \cdot 0 \cdot 995 = 39 \cdot 8 \text{ A}.$$

$$I_p = \text{Fe loss}/E_1 = \frac{23 \cdot 7}{3} \Big/ 3 \cdot 614 = 2 \cdot 2 \text{ A}.$$

The magnetising branch impedance components are:

$$X_m = 3614/39 \cdot 8 = 91 \, \Omega \quad \text{and} \quad R_m = 3614/2 \cdot 2 = 1650 \, \Omega.$$

In the following solution, the approximate circuit is used, though the impedance values deduced from the tests could of course be used in the exact circuit also.

To find the full-load slip:

total mechanical output $= (1-s) \cdot 3P_g$

$$1330 \times 746 + 15,000 = (1-s) \cdot 3 \cdot \frac{(6600/\sqrt{3})^2}{(1 \cdot 31 + 1 \cdot 31/s)^2 + (9 \cdot 8)^2} \cdot \frac{1 \cdot 31}{s}.$$

Dividing by 10^6 and simplifying:

$$1 \cdot 009 = \frac{(1-s)}{s} \times \frac{43 \cdot 6 \times 1 \cdot 31}{(1 \cdot 71 + (3 \cdot 42/s) + (1 \cdot 71/s^2) + 96}.$$

By further simplification the quadratic:

$$\frac{1}{s^2} - \frac{31 \cdot 1}{s} + 90 \cdot 5 = 0$$

is obtained. Solving for $(1/s)$ which is more accurate since s is small gives, as the larger value, $(1/s) = 27 \cdot 8$. Therefore s at full-load is equal to $0 \cdot 036$. The other value of s corresponds to the unstable portion of the curve.

At full load, the rotor current as reflected in the primary is:

$$\mathbf{I_2'} = \frac{(6600/\sqrt{3})}{(1 \cdot 31 + 1 \cdot 31/0 \cdot 036) + j9 \cdot 8} = 95 \cdot 5 - j24 \cdot 6 = 99 \text{ A};$$

adding $\mathbf{I_0} = 2 \cdot 2 - j39 \cdot 8$ gives $\mathbf{I_1} = 97 \cdot 7 - j64 \cdot 4 = 117$ A.

The total copper loss is $(99^2 + 117^2)1 \cdot 31 \times 3 =$	92 kW
Iron loss + mechanical loss	$=$ 38·7 kW
	130·7 kW
Useful output $= 1330 \times 0 \cdot 746$	$=$ 995 kW
$\therefore$ Input	$=$ 1125·7 kW

From eqn. (3.6): full-load efficiency $= 1 - 130 \cdot 7/1125 \cdot 7 = \underline{88 \cdot 4 \%}.$

Note. The data in this question are to be used for a circle diagram problem (E.7.5). For clarity in the construction, the copper losses have been exaggerated and this explains the low value of efficiency. For a machine of this size, the full-load efficiency would be well over 90%.

7.5 CIRCLE DIAGRAM

As the slip varies over the range $\pm \infty$, the primary current phasor will be shown to trace out a circular locus, if the impedance parameters are assumed to remain constant. From this fact, it is possible to construct the complete locus knowing only two current phasors. Hence the performance of the machine over the operating range can be found. With modern aids to computation, the virtues of the circle diagram, which are mainly associated with speedy estimation of machine behaviour, are rather diminished. Any information which can be deduced from the diagram can be obtained more accurately from the equations or equivalent circuit, of which the diagram is only a graphical representation.

Current Locus Diagram for Circuit with Variable Resistance,
Constant Reactance and Constant Voltage

The current in the circuit of Fig. 7.27 varies from V/X at zero power-factor when $R = 0$, to 0 at "unity power-factor" when $R = \infty$. In between these two extremes, the current phasor can be obtained by rearranging the circuit equation:

$$\mathbf{V} = R\mathbf{I} + jX\mathbf{I}.$$

Dividing by jX; $\qquad \mathbf{V}/(jX) = R\mathbf{I}/jX + \mathbf{I}$

from which: $\qquad -j\mathbf{V}/X = \mathbf{I} - j(R/X)\mathbf{I}.$

This equation states that a constant current V/X lagging 90° behind $\mathbf{V}$ is made up of two components; the current in the circuit itself plus a variable current $(R/X)\mathbf{I}$, lagging $\mathbf{I}$ by 90°. From the geometry of the phasor diagram shown in Fig. 7.27, where a right-angled triangle is

416

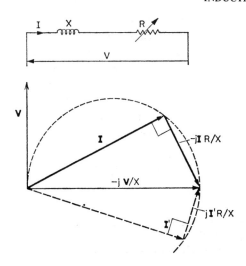

Fig. 7.27. Locus of current phasor with varying resistance.

formed over a constant diameter V/X, it can be seen that the current phasor $\mathbf{I}$ traces out a semicircle as R varies from 0 to ∞. If R was able to take negative values, the remainder of the circle would be traced out by $\mathbf{I}'$ as indicated.

Application to Induction Machine

In the approximate equivalent circuit, the current component $\mathbf{I}_0$ is constant and the component $\mathbf{I}_2'$ flows in a circuit of fixed reactance $x_1 + x_2'$ and a resistance $R_1 + R_2'/s$ varying between infinity when $s = 0$, and $R_1 + R_2'$ when $s = 1$. A constant voltage V_1 is applied and so the locus of $\mathbf{I}_2'$ must lie on a circle. In fact if s assumes all positive and negative values, the whole circle is traced out. Since $\mathbf{I}_1$ is equal to the vectorial summation of $\mathbf{I}_2'$ and the constant component $\mathbf{I}_0$, the locus of $\mathbf{I}_1$ too is a circle, Fig. 7.28. Geometrically, a circle can be defined by two points, if the position and direction of a diameter are known. For example if $\mathbf{I}_0 = I_p - jI_m$ is given, this establishes the

417

horizontal diameter of Fig. 7.28 and only one other current phasor is required. This is usually obtained from the short-circuit test. By scaling up the value I_{sc} in the ratio V_1/V_{sc}, the short-circuit current at full voltage, I_{1sc} is obtained. The corresponding referred rotor current I'_{2sc} is the phasor joining I_0 to I_{1sc}. Bisecting I'_{2sc}, which is a chord of the circle, locates the centre from which the circle can be drawn. It could of course be found by trial and error.

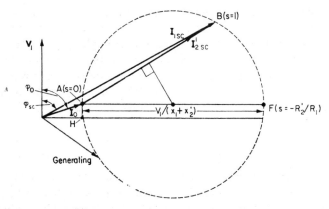

FIG. 7.28. Circle diagram for approximate circuit.

The horizontal diameter represents the referred rotor current when the effective resistance $R_1 + R'_2/s$ is zero, i.e. it is equal to $V_1/(x_1 + x'_2)$. At the point F, the slip must therefore have a value of $-R'_2/R_1$, but this has no practical significance because it is on the unstable portion of the torque characteristic for negative slips at a speed of approximately $2n_s$ (since $R_1 \simeq R'_2$; $s = -1$). When the I_1 phasor is below the zero voltage axis, the machine must be generating; i.e., it is a "negative" motor, because the current has components in antiphase with the supply voltage. The slip must be negative for this condition and R_2/s small enough to give a power component of current equal and opposite to I_p. The portion of the circle between its horizontal diameter and the zero voltage line is also for negative slips, but although there is some

418

conversion of mechanical to electrical power here, it is not sufficient to overcome the internal losses and give a net generating output. The portion of the circle between B and F is the reverse-current braking region and covers both positive and negative slips, but only a small portion of the positive region is of practical interest. The region between A ($s = 0$, $n = n_s$) and B ($s = 1$, $n = 0$) is the current locus of I_1 and I_2' when the machine is motoring and covers the vast majority of practical applications.

If the diagram is drawn in terms of line currents, whether the primary is star or delta connected, it should be clear that vertical, or in-phase, components represent power to scale. For example, $\sqrt{3}V_1$ AH is the total stator iron loss. It should be remembered that all phasors on the diagram are current phasors; V_1 only being drawn in for reference purposes. Nevertheless, it is the possibility of converting in-phase currents to power values that makes the diagram useful.

Analysis of Circle Diagram

The diagram is redrawn on Fig. 7.29 with some additional construction lines. Consider point B, $s = 1$ and rotor stationary. The total input is $\sqrt{3}V_1$ BE to scale and contains a component DE representing the iron loss. The remaining component BD can only be the stator and rotor copper loss since there is no mechanical output. Choosing another operating point G, arbitrarily:

$$\frac{bd}{\text{BD}} = \frac{Ad}{\text{AD}} = \frac{\text{AG} \cos \text{GAF}}{\text{AB} \cos \text{BAF}} = \frac{\text{AG} \cdot \text{AG}/\text{AF}}{\text{AB} \cdot \text{AB}/\text{AF}} = \frac{\text{AG}^2}{\text{AB}^2}$$

$$\therefore \quad \frac{bd}{\text{BD}} = \frac{\text{Total } I_2'^2(R_1 + R_2') \text{ at current AG}}{\text{Total } I_2'^2(R_1 + R_2') \text{ at current AB}},$$

i.e. vertical intercepts such as bd are proportional to $3I_2'^2(R_1 + R_2')$ the constant of proportionality being $\sqrt{3}V_1$ if bd is scaled in amperes.

If BD is so divided that $\sqrt{3}V_1$ BC is $3I_2'^2R_2'$ and $\sqrt{3}V_1$ CD is $3I_2'^2R_1$, then at any current such as I_1, the rotor copper loss is $\sqrt{3}V_1 \cdot bc$

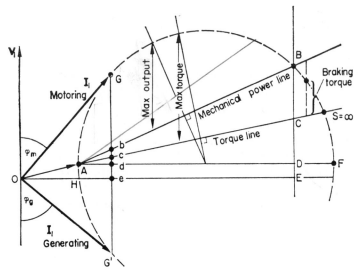

FIG. 7.29. Analysis of circle diagram.

and $\sqrt{3}V_1 \cdot cd$ is $3I_2'^2 R_1$. The stator copper loss is $3I_1^2 R_1$ which is slightly greater than $3I_2'^2 R_1$ but for the purpose of problems, bearing in mind the errors inherent in a graphical method, CD will be taken as the stator copper loss if BD is divided in the ratio $BC/CD = R_2'/R_1$. For a squirrel-cage rotor machine it would be necessary to calculate the stator copper loss as $3I_{1sc}^2 R_1$ and set this off to scale as CD. The effective value of R_1 when carrying normal-frequency currents should be increased by perhaps 5–20% over the measured d.c. value to allow for copper eddy-current losses.

The line AB is called the mechanical-power line, since vertical distances from it to the circle represent the mechanical output (including mechanical loss). Vertical distances below this line to OE represent the other machine losses.

The line AC is called the torque line since vertical distances from it to the circle represent the rotor input $3P_g$ to which the torque is proportional. This is so because Ge represents the total input and ce

420

represents the total stator loss. Therefore Gc represents the rotor input, the power across the air gap for the motor case.

At any working point such as G

$$\text{Input} = \sqrt{3}V_1 I_1 \cos \varphi_m = \sqrt{3}V_1 \; Ge \text{ to scale}$$

$$= \sqrt{3}V_1 (\quad Gb \quad + \quad bc \quad + \quad cd \quad + \quad de \quad)$$

$$= \text{mech. output} + \text{rotor Cu loss} + \text{stator Cu loss} + \text{Fe loss}.$$

From this equation the following information is obtained:

$$\text{Efficiency} \quad = \frac{\sqrt{3}V_1 Gb; \text{ less mechanical loss if known}}{\sqrt{3}V_1 Ge}$$

Slip $\qquad s = \text{rotor Cu loss/power across the gap} \Rightarrow bc/Gc.$

Torque $\quad T_e = 3P_g \text{ sync watts}$

$$= \sqrt{3}V_1 Gc \quad \text{to scale or} \quad \sqrt{3}V_1 Gc/2\pi n_s \text{ Nm}.$$

Starting torque $= \sqrt{3}V_1 BC \quad \text{to scale}.$

The maximum torque and the maximum power correspond to the maximum vertical distances between the circle and the torque- and mechanical-power-lines respectively. By drawing perpendicular bisectors through these lines from the circle centre as shown, their intersections with the circle determine the operating points directly for these two conditions.

Normal motoring operation is somewhere near the point of maximum power factor where a current phasor from the origin is tangential to the circle. A typical current phasor for generating operation is also shown. The electrical output would be $\sqrt{3}V_1 G'e$ to scale at a power factor $\cos \varphi_g$. The power across the air gap, which is equal to the stator loss plus the output, is still measured from the torque line, downwards this time, giving an electromagnetic resisting torque of $\sqrt{3}V_1 G'c/2\pi n_s$ Nm. The mechanical loss-torque must be supplied from the mechanical source and does not appear on the circle diagram, because it is not involved in the electromagnetic power conversion in the air gap. The slip, as for the motor, see p. 396, is: Rotor Cu loss/power across the

421

gap = bc/cG'. The total mechanical input power is $\sqrt{3}V_1bG' +$ mechanical loss.

For reverse-current braking, the slip is greater than unity so the rotor copper loss is greater than the total rotor input obtained from the stator supply, as shown by a typical operating point further round the circle from point B.

Accuracy

It is important to realise that the circle diagram is only a graphical representation of the equivalent circuit and is bound by the same limitations. For example, the power input $\sqrt{3}V_1$AH is the iron loss and from the approximate circuit this is constant. In practice the flux distribution changes, particularly beyond maximum torque towards standstill and the increasing rotor frequency means that the rotor iron losses are no longer negligible. The horizontal diameter in some treatments of the subject is drawn from the end of the no-load I_{n1} phasor which therefore includes the mechanical loss automatically even at standstill! Sometimes the diameter is drawn along the zero voltage axis OE on the ground that part of the iron loss gives rise to a reaction torque. These are points for the specialist and are mentioned here to indicate that there are other interpretations of the data. A major source of error occurs at high currents, which are sufficient to cause saturation in the leakage flux paths, particularly in the teeth, and so reduce x_1 and x_2, making the current locus elliptical. Allowance must be made for this in practice, but apart from starting and braking, normal operation is at low slips where rotor iron-loss is small and x_1 and x_2' are unsaturated. A further error results from resistance changes, not just due to temperature, but in the case of the rotor, from the d.c. value on no load to a.c. values at higher slip frequencies. However, a 15 cm circle will give a good estimate of full-load current after locating the rated hp point and hence full-load torque and efficiency follow. Care must be exercised to get a reasonably accurate value of slip and the use of similar triangles is helpful here; e.g. $bc = $ BC(Ac/AC). The values of maximum torque and maximum power are nearly as accurate as those obtained from the equivalent circuit, bearing in mind errors in the parameters under these conditions. Sometimes additional scales are added for power, slip and power factor, using more construction lines, but this tends to make the diagram confusing. It is simplest to regard the circle as a current diagram and make the necessary conversions to power as described.

"Exact" Circle Diagram

It can be proved that the current locus for the "exact" circuit is also circular. On small machines up to a few hp, the calculated circle

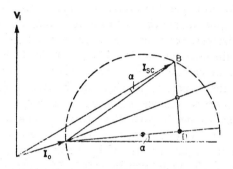

FIG. 7.30. Circle diagram for "exact" circuit.

based on the approximate circuit is noticeably erroneous when compared with the measured locus. This is due to neglecting the relatively high $I_0 z_1$ drop. To draw the "exact" circle, the diameter must be set off at an angle α from AF, where α is the angle between stator and rotor currents on short circuit, as shown in Fig. 7.30. Torque and rotor copper loss are measured perpendicular to this diameter, the current components being multiplied by $\sqrt{3}V_1$ as before. The torque line is found by dividing BD in the ratio R'_2/R', where R' is R_1 reduced in the ratio of no-load e.m.f. E_1 to applied voltage V_1. The proof of the above statements is tedious and still involves minor approximations. The information given is adequate for checking the measured performance

423

of small machines against figures deduced from this "exact" primary current locus. Note that I_0 as shown is the current through the magnetising branch only at synchronous speed. The full diagram for the "exact" circuit makes allowance for the variation of E_1, I_m and I_p as the mechanical load increases from zero.

Problems

The effects of changes to various parameters can be studied quite easily on the circle diagram, though since x_1 and x_2' are not often changed, variations of these will not be considered.

A reduction of voltage will diminish all current phasors and the circle itself in direct proportion. All powers would be reduced as V^2, since they are obtained by multiplying a current and voltage reduced by the same fraction.

Changes to R_1 or R_2' are accommodated by extending the vertical line DB to DB' say, in proportion to the increase of resistance. This line is then divided in the new ratio of the resistances, say B'C'/C'D; B' and C' being joined to A to form the new mechanical-power and torque lines respectively. The torque line will intersect the circle at a lower value of current to give the new value of I_{sc} with the increased resistance. The current locus follows the same circle since it is determined by the diameter $V_1/(x_1+x_2')$ and these parameters are unchanged. Movement of the current phasor around the circle towards A is due to an apparent increase in resistance viewed from the primary terminals, brought about by changes in R_1, R_2' or s, alone or in combination. A practical problem in which this knowledge is applied, occurs when extra rotor resistance for a given starting torque is required. It is only necessary to find the nearest point to B where the vertical distance to the torque line from the circle gives the stipulated value of starting torque. This point defines the new starting current and producing through it from A to the vertical along DB, will give the extra resistance as a fraction of the natural rotor resistance.

To construct the circle, I_0 and a second current phasor are required. This other current need not be the short-circuit current. For example, it could be the current at maximum power factor, or at maximum torque, or the current with extra rotor resistance inserted at starting. The constructional procedure is the same as before and is the more accurate, the larger the second current phasor relative to I_0.

EXAMPLE E.7.5

From the data of E.7.4 construct the approximate circle diagram and check the full-load efficiency, input current, and speed. Determine also the starting torque, maximum torque and the extra rotor resistance to be connected in series with the slip rings to give maximum torque at starting. What would be the output and efficiency of the machine while generating with the same value of line current as at full load. The rotor winding is delta-connected with a phase turns ratio, stator/rotor, of 3/1.

Using the value of I_0 and I_{1sc}, calculated in E.7.4, the circle diagram is constructed as already described. The location of the full-load point is determined by a line vertically displaced from the mechanical-power line by a distance

$$\frac{1330 \times 746 + 15,000}{\sqrt{3} \times 6600} = 86\cdot7 + 1\cdot3 \text{ A} = 88 \text{ A to scale.}$$

The various measured current components are shown on the circle diagram, and the nomenclature of Fig. 7.29 is used. BD is divided equally ($R_1 = R_2'$) which gives rise to small errors, because the stator and rotor copper losses are not equal. Such anomalies arise because of the errors in the approximate circuit.

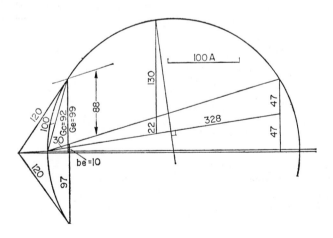

The full-load current checks, within 2–3%.

The full-load efficiency = 86·7/99 = 87·5% which checks within 1%. If the mechanical loss had been neglected, Gb/Ge = 88/99 = 88·8%.

For the slip, bc = 47(30/328) = 4·3 giving bc/Gc = 4·3/92 = 0·047, which is rather different from the value obtained more accurately by calculation with unequal stator and rotor copper losses. The speed at full load is:

$$(1-s) \times 60f/p = (1-0\cdot047)60 \times 50/4 = 715 \text{ rev/min.}$$

The starting torque $= \sqrt{3} \times 6\cdot6 \times 47$ $= \underline{536}$ synchronous kW.

The maximum torque $= \sqrt{3} \times 6\cdot6 \times (130+22) = \underline{1740}$ synchronous kW.

The extra rotor resistance per phase to get maximum torque at starting must cause the operating point to coincide with that for maximum torque. The internal rotor resistance consumes a loss here which is proportional to 22 A. The external resistance must consume the remainder, proportional to 130 A., i.e. extra resistance per phase = $(130/22) \times 1\cdot31 \times (1/3)^2 = 0\cdot86 \, \Omega$.

Note that this is referred back to the rotor, but it is not possible to insert it in series with each phase because the rotor is delta-connected with only three connection points available, see Fig. 7.7. Using the delta/star transformation, the star-connected impedances giving the same line/line impedance are each 1/3 of the individual delta components, i.e. in the leads to the slip rings, a resistance of $0\cdot86/3 = 0\cdot287 \, \Omega$/phase will have the same effect as $0\cdot86 \, \Omega$ in series with each phase winding. This can be checked readily by calculating the I^2R loss for the two cases which will be found to be the same.

With the same line current of 120 A, the active component when generating is 97 A, giving an output of $\sqrt{3} \times 6\cdot6 \times 97 = \underline{1110}$ kW at a power factor of 97/120 = 0·81 leading.

The generator efficiency is 97/(97+10+1·3) = 89·5%.

7.6 SPEED CONTROL: STARTING PERFORMANCE

The induction motor is essentially a constant-speed machine and much thought has been and is being exercised on the problem of overcoming this limitation, so that it can compete with the d.c. machine and thus avoid the expense of a.c./d.c. conversion. Although the majority of industrial drives run at substantially constant speed, there are many applications in which variable speed is a necessity.

The appropriate equation to be examined is:

$$n = (1-s)n_s = (1-s)f/p$$

and some possibilities suggest themselves immediately.

Frequency Change

The attractiveness of variable-frequency supplies lies in the possibility of using the simple, robust, squirrel-cage motor for which maintenance requirements are much less than for machines with rubbing contacts. This is particularly important for motors operating in difficult environmental conditions. Figure 7.24 has shown that the characteristics permit a wide range of speed control. With maximum flux maintained, maximum torque is available at all frequencies, and with high efficiency since low-resistance rotors can be used. High-speed, low-torque operation is possible as on the d.c. machine but in this case, flux control is via the voltage/frequency relationship, see p. 404. If the characteristics of Fig. 7.24 are continued into the generating quadrant (T_e negative, ω_m positive), it will be seen that very efficient braking results from a frequency reduction, providing that regenerated power can be accepted by the supply system. The expense of the special variable-frequency supply equipment may be justified in the same way that in many cases, the advantages of d.c. motors justify the provision of a special, d.c. supply. Large marine-propulsion units supplied from variable-speed turbine generators have already been mentioned. There are many small-power applications, particularly at low frequency, where a rotating frequency changer offers an economic solution to a speed-control problem. The most likely possibility for general use however, is the static frequency-changer using controlled rectifiers. Such schemes have been used in the past with grid-controlled, mercury-arc converters, but it is the increasing pace of semi-conductor development which has made the adoption of such systems more common. The thyristor is smaller than a mercury discharge device of comparable current rating and though this brings its own problems in terms of surge capacity, these can be overcome by more elaborate protective schemes. The control equipment is relatively complicated and costly, but once the best schemes have been found and tested thoroughly, it can be expected that they will prove to be as reliable as rotating machines, or perhaps moreso and have the normal advantages of static equipment. With increasing usage of such schemes, the cost disadvantage may well disappear, and in any case this is offset to some extent by possible savings on maintenance, space and heavier foundations which would be needed for rotating motor-generator sets.

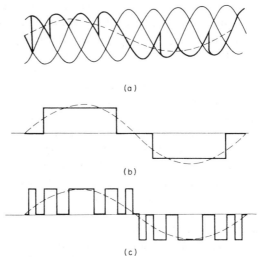

FIG. 7.31. Static frequency-changer output-voltage waveforms:
(a) gate-controlled cycloconverter (a.c. input);
(b) quasi-square wave inverter (d.c. input);
(c) pulse-width-modulated inverter (d.c. input).

The circuitry required to produce variable frequency is very simply supplemented to control the machine for maximum performance, for example to increase the starting torque. At zero speed, slip frequency is the same as line frequency and the motor impedance is low. At a low frequency therefore, extra current could be forced in for a short time by increasing the voltage—up to the maximum. The excess current will,

428

of course, drive the machine into saturation eventually, and much of it will be required to provide a disproportionate magnetising current through a reduced value of X_m. A 5 times increase in current, say, might only increase the rotor current by about 4 times and the flux might increase by say 50%. The torque would therefore be about 6 times and in practice is little more than proportional to current, instead of being proportional to (current)^2 approximately, as in the unsaturated condition. The starting torque per ampere for a d.c. machine would be better than this because the field current is not dependent on induced current from the other winding as in the case of the induction motor.

For conditions beyond starting, i.e. acceleration and running, it is possible to design the control scheme so that the input current is limited to some adjustable figure, and that f_2, the slip frequency, can also be set. This can be achieved by measuring the shaft speed n, and controlling the stator frequency to the appropriate value given by the equation on p. 404, $f = f_2 + pn$. It was also explained here that at any particular slip frequency f_2 and flux ϕ; z_2', I_2', I_m, I_1, and T_e are also fixed. This is easily checked by getting expressions for z_2', etc., for any particular line frequency f, the inductive parameters being known at some base frequency, f_{base}; see Appendix E, Problem (18) for Chapter 7. The required voltage, within the maximum value available, depends on I_1 and the input impedance. The control system can therefore process the measured information on current and speed to give maximum possible torque and so achieve final speed with rapid acceleration, within prescribed current, voltage and frequency limits.

Two main, static, variable-frequency systems are being used. The *cycloconverter*[12] employs direct a.c./a.c. conversion, synthesising a lower-frequency wave from the higher-frequency polyphase supply, by switching the motor terminals sequentially to successive supply phases. The main circuit for each phase is similar to that of Fig. 6.42, the two banks of anti-parallel thyristors supplying the positive and negative half-cycles alternately to the common connected load (the motor phase) at the required frequency. Figure 7.31a shows the voltage waveform derived from a simple cycloconverter circuit. The waveform varies with the voltage and power factor, being rich in harmonics and the circuit can produce sub-harmonics which would be troublesome if they were to excite a mechanical resonance in the system. A better waveform than that shown is normally required, though at the cost of increasing the number of thyristors. The amplitude of the positive and the negative low-frequency voltage swings is determined by the variation of firing-delay angle about the zero-voltage condition at $90°$ (see Fig. 6.43c). Figure 7.31a shows a condition where α varies from $0°$ to $180°$, giving maximum output voltage.

The *d.c. link inverter*[14] operates from a d.c. supply which is normally obtained by rectifying the a.c. mains-supply and forms the link between the constant- and variable-frequency systems. The inverter circuit is basically the same as that shown in Example E.6.7, p. 341, but operating in reverse, from d.c. to a.c.; direct current being routed through each load phase in sequence, and in positive and negative senses by appropriate thyristor switching. As explained on p. 404, it is necessary to incorporate a volts/cycle control to maintain the motor flux-per-pole substantially constant and there are two ways of achieving this electronically with the same main inverter circuit. The d.c. bus voltage can be controlled via the rectifying circuit, in which case the output line-voltage waveform takes the shape shown on Fig. 7.31b. This is the quasi-square inverter and typically has a frequency range of about 5 Hz to 200 Hz, though operation into the kHz region is feasible. The lower limit is partly related to problems of thyristor switch-off at low voltages, but is also influenced by the motor behaviour, which sometimes tends to result in pulsating or "jerky" torque at very low frequencies and mechanical resonances are more likely.

If the d.c. bus voltage is constant, e.g. if battery supplied, then a chopper control, see Fig. 6.52, switching the thyristors ON and OFF at frequencies of perhaps 500 Hz, reduces the average voltage applied to the load. By modulating the pulse width in addition, a waveform with a larger fundamental component can be obtained and Fig. 7.31c shows a typical waveform for a pulse-width-modulated (PWM) inverter. The requirement for high-frequency chopping, has previously imposed limits on the upper frequency of the output in this mode to about 100 Hz, but very low frequencies including d.c., are easily obtained. Note that although these various waveforms are far from sinusoidal, the machine inductance smooths the current waveshape somewhat, damping the harmonics considerably. Refer also to Appendix A, Fig. A.6.

The *cycloconverter* has the built-in advantage that it permits reverse-power flow for regeneration and reactive current. It gives a working frequency range from zero to about half of the supply frequency, with readily reversible phase sequence for reversal of motor rotation. On aircraft electrical systems, where the supply generator is driven by engines at variable speed, the cycloconverter can be used as the link between this variable-frequency source and the "constant" frequency distribution system. The *inverter* has the wider frequency range but requires additional facilities to provide for regenerative power flow. A useful feature of the inverter drive is the increased value of frequency over normal power supplies so that higher synchronous speeds are possible. Better overall performance will be obtained with gate turn-off thyristors or power transistors, which permit high switching frequencies and improved waveforms through PWM techniques. Departures from the ideal sinusoidal waveshape lead naturally to increased motor losses. Time harmonics in the current will add to the heating but in general tend to reduce the torque since the 5th harmonic, for example, produces a reversed rotating field; see Fig. 5.28

Change of Number of Poles

If a machine is provided with two stator windings arranged for different numbers of poles, and having a squirrel-cage rotor preferably, so that no connection change is required on the secondary, two synchronous speeds become available. It is possible with special connections to have one winding which can be reconnected simply to give two or even three different numbers of poles. Figure 7.32 shows one phase of a four-pole winding having two coils per phase. Reversing the current in one of the coils changes the pattern to give a two-pole m.m.f., Fig. 7.32b. This change is accomplished quite simply as indicated on the schematic diagrams. Each phase has two equal sections, which in series-delta absorb half the line voltage per section. When re-formed with the two sections in parallel and star connected, the voltage is increased slightly, to 0·577 of the line voltage, and the m.m.f. pattern is changed. This system of pole changing is in fact a special case of *pole amplitude modulation* and to achieve it, the top coil m.m.f. has effectively been multiplied by $+1$ and the bottom coil m.m.f. by -1. This in turn is equivalent to modulating the original m.m.f. wave by a

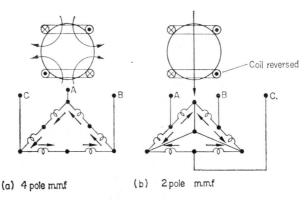

(a) 4 pole m.m.f   (b)   2 pole m.m.f

FIG. 7.32. Pole-changing winding.

square space wave of unit amplitude and of period equal in length to the
stator periphery. When the number of pole pairs is greater than two,
the effect of thus reversing one half of the winding is to change the
number of pole pairs by $+1$ or -1 as desired; Fig. 7.33. Refinements
to this simple method of coil reversal are necessary in practice to
improve the balance and performance in the modulated condition.
This may involve cutting out certain coils to give a stepped modulating

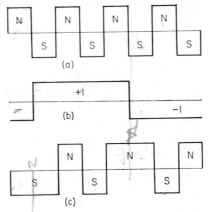

FIG. 7.33. Pole-amplitude modulation: (a) 8-pole wave;
(b) modulating wave; (c) modulated wave, 6-pole.

432

wave, or, for example, to get a ±2 change in the number of pole pairs, the modulating wave must have two cycles over the stator periphery.[20]

Many ingenious methods of varying the number of poles in an infinitely finer manner have been tried recently. They depend on the ability to reverse the current zones round the periphery at any desired points. At every point where there is a current reversal, see Fig. 5.15b for example, there is a peak of m.m.f., the next reversal giving a peak of opposite polarity. By controlling the time phase of the currents supplied to the coils, *phase modulation*, the spatial positions of the reversals can be controlled. For fine speed-control, extra apparatus is required, usually of full rating and this is an economic handicap. Indeed all methods of speed control require a sacrifice in performance, cost or simplicity to a greater or lesser extent and these disadvantages must be considered carefully against the advantages offered.

Slip Control and Slip-power Recovery Schemes

The remaining term s varies with load, but load variation does not offer a practicable method of speed control. However, it is possible to change the speed/torque characteristic by various means so that a different value of s is required for any given load torque. These methods

(a)   (b)

FIG. 7.34. ON/OFF voltage control: (a) phase control; (b) integral-cycle control.

are often wasteful in power and machine capacity and give poor speed regulation but the control may be simple and justifiable in certain cases. Two such techniques have already been illustrated in Figs. 7.22 and 7.23. In the case of voltage variation, thyristors in series with the stator leads can be used to switch on the supply for only part of the time so that the the average applied voltage is reduced. Switching can be performed in two different ways as illustrated in Fig. 7.34. Phase control gives rise to high-order harmonics in the frequency spectrum whereas integral cycle control produces less interference of this kind while raising the possibility of low-order subharmonics to which the mechanical system

EM-O*

433

might respond with sympathetic vibration. Thyristors have also been used in the rotor circuits to give similar machine-control characteristics. Since thyristors, triacs etc. require a control circuit for firing, it is quite easy to incorporate a feedback loop, see p. 621, which would permit the speed to be held substantially constant at any desired setting.

Added rotor resistance, which requires a phase-wound secondary, can be regarded as a special case of rotor voltage-injection. If instead the resistance is replaced by an active element, the slip power is not wasted and in addition, super-synchronous speeds and power-factor correction become possible. The external source across which the voltage drop is say V_3 can be regarded as an additional impedance $Z_3 = V_3/I_2$, which may have a leading or lagging reactive component, and may have a positive or negative resistance component. If positive, the external device absorbs power from the rotor circuit, tending to reduce the rotor current, torque and speed; the slip increases. If negative, the device is a source providing power to the rotor circuit, tending to increase the current and torque, so that the speed increases and the slip eventually becomes negative. The phase and magnitude of the injected voltage may or may not be controllable, but it is essential that the frequency is precisely equal to the slip frequency at all times if wattful power is to flow. This might appear to be a very severe requirement, but there are various ways of meeting it; see Chapter 9 and Reference 4. One way has already been discussed; i.e. to use a resistance R_3 across which the voltage drop I_2R_3 is automatically at the same frequency as the rotor current. Another way is to mechanically couple a second induction motor and supply it from the slip rings of the first motor at frequency s_1f_1, Fig. 7.35. This is called *cascade control*. Since the two motor speeds must be the same;

$$n = \frac{f_1}{p_1}(1-s_1) \simeq \frac{s_1 f_1}{p_2},$$

s_2 being small. From this equation:

$$s_1 \simeq \frac{p_2}{p_1+p_2} \quad \text{and} \quad n \simeq \frac{f_1}{p_1+p_2}.$$

The speeds available correspond to p_1 or p_2 using one machine. Using cascaded machines, the speeds would correspond to p_1+p_2, or

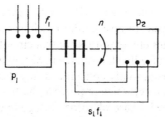

FIG. 7.35. Cascade speed control.

$p_1 - p_2$ if machine 2 is connected with reversed phase sequence. In this case, the value of s_2 may no longer be neglected.

The more usual method of regulating the slip power and the speed is to employ a controllable slip-frequency supply. The effects of an injected voltage from such a source can be investigated quite simply by modifying the equivalent circuit as shown in Fig. 7.36. The power supplied to the rotor circuit from the primary is still $sE_1 I_2'$ cos φ_2 and this is balanced by the electrical power absorbed in rotor copper loss and in the external element, $P_3 = V_3' I_2'$ cos φ_3, i.e.

$$sE_1 I_2' \cos \varphi_2 = I_2'^2 R_2' + P_3. \tag{7.10}$$

The copper loss is always positive, but P_3 and s may be negative. In any case, the electromagnetic power $(1-s)E_1 I_2'$ cos φ_2 (not shown on the diagram) is converted to mechanical power and thus the torque is still given, as derived from eqn. (7.2), by the expression:

$$T_e = (3E_1 I_2' \cos \varphi_2)/\omega_s \text{ Nm.}$$

If the rotor-circuit parameters are divided throughout by s, the

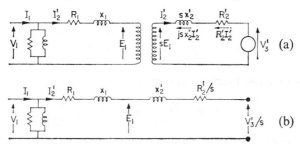

FIG. 7.36. Approximate equivalent circuit with injected secondary voltage.

435

whole of the power across the gap is shown in electrical terms and the balance equation becomes:

$$E_1 I_2' \cos \varphi_2 = I_2'^2 R_2'/s + P_3/s = P_g.$$

The torque is: $\qquad T_e = 3(I_2'^2 R_2' + P_3)/s\omega_s,$

an equation which is easily verified against previous knowledge by considering the case where P_3 is the loss in an external resistance R_3 so that $T_e = I_2'^2(R_2' + R_3')/s\omega_s$. In general, however, P_3 may be either a positive sink of energy as in this case, or a negative sink, i.e. a source.

For the special case where $P_3 = -I_2'^2 R_2'$, the slip is zero from eqn. (7.10) and the rotor carries a d.c. (zero frequency) current. The speed is therefore synchronous, the rotor copper loss being supplied externally. The rotor circuit equation for any slip is:

$$sE_1 = R_2' I_2' + jsx_2' I_2' + V_3', \qquad (7.11)$$

and when s is zero, the required voltage drop across the element is $V_3' = -I_2' R_2'$; i.e. at the terminals of the external device, the characteristic of a pure negative resistance is exhibited, indicating a source of power.

Speed/Torque Curves. From the approximate circuit of Fig. 7.36b:

$$I_2' = \frac{V_1 - V_3'/s}{(R_1 + R_2'/s) + j(x_1 + x_2')}. \qquad (7.12)$$

The impedance of the externally connected device could be included with the rotor circuit, though in the above expression it is implicitly included in V_3' because this is the terminal voltage.

For a stipulated value of V_3' the values of I_2', φ_2 and the torque could be calculated for various values of s, though each point would take some time to work out. Speed/torque curves could be plotted and the primary input and power factor obtained by adding I_0 vectorially to I_2'.

Figure 7.37a shows three typical curves for different values of V_3'; one positive, causing secondary power to be absorbed externally and the speed to fall, one negative giving increased speed, and one with V_3' zero (slip rings short-circuited), to compare with the other two. Using s from eqn. (7.11), the speed can be expressed as:

$$n = n_s(1-s) = n_s(1 - V_3'/E_1 - I_2' z_2'/E_1). \qquad (7.13)$$

Although $\mathbf{z}_2 = R_2 + jsx_2$ is itself a function of slip, the effect of the $I_2'\mathbf{z}_2'$ term is small, causing only a moderate fall in speed as the load increases. The middle term however has a pronounced effect on the speed at all loads since the slip at no load starts from a value $s = V_3'/E_1$, with $I_2 = 0$.

Figure 7.37b shows how the electromagnetic power, P_g per phase, is apportioned. If the mechanical load demand can be expressed as a simple power function of the speed, x say, so that $P_m = \hat{P}_m(1-s)^x$ then the required rating of the external slip-frequency source, neglecting

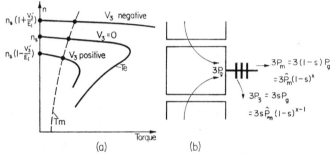

(a)   (b)

FIG. 7.37. Slip control by injected secondary voltage: (a) speed/torque characteristics; (b) power distribution neglecting rotor losses. Mechanical power assumed proportional to (speed)^x.

the rotor losses, is seen to be $P_3 = s.\hat{P}_m.(1-s)^{(x-1)}$ per phase, where $3\hat{P}_m$ is the mechanical power demand if running at synchronous speed. Thus the external source power-rating depends on both the speed variation required and on the mechanical load characteristic. It is a straightforward matter to show, by differentiation, that the maximum value of P_3 occurs at a slip of $s = 1/x$. For a constant-torque load therefore, for which the power varies directly as the speed, i.e. $x = 1$ and $P_3 = s\hat{P}_m$, a variation of speed down to zero would require a large value of P_{3max} equal to $\hat{P}_m$. For a propellor-type load, for which power varies as the cube of the speed approximately, P_{3max} occurs at $s = 1/3$ and by substitution, has a value of only $4\hat{P}_m/27$, though the source must still be capable of providing a voltage V_3 equal to E_2, if speed control down to zero is required.

The power P_3 through the slip-frequency source may be processed in various ways. It could be rectified and used to supply a d.c. motor coupled on the same shaft. In this way, neglecting losses, the system as a whole is converting all the electrical power to mechanical power as speed changes. This is therefore a nominally *constant-power* drive. Alternatively, the power can be fed back to the supply mains through a frequency converter, often mechanically coupled to the motor shaft, in which case the mechanical power available varies with speed; suitable for a nominally *constant-torque* characteristic. Various rotating frequency changers for this purpose are described in Chapter 9, together with machines which include both motor and frequency-converter functions in the one device. The conversion from slip frequency to line frequency can also be achieved using either cycloconverters or a rectifier and inverter with d.c. link. Such schemes are sometimes referred to as static Kramer or Scherbius drives, from the original schemes which used rotating machinery for frequency conversion.

Phasor Diagrams for Various Conditions. If a constant-torque load is assumed, then $I_2 \cos \varphi_2$ is constant, neglecting the slight changes in E_1 due to variations of $I_1 z_1$. The locus of $\mathbf{I}_2$ under these conditions, but with $\mathbf{V}_3$ injected, is horizontal, Fig. 7.38. The phasor diagrams are drawn by assuming certain values of slip and rotor power-factor. The required value of $\mathbf{V}_3$ follows from eqn. (7.11) as:

$$\mathbf{V}_3 = s\mathbf{E}_2 - R_2\mathbf{I}_2 - \mathrm{j}sx_2\mathbf{I}_2$$

Starting from normal induction motor operation, Fig. 7.38a, three other cases are shown:

Fig. 7.38b slip increased (sub-synchronous) with no change to φ_2;

Fig. 7.38c s unchanged but φ_2 leading, to give improved power factor;

Fig. 7.38d s decreased and made negative (super-synchronous), φ_2 unchanged.

Note that in the last figure, all phasors involving s are reversed from the normal condition because s is negative as for the induction generator. $\mathbf{I}_2$ is not reversed in phase however, because a motoring condition is shown here. A variable-speed generating condition is possible of course. Note also that to change slip alone, the voltage $\mathbf{V}_3$ must be in phase or antiphase with the rotor current, because sE_2

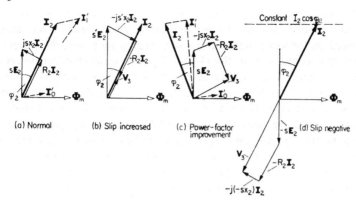

FIG. 7.38. Speed and power-factor control by injected secondary voltage.

and sx_2 change in the same proportion with speed. If, for example, V_3 was in line with sE_2, sub-synchronous speeds would give rise to a worsening of power factor, whereas super-synchronous speeds would cause a power-factor improvement. To change power factor alone, V_3 must have a strong component in quadrature with I_2. If the injected voltage V_3 and the slip are specified, the following rearrangement of eqn. (7.11) will give the phasor I_2 and hence the power factor and torque:

$$I_2 = sE_2/(R_2 + jsx_2) - V_3/(R_2 + jsx_2).$$

Starting; from Constant-voltage, Constant-frequency Supplies

Where high starting torques are required, it is sometimes necessary to have a wound rotor, so that extra rotor resistance can be inserted and maximum torque becomes available. The external starting resistance on the medium and larger machines often consists of a tank, with three insulated electrodes, one for each phase, dipping into treated water as the electrolyte. A star-connected resistance is formed and the three electrodes are shorted out by switch contacts when completely immersed.

For cage rotors, the problem is to keep down the starting current while maintaining adequate starting torque, a 50% reduction in motor voltage and starting current giving rise to a 75% reduction of starting torque; eqn. (7.9). The use of stator impedance is permissible if the starting requirements are not severe, but the starting torque per line ampere is greatly improved if a transformer is used to reduce the motor voltage. Another common method of achieving the same result is to design the primary winding so that it operates normally in delta connection but all the phase ends are brought out so that it can be connected in star for starting.

Four methods of starting a cage-rotor machine are compared on Fig. 7.39, a delta primary being maintained throughout, though this is only essential for case (b).

For direct-on-line (a), the supply line current is I_d and phase current $I_d/\sqrt{3}$.

For star/delta, (b), the voltage per phase is reduced by a factor of $1/\sqrt{3}$ and so the phase current is $I_d/3$, which is also the line current.

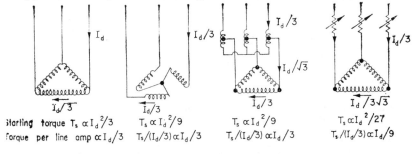

Starting torque $T_s \propto I_d^2/3$	$T_s \propto I_d^2/9$	$T_s \propto I_d^2/9$	$T_s \propto I_d^2/27$
Torque per line amp $\propto I_d/3$	$T_s/(I_d/3) \propto I_d/3$	$T_s/(I_d/3) \propto I_d/3$	$T_s/(I_d/3) \propto I_d/9$
a) Direct-on-line	(b) Star connection for starting	(c) Auto transformer start	(d) Stator impedance starting

FIG. 7.39. Starting methods for single-cage rotor machines.

In Fig. 7.39c the voltage is reduced by an auto-transformer, which is economical and quite satisfactory for the duty. For comparison with (b), the supply line current is set at $I_d/3$ so with an h.v./l.v. turns ratio N_1/N_2, the line current at the motor terminals is $(N_1/N_2) \times I_d/3$. An alternative expression for this current is obtained by considering the reduction of motor voltage. With full voltage, the *motor* line current is I_d, so with reduced voltage it must be $(N_2/N_1)I_d$. Equating these expressions:

$$\frac{N_1}{N_2}\frac{I_d}{3} = \frac{N_2}{N_1} I_d$$

440

from which the required turns ratio for this line current is $\dfrac{N_1}{N_2} = \sqrt{3}$.

It will be noticed that the auto-transformer with this ratio, gives the same performance figures as the star/delta connection; they both give an improvement over the stator impedance case, Fig. 7.39d, of three times the starting torque per line ampere, when the external impedance is set to give the same line current.

When the supply system can withstand the large low-power-factor currents associated with direct-on-line starting and without endangering system stability, p. 403, this method is preferred, there being no insuperable design difficulty as far as the motor is concerned.

EXAMPLE E.7.6

An induction motor for which the starting current is 6 *per unit* and the full-load slip is 0·04 *per unit* is to have an auto-transformer starter. If the minimum starting torque must be 0·3 *per unit*, determine the required tapping on the transformer and the *per-unit* line current at starting.

1 *per-unit* current and torque correspond to full load $I_{f.1.}$ and $T_{f.1.}$.

From eqn. (7.9) written in per unit values:
$$T_s = I_s^2 . s_{f.1.}$$
Substituting values, $0·3 = I_s^2 \times 0·04$ from which:
$$I_s = \sqrt{0·3/0·04} = 2·74 \text{ p.u.}$$
Hence the applied voltage to the motor must reduce the current from 6 *per unit* to 2·74 *per unit*, i.e. the tapping must reduce the voltage to $\dfrac{2·74}{6} = \underline{0·456}$ of the full value.

The line current will be 2·74 multiplied by the secondary/primary turns ratio, i.e. $2·74 \times 0·456 = \underline{1·25}$ *per unit*.

Although methods (b) and (c) reduce the starting current, there is usually a very high, short-lived switching transient on changing over to the full-voltage connection. This excessive current must be avoided on the larger machines, one method being shown on Fig. 7.40 together with the switching sequence. In the first connection, the situation is the same as for transformer starting, but when the neutral switch is opened, the line portion of the transformer still in circuit is just a series reactance. Finally, this portion is shorted out, putting the machine directly across the supply. At no stage is the circuit opened and so the transition switching peaks, see page 126, are absent.

441

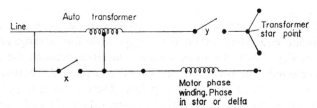

FIG. 7.40. Korndorffer method for starting large a.c. motors
(one phase only shown).
Switching sequence 1. y closed   x open
2. y open   x open
3. y open   x closed.

Double-cage Rotors

By having more sophisticated cage arrangements, starting torques can be greatly increased. For example, very deep conductors in the cage give rise to considerable eddy current effects at line frequency. At starting therefore, with slip frequency equal to line frequency, the effective cage resistance is very high and this improves the rotor power-factor and torque. As the slip falls, so the a.c. resistance falls, till at normal slips it is little different from the d.c. value, thus giving low loss and high efficiency. Another method is to use two or even three cages.

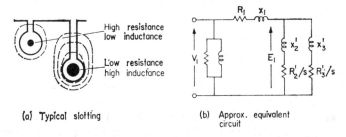

(a) Typical slotting

(b) Approx. equivalent circuit

FIG. 7.41. Double-cage rotor.

The outer cage is made with high resistance but has low leakage inductance, whereas the inner cage or cages are made with relatively low resistance, but have a high leakage flux per ampere, Fig. 7.41a. At starting, most of the rotor m.m.f. is provided by the outer cage, which has the lowest impedance $R_2' + jsx_2'$. When running at low slips

442

the inner-cage reactance becomes negligible, so that its impedance $R'_3 + jsx'_3$ permits a larger current to flow than in the outer cage. The approximate equivalent circuit for a typical slotting arrangement is shown in Fig. 7.41b, the small leakage reactance between the cages being neglected. The two cages are virtually in parallel, because they experience very nearly the same gap flux. Their referred impedances are therefore combined in parallel at various values of s and from the stator point of view, the only change from the single-cage analysis is the different behaviour of the apparent rotor resistance.

The impedance is given by:

$$Z = R_1 + jx_1 + \cfrac{1}{\cfrac{1}{R'_2/s + jx'_2} + \cfrac{1}{R'_3/s + jx'_3}}$$

and the current flowing through the combined rotor impedance is V_1/Z. The torque, as before, is obtained from the rotor copper loss as:

$$T = (V_1/Z)^2 R_{eff}$$

where R_{eff} is the "real" part of the two rotor complex-impedances combined in parallel and compares with R'_2/s for a single rotor winding.

The performance of the double-cage rotor can be estimated approximately by superimposing, at the same values of slip, two torque/slip characteristics corresponding to the different x/R ratios, Fig. 7.42. The starting torque can be made as high or even higher than the pull-out torque. However, this peak torque is less than would occur with a single cage, as can be understood by considering the effect of increased reactance on the diameter of the circle diagram for a single cage. The current locus for a double cage is not a circle. The efficiency and power factor are also reduced, the former due to the deliberate increase in resistance and the latter due to the increase in reactance. It must be remembered, though, that a single cage would have to be modified anyway to meet the starting duty. In addition to this, the multi-cage and deep-bar rotors give more rapid acceleration due to the higher torque from $s = 1$ to $s = 0$. This in turn reduces the energy wasted in stator copper losses during the starting period and virtually reduces the size of motor required if starts are frequent. Double-cage rotors can be designed with a higher ratio of effective rotor-winding a.c. to d.c. resistance than the deep-bar rotors.

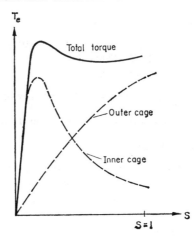

FIG. 7.42. Torque/slip curve.

EXAMPLE E.7.7

A double-cage induction motor has the following equivalent circuit parameters, all of which are phase values referred to the primary:

primary	$R_1 = 1\,\Omega$	$x_1 = 3\,\Omega$
outer cage	$R_2' = 3\,\Omega$	$x_2' = 1\,\Omega$
inner cage	$R_3' = 0\cdot6\,\Omega$	$x_3' = 5\,\Omega$.

The primary is delta connected and supplied from 440 V. Calculate the starting torque and the torque when running at a slip of 4%. The magnetising branch can be assumed connected across the primary terminals.

$$\text{Impedance at starting} = 1 + \mathrm{j}3 + \cfrac{1}{\cfrac{1}{3+\mathrm{j}1} + \cfrac{1}{0\cdot6+\mathrm{j}5}}$$

$$= 1 + \mathrm{j}3 + \frac{(3+\mathrm{j}1)(0\cdot6+\mathrm{j}5)}{3\cdot6+\mathrm{j}6}\frac{3\cdot6-\mathrm{j}6}{3\cdot6-\mathrm{j}6}$$

$$= 1 + \mathrm{j}3 + 1\cdot68 + \mathrm{j}1\cdot538$$

$$= 2\cdot68 + \mathrm{j}4\cdot538\ \Omega.$$

$$\text{(Current)}^2\ \text{per phase} = \frac{440^2}{(2\cdot68)^2 + (4\cdot538)^2} = 6960\ \text{A}^2.$$

$$\text{Torque} = 3 \times 6960 \times 1\cdot68 = \underline{35}\ \text{synchronous kW}.$$

An estimate of the torque at 4% slip can be made by neglecting the outer-cage impedance altogether. However, it does carry some current, but this is determined almost entirely by its resistance.

444

Hence, total impedance $= 1+j3+\dfrac{1}{\dfrac{1}{3/0\cdot04}+\dfrac{1}{j5+0\cdot6/0\cdot04}}$

$$= 1+j3+12\cdot65+j3\cdot45$$
$$= 13\cdot65+j6\cdot45.$$

Torque $=\dfrac{440^2}{(13\cdot65)^2+(6\cdot45)^2}\times3\times12\cdot65 = \underline{32}$ synchronous kW.

The starting torque is higher in this case than the full-load torque.

Electromechanical Transients

The same basic electromechanical equation applies to the induction machines as to the d.c. machine; viz., eqn. (6.9), p. 348. The expression for T_e cannot, however, be so simply stated as in eqn. (6.4), because the torque angle δ is not independent of load as on the d.c. machine. When all the primary and secondary winding inductances are included and the equations expressed in their simplest form, the calculation of torque requires the simultaneous solution of four differential equations having coefficients which are functions of speed, see Chapter 10. The assistance of a digital or analogue computer is required. It is possible to get an approximation to the transient performance in the form of a second-order (quadratic) equation, by just considering the overall effect of the stored energy in the leakage fields and in the mechanical-system rotational-inertia. In general, the transient response is fairly heavily damped, but becomes less so as the frequency is reduced. The oscillatory tendencies are also aggravated when thyristor inverters, p. 430, are used, since such circuits require extensive use of capacitors, thus providing further energy storage. Even with normal mains supplies, there may be considerable oscillations of electro-magnetic torque during fast speed changes. For example, when starting, the speed may not build up in a steady fashion but may have clearly-defined speed pulsations, even overshooting synchronous speed under certain conditions. The dynamic speed/torque curve, for which instantaneous values of speed against torque are plotted during the speed build up, is quite different from the steady-state characteristic shown by Fig. 7.13, for which slow speed-changes have been assumed and the electrical time constants have been neglected. The dynamic curve exhibits more or less severe torque oscillations, mostly at the lower end of the curve. The mean of this curve may be quite close to the steady-state curve and could then give a good estimate of the overall time for speed changes. Consequently, if the electrical time-constants are relatively small and are neglected, the examples discussed for the d.c. machine; starting, reverse-current braking, dynamic braking and pulse loading, can be solved for the induction machine too, either analytically or graphically as convenient and based on the steady-state speed/torque characteristics. Reference 33 covers these problems.

7.7 OTHER MODES OF OPERATION

With nearly a century of development, many variants on straightforward polyphase induction-machine operation have been evolved. Some have already been discussed in Section 7.6 and a brief review of other operational modes is given in this section. For a more detailed treatment of these and other special induction machine applications, see Reference 21, for example. Discussion of the *Synchronous Induction Motor* will be deferred till Section 8.6.

Operation in Dynamic Braking and Constant-current Modes

As an induction motor, the field produced by polyphase currents moves at n_s in the same direction, but faster than the rotor conductors. If, when operating with an initial slip s, the primary winding is switched over to d.c. excitation, usually in accordance with Fig. 5.30b for simplicity, the field is now stationary in space, so that the conductors are moving faster than the field at a speed $(1-s)n_s = Sn_s$, Fig. 7.43. The induced current is therefore reversed and a generating torque is produced. Retaining the induction machine viewpoint, the rotor referred reactance, frequency and e.m.f. become $(1-s)x'_2$, $(1-s)f$ and

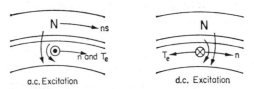

a.c. Excitation   d.c. Excitation

FIG. 7.43. Changeover from a.c. to d.c. excitation.

$(1-s)E_1$ respectively, where E_1 is the e.m.f. for a flux/conductor relative motion corresponding to synchronous speed, i.e. the E_1 of the induction motor equivalent circuit. For convenience in the equations for this operational mode, the variable $(1-s)$ is replaced by S, which is proportional to the flux/conductor relative motion, now directly varying as the speed n, where $n = Sn_s$.

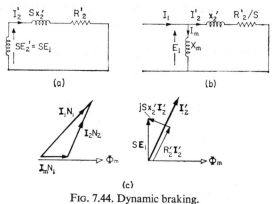

(a)   (b)

(c)

FIG. 7.44. Dynamic braking.

The rotor equivalent circuit is shown in Fig. 7.44a, and on dividing throughout by S, is modified to Fig. 7.44b. The stator circuit no longer carries alternating current so the reactance is of no interest, except during transients, and the resistance is only required to determine the d.c. excitation voltage. Further, there will be no iron loss in the stator core and the rotor iron loss will cause an effective increase in the rotor referred resistance, so that it could be included in the torque calculation. Consequently, the magnetising branch of the equivalent circuit will be just X_m, the current through it being I_m and the e.m.f. at synchronous speed E_1 being equal to $I_m X_m$.

Referring to the phasor diagrams of Fig. 7.44c, the rotor circuit has a.c. quantities, but they are produced by a steady flux Φ_m and a net m.m.f. $I_m N_1$, which are stationary in space. Nevertheless, viewed from the rotor, the flux and stator m.m.f. are alternating since flux variations are experienced by the winding. Further, since the rotor

447

m.m.f. $I_2 N_2 = I_2' N_1$ must be cancelled by stator ampere-turns in order that $I_m N_1$ may remain, the stator m.m.f. $I_1 N_1$ viewed from the rotor is the "vector" sum of $\mathbf{I}_m N_1$ and $\mathbf{I}_2' N_1$, expressed in terms of an r.m.s. sinusoidal a.c. current I_1. As discussed in connection with Fig. 5.30b, a d.c. current of $1 \cdot 225 \, I_1$ would produce the same m.m.f. Hence, the solution of the equivalent circuit can be carried out using a.c. theory, and the d.c. excitation found later by the conversion just explained. Actually, the machine under these conditions is a variable-frequency synchronous generator, for which this circuit is the third interpretation of Fig. 8.10 (q.v.), ($I_1 = I_f' \times$ turns ratio), but it is more convenient to use induction machine theory in which the variable frequency is a normal operating condition. In practice, too, saturation must be allowed for, and this causes the value of X_m to fall considerably over the operating range as I_1 and the flux increase.

The braking, or generating torque can be obtained from eqn. (7.3), or more conveniently by considering that the generated power is all dissipated in the rotor-circuit resistance. Hence the torque must be:

$$T_e = 3P/\omega_m = 3 I_2'^2 R_2'/S \omega_s \text{ Nm.}$$

In terms of the equivalent circuit parameters:

$$T_e = \frac{3 \cdot I_1^2}{2\pi n_s} \cdot \frac{X_m^2}{(X_m + x_2')^2 + (R_2'/S)^2} \cdot \frac{R_2'}{S} \text{ Nm.} \tag{7.14}$$

Equation (7.14) permits the calculation of torque at any speed, given the circuit parameters and the d.c. excitation for which I_1 is the equivalent r.m.s. value. I_1 is divided between the two parallel branches in accordance with the appropriate relationships for parallel circuits. By comparison of eqns. (7.14) and (7.5), or by differentiation, it will be seen that a maximum torque of value $3 I_1^2 X_m^2 / (2X_m + 2x_2') \, \omega_s$ occurs at a "slip" $\hat{S} = R_2'/(X_m + x_2')$.

The torque/speed curves, Fig. 7.45, are similar to those of the induction motor with the speed scale reversed. The curves are in the quadrants shown and the effect of variations in I_1 and R_2 are indicated. Higher values of I_1 lead to greater saturation. X_m falls so the "slip" for maximum torque increases and the torque itself is not directly proportional to I_1^2 as indicated by eqn. (7.14).

This method of braking is very useful both for stopping quickly from

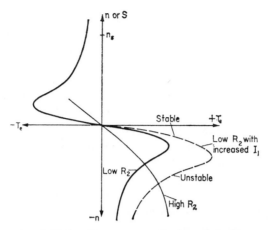

FIG. 7.45. Speed/torque curves for dynamic braking.

a normal motoring condition and also for lowering a hoist-type load, for which operation would be in the bottom right-hand quadrant. The machine can be controlled to give operation on the stable section of the curve in this last case, if I_1 and/or R_2 are adjusted automatically with speed, thus ensuring that the restraining torque T_e will increase if T_m and the speed tend to increase. The development of dynamic braking with closed-loop control has permitted the induction motor to usurp some traditional d.c. motor applications, particularly for hoists. Here, operation is at constant speed for most of the time; starting is by resistance control in the rotor circuit and braking is by the method just described.

It is possible to get a dynamic braking torque using a.c. instead of d.c. excitation, but this is only suitable in general for small machines because the characteristic is inferior.

If S is replaced by s, the equivalent circuit and torque expressions derived above, apply also to the case of an induction motor supplied from a current source, maintaining a preset stator current, instead of the normal, nominally-constant-voltage source. The stator impedance does not come into the calculation of rotor-current and torque, but must be included in the data if the voltage required to maintain a particular stator current is being calculated. The speed/

torque curve for a constant-current drive appears, superficially, to be attractive because the slip for maximum torque is low, giving good speed regulation, and the maximum torque itself is higher at rated current. However, these features are not attainable in practice since the supply voltage would have to be excessive; see Appendix E, Problem 19 for Chapter 7. But with variable-frequency control, as described on p. 428, higher currents and torques can be obtained at the lower frequencies to get maximum performance. Within the limits of maximum voltage, the r.m.s. current can be maintained constant at a particular limit during acceleration and so ensure maximum flux and torque per ampere are maintained.

The current-source equivalent circuit is useful for calculations under these conditions. For example, if the impedance parameters are known at some reference frequency f_{base}, then at any other frequency f, they have to be multiplied by the ratio f/f_{base}. Hence the required slip frequency $\hat{f}_2$, for maximum torque, from p. 448 is:

$$\hat{s} = \frac{\hat{f}_2}{f} = \frac{R'_2}{(f/f_{base})(X_m + x'_2)} \quad \text{or:} \quad \frac{\hat{f}_2}{f_{base}} = \frac{R'_2}{X_m + x'_2} \quad (7.15)$$

The torque, from eqn. (7.14), allowing for the frequency change is:

$$T_e = \frac{3 . I_1{}^2}{2\pi . f_{base}/p} \left\{ \frac{X_m{}^2}{(X_m + x'_2)^2 + [R'_2/(f_2/f_{base})]^2} \right\} \frac{R'_2}{f_2/f_{base}} \quad (7.16)$$

and with the ratio $\hat{f}'_2/f_{base}$ substituted, the maximum torque is:

$$\frac{3 . I_1{}^2}{2\pi . f_{base}/p} \left\{ \frac{X_m{}^2}{2(X_m + x'_2)} \right\} \text{Nm.} \quad (7.17)$$

The detailed proof of these expressions is left as a simple exercise and it is interesting to note that they are all independent of supply frequency f. The ratio f_2/f_{base} appears in place of the slip s in the normal expressions and it can be seen that at normal frequency, the equations revert to the ones derived earlier. See also Reference 33.

If currents are increased for starting, then saturation effects cannot be ignored and show up in the same way and for the same reason as on Fig. 7.45. If, for various constant currents, frequency is varied and starting torque calculated, $\hat{f}_2$ for maximum torque increases at higher currents due to the fall of X_m and the maximum torque is less than proportional to the increase in $I_1{}^2$.

The allowance for saturation, though it neglects the waveform distortion, can be incorporated in the equations in a manner similar in principle to that used when calculating the speed/torque curves of the d.c. series motor. A curve of X_m/I_m at base frequency is required; see p. 411. Choosing various values of I_m will give the corresponding values of X_m. Now $I_m X_m$ is proportional to the flux and is equal to E_1 at base frequency. The actual e.m.f. at slip frequency f_2 is reduced to $E_1 . (f_2/f_{base})$ and the same reduction factor applies to the rotor leakage reactance x_2'. Hence I_2' can be calculated from the e.m.f. and the rotor impedance, and the torque follows. If only the maximum torque is required, then $\hat{f}_2$ is obtained from eqn. (7.15) and the value of I_1 is obtained from I_2 (or I_m) and the parallel-circuit relationships. Values of maximum torque, from eqn. (7.17), can be plotted against I_1 and/or slip frequency. If other curves are required, e.g. starting-torque/slip-frequency or speed/torque, the same basic technique can be used for various values of I_m and various values of f_2 and using eqn. (7.16). However, it will be noticed that since I_1 is an end product of the calculation, a curve at constant I_1 is not obtained directly but only after calculating many points. For a fixed I_1, an iterative solution and a short computer program are really required. Once the torque/slip-frequency curve is available for a particular I_1, the speed/torque curve at any particular line frequency follows from the relationship: $n = (f - f_2)/p$. Note that the above remarks apply, with slight modification, to the calculation of the dynamic braking curves—with allowance included for saturation of X_m. This time, however, since the stator iron is magnetised with steady d.c. flux, the actual not the average slope of the magnetising curve ($X_m = dE_1/dI_m$, at the operating point) is required.[21]

EXAMPLE E.7.8

The machine of E.7.1 is to be braked from full speed (5 % slip), by changing the stator connections and inserting an external rotor-circuit resistance which in primary terms is 1·5 Ω per phase. Determine the braking torque initially:

(a) when the stator is disconnected from the a.c. supply and d.c. is fed into one phase, taken out of another and adjusted to give the same air-gap flux,

(b) when two stator leads from the a.c. supply are interchanged.

451

In case (a) what is the required d.c. excitation to maintain the same gap flux as when motoring at 5% slip?

Use the approximate equivalent circuits.

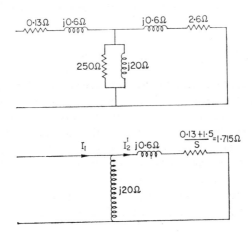

(a) The circuit is shown for the motoring condition at 5% slip, with the magnetising admittance converted to ohmic values.

$$E_1 \simeq \frac{|Z_2'|}{|z_1 + Z_2'|} \, V_1 = \sqrt{\left(\frac{2 \cdot 6^2 + 0 \cdot 6^2}{2 \cdot 73^2 + 1 \cdot 2^2}\right)} \times \frac{500}{\sqrt{3}} = 258 \text{ V}.$$

The *per-unit* relative speed $S = 1 - s = 0.95$ and, with the additional rotor resistance, the equivalent circuit when dynamic braking is as shown on the figure; therefore:

$$I_2' = 258/\sqrt{(0 \cdot 6^2 + 1 \cdot 715^2)} = 142 \text{ A}.$$

Initial torque $= 3 . I_2'^2 R_2'/S \omega_s = 3 \times 142^2 \times 1 \cdot 715/25\pi = \underline{1320} \text{ Nm}.$

The relationship between I_1 and I_2' used in eqn. (7.14) gives:

$$I_2' = I_1 \frac{X_m}{\sqrt{[(x_2' + X_m)^2 + (R_2'/S)^2]}} = I_1 \frac{20}{\sqrt{[20 \cdot 6^2 + 1 \cdot 715^2]}} = 142 \text{ A}$$

$$\therefore I_1 = 147 \text{ A}.$$

With the circuit arrangement of Fig. 5.30b the d.c. excitation must be $1 \cdot 225 \times 147 = \underline{180} \text{ A}.$

(b) The slip on reverse current braking is 1·95 initially.

The current $I_2' = (500/\sqrt{3})/[(0 \cdot 13 + 1 \cdot 63/1 \cdot 95)^2 + 1 \cdot 2^2]^{1/2}$

$$= 187 \text{ A}$$

452

and the torque initially $= \dfrac{3 \times 187^2 \times 1 \cdot 63/1 \cdot 95}{25\pi} = \underline{1120 \text{ Nm}}$.

The interested reader will notice that for both conditions, the torque per stator ampere is inferior to the motor case, because the rotor reactance is higher.

The Induction Frequency Changer

The secondary frequency varies with speed. Therefore, an induction machine can be used to obtain a frequency, different from that of the supply if driven at a suitable speed. The output voltage sE_2 varies directly as the frequency but this is not necessarily a disadvantage, particularly if the output is for use in controlling the speed of other induction motors. In this case, from eqn. (5.8a) an approximately constant motor-flux would result. There are many schemes for obtaining either an increase or a decrease from the line frequency. For the former case, the slip must be greater than unity, i.e. rotor driven backwards against rotating field. For the latter, the slip must be less than unity so the motor must be loaded mechanically. If this mechanical energy is converted to electrical energy through a coupled synchronous generator of correct design, it can be used to supplement the slip-frequency output from the rotor slip rings. When rotor and generator terminals are paralleled, the rotor frequency $f_2 = sf_1$, must be the same as the generator frequency $p_g n$. Since $n = n_s(1-s)$ and $n_s = f_1/p_m$, then, by substitution, $f_2 = f_1 p_g/(p_g + p_m)$.

Power Selsyns

Consider two identical 3-phase motors having their primary windings paralleled to the same supply. When running at the same speed, their secondary e.m.f.s will be of the same magnitude but may differ in phase depending on the rotor positions relative to the rotating fields of the two machines. If the secondaries are now connected together with the same sequence, no current will circulate between the rotors if they maintain identical relative positions, so that the e.m.f.s are equal and opposite at all times. No driving torque can then be developed of course, unless a separate current path is provided for each rotor in the form of a common, 3-phase rotor resistor. If one rotor is in advance of the other, there will be a circulating current, in a direction to cause a braking effect on this rotor and an accelerating effect on the other, so that they will tend to remain in step. This self-synchronising action is similar to that occurring with paralleled synchronous machines as will be discussed in connection with Fig. 8.19. Again the variable rotor-frequency and reactance are more conveniently treated in terms of the induction-machine equivalent circuit.

The voltage difference between the two secondary e.m.f.s is a measure of the angular error, so the arrangement is also useful as an angular position indicator. Small angular-position transducers of this kind, "*synchros*", are used for control systems. For this purpose it is only necessary to have single-phase excitation on the primary windings.

Single-phase Induction Motor

The number of machines operating from single-phase supplies is greater than all other types taken in total. For the most part, however, they are only used in the smaller sizes, less than about 5 kW and mostly in the small-power range. They operate at lower power-factors and are relatively inefficient when compared with polyphase motors. Though simplicity might be expected in view of the two-line supply, the analysis is quite complicated[2] and will not be detailed here. However, enough work has been done to permit of a simple explanation although the theory will not be proved rigorously.

If a 3-phase induction motor, when running normally, has one supply line opened, it will continue to rotate even though its supply is then single-phase virtually, and produces only a pulsating stator m.m.f. One approach to the analysis, the rotating field theory, has already been introduced in connection with Fig. 5.31. Here, on the assumption of sinusoidally distributed m.m.f.s, a pulsating field was considered as two equal m.m.f. components of half magnitude rotating in opposite directions at synchronous speed. With the rotor stationary, each would be responsible for half of the flux and stator e.m.f. If further, each

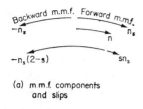

(a) m.m.f. components and slips

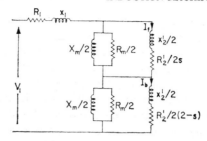

(b) Equivalent circuit

FIG. 7.46. Single-phase induction motor.

component is considered to act on one-half of the rotor impedance, equal and opposite torques would be produced to give a zero resultant at standstill. With rotation in either direction somehow initiated, since the half rotor impedances are affected by different slips, s for forward and $2-s$ for the backward component, Fig. 7.46a, the forward half with the higher impedance absorbs more of the available voltage and produces a strong forward torque. The backward component produces a smaller, reverse (plugging) torque.

This is not a rigorous proof of the equivalent circuit of Fig. 7.46b but it is reasonable and is easily checked. Consider, for example, the locked rotor condition, $s = 1$:

$$z_{sc} = R_1 + jx_1 + \frac{1}{1/R_m + 1/jX_m + 1/(R'_2 + jx'_2)}$$

as for the exact circuit of the single-phase transformer; Fig. 4.8, when the secondary is short-circuited.

The parameters of the circuit can be measured by o.c. and s.c. tests as for the 3-phase induction motor taking due account of the different arrangement of Fig. 7.46b, when making approximations at small slips. A rough idea of the performance of a 3-phase motor with one line opened can be obtained from the following considerations. For the same line current, since the stator windings are only used effectively at 2/3 of the rating, the copper loss is only about 2/3 of that occurring when operating as a 3-phase machine. Hence, the value of $(R_1 + R'_2)$ is obtained on dividing this reduced copper loss by (line current)^2, giving twice the resistance/phase of the 3-phase equivalent circuit.

455

Further, since the ratio of resistance to leakage reactance is about the same, $(x_1 + x_2')$ can be found. When running, of course, the a.c. resistance of the rotor for the backwards component at nearly 100 Hz is perhaps 50% higher than the value at 50 Hz and allowance can be made for this in calculation. Consider now the magnetising branch. For a given terminal voltage, the winding e.m.f. will only be slightly reduced so the flux and iron loss will be little different from three-phase operation. However, with fewer active turns and interference from the backwards rotating field, the magnetising current will be increased by as much as 100%, so X_m will be about the same as in the 3-phase equivalent circuit.

By calculating the copper loss in each rotor branch for various values of slip, the torque follows from:

$$T_e = \left[\frac{I_f^2 R_2'/2}{s} - \frac{I_b^2 R_2'/2}{2-s} \right] \Big/ \omega_s \quad \text{Nm.}$$

The forwards and backwards components are in opposition, of course, and it can be seen that manual calculation would be quite tedious, the two currents I_f and I_b being unequal except at standstill.

The speed/torque characteristic obtained by combining the forwards and backwards components is shown on Fig. 7.47. The starting torque is zero and for starting purposes an auxiliary winding in space quadrature with the main winding is provided. The auxiliary current must be displaced in time-phase from the main current to produce a rotating-field component. Sufficient starting torque may be obtained merely by having a relatively high-resistance short-time-rated auxiliary winding circuit, which is opened by a centrifugally operated switch when approaching synchronous speed. The phase shift can however be increased to 90° using the capacitor phase-splitting circuit of Fig. 5.32. A large capacitor, which may be more than 100 μF on the larger machines, is required to give the best starting performance, but a smaller value is adequate for good running performance if left in circuit. On the *Capacitor-run Motor*, a centrifugal switch is employed to make this changeover and so there is an approximation to 2-phase operation for both starting and running conditions.

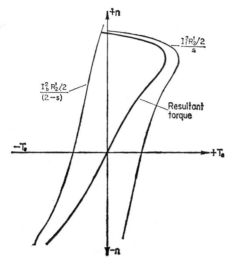

FIG. 7.47. Speed/torque curve for single-phase induction motor.

Shaded-pole Motor

This is a simple, robust, single-phase motor suitable for the lower end of the small-power range. It has a squirrel-cage rotor and an elementary form of moving field is produced by the interaction of the stator primary currents and the currents induced in a stator "secondary" circuit. One side of each stator pole is enclosed by a solid ring of copper or brass called a shading ring, see Fig. 7.48. As explained in Section 3.10,

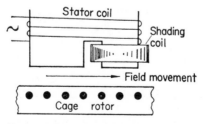

FIG. 7.48. Shaded-pole induction motor.

induced currents in a coupled coil, delay the rise of flux linking the coil. In this particular case, the flux in the shaded portion of the pole will reach its peak value later in time than the flux in the unshaded portion, due to the shading-coil eddy currents. The flux-density distribution moves across the pole face as a result and induces torque-producing currents in the squirrel cage. Although the motor efficiency is low, this has generally been considered of no great consequence when ratings are usually less than 50 watts, but see final remarks in Appendix A.

Unbalanced Operation

By deliberately unbalancing the polyphase supply to an induction motor in a controlled fashion, the speed/torque curves can be modified so that a wide range of forward and reverse operating-speeds are possible by voltage control, though with some sacrifice in efficiency. Not only is this method used for 3-phase motors of modest size, but small, 2-phase servo motors are operated in a similar way. A single-phase supply is employed but with some form of phase-splitting arrangement, through which a fixed voltage is applied to one phase. From the supply is also derived a variable voltage for the other phase, which is controlled in accordance with the error between the desired shaft position and the actual position, to produce the necessary correcting torque.

The rotor on very small servo-control motors may consist only of a thin metal cylinder, having very low inertia, rotating in the air gap between the stator core and a fixed cylindrical inner iron-core. This "drag-cup" rotor construction is also used on certain a.c. tachometer generators which are a.c.-excited on one phase-winding only. By virtue of the induced rotor currents, the other stator phase in quadrature experiences flux variations and its induced voltage is very nearly proportional to the rotational speed. The tachometer function of converting speed into a linearly related voltage is therefore achieved.

Another form of unbalanced operation is the single-phase/3-phase converter.[2] This consists of a 3-phase motor driven from a single-phase supply. The arrangement utilises the forward rotating-field component to provide time-phase displacement between the potentials at the three phase terminals, from which a 3-phase supply can be obtained.

Induction Motors with Solid-iron Rotors

These machines are sometimes made, even though the efficiency is much lower than with a cage rotor. The effective solid-rotor resistance may be more than 5 times greater. It can be shown[22] that for any value of slip s, the effective rotor impedance is of the form $R/s + jR/2s$, where R is the standstill effective resistance multiplied by the ratio of actual air-gap flux to the flux corresponding to normal frequency and terminal voltage. The parameter R can be measured from a standard short-circuit test if the stator leakage impedance is known from a rotor-removed test, and calculations of performance can then be made using the equivalent circuit in the normal way, bearing in mind the form of the effective rotor leakage reactance; $x_2' = R/2s$.

Eddy-current Couplings and Brakes

Consider a d.c. excited salient-pole "stator" which is mechanically rotated about an induction-machine-type rotor. The rotor reaction will be identical to that occurring if it experienced a rotating field produced by a stationary, a.c.-excited polyphase winding; i.e. it will have induced currents producing a torque in the direction of the field. The slip and speed will be determined by the rotor-circuit characteristics, the mechanical load to which it is connected, and the strength of the stator field, as on an induction motor. Such a device[14] is often used to couple a nominally constant-speed motor to a variable-speed load. The motor drives the d.c.-excited "stator" of the coupling and the mechanical load is driven by the rotor of the coupling, the torque corresponding to the coupling air-gap power. The slip is controlled by varying the d.c. excitation. As for the induction motor, $s \times$ (air-gap power) is wasted as heat.

Similar devices can be used as eddy-current brakes, dissipating the mechanical power as heat in the a.c. winding circuits. One member is mechanically coupled to the motor which is to be braked or loaded and the other member is locked, to give a slip of unity. The device can therefore be used in a similar way to the d.c. dynamometer, Fig. 6.51, though all the power is now dissipated in the brake itself.[34] Both eddy-current couplings and brakes may employ drag-cup-type rotors on the smaller sizes.

CHAPTER 8

SYNCHRONOUS MACHINES

8.1 BASIC THEORY AND CONSTRUCTION

Synchronous Operation

It was explained on p. 373 that induction torque cannot be produced when the slip is zero, i.e., $n = n_s$. It is possible to produce synchronous torque however and a synchronous machine, as such, can *only* run at a speed $n_s = f/p$ in the steady state. For the polyphase machine, the armature winding, usually on the stator, produces an m.m.f. wave moving at synchronous speed relative to the winding as for the induction machine. The field winding, usually on the rotor, is d.c. excited so that it produces an m.m.f. wave which is stationary with respect to the rotor. For the machine to develop a uniform torque, the armature- and field-m.m.f. axes must be stationary with respect to one another, see Section 1.3. Consequently, there must be a physical movement of the rotor at speed n_s, so that its m.m.f. wave travels in step with the armature m.m.f. axis, see Fig. 8.1. If the rotor were to run at a different speed, any particular rotor pole would be approaching alternately a stator north pole and than a stator south pole. The torque would change from positive to negative values at a frequency corresponding to the difference of speeds. Clearly such a running condition could not be sustained, although it can occur during transient processes; see also under *Synchronous Induction Motor*, p. 537.

It will be assumed that the armature is wound for three phases and is on the stator. The theory developed is quite general, however, providing the winding is polyphase. The same fundamentals of m.m.f.

461

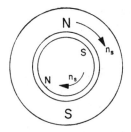

FIG. 8.1. Synchronous operation.

and e.m.f. could be developed by a two-phase winding, for example, with suitable adjustments to the number of turns, or to the current or to both. On small machines, the armature is sometimes wound on the rotor, i.e. the *inverted* arrangement.

Motoring and Generating Operation

Consider a conductor on the stator as shown in Fig. 8.2, and assume that the north pole is moving past it in a counterclockwise direction. Applying the "whiplash rule", it is seen that the direction of the induced e.m.f. and current would be out of the paper and the torque on the conductor would be counterclockwise; see also Fig. 2.15c. The reaction force on the rotor would be in a clockwise sense. Consequently

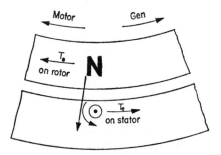

FIG. 8.2. Motoring and generating operation.

462

for this condition to be sustained, a counterclockwise mechanical torque would have to be applied to the rotor shaft to counterbalance T_e. There would be a conversion of mechanical power to electrical power; the generating condition. Conversely, if the conductor current was sustained in the direction shown by an external electrical source, T_e would be available for driving a coupled mechanical load in the clockwise direction; the motoring condition. Because of the reversal in rotation, the induced e.m.f. would be in the opposite sense to the direction of current shown, another indication of the motoring mode.

For a generator, sometimes called an alternator, the frequency of the conductor currents would vary directly as speed n, being equal to $f = pn$ cycles/sec. The rotating armature m.m.f. wave due to currents at this frequency would move in space at a speed $n_s = f/p = pn/p$, which, as might be expected, is the same speed as that of the rotor. For a motor, the rotating armature-field fixes the motion of the rotor at the synchronous speed. On some hydro-electric pumped-storage plants, the one machine may act as a motor during the night, pumping water up to the reservoir, and as a generator during the day, peak-load periods.

Cylindrical Rotor Construction

Figure 8.1 shows a rotor which is cylindrical in form. The term *cylindrical rotor*, or *round rotor*, has a special meaning when applied to synchronous machines. It implies not only that the air gap is uniform, but that the rotor m.m.f. may be taken to be distributed sinusoidally in space. It can therefore be combined vectorially with the stator m.m.f. as on the induction machine, to give a sinusoidally distributed resultant m.m.f., which, applied to the air gap, produces a sinusoidally distributed mutual flux. Such a rotor has a suitably distributed and chorded 3-phase winding and is excited with a direct current, usually with one phase carrying the full value, and the other two, half negative value each, see Fig. 5.30a. Thus both rotor and stator appear like those of an induction motor, though on all but the smaller sizes, open slots and two-layer windings are used for preference and the radial air gap is longer.

463

Reactive current is not required to magnetise the air gap of synchronous machines, so there is no restriction on the radial-gap length from this point of view as in the case of the induction motor. Apart from field heating limitations, the balance between the field-winding m.m.f., and full-load armature-winding m.m.f. (which determines the control exerted by each of them), is the deciding factor in fixing the radial-gap dimension, as for the d.c. machine. For the largest turbo-alternators envisaged, it may be longer than 150 mm, but apart from these very high-power units, air gaps are usually a good deal less than 25 mm, decreasing with the physical size of the machine.

Turbo-generator Rotor Construction

Bulk power supplies are mostly generated by steam-turbine driven alternators where water power is not plentiful. For reasons of economy and efficiency, the speeds are as high as possible, and for a 50-Hz supply using only one pair of poles the maximum speed is $f/p = 50$ rev/sec = 3000 rev/min. The rotor, though cylindrical in form, is not phase wound and the coils only occupy a portion of the periphery, see Fig. 8.3a and Fig. 8.3c. The m.m.f. wave is not necessarily such a good sinusoid as that produced with a *cylindrical rotor*, and saliency effects are noticeable, see below. The material is high-tensile steel, mechanical properties at these high speeds being the first consideration rather than the magnetic properties.

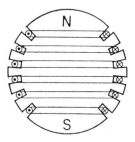

Fɪɢ. 8.3a. 2-pole Turbo-generator rotor construction.

464

Salient-pole Rotor Construction

The lower-speed synchronous machines, hydro-electric generators for example, utilise salient-pole rotors as illustrated by Figs. 1.11b and 8.3b. In general use, 4-pole construction is quite common, but 2-pole construction is only used on very small machines. Salient poles leave a large space for field windings, but in general give a flux wave with pronounced lower-order harmonics, though these can be reduced by grading the air gap so that it is longer under the pole tips than under the centre of the pole. The two magnetic axes of symmetry per pole pitch, and their different reluctances, result in a natural tendency for the field poles to line up with the armature poles. This *reluctance torque* permits the motor to run synchronously with zero excitation up to moderate loads. With normal excitation it supplements the torque produced by the interaction of armature and field m.m.f.s and improves the transient stability; i.e. the capacity of the machine to withstand sudden changes of load without losing synchronism. Cylindrical-rotor theory gives quite a good estimate of the normal steady-state performance, but for transient behaviour, *two-axis* theory must be used and this is introduced in Section 8.2. Unless specifically stated otherwise, however, the equations in this chapter are based on cylindrical-rotor theory.

Excitation Supplies: Cooling Arrangements

The power required to provide the field excitation is not very large relatively; perhaps only up to 2% of the power handled by the armature winding, even on machines of low rating. It is for this reason that the field is on the rotor on all but the small machines so that the larger amount of power is associated with the stationary winding, easing the problems of insulation and current collection. However, with the development of turbo generators with ratings of 1,000,000 kW (1000 megawatts), and with possible increases beyond 2000 MW, the excitation requirements may be more than 3000 kW. The problem of conveying such amounts of power through a high-speed sliding contact becomes formidable. The present tendency, even on the highest ratings, is to have a direct-coupled a.c.-generator exciter, with rectifiers for a.c./d.c. conversion. If the rectifiers are mounted on the rotating shaft, Fig. 8.3b, the exciter armature must be on the rotor and the rectified output can be connected directly to the field winding, thus eliminating the sliding contact. Otherwise, as when coupled d.c.-generator exciters are used, slip-ring collectors and brushes must be provided to receive the main-generator field power. The turbo generator of Fig. 8.3c is arranged this way, though in this case, the mechanically coupled exciter was of an a.c. type with stationary rectifiers. Large d.c. exciters are sometimes separately driven, at a speed of 1000 rev/min or less because of speed limitations on the d.c. machine. The d.c. required for the field of the exciter itself, is sometimes provided by a permanent magnet generator, see Appendix A, which if mechanically coupled to the main machine as on some hydro-electric plant, gives complete independence from external supplies. The machine of Fig. 8.3b has such a pilot-exciter generator, the rotor poles being visible between the cooler fan and the main-exciter armature.

The construction shown by Fig. 8.3c is typical of high-power turbogenerators up to about 1000 MW. The cooling ventilation paths are shown, hydrogen being used for the rotor, and for the stator, water is circulated through the conductors permitting very high current loadings without excessive temperature rise. Superconducting machines would have differences in construction, see p. 529. For hydro-electric generators, speeds are slower and diameters larger so that natural fanning action is generally adequate for cooling purposes. Maximum ratings are approaching 1000 MVA, at speeds somewhat lower than 100 rev/min.

8.2   EQUIVALENT CIRCUIT

General Notes on Treatment

The following treatment will be based largely on the 2-pole round-rotor machine but is applicable directly to multipolar construction with the change of synchronous speed and with allowance for the mechanical-electrical angular relationship. At 50 Hz, the speed is there-

FIG. 8.3b. Salient-pole rotor with a.c. main exciter and rectifiers and permanent-magnet a.c. pilot exciter for 3000-kVA, 11-kV, 500-rev/min diesel-drive brushless generator. (By courtesy of G.E.C. Machines Ltd.)

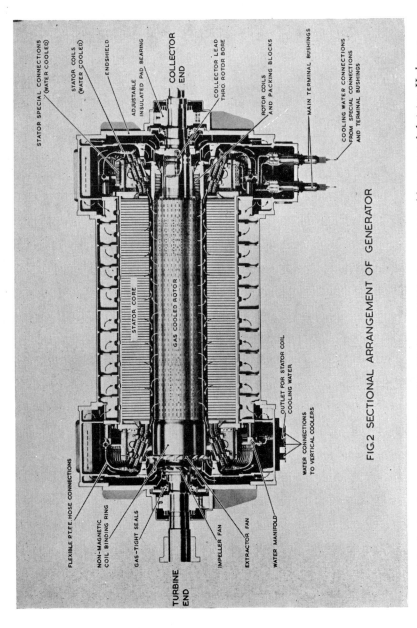

STATOR SPECIAL CONNECTIONS (WATER COOLED)

STATOR COILS (WATER COOLED)

ENDSHIELD

ADJUSTABLE INSULATED PAD BEARING

COLLECTOR END

COLLECTOR LEAD THRO ROTOR BORE

ROTOR COILS AND PACKING BLOCKS

MAIN TERMINAL BUSHINGS

COOLING WATER CONNECTIONS FROM SPECIAL CONNECTIONS AND TERMINAL BUSHINGS

STATOR CORE

GAS COOLED ROTOR

OUTLET FOR STATOR COIL COOLING WATER

WATER CONNECTIONS TO VERTICAL COOLERS

FLEXIBLE P.T.F.E. HOSE CONNECTIONS

NON-MAGNETIC COIL BINDING RING

GAS-TIGHT SEALS

TURBINE END

IMPELLER FAN

EXTRACTOR FAN

WATER MANIFOLD

FIG.2 SECTIONAL ARRANGEMENT OF GENERATOR

FIG. 8.3c. Sectional arrangement of 776-MVA, 660-MW, 23-kV turbo-generator with water-cooled stator. Hydrogen pressure 60 p.s.i. (By courtesy of G.E.C. Turbine Generators Ltd.)

fore $60f/p = 3000$ rev/min. As the number of poles is increased two at a time, the synchronous speeds fall to 1500, 1000, 750, 600, 500, 428, etc. On low-head hydro-electric installations the water turbine speed may be 100 rev/min or lower, so a very large number of poles is required.

Figure 8.4 shows some important features of the representation which is going to be used. The resultant m.m.f. of armature- and

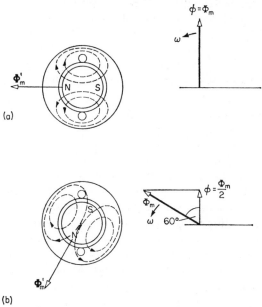

FIG. 8.4. Flux space-phasor Φ_m', and flux time-phasor Φ_m.

field-windings, will establish the mutual flux Φ_m', the axis being indicated by N and S on the rotor. This is the same as the axis of the rotor *winding* m.m.f. only at zero power; i.e. zero armature current or zero power factor.

Figure 8.4a shows the rotor in a position where the stator coil is linked by all the flux lines. The flux time-phasor Φ_m is therefore in the vertical position to indicate maximum linkage.

Figure 8.4b shows the rotor position after 60° of movement. The flux is unchanged but the flux actually linking the coil is $\phi = \hat{\phi}_m \cos 60° = \hat{\phi}_m/2$. This expression follows from the sinusoidal spatial distributions assumed in round-rotor theory.

When both field and armature windings carry current, each has a leakage flux, which for the moment will be neglected, and each winding contributes to the mutual flux. In the general case, this is found by combining the two m.m.f.s "vectorially" and applying the resultant to the permeance of the air gap and the rest of the magnetic circuit. Alternatively, if saturation is neglected, each m.m.f. can be considered as producing its own flux, some of which will be leakage and the two mutual components of flux crossing the air gap can be combined "vectorially" to give the same resultant mutual flux ϕ_m as before.

Development of Phasor Diagram for a Generator

Neglecting leakage and saturation, the field m.m.f. F_f acting alone will produce a flux ϕ_f, the maximum value linking the armature winding being represented by the time phasor $\mathbf{\Phi}_f$. Similarly, the flux ϕ_a, produced by the armature m.m.f. F_a acting alone, has a maximum value linking the armature winding which is represented by the time phasor $\mathbf{\Phi}_a$. When these two time-phasors are combined, the resultant $\mathbf{\Phi}_m$ represents the maximum value of the resultant mutual flux ϕ_m crossing the air gap and linking both armature and field. The three time-phasors are shown on Fig. 8.5a together with the induced armature winding e.m.f. E, $(N_a.d\phi_m/dt)$, leading $\mathbf{\Phi}_m$ by 90°. Note that the sinusoidal space distribution of flux assumed, gives rise to a sinusoidal time variation of flux and induced e.m.f.

It will be noticed that $\mathbf{\Phi}_a$ is drawn in *antiphase* with $\mathbf{I}_a$, unlike Fig. 5.29 for a power sink, which showed a co-phasal relationship between the time phasors of armature current and flux. To determine the relationship for the generator, it will be helpful to consider the physical situation by drawing the rotor in the position corresponding to the instant depicted by Fig. 8.5a. Here, the flux linkage of ϕ_m with the phase is passing through zero and maximum e.m.f. is induced since, for a sinusoidal flux variation, $d\phi_m/dt$ is then at its maximum. Figure 8.5b shows just the one phase under consideration, phase A, say. The resultant mutual-flux axis as indicated by N and S on the rotor, is so arranged that ϕ_m is symmetrically disposed about the centre of the phase winding. The net flux linkage is therefore zero. Further confirmation that the rotor position is correct for the instant shown by Fig. 8.5a, is given by applying the flux cutting rule (*Blv*) from which the centre coils of the phase should be in a position to experience maximum flux density and hence give maximum e.m.f. This is seen to be true.

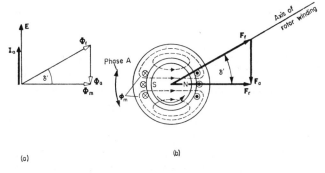

(a)

(b)

FIG. 8.5. Relationship of time phasors and space phasors for a generator.

Figure 8.5a is drawn for a unity power-factor condition with $\mathbf{I_a}$ in phase with $\mathbf{E}$. Consequently, when the e.m.f. is maximum, the current is maximum too, so that the direction of conductor current follows from that of the induced e.m.f., determined, for example, by using the "whiplash rule", see Fig. 2.15c; the rotation being anticlockwise. From Fig. 5.27 it was found that the armature m.m.f., due to all phases, is in alignment with the particular phase which is carrying maximum current. Therefore, on Fig. 8.5b, the direction of the whole armature m.m.f. is the same as that of phase A, i.e., $\mathbf{F_a}$ is vertically downwards at the instant shown. Since the resultant m.m.f. $\mathbf{F_r}$, which is in line with the mutual-flux axis, is itself due to the space-phasor summation of $\mathbf{F_f}$ and $\mathbf{F_a}$, the axis of the rotor winding must be as indicated to fulfil this condition. Mathematically:

$$\mathbf{F_r} = \mathbf{F_f} + \mathbf{F_a}. \qquad (8.1a)$$

Saturation is being neglected for the moment so that each of these m.m.f. components produces a proportionate flux component, the time variations of which are represented on Fig. 8.5a by $\mathbf{\Phi_m}$, $\mathbf{\Phi_f}$ and $\mathbf{\Phi_a}$ respectively. The space phasors of m.m.f. and the time phasors of flux form similar triangles, oriented in the same way. The shape similarity follows from the neglect of saturation but the similarity of position arises accidentally from the arbitrary choice of rotation and of phase position. No alignment can be expected since time- and space-phasors are different in nature. Nevertheless it is convenient to take advantage of the shape similarity when drawing the time-phasor diagram. The space-phasor diagram can be superimposed. $\mathbf{F_f}$ leads $\mathbf{F_r}$ in the direction of rotation by the same angle that $\mathbf{\Phi_f}$ leads $\mathbf{\Phi_m}$. For a generator, $\mathbf{\Phi_a}$ must be drawn in antiphase with $\mathbf{I_a}$, since $\mathbf{F_a}$ is in antiphase with $\mathbf{I_a}$ as explained above. The flux equation follows from eqn. (8.1a) as:

$$\mathbf{\Phi_m} = \mathbf{\Phi_f} + \mathbf{\Phi_a}. \qquad (8.1b)$$

Where $\mathbf{\Phi_f}$ is the unknown, the equation can be rearranged as:

$$\mathbf{\Phi_f} = \mathbf{\Phi_m} + (-\mathbf{\Phi_a}). \qquad (8.1c)$$

Looked at in this way, $\mathbf{\Phi_f}$ is composed of $\mathbf{\Phi_m}$ to generate $\mathbf{E}$ and $-\mathbf{\Phi_a}$ to cancel armature reaction.

Equations (8.1a, b, c) should be compared with eqn. (4.2) for the transformer. The transformer secondary is also a source like the generator and the resultant mutual flux $\mathbf{\Phi_m}$ is the combination of the primary and secondary mutual components $\phi_{21} + \phi_{12}$ (corresponding to $\phi_f + \phi_a$) and shown on Fig. 4.6a. Further, it will be noticed that for the m.m.f.s, $\mathbf{F_s}$ is drawn in antiphase with $\mathbf{I_2}$, as $\mathbf{F_a}$ must be drawn in antiphase with $\mathbf{I_a}$.

Load Angle

Figure 8.5b also shows that because of the armature flux ϕ_a, there is an angular advance of the rotor from the position it would have had on no load at the same instant in the terminal voltage cycle. The angle δ' which corresponds to the load angle δ_{fr} discussed in Section 1.4, excludes the effects of leakage. Consequently, the actual angular shift, δ, from no load is slightly different. It can be measured by using a stroboscope with its light flashing in step with the machine frequency and supplied from the machine terminals. A fixed mark on the rotor is, therefore, sighted at a particular instant in the terminal voltage cycle. As the prime-mover applies more torque to the shaft, there is a momentary acceleration until the machine takes up a new angular position relative to the rotating armature m.m.f., where it can develop an equal counterbalancing torque electromagnetically, see Section 1.3. Starting from no load, the movement would be the angle δ, the measurement actually being in mechanical degrees and requiring the appropriate conversion factor. As shown by eqn. (1.2b) the load angle is a function of the load torque, the field current and the resultant air-gap m.m.f.

Conditions at Zero Power-factor

For zero power-factor lagging, by the time that the current has reached maximum value, the mutual flux, e.m.f. and flux time-phasors will be 90° ahead of the positions shown in Fig. 8.5. The change is illustrated in Fig. 8.6a. F_a is seen to be directly demagnetising and consequently F_f is in line with and supporting F_r. There is no load angle because there is no torque or kW load, Fig. 8.6b. For zero power-factor leading, Fig. 8.6c, F_a would be directly magnetising and F_f therefore relatively small, still in line with F_r but not necessarily supporting it, depending on the magnitude of F_a. This shows that synchronous machines too can be self-excited if the load is large enough and at a low enough leading power-factor. The phasor relationships compare with those for the self-excited induction generator, not a very surprising fact in view of the similarity in both the primary

471

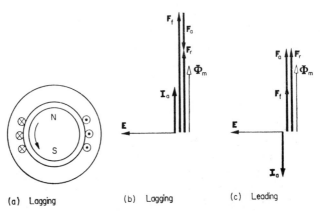

FIG. 8.6. Zero power-factor conditions; generator conventions.

(armature) winding and primary m.m.f. This phenomenon is sometimes an embarrassment and can lead to instability since the armature m.m.f. tends to take control. A practical example of this situation occurs when a generator is supplying a long unloaded transmission line. The line itself forms a predominantly capacitive load, see Example E.8.6.

Effect of Leakage

Figure 8.7 shows the flux pattern diagrammatically on the basis of the individual components produced by field and armature acting alone. That part ϕ_a of the total armature flux due to all phases, which crosses the gap to link the field winding, and that part ϕ_f of the total field flux, which links the armature winding, are shown separately, though they cannot exist like this and resolve into the mutual flux ϕ_m which is susceptible to saturation. That part ϕ_{a1} of the total armature flux which previously has been neglected, "leaks" between the windings and links the armature alone. It is substantially proportional to current because it is relatively small and most of the reluctance in its path is provided by air or some other non-magnetic medium. ϕ_{a1} is responsible for an induced armature winding e.m.f. of value $N_a.d\phi_{a1}/dt$. For sinusoidal

472

variations this can be expressed as $jx_{al}\mathbf{I}_a$, where x_{al} is the armature leakage reactance and $\mathbf{I}_a$ the armature current. Although the field leakage flux ϕ_{fl} *does* affect the performance on steady state as a result of saturation at high field currents, there is no field induced-e.m.f. except under transient conditions. $\mathbf{\Phi}_{fl}$ will not be shown on the phasor diagrams.

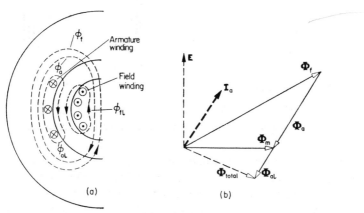

FIG. 8.7. Flux components on load: (a) flux pattern;
(b) phasor diagram.

The phasor diagram of Fig. 8.7b shows the time-phase relationship of the flux components for the condition when the machine is generating at a lagging power factor. The total induced armature voltage will be $N_a \cdot d(\phi_{total})/dt$, where:

$$\phi_{total} = \phi_f + \phi_a + \phi_{al}$$

corresponding to:

$$\phi_s = \phi_{21} + \phi_{12} + \phi_2 \tag{8.2}$$

when the armature, as a source, is compared with the secondary winding of a transformer; cf. the secondary-winding phasor diagrams of Figs. 4.6a and 4.6b. In the equations above, the positive directions of the flux components have been taken in the same magnetic sense, though for the particular condition illustrated by Fig. 8.7, the field and armature fluxes will be in opposition for most of the cycle. When saturation is

473

neglected, it is permissible, because of the system linearity, to deal with the flux components either individually or in any convenient groupings. To allow for saturation, however, it is necessary in some manner to combine ϕ_f with ϕ_a to get ϕ_m, because this determines the flux in the iron and the magnetic condition. It is usually more convenient to use the air-gap e.m.f. E for this purpose, since E is readily found from the circuit equation and is directly proportional to ϕ_m at any particular speed.

The induced e.m.f. components can be expressed in terms of inductances and the appropriate rates of current change. Although ϕ_f may be due to a steady d.c. current, there *appears* to be a rate of change with rotation when viewed from the armature winding. However, until Chapter 10, the voltage due to ϕ_f will just be designated E_f and will be left as a function of field current and speed. When ϕ_{a1} and ϕ_a are grouped together as the total self-produced flux of the armature, the corresponding self-inductance, when multiplied by the angular frequency of the armature-current changes, ω, is called the synchronous reactance $X_s = \omega . N_a . d(\phi_{a1} + \phi_a)/di_a$. It has two components, x_{a1}, the leakage reactance due to leakage flux ϕ_{a1} and X_m due to that portion of the armature flux ϕ_a crossing the air gap as part of the total mutual flux. X_m is sometimes called the armature reactance though it is directly analogous with the magnetising reactance of the transformer and the induction machine.

$X_s = x_{a1} + X_m$ could be measured by supplying normal-frequency currents to the armature winding, meanwhile driving the rotor at synchronous speed so that the field was in step with the armature m.m.f. wave. This would ensure that ϕ_f was zero since there would be no relative motion to give rise to induced currents in the field system, and the field winding could be left unexcited. Hence the only flux and voltage components would be as shown on Fig. 8.8a. The applied voltage **V** would be balanced by $[R_a + j(x_{a1} + X_m)]\mathbf{I}_a$ which in fact is very nearly equal to $j\mathbf{I}_a X_s$ because R_a is relatively small. The synchronous impedance is therefore $Z_s = V/I_a$ for the stated conditions and is little different from X_s.

An alternative method of measuring X_s is to run the machine in a

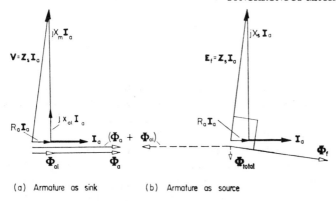

(a) Armature as sink   (b) Armature as source

FIG. 8.8. Synchronous impedance measurement.

generating mode at synchronous speed with the armature terminals
short circuited. If the field is excited, the flux component ϕ_f and the
corresponding voltage E_f will be produced. $\mathbf{E_f}$, circulating a current
$\mathbf{I_a}$, will be absorbed in the internal impedance drop $[R_a + j(x_{a1} + X_m)]\mathbf{I_a}$.
Figure 8.8b is the phasor diagram, which, since the machine is a source,
shows the armature flux components in phase opposition to $\mathbf{I_a}$. The
field flux $\mathbf{\Phi_f}$ lags $\mathbf{E_f}$ by 90° so that the total flux linking the armature,
obtained by summing the phasors $\mathbf{\Phi_{a1}} + \mathbf{\Phi_a} + \mathbf{\Phi_f}$, is very small, just
sufficient to provide the resistance drop. I_a can be measured during this
test and E_f can be read from the open-circuit curve at a point corresponding to the field excitation, as will be described in connection
with Fig. 8.11a. Hence, Z_s is given by E_f/I_a.

Complete Phasor Diagram for a Generator

Figure 8.9 completes the generator diagram given on Fig. 8.7b,
by adding the m.m.f. space phasors and the imperfections manifested
by the armature leakage impedance drop $(R_a + jx_{a1})\mathbf{I_a}$. It also shows the
relationship of all the voltage and flux components. Saturation is still
being neglected and this is indicated by designating the unsaturated
magnetising reactance as X_{mu}. To draw the diagram, V, I_a, F_a, φ and

475

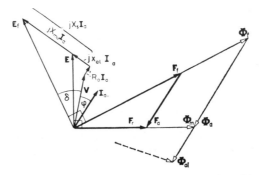

FIG. 8.9. Generator phasor diagram; unsaturated machine.

$R_a + jx_{al} + jX_{mu}$ are assumed to be known. E follows from $V + (R_a + jx_{al})I_a$, for a generator and $\mathbf{\Phi}_m$ lags E by 90°. $\mathbf{\Phi}_m$ requires a resultant m.m.f. F_r and from eqn. (8.1a), $F_f = F_r - F_a$. Considering F_f alone, this would produce a flux component $\mathbf{\Phi}_f$ which could be obtained directly from eqn. (8.1c) if $\mathbf{\Phi}_a$ was known. $\mathbf{\Phi}_f$ in turn will generate the e.m.f. E_f, leading $\mathbf{\Phi}_f$ by 90°. Alternatively, in circuit terms, E_f follows after the addition of voltage $jX_{mu}I_a$ to the $jx_{al}I_a$ phasor. A little consideration at this stage should make it clear that although the m.m.f. and flux phasors help to understand the physical operation of the machine, they could in fact be dispensed with now, since they are reflected as proportional voltage components. Note that E_f leads V by the load angle δ which is slightly greater than the angle between E_f and E (or between F_f and F_r as on Fig. (1.10)), due mainly to the leakage flux.

For a qualitative analysis of the diagram, the relationship between the armature induced voltage and the m.m.f. must be known. This relationship is conveniently obtained by driving the machine at synchronous speed with the armature terminals open circuited. At various field currents, the armature voltage is measured and the o.c. curve plotted as on Figs. 2.10 and 8.11a. Note, that on open circuit, I_a and therefore F_a are zero. Consequently, $E_f = E = V$ so that the terminal voltage is a direct measure of the induced e.m.f. Further, $\phi_f = \phi_m$ and $F_f = F_r$ since $F_a = 0$ and therefore F_r is measured directly by F_f. The m.m.f.s will be in terms of the field turns, i.e. ampere-turns/ field turns, but will be designated by F with an appropriate suffix. Since saturation is being neglected for the moment, the induced e.m.f.s can be expressed as $k_f.F$, where k_f is the slope of the air-gap line of the o.c. curve as for the d.c. machine. The voltage components will be:

$$E = k_f F_r, \quad X_{mu}I_a = k_f F_a \quad \text{and} \quad E_f = k_f F_f,$$

so that after constructing the voltage diagram from a knowledge of the impedances, the required excitation follows from above as:

$$F_f = E_f/k_f. \tag{8.3a}$$

Equivalent Circuit

The equivalent circuit has already been derived in Section 3.9 and was shown on Fig. 3.27. The preceding discussion should also have clarified the circuit equations so that Fig. 8.10 follows naturally. The voltage and current conventions have been chosen to illustrate the

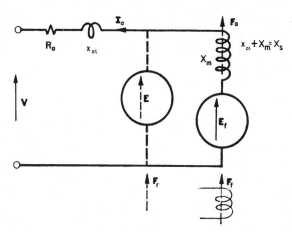

Fig. 8.10. Equivalent circuit; generator conventions.

477

generator equation though the *circuit* is the same whether the machine
is motoring or generating. Two possible interpretations are indicated:

(a) the machine has an internal impedance $R_a + jX_s$ per phase and an
internally generated e.m.f. E_f per phase which is a function of
field current alone, i.e. due to $N_a . d\phi_f/dt$;

(b) the machine has an internal impedance $R_a + jx_{a1}$ per phase and an
internally generated e.m.f. E per phase which is a function of both
field and armature currents, i.e. due to $N_a . d(\phi_a + \phi_f)/dt$.

E_f is the voltage behind synchronous impedance. E is the voltage
behind leakage impedance and is sometimes called the gap e.m.f.
because it is due to the mutual air-gap flux. The voltage components
$x_{a1}I_a$ and $X_m I_a$ are respectively due to $N_a . d\phi_{a1}/dt$ and $N_a . d\phi_a/dt$.

A third interpretation (cf p. 87) would replace the voltage "drive" E_f in
series with X_m, by a current source $E_f/jX_m = I'_f$, which is the field current
referred to the armature turns. X_m is now in parallel with this source and
carries $I_m = I'_f - I_a = F_r/N_a$, see also Fig. 8.9 and Reference 33.

Effect of Saturation; Calculation of Load Excitation

Since the iron parts of the magnetic circuit are saturable, superposition of the mutual-flux components produced by each winding
acting alone, is not permissible. However, the techniques explained in
Section 3.3, with particular reference to Fig. 3.17b, can be used to
give answers which only exclude second-order effects like the harmonic
distortion illustrated in Fig. 4.5. The traditional method of calculating
the excitation under saturated conditions is to find the gap e.m.f. E
from the circuit equation as for Fig. 8.9, and read the required resultant
m.m.f. F_r from the o.c. curve, Fig. 8.11a. The armature m.m.f. F_a
must be known, and in accordance with eqn. (8.1a), this is combined
"vectorially" at the appropriate angle to find F_f, the field m.m.f. Up to
this point, the phasor diagram is constructed in just the same way as for
Fig. 8.9, saturation having been accounted for when reading F_r from the
o.c. magnetisation curve. The values of F_r and F_f are, of course, larger
than before, as indicated on Fig. 8.11b.

It would be convenient to retain the equivalent-circuit viewpoint

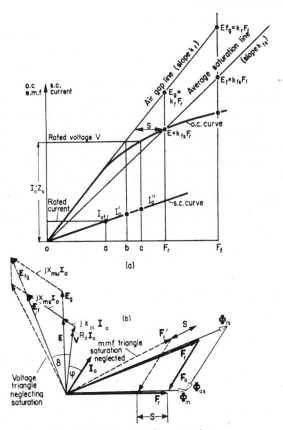

FIG. 8.11. Allowance for saturation.

so that calculations other than those concerned with determining the excitation could be carried out for saturated conditions. This means finding the appropriate value of the saturated magnetising reactance X_{ms}. The armature leakage flux remains virtually proportional to current because of the high-reluctance air-path and it does not affect appreciably, the level of core saturation, because in general Φ_{al} is out of phase with Φ_m. Except at very high overloads, then, x_{al} can be taken as constant. If now, the operating point E on Fig. 8.11a is considered,

479

it appears that the sensitivity of the electromagnetic circuit in terms of induced voltage per unit of m.m.f. has been reduced by saturation in the ratio E/E_g, from k_f to k_{fs} say. The mutual flux components which make up Φ_m must be reduced in the same ratio, the Kingsley factor, to Φ_{fs} and Φ_{as}, since they are associated with the same low-reluctance path through the saturable iron. It must be remembered, however, that these fluxes do not exist separately, though they are reflected in the reduced voltage components which can be expressed as:

$$E = k_{fs}F_r, (\phi_m); \quad X_{ms}I_a = k_{fs}F_a, (\phi_{as}); \quad E_f = k_{fs}F_f, (\phi_{fs}).$$

The middle term can be further explained if the ratio $F_a/I_a = X_{mu}/k_f$ is extracted from the corresponding unsaturated expression on p. 477. By substitution this gives:

$$X_{ms} = X_{mu}(k_{fs}/k_f) = X_{mu}(E/E_g). \tag{8.3b}$$

The reduced value of magnetising reactance is a reflection of the fact that inductance is proportional to $d\phi/di$, and the average saturation line shown on Fig. 8.11a corresponds to the average permeance over a complete magnetisation cycle as discussed in Section 3.3. Figure 8.11b shows how the saturated voltage triangle E_f, E, $X_{ms}I_a$, is still proportional to the m.m.f. triangle and the excitation can be calculated from the voltage triangle alone as before, if the appropriate values of k_{fs} and X_{ms} are used. In general then, from the above expressions:

$$F_f = E_f/k_{fs}. \tag{8.3c}$$

Note that when compared with the unsaturated machine, E_f and the load angle δ are reduced a little.

This direct method of excitation calculation will now be summarized. The gap e.m.f. E is determined from two independent considerations which must, of course, give the same answer. E is first found from the modulus of the linear circuit equation $\mathbf{E} = \mathbf{V} \pm (R_a + jx_a)\mathbf{I}_a$, but it is also proportional to ϕ_m at any particular speed, where $\phi_m = f(F_r)$. E determines the operating magnetic condition and when located on the o.c. curve gives the appropriate value of k_{fs}. The reduction factor k_{fs}/k_f is now calculated and applied to the unsaturated value of X_{mu} to give X_{ms} so that E_f follows by combining "vectorially", the phasor

$jX_{ms}I_a$ with E. F_f follows from eqn. (8.3c); (cf. Fig. 8.35). The method of finding the unsaturated value of synchronous reactance will be described shortly, and the leakage reactance must be deducted to get X_{mu}. Excitation calculations will be discussed again in Section 8.5.

This use of saturated reactance is quite accurate at one particular value of mutual flux and e.m.f. E. However, even if the terminal voltage V is maintained constant by automatic means, E is not constant at all loads. The average magnetic saturation condition changes and with it the value of X_{ms}. A median value of X_{ms} can be chosen so that the errors over a moderate operating range of currents and power factors are tolerable for many purposes.

On Fig. 8.11 is also indicated an alternative method of allowing for saturation when calculating the field ampere-turns. F_f' is the value if saturation is neglected altogether and this must be supplemented by an amount S corresponding to the difference between the air-gap line and the o.c. curve at the operating flux. $F_f' + S$ is not quite the same as F_f for which S was added at an earlier stage in the triangle construction, see Fig. 8.11b.

Per-unit Synchronous Impedance and Short-circuit Ratio

The standard definition of synchronous impedance as in BS 4727 is based on the unsaturated value. It can be measured, see Fig. 8.8b, by generating the armature current under short-circuit conditions, which, since mutual flux is small, give a linear s.c. curve (I_a/I_f). Using also the air-gap line of the o.c. curve, E_f and I_{sc} at any particular field current can be found. Hence, referring to Fig. 8.10, interpretation (a), with $V = 0$, the synchronous impedance is given by the quotient of the o.c. voltage and the s.c. current at any convenient value of field current. For example, referring to Fig. 8.11a and using the field current at rated voltage:

$$Z_{s(unsaturated)} = (\text{rated voltage } V)/I_a'$$

In *per-unit*, with rated terminal voltage as V_{base}:

$$Z_{sp.u.} = \frac{I_{a.fl}Z_s}{V} = \frac{I_{a.fl}}{V/Z_s} = \frac{I_{a.fl}}{I_a'} = \frac{ao}{bo}$$

Another related term defined in BS 4727 is the *short-circuit ratio* and this is equal to co/ao. co is proportional to the armature current which would flow if the field current giving rated voltage on open circuit was applied with the generator short-circuited. Since the s.c. characteristic is straight, ao is proportional to full-load current so the short-circuit ratio is really the steady-state *per-unit* short-circuit current when the excitation is sufficient to generate *V* on open circuit. It will be noticed that co takes the open-circuit voltage beyond the air-gap line and therefore the definition includes effects of saturation. The inverse of the short-circuit ratio, ao/co is sometimes used as a near approximation for the saturated synchronous impedance in *per-unit*. Expressed in ohms, the value would be equal to (rated voltage)/I_a''; see problem 18, p. 648. The short-circuit ratio is a measure of the relative strengths of the field m.m.f. and the full-load armature m.m.f. It has a value rather less than 0·5 for highly rated turbo-generators, where the greatest possible armature m.m.f. and current rating are required for any given rotor, designed to its maximum m.m.f. capacity. Values greater than unity may be employed on other types of synchronous machine where operational stability considerations may necessitate a strong field.

Motor Phasor Diagram

The characteristic of motoring operation is that current is made to flow through the armature conductors in opposition to the e.m.f. generated in them. The motor diagram at unity power-factor, corresponding to the generator diagram of Fig. 8.5b, must show a reversal of current if the same instant in the e.m.f. cycle is portrayed. Figure 8.12b shows that the space phasor of armature m.m.f. $\mathbf{F_a}$ is now directed vertically upwards. $\mathbf{I_a}$ remains in phase with $\mathbf{E}$ if motoring conventions are adopted, because the positive sense of current on the circuit is reversed with the machine function; cf. the change from Fig. 3.7a to Fig. 3.7b. For a motor then, using motor conventions, the space phasor $\mathbf{F_a}$ must be drawn in phase with the time vector $\mathbf{I_a}$, this being true for any power factor. $\mathbf{\Phi_a}$ must also be drawn in phase with $\mathbf{I_a}$. The situation has its analogy with the transformer as for the generator case. This time, referring to Figs. 4.6a and 4.6b, the transformer primary must be considered since this is a power sink like the motor. Φ_{21} is drawn in phase with I_1 as Φ_a is drawn in phase with I_a. Φ_{12} is the flux linkage due to the other winding and must be compared with Φ_f. In each case, the two flux components sum to Φ_m.

The phasor diagram of Fig. 8.12a shows that the rotor must now lag by an angle δ' to permit the correct alignment of the m.m.f. phasors.

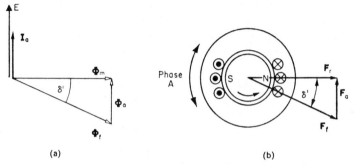

(a)   (b)

FIG. 8.12. Time phasors and space phasors for motor.

This again is only to be expected, since the application of load torque will decelerate the rotor until it has slipped back relative to the flux wave and developed a counterbalancing torque electromagnetically. This also causes E to lag V.

To construct the phasor diagram and calculate the excitation, the procedure is similar to that for the generator and the same equivalent circuit is used. The motor equation gives $E = V - (R_a + jx_{a1})I_a$ which determines F_r and X_{ms}, the saturated magnetising reactance. The e.m.f. due to field current follows from $E_f = E - jX_{ms}I_a$ and eqn. (8.3c) yields F_f. Alternatively, $F_f = F_r - F_a$ as for the generator, noting that F_a must now be drawn in phase with I_a from Fig. 8.12. The equations themselves are sufficient, without drawing the phasor diagram, but this has been shown on Fig. 8.13 for a lagging power-factor condition. It can be seen that the excitation requirements are relatively low. The diagram for a leading power factor would show that E_f would be relatively high.

These excitation conditions can be confirmed by reference to Fig. 8.6a, which if drawn for a motor, would require a reversal of conductor current for the same instant in the e.m.f. cycle. F_a would therefore be directly magnetising on lagging power factor, so that F_f would only need to be small. Figure 8.6c would be the appropriate phasor diagram

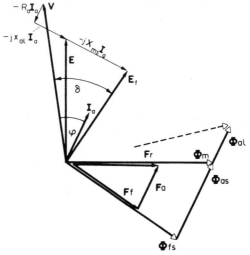

FIG. 8.13. Motor phasor diagram.

if the current phasor was reversed. Figure 8.6b, with the current phasor reversed, would be the diagram for a motor accepting a current at zero leading power-factor.

The effect of excitation changes on a motor is opposite to the effect on a generator only because of the change of viewpoint from a *source delivering power*, to a *sink accepting power*. Assuming constant terminal voltage, the effects of departures from unity power-factor can be summarised as below:

For a generator, a lagging power-factor causes the armature m.m.f. to weaken the flux and the field current must be increased to maintain the voltage; i.e. an over-excited generator exports lagging kVAr to the system connected to its terminals. Refer to Fig. 8.19d. A leading power-

factor causes the armature m.m.f. to strengthen the flux. The field current must therefore be reduced; i.e. an under-excited generator exports leading k V Ar to the system.

For a motor, the relative current reversal means that the acceptance of power at lagging power-factor causes an armature magnetising m.m.f. so that the machine must be under-excited. To accept a leading power-factor current requires over-excitation to cancel the demagnetising effects of armature reaction.

For any condition, the armature m.m.f. can be resolved into two perpendicular components, one of which is directly in line with the poles, the other being in quadrature with the poles. The direct-axis component gives rise to magnetisation or demagnetisation depending on the power factor. The quadrature component gives rise to torque, its demagnetising action due to saturation often being neglected. This two axis treatment has already been applied to the d.c. machine and will now be considered in more detail for the salient-pole synchronous machine.

Two-axis Theory

Figure 8.14a shows a salient-pole rotor drawn to correspond with

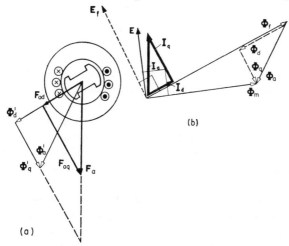

FIG. 8.14. Space phasors and time phasors for salient-pole generator.

485

Fig. 8.5b, i.e. the machine is generating with the armature m.m.f. directed vertically downwards at the instant chosen. The sinusoidally distributed m.m.f. $\mathbf{F_a}$ is resolved into perpendicular components, $\mathbf{F_{ad}}$ along the direct axis and $\mathbf{F_{aq}}$ along the quadrature axis. It can readily be seen that the permeances offered to the two components are different. Thus, the component fluxes Φ_d' and Φ_q' are not in the same ratio as the component m.m.f.s. Further, since the permeance is not uniform, there will be space harmonics in the flux wave. In the following, only the fundamentals of the flux space wave will be considered, and on this basis, the phasor diagrams superimposed on Fig. 8.14a show that the resultant armature flux Φ_a' is not in space-phase with the armature m.m.f. $\mathbf{F_a}$.

It is convenient also to resolve the time phasor of armature current along and in quadrature with the direct-axis voltage $\mathbf{E_f}$. Thus, $\mathbf{I_d}$ and $\mathbf{I_q}$ are responsible in turn for $\mathbf{F_{ad}}$ and $\mathbf{F_{aq}}$. Note that the 90° shift between $\mathbf{E_f}$ and Φ_f means that the quadrature component of current $\mathbf{I_q}$ is in phase with the direct-axis voltage $\mathbf{E_f}$. $\mathbf{I_d}$ is in time-quadrature with $\mathbf{E_f}$. If the reactive e.m.f. produced by the armature flux is considered in terms of the components Φ_d and Φ_q, the values will be $jX_d\mathbf{I_d}$ and $jX_q\mathbf{I_q}$ respectively, X_d being the direct-axis synchronous reactance and X_q being the quadrature-axis synchronous reactance. X_d and X_q can be measured in a similar way to that first described for the synchronous reactance of a cylindrical rotor machine, p. 474, though with the speed held sufficiently less than synchronous, to ensure that induced rotor currents are small, but with a large enough slip to avoid pulling into step on reluctance torque. The armature m.m.f. wave is then alternately presented with the direct-axis and quadrature-axis permeance so that the measured reactance varies between X_d and X_q. Oscillographic records of applied voltage and current would be necessary to follow the variations of impedance.

The complete phasor diagram of the salient-pole generator is shown on Fig. 8.15. Presuming for the moment that $\mathbf{I_d}$ and $\mathbf{I_q}$ are known, the reactive voltages due to these component currents are shown in terms of the voltage drops $jX_d\mathbf{I_d}$ and $jX_q\mathbf{I_q}$. For a cylindrical-rotor machine, the permeances on the two axes are the same so $X_q = X_d = X_s$, and the

E_f phasor would be drawn to the end of the $jX_dI_a = jX_sI_a$ phasor. The reduction of X_q due to saliency means that the closing phasor E_f is at a smaller load angle δ to V. The smaller load angle indicates that the salient-pole machine develops its electromagnetic resisting torque more readily than the unsaturated cylindrical-rotor machine. It is "stiffer" and more difficult to pull out of synchronism.

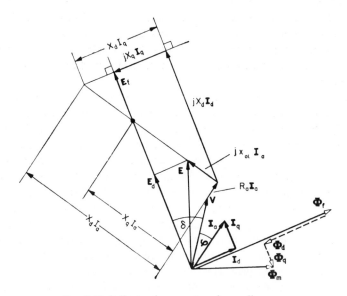

FIG. 8.15. Salient-pole generator phasor diagram.

To construct the phasor diagram, X_d and X_q must be known. The technique is demonstrated by Fig. 8.15. The phasor jX_dI_a is divided in the ratio X_q/X_d which defines a point through which E_f is drawn. A perpendicular to the direction of E_f is drawn from the end of the phasor jX_dI_a which defines both the end of the E_f phasor and the orthogonal components jX_dI_d and jX_qI_q. The components I_d and I_q, in quadrature and in phase respectively with E_f, can now be drawn if desired. Enough data is given on Fig. 8.15 to justify the construction which reduces only the component $X_d I_q$ of the phasor $X_d I_a$. The construction could be applied with equal facility to the motor phasor diagram, this being left as an exercise. In the general case, the diagram represents the phasor equation:

$$\mathbf{V} = \mathbf{E_f} \mp R_a \mathbf{I_a} \mp j X_d \mathbf{I_d} \mp j X_q \mathbf{I_q},$$

but the components $\mathbf{I_d}$ and $\mathbf{I_q}$ are not known until the construction has been carried out graphically or calculated as in Example E.8.1.

The armature leakage flux paths are not greatly changed with the salient-pole construction so that the mutual flux still corresponds to

$$\mathbf{E} = \mathbf{V} \pm R_a \mathbf{I_a} \pm j x_{a1} \mathbf{I_a}$$

The peak of the mutual flux does not occur on the direct axis and the saturation correction S of Fig. 8.11a corresponds to the component $\mathbf{E_d}$, from $\mathbf{E}$ resolved along $\mathbf{E_f}$. To find the total excitation, these extra ampere-turns are added to F_f read off the air-gap line at voltage E_f. The alternative method is to find the appropriate saturated values of X_d and X_q. Since the quadrature-axis permeance includes a long air-path, X_q is much less affected by saturation than X_d. Equation (8.3b) can be used to find the corrected value of X_d if the operating point is based on E_d.

Note that although the field ampere-turns are not sinusoidally distributed, and the air gap is not uniform, the test o.c. curve relates the fundamental of generated e.m.f. to the d.c. field current. At the design stage, the fundamental of the flux wave-form would have to be extracted by Fourier analysis to get this relationship. As already pointed out in Chapter 5, distribution and chording of the winding bring about significant reductions in the harmonic content of the voltage. This is particularly important on the salient-pole machine which may have substantial low-order harmonics in the no-load flux wave.

The effective stator-winding reactance under transient conditions is modified by the presence of induced rotor currents, as discussed briefly in Section 3.10. For the salient-pole machine, these transient reactances must be obtained for both direct and quadrature axes, see p. 531.

EXAMPLE E.8.1

On a certain synchronous generator, $X_d = 0.9$ *per-unit* and $X_q = 0.6$ *per-unit*. The machine is operating at full load, 0.8 p.f. lagging. Calculate the value of excitation in terms of the terminal voltage. Calculate also the load angle and the values of direct-and quadrature-axis currents. Neglect resistance and saturation.

Taking V as 1 *per-unit* voltage, full-load current as 1 *per-unit* current and referring to the figure:

$$V \sin \delta = X_q I_q = 0.6 \times I_a \cos(\varphi + \delta).$$

In *per-unit*, $\sin \delta = 0.6(\cos \varphi . \cos \delta - \sin \varphi . \sin \delta)$, since $V = 1$ and $I_a = 1$,
$$= 0.6(0.8 \cos \delta - 0.6 \sin \delta)$$

from which:

$$1.36 \sin \delta = 0.48 \cos \delta$$

and:

$$\tan \delta = 0.48/1.36 = 0.353 \quad \underline{\delta = 19.4°}.$$

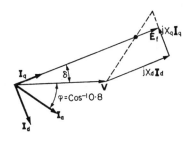

$$E_f - V \cos \delta = X_d I_d = 0.9 \times I_a \sin (\varphi + \delta) \quad \text{where} \quad \varphi = \cos^{-1} 0.8.$$

In *per-unit:* $E_f = \cos 19.4° + 0.9 \sin (36.9° + 19.4°)$
$$= 0.942 + 0.9 \times 0.833 = \underline{1.69 \ per\text{-}unit}$$
$$I_q = I_a \cos 56.3 = 1 \times 0.555 = \underline{0.555 \ per\text{-}unit}$$
$$I_d = I_a \sin 56.3 = 1 \times 0.833 = \underline{0.833 \ per\text{-}unit}$$

External Load Characteristics by Synchronous Impedance Method

With the aid of the equivalent circuit, using $\mathbf{Z_s} = R_a + jX_s$ and the equation $\mathbf{V} = \mathbf{E_f} - \mathbf{I_a Z_s}$, the behaviour of terminal voltage or e.m.f. as load varies, can readily be estimated from cylindrical-rotor theory. Usually a generator terminal voltage is maintained constant by controlling the field current manually or by using an automatic voltage regulator, which senses the departure of the voltage from a desired figure and applies a correction to the field current. To illustrate the required variation of E_f, consider a simple case where resistance is negligible and the synchronous reactance is 1 *per-unit*, i.e. at full load the reactance drop is equal to the terminal voltage, quite a representative figure. The three phasor diagrams of Fig. 8.16 show that even with constant terminal voltage and current, E_f varies from 0 at zero power-factor leading, to $\sqrt{2}V$ at unity power-factor, and at zero power-factor lagging, it rises to $2V$. Such a wide change of power-factor is not a

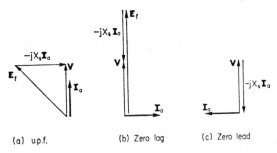

(a) u.p.f.   (b) Zero lag   (c) Zero lead

Fig. 8.16. Effect of power-factor on generator excitation requirements.

normal condition but it illustrates the considerable effect of armature reaction on the excitation requirements. For load changes alone, E_f would have to vary from V on no-load to $\sqrt{2}V$ on full-load u.p.f.

With constant field-current and therefore constant E_f, the *voltage regulation* of a single generator can be obtained from the transformer eqn. (4.3a), since the equivalent circuit of Fig. 8.10 is substantially the same as Fig. 4.10 apart from the different symbols.

Hence, numerically:

$$V = E_f - I_a X_s \sin \varphi - (I_a^2 X_s^2 \cos^2 \varphi)/2E_f \quad \text{neglecting resistance.}$$

Typical curves are shown on Fig. 8.17. The voltage falls on lagging power-factors and may rise on leading power-factors, though in practice undue increases would be limited by saturation. Note that

490

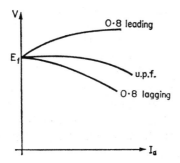

FIG. 8.17. Generator external characteristics.

eqn. (4.3b) is not accurate enough due to the large difference between E_f and V.

8.3 PARALLEL OPERATION: MACHINE PERFORMANCE EQUATIONS

The vast majority of synchronous machines are connected across a supply system of much greater capacity than themselves. So much of synchronous machine theory is concerned with the behaviour under these conditions that it is appropriate to consider parallel operation at an early stage.

Synchronising

Referring to Fig. 8.18, the incoming machine 1 has been brought up to speed and is to be paralleled to the supply system represented by machine 2. Lamps or other indicators are connected across the switches. Figure 8.18b is the phasor diagram representing, for each machine, the line potential variations about a common neutral potential. Before closing the switches, A_1 must be in phase with A_2, or nearly so, and the phase sequence must be checked so that B_1 is in phase with B_2 and

491

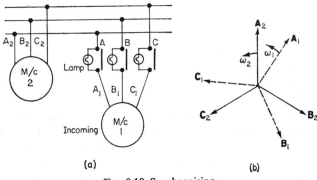

Fig. 8.18. Synchronising.

C_1 is in phase with C_2 at the same instant. It will be found that kW and kVA load sharing can be controlled by adjustments to mechanical power input and field current respectively, so that the relative values of *per-unit* impedances are not restricted in the same way as for transformers. Further, although the e.m.f.s must be adjusted to near equality before paralleling, thereafter they may be changed to quite different values.

The voltage across switch A and its lamp is represented by $A_1 - A_2$, a phasor joining A_1 and A_2. This phasor can be considered as rotating at supply frequency but becoming longer and shorter in time in accordance with the difference frequency. Each time A_1 passes A_2 when $\omega_1 \neq \omega_2$, the switch and lamp voltages pass through a zero. Since the phasors for machine 2 travel at $\omega_2/2\pi$ rev/sec, and those for machine 1 at $\omega_1/2\pi$ rev/sec, A_1 passes A_2 $(\omega_1 - \omega_2)/2\pi$ times every second, which is the beat frequency at which the lamps go on and off. If the speed of the prime-mover driving machine 1 is so adjusted that this beat frequency is quite low, the middle of the dark period can be estimated and this is the moment to close the switches. Sometimes two lamps are cross connected to record say the voltages A_1B_2, B_1A_2 and C_1C_2. With $\omega_1 > \omega_2$, the voltage between A_1 and B_2 would be approaching its maximum, whereas that between A_2 and B_1 would have passed its maximum for the instant shown in Fig. 8.18b. Meanwhile, the voltage between C_1 and C_2 would be practically zero. A little consideration will show that if ω_1 were less than ω_2 the lighting sequence of the lamps would be reversed. The cross connection results in an indication as to whether the incoming machine is fast or slow, and the switches can be closed when the lamp across switch C is dark. Note that across each switch of Fig. 8.18 the voltage rises to twice the line/neutral value. Lamps must be rated accordingly.

More elaborate synchronising arrangements are necessary for the larger machines. Special instruments called synchroscopes sense the difference frequency and relative phase of the two machine voltages which are used to supply the wound rotor and stator of the instrument to give contra-rotating fields. When the pointer, which is attached to the rotor, is steady, indicating identical frequencies, and in a vertical position, indicating correct phase, the switch may be closed. The whole synchronising operation can also be carried out automatically using a suitable logic and control circuit which determines the correct instant to close the main switch.

The behaviour subsequent to closing the switch will be explained on the basis of the equivalent circuit using synchronous impedance. Initially, two machines only will be considered, with synchronous impedances Z_1 and Z_2 and with the e.m.f.s behind synchronous impedance having values E_1 and E_2 respectively. The usual star connection of the phases will be assumed and for balanced conditions the treatment of each phase can be identical. Around the local circuit, the two e.m.f.s will be in opposition though not necessarily in direct antiphase as at synchronising. Any additional electrical load circuit will be assumed, for the present, to be disconnected. It will be found that the behaviour is different from that of d.c. machines in parallel due to the effect and predominance of inductive reactance in the local circuit impedance made up of $\mathbf{Z_1 + Z_2}$.

Two Machines in Parallel. Effect of Changing Mechanical Torque

If the two machines are driven at the same speed and with their e.m.f.s adjusted to equality, Fig. 8.19a will be the phasor diagram

493

for this condition. If Machine 1 has extra driving torque applied, the rotor will tend to accelerate and advance the phase of E_1, Fig. 8.19b. Assuming generator conventions for Machine 1, it will drive a circulating current $I_c = (E_1 - E_2)/(Z_1 + Z_2)$ through the local circuit impedance. Though this current will be lagging by nearly 90° behind $E_1 - E_2$, Fig. 8.19b shows that it is nearly in phase with E_1 so that Machine 1 will experience a generating, retarding torque. Machine 2

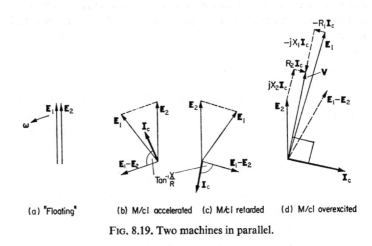

(a) "Floating"   (b) M/cl accelerated   (c) M/cl retarded   (d) M/cl overexcited

FIG. 8.19. Two machines in parallel.

has been assumed to be a motor in the above equation and I_c is seen to have a strong component in phase with E_2. Machine 2 receives power and produces a motoring, accelerating torque. Thus the electromagnetic torques exert a synchronising action tending to return the e.m.f.s to their cophasal relationship. If the shaft of Machine 1 had been loaded mechanically so that it tended to decelerate, the phase of E_1 would have been retarded. Figure 8.19c shows that the circulating current calculated from the above equation has antiphase components with both E_1 and E_2 which indicates that both machine functions have reversed. Machine 1 becomes a motor temporarily and Machine 2 becomes a generator. The electromagnetic torques again cause the

494

retarded machine to be accelerated and the advanced machine to be retarded. Thus, within the limits of maximum electromagnetic torque, there is a tendency for both machines to run in synchronism in spite of applied disturbances. The generating machine provides a synchronising power which is the real part of $\mathbf{E}^*\mathbf{I}_c$. It will be noticed that in order to get the circulating current in the right phase for corrective action, it is necessary that the local circuit impedance should be highly inductive and this is the natural characteristic of synchronous impedance. A predominantly capacitative reactance would not permit this stable situation.

Two Machines in Parallel. Effect of Changing Excitation

Starting again from the no-load condition, Fig. 8.19a, an increase in the excitation (E_1) of machine 1 will cause an interchange of current $\mathbf{I}_c = (\mathbf{E}_1 - \mathbf{E}_2)/(\mathbf{Z}_1 + \mathbf{Z}_2)$, Fig. 8.19d. The kVAs $\mathbf{E}_1^*\mathbf{I}_c$ or $\mathbf{E}_2^*\mathbf{I}_c$ will be almost entirely reactive, the active components corresponding to the copper losses $I_c^2 R_1$ and $I_c^2 R_2$ respectively, presuming the losses are proportionately shared by the respective prime movers. It follows that neither machine requires an electrical input, so the terminal power factor will be zero. Machine 1 has been regarded as a source and it is seen that it is operating at a lagging power factor because it is over-excited. Since Machine 2 is relatively under-excited, and because it has been considered as a sink when formulating the equation, it also operates at a lagging power factor. This is in accordance with the deductions made in Section 8.2, p. 484. The meaning of leading and lagging power-factors when referred to sources and sinks was also discussed in Section 3.2. Actually, since the real part of $\mathbf{E}_2^*\mathbf{I}_c$ is seen to be *negative* from Fig. 8.19d, *both* machines are sources, supplying their own I^2R losses.

If the machines were supplying an external load, any circulating current $\mathbf{I}_c$ would be superimposed on the machine load currents, power being added to one machine and subtracted from the other. The no-load expression for $\mathbf{I}_c$ is slightly modified in the presence of external impedance, as can be seen from the second term in eqn. (4.8), p. 190.

495

Load Sharing of Two Machines in Parallel

Unlike d.c. machines in parallel, disturbing the balance between the e.m.f.s does not change the power of either machine, only the reactive kVA. Furthermore, the speeds must be identical, because of the synchronising action which ties together the speed/torque characteristics of the prime movers in a more rigid manner. The construction of Fig. 8.20 explains how the characteristics determine the power supplied

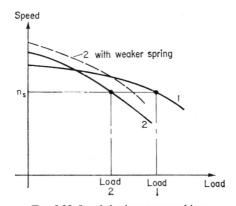

FIG. 8.20. Load sharing; two machines.

through each shaft at any particular speed. To change the load sharing necessitates a change of mechanical characteristic, see broken line, and for a mechanical type of drive this involves an alteration to the governor spring setting.

EXAMPLE E.8.2

Two 3-phase, 6·6-kV, star-connected alternators supply a load of 3000 kW at 0·8 p.f. lagging. The synchronous impedance per phase of machine A is $0·5 + j10\,\Omega$ and of machine B is $0·4 + j12\,\Omega$. The excitation of machine A is adjusted so that it delivers 150 A at a lagging power-factor, and the governors are so set that the load is shared equally between the machines. Determine the current, power factor, induced e.m.f. and load angle of each machine.

496

For machine A, $\cos \varphi_A = \dfrac{1500}{\sqrt{3} \times 6 \cdot 6 \times 150} = 0 \cdot 874, \quad \varphi_A = -29°.$

$\qquad \sin \varphi_A = -0 \cdot 485.$

Total current $= \dfrac{3000}{\sqrt{3} \times 6 \cdot 6 \times 0 \cdot 8} = 328 \text{ A} = 328(0 \cdot 8 - j0 \cdot 6)$

$\qquad\qquad\qquad\qquad\qquad\qquad = 262 - j195 \text{A}.$

$I_A = 150(0 \cdot 874 - j0 \cdot 485) \qquad = 131 - j72 \cdot 6 \text{ A}$

$\qquad\qquad\qquad\qquad I_B \quad = 131 - j124 \cdot 4 \text{ A}$

$\qquad\qquad\qquad\qquad\qquad = 181 \text{ A}$

$\cos \varphi_B = \dfrac{131}{181} = 0 \cdot 723$ lagging.

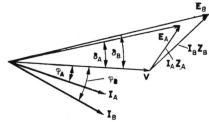

Taking V as reference time-vector and working in phase values:

$E_A = V + I_A Z_A = (6 \cdot 6 / \sqrt{3}) + (131 - j72 \cdot 6)(0 \cdot 5 + j10) \times 10^{-3}$

$\qquad\qquad = 4 \cdot 6 + j1 \cdot 27 \text{ kV}$

$\qquad$ load angle $\delta_A = \tan^{-1} 1 \cdot 27 / 4 \cdot 6 = \underline{15 \cdot 4°}$

$\qquad$ line value of e.m.f. $= \sqrt{3} \sqrt{[(4 \cdot 6)^2 + (1 \cdot 27)^2]} = \underline{8 \cdot 26 \text{ kV}}.$

$E_B = V + I_B Z_B = (6 \cdot 6 / \sqrt{3}) + (131 - j124 \cdot 4)(0 \cdot 4 + j12) \times 10^{-3}$

$\qquad\qquad = 5 \cdot 35 + j1 \cdot 52 \text{ kV}$

$\qquad$ load angle $\delta_B = \tan^{-1} 1 \cdot 52 / 5 \cdot 35 = \underline{15 \cdot 9°}$

$\qquad$ line value of e.m.f. $= \underline{9 \cdot 6 \text{ kV}}.$

Parallel Operation with an "Infinite" Number of Machines

Electrical power is mostly obtained from a power system composed of a very large number of interconnected synchronous machines. A local load applied to, or removed from such a large-capacity system causes a hardly detectable change of voltage or frequency; the system behaves like a large generator having virtually zero internal impedance and infinite rotational inertia. When constant voltage and frequency can be assumed, the arrangement is called an *infinite-busbar* system. The behaviour of an incoming generator is not quite the same as just discussed for two machines in parallel, since the steady-state speed is fixed at a value corresponding to the line frequency. Changes to the governor setting of one machine alter the load only, without perceptible change in the common frequency as in the two-machine case.

The phasor diagrams are similar to the two-machine case, though with certain different features. E_2 will be replaced by V and E_1 by E_f, the voltage behind synchronous impedance Z_s. Under all load conditions, not just on no load, the total machine generating current is given by the simple expression $I_a = (E_f - V)/Z_s$ and $V*I_a$ is the machine kVA per phase. These expressions apply throughout, because the impedance of the infinite system connected to the terminals is zero. Figure 8.21b shows the conditions just after synchronising but with E_f increased. Because of the constant speed, the mechanical power input remains constant, so the current cannot have an active component in phase with the machine e.m.f. unless the governor setting is changed; the copper losses must all be supplied from the external electrical

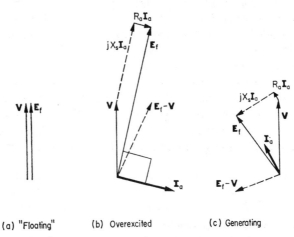

(a) "Floating"   (b) Overexcited   (c) Generating

FIG. 8.21. Generator in parallel with "infinite" system.

498

system. Increasing the excitation has not brought about any increase in terminal voltage. The additional field ampere-turns have been cancelled by the armature m.m.f. which is wholly demagnetising since the current is in lagging quadrature with the e.m.f. If under-excited, the armature current would lead the e.m.f. and the resulting ampere-turns would be magnetising and make up the deficiency in the field. Thus, excitation control changes the magnitude and sense of the reactive current so that any desired sharing of the load kVAr can be provided by the machine within its capacity.

In the condition of Fig. 8.21a, the machine is electrically "floating" on the line, neither motoring nor generating. Similarly for Fig. 8.21b, since there is no electromagnetic conversion of power; the line merely supplies copper loss to an "impedance" with a resistive component. If the supply of mechanical energy which provides the torque losses is withdrawn, the machine will tend to slow down. However, the resulting delay in the phase of $\mathbf{E_f}$ and its effect on the phasor $(\mathbf{E_f} - \mathbf{V})$ will cause the current to have a negative real component, which means that power is being taken from the supply to provide the mechanical losses by electromechanical conversion. The machine, still over-excited, would be a motor on no load taking a leading current, with the armature m.m.f. still demagnetising. A reduced excitation would cause the motor to take a lagging current from the supply with magnetising effects from the armature m.m.f.

If mechanical load torque were to be applied to the shaft, the further retardation of $\mathbf{E_f}$ would cause the current to have a greater negative real component, a similar situation being shown on Fig. 8.19c. Alternatively, if a mechanical driving torque is applied, $\mathbf{E_f}$ will advance in phase and the active component of current will reverse so that it has a component in phase with $\mathbf{E_f}$ to give an electrical output, Fig. 8.21c, see also Fig. 8.19b.

Power in Terms of Load Angle

A deeper insight into synchronous machine operation is provided by expressions for power and torque as functions of the load angle δ.

These relationships can be derived concisely by using the motor or generator equations alone. Using motor conventions and adapting eqn. (3.5):

$$\mathbf{I}_a = (\mathbf{V}/\mathbf{Z}_s) - (\mathbf{E}_f/\mathbf{Z}_s) \qquad (8.4)$$

Defining $\sin \alpha$ as R_a/Z_s, the two components of the current lag $(90 - \alpha)°$ behind the voltages $\mathbf{V}$ and $\mathbf{E}_f$ respectively. The equation is applied in the phasor diagrams of Fig. 8.22 for the case where $\mathbf{E}_f$ leads $\mathbf{V}$ and also

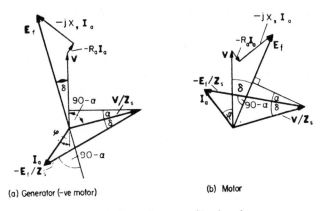

(a) Generator (−ve motor) (b) Motor

FIG. 8.22. Power in terms of load angle.

for $\mathbf{E}_f$ lagging $\mathbf{V}$. For the first condition, Fig. 8.22a, the current phasor is seen to have a negative real part so that the machine must be generating. However, a motor convention has been chosen so the current components will be resolved along $+\mathbf{V}$ giving:

$$\text{input per phase} = VI_a \cos \varphi = V\left(\frac{-E_f}{Z_s}\sin(\delta + \alpha) + \frac{V}{Z_s}\sin \alpha\right)$$

$$= \frac{-VE_f}{Z_s}\sin(\delta + \alpha) + \frac{V^2 R_a}{Z_s^2}. \qquad (8.5)$$

For a generator, the load angle δ is advanced with load, in the direction of rotation and so will be positive. Hence the expression will be negative, correctly indicating that the electrical terminal power is an output, not an input. For a motor, however, δ is negative and so the expression will yield a positive value when the machine is motoring. The expression can also be derived directly from Fig. 8.22b, noting that if the symbol δ is to represent a positive number, it must be preceded by a negative sign in any equation derived from this figure, for which δ is a negative quantity.

To obtain an expression for the mechanical output, rewrite input as:

$$VI_a \cos \varphi = I_a(V \cos \varphi)$$

Now resolve the components of V along I_a giving:

$$= I_a(E_f \cos(\varphi - \delta) + I_a R_a)$$
$$= E_f I_a \cos(\varphi - \delta) + I_a^2 R_a$$

Electrical input power = mechanical output + copper loss

This corresponds to eqn. (6.3) for the d.c. machine and since it is derived from Fig. 8.22b, there is a negative sign in front of δ so that for motoring, the numerical value of $-\delta$ will be positive. Hence, resolving the components of current along E_f, the mechanical output per phase P is:

$$E_f I_a \cos(\varphi - \delta) = E_f\left(\frac{V}{Z_s} \sin(-\delta + \alpha) - \frac{E_f}{Z_s} \sin \alpha\right)$$

$$= -\frac{VE_f}{Z_s} \sin(\delta - \alpha) - \frac{E_f^2 R_a}{Z_s^2}. \tag{8.6}$$

Equation (8.6) applies to the generator also, but with δ positive for generating the mechanical output will be negative.

These two equations, although deduced with the aid of phasor diagrams showing E_f numerically greater than V, are of general application, with any value of E_f. They are not phasor equations, r.m.s. values of voltage and the impedance modulus being used. Figure 8.23 shows the equations plotted with load angle as the independent variable. The expression $3E_f I_a \cos(\varphi - \delta)$ is the air gap power and is a direct measurement of torque in synchronous watts. Note that since speed is constant, the speed/torque curve for a synchronous motor is a horizontal line extending to the pull-out torque.

The maximum electrical output of a generator and the maximum mechanical output of a motor occur at angles less than 90 electrical degrees due to the effects of resistance. The maximum inputs have less

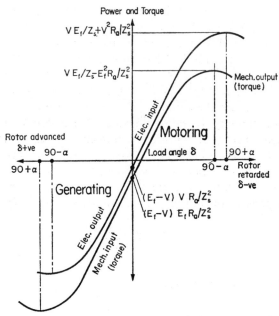

Fig. 8.23. Power/angle and torque/angle curves for constant voltage and excitation.

practical significance; a motor will have passed its stalling-torque position and probably pulled out of step before reaching this point. In the case of a generator, as the mechanical input increases, the electrical output will begin to fall off before the machine develops its maximum electromagnetic resisting torque. At this point, overspeeding out of synchronism can occur. It can be seen from the equations that all ordinates of the curves would be increased proportionately to any increase in either voltage V or excitation E_f, saturation affecting both E_f and Z_s in about the same proportion. If the relatively small effect of resistance is neglected entirely, all equations reduce to the same form and the torque is:

$$T_e = -\left(3\,\frac{VE_f}{X_s}\sin\delta\right)\bigg/2\pi n_s = -3\,\frac{\hat{P}}{\omega_s}\sin\delta \text{ Nm.} \tag{8.7}$$

502

Equation (8.7) can be put in the same form as eqn. (6.4a) for the d.c. machine and eqn. (7.3) for the induction machine. These equations are in terms of a flux, a current, a spatial displacement-angle and various constants and are similar in form to eqns. (1.2a, 1.2b, and 1.2c) expressed in terms of m.m.f.s. The flux corresponds to one of these m.m.f.s and the current corresponds to the other. The angle corresponds to the angle between these m.m.f.s. For the synchronous machine for which it is customary to choose a load angle between V and E_f, the flux can conveniently be taken to correspond with E_f as determined from eqn. (5.8), i.e. $E_f = 1 \cdot 11 \times 2 p \phi_f n_s z_s \times k_d k_p$. The current in the torque equation must therefore be taken, in a manner explained by eqn. (8.17), p. 540, to correspond with the resultant m.m.f. or flux which is responsible for V. Hence by substitution in eqn. (8.7):

$$T_e = -3 \cdot \frac{V}{X_s} \cdot \frac{1 \cdot 11 \times 2 p \phi_f n_s z_s \times k_d k_p}{2 \pi n_s} \cdot \sin \delta$$

$$= -1 \cdot 11 k_d k_p \cdot \frac{p \times 3 z_s}{\pi} \cdot \phi_f \cdot \frac{V}{X_s} \cdot \sin \delta. \qquad (8.7a)$$

This is another way of expressing eqn. (1.2b) and is similar in form to eqns. (6.4a) and (7.3) so that the three main types of machine are now on a comparable basis. These new equations explain the constants of proportionality omitted from eqns. (1.2a, 1.2b and 1.2c).

EXAMPLE E.8.3

A 3-phase, 100-hp, 440-V, star-connected synchronous motor has a synchronous impedance per phase of $0 \cdot 1 + j1 \; \Omega$. The excitation and torque losses are 4 kW and may be assumed constant. Calculate the line current, power factor and efficiency when operating at full load with an excitation equivalent to 400 line volts.

$$\alpha = \sin^{-1} R_a/Z_s = \sin^{-1} 0 \cdot 1/\sqrt{[(0 \cdot 1)^2 + 1^2]}$$

$$= \sin^{-1} 0 \cdot 1/1 \cdot 005 = 5 \cdot 7°.$$

Gross output $3P = 100 \times 746 + 4,000 = 78,600$ watts.

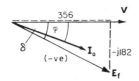

Using eqn. (8.6) and with reference to the figure:

$$78{,}600 = -3 \left[\frac{(440/\sqrt{3})(400/\sqrt{3})}{1 \cdot 005} \sin(\delta - 5 \cdot 7)^\circ + \frac{(400/\sqrt{3})^2 \times 0 \cdot 1}{(1 \cdot 005)^2} \right]$$

$$= -\frac{400 \times 440}{1 \cdot 005} \left[\sin(\delta - 5 \cdot \hat{7})^\circ + \frac{400 \times 0 \cdot 1}{440 \times 1 \cdot 005} \right]$$

$$0 \cdot 45 = -\sin(\delta - 5 \cdot 7)^\circ - 0 \cdot 0905$$

from which $\delta = (\sin^{-1} - 0 \cdot 5405) + 5 \cdot 7^\circ = -27 \cdot 1^\circ$.

Taking **V** as reference phasor:

$$\mathbf{E}_f = (400/\sqrt{3})(\cos\delta + j\sin\delta) = 231(0 \cdot 89 - j0 \cdot 455)$$
$$= 206 - j\,105$$

$$\mathbf{I}_a = \frac{(\mathbf{V} - \mathbf{E}_f)}{\mathbf{Z}_s} = \frac{254 - 206 + j105}{0 \cdot 1 + j1} = 109 - j38 \cdot 1 = \underline{115 \cdot 5 \text{ A.}}$$

$$\cos\varphi = 109/115 \cdot 5 = 0 \cdot 945 \text{ lagging.}$$

The answer can be checked by calculating $\sqrt{3}VI_a \cos\varphi$ — copper losses and comparing with the gross mechanical output $3P$; i.e.

$$\sqrt{3} \times 440 \times 109 - 3 \times 115 \cdot 5^2 \times 0 \cdot 1 = 83{,}000 - 4000$$
$$= 79{,}000 \text{ watts.}$$

The discrepancy is only about 0·5% and is due to rounding-off errors.

$$\text{Efficiency} = \frac{746 \times 100}{\sqrt{3} \times 440 \times 109} = \underline{89 \cdot 6\%.}$$

Synchronising Power and Torque

The inherent tendency of synchronous machines to remain in step has already been explained. The restoring torque at any particular excitation is a function of the load angle and hence of the load. A change of load angle occurs whenever there is unbalance between the applied mechanical torque T_m, and that developed electromagnetically, T_e. This in turn can be caused by a change to the mechanical load of a motor or to the mechanical input from the prime mover; alternatively, by a change of armature current due to circuit conditions or by a change of excitation. For any particular excitation, the relationship between a small change of load angle $\Delta\delta$ and the additional power ΔP can be obtained from:

$$\frac{\Delta P}{\Delta\delta} \simeq \frac{\mathrm{d}P}{\mathrm{d}\delta},$$

from which:

$$\Delta P \simeq \frac{\mathrm{d}P}{\mathrm{d}\delta}.\Delta\delta,$$

see Fig. 8.24. The slope of the curve $\mathrm{d}P/\mathrm{d}\delta$, is called the synchronising power per radian (elec.) and by differentiating eqn. (8.7):

$$\frac{\mathrm{d}P}{\mathrm{d}\delta} = -3(VE_f/X_s)\cos\delta = -3P\cos\varphi \text{ (total)}. \tag{8.8}$$

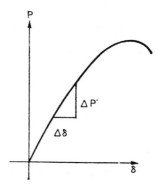

FIG. 8.24. Synchronising power.

This has a maximum value on no load with $\delta = 0$. It falls to zero when $\delta = 90°$ and the machine is developing its maximum torque $3\hat{P}/\omega_s$. The synchronising power, ΔP, is called into play in the event of a load-angle change $\Delta\delta$.

EXAMPLE E.8.4

A 3-phase, 5000-kVA, 11-kV, 50-Hz, 1000 rev/min, star-connected synchronous motor operates at full load at a power factor of 0·8 leading. The synchronous reactance is 60% and resistance may be neglected. Calculate the synchronising power per mechanical degree of angular displacement. What is the ratio of maximum to full-load torque and the value of the maximum torque?

Taking rated terminal voltage, full-load current and hence rated kVA each as 1 *per-unit*:

$$E_f = V - jX_sI_a = 1 - j(0·6) \times 1(0·8 + j0·6)$$
$$= 1 - j0·48 + 0·36 = 1·36 - j0·48$$
$$= 1·44 \text{ } per\text{-}unit$$

load angle $\delta = \sin^{-1} - 0·48/1·44 = -19·4°$

$$\frac{dP}{d\delta} = \frac{VE_f}{X_s} \cos \delta = \frac{1 \times 1·44}{0·6} \cos 19·4°$$
$$= 2·26 \times \text{rated kVA per electrical radian.}$$

From the data, the machine must have 6 poles to operate at 1000 rev/min and 50-Hz. Therefore in this case, one electrical radian is equal to:

$$\frac{180}{\pi} \times \frac{2}{6} = \frac{60}{\pi} \text{ mechanical degrees.}$$

∴ synchronising power per mechanical degree $= 2·26 \times \pi/60$
$$= 0·1185 \text{ } per\text{-}unit$$

in synchronous watts $= 0·1185 \times 5000 \times 10^3 = 593 \times 10^3$;

in terms of torque $= \dfrac{7·04}{1000} \times 593,000 = \underline{4170}$ lbf ft per mechanical degree.

Maximum torque $= VE_f/X_s = 1 \times 1·44/0·6 = 2·4 \times$ rated kVA

$$= \frac{7·04}{1000} \times 2·4 \times 5000 \times 10^3 = \underline{84,200 \text{ lbf ft.}}$$

Full-load torque = maximum torque $\times \sin \delta$ = maximum torque $\times \sin 19·4°$.

∴ maximum torque/full-load torque $= 1/\sin 19·4° = \underline{3·01}$.

Electromechanical Transient Stability

The previous example has illustrated the steady-state stability limit of the synchronous machine in terms of the maximum mechanical torque which can be applied, gradually, without causing the machine to fall out of synchronism. For suddenly applied changes, either to T_m or T_e, the transient stability performance must be studied. Since the electromagnetic "coupling" between stator and rotor fields is a flexible one, any change of either the load or the electromagnetic torque is liable to cause a mechanical oscillation. Consider a synchronous motor operating against a mechanical torque T_{m1}, which is suddenly reduced to T_{m2}, Fig. 8.25a. The electromagnetic torque T_e is now greater than the mechanical load torque and the rotor will accelerate, reducing the

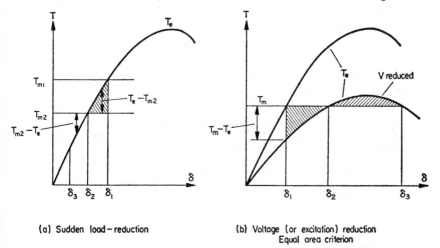

(a) Sudden load–reduction

(b) Voltage (or excitation) reduction
Equal area criterion

FIG. 8.25. Transient stability of synchronous motor.

load angle. On first reaching the new steady-state balance point δ_2, the rotor cannot suddenly decelerate back to n_s; energy extracted from the magnetic field represented by X_s has increased the stored energy in the rotating parts by an amount equal to the work done:

$$\Delta w = \int_{\delta_1}^{\delta_2} (T_e - T_{m2}) \, d\delta,$$

see shaded area. As the rotor is carried beyond the balance point, and the load angle continues to decrease, a retarding torque sets in since T_{m2} is now greater than T_e. At some angle δ_3, when the additional stored energy has been partly dissipated and partly transferred back to the magnetic field, the speed is again synchronous, but the retarding

507

torque which is still acting, causes speed to fall further, allowing the rotor to swing back and reduce the load angle towards δ_2 again. This interchange of energy between the inertia of the rotating parts and the magnetic field causes a superimposed speed oscillation about synchronous speed, in a manner analogous to the oscillations in the LCR circuit. The disturbance continues until the initial excess energy Δw is dissipated in losses.

It does not always follow that all disturbances will cause a decaying oscillation, and transient stability studies are concerned with whether the oscillation could reach such a magnitude that the machine would "swing" out of synchronism. Figure 8.25b shows the torque/angle curves for a motor which is subjected to a sudden fall in terminal voltage V, which could be brought about by a supply-line fault causing a large line-impedance drop. Momentarily, the load angle is unchanged but immediately there arises a decelerating torque $T_m - T_e$. The rotor slips backwards but after an amount of energy corresponding to the lower shaded area is extracted from the rotating inertia, torque balance is restored at angle δ_2. Beyond this point, T_e becomes greater than T_m so that acceleration towards synchronism commences. However, if the upper shaded area does not give an energy increment equal to the lower area, synchronism will be lost as T_m again becomes greater than T_e at angle δ_3 and the motor will stall unless the protective equipment has cleared the fault and restored the voltage in time. A similar situation was discussed on p. 403 when dealing with induction motors on reduced voltage. The condition on the synchronous machine is rather better since torque T_e is proportional to V, whereas for the induction machine, torque is proportional to V^2. In addition, E_f can be boosted by quick-response excitation to offset in part, the effect of any voltage reduction.

Instability and hunting can occur in generators also and Fig. 8.3b shows the measures taken to suppress the oscillations. Damper bars of copper or brass are inserted in the pole-face slots and connected at the ends to form closed circuits as in a squirrel-cage winding. At synchronous speed this damper winding moves in step with the rotating field and is inactive. A departure from synchronous speed will cause relative movement, and thus induced voltages and currents which, as in the induction machine, produce a torque opposing the relative motion. This action, which restricts the transfer of energy to the magnetic field, is directly assisted by the energy dissipated in damper I^2R loss. The resistance must be relatively low if strong electromagnetic damping is required. However, a compromise on resistance must be made if, in addition, these dampers are to be used on a motor to give induction type starting and accelerating torque, for the purposes of running up to speed. A high resistance is preferable for good starting torque, see Fig. 7.22, p. 401.

The equation of motion can be derived from the general eqn. (6.9), p. 348, though in application it is convenient, as a near approximation, to consider small angular oscillations about the steady-state operating speed ω_s. The electromagnetic torque has two components, one proportional to the departure $\Delta\delta = \beta$ say, from the steady-state load angle, and the other proportional to $d\beta/dt$ due to relative movement between the air-gap field and the rotor damping circuits. The inertia torque due to the superimposed oscillation will be $Jd^2\beta/dt^2$ and the characteristic equation for small oscillations will therefore be a quadratic (see p. 128) in β. The oscillation frequency is little different from the undamped natural frequency ω_n. This in turn can be found directly from the general expression developed at an early stage in dynamics, i.e. $\omega_n = \sqrt{(\text{stiffness/inertia})}$. The stiffness must be in Nm per mechanical radian and can be obtained from the torque/angle curve at the operating point. The total inertia must then be expressed in kg m^2. The stiffness is proportional to $\hat{P}$, eqn. (8.8), and is therefore related to the machine rating. The inertia is related to the stored mechanical energy $\frac{1}{2}J\omega_s^2$. Consequently, the ratio of stored energy (in watt seconds), to the machine VA rating (in volt amperes), is significant from the viewpoint of transient stability and is called the stored-energy constant $H = \frac{1}{2}J\omega_s^2/$ rated VA. The value of H varies from less than 1 second up to perhaps 10 seconds depending on the machine type. See earlier reference on p. 348.

Damper windings can be designed to limit any forced oscillations arising due to coupled mechanical apparatus such as reciprocating engines and compressors; flywheel sizes can thus be reduced. Dampers also tend to suppress any backwards rotating m.m.f. component due to single-phase or unbalanced loading; overheating if this is excessive.

Power/Load-angle Relationship for Salient-pole Machine

Figure 8.26 is the phasor diagram for the salient-pole motor neglecting resistance, but the expression derived will apply to both the motor and generator, remembering that δ is negative for a motor. Further, since for a motor at leading power factor as shown, I_d has a demagnetising effect, the equations will be arranged to yield I_d negative for this condition. Hence, from the phasor diagram:

$$X_q I_q = - V \sin \delta \qquad \therefore \quad I_q = \frac{- V \sin \delta}{X_q}$$

and

$$X_d I_d = V \cos \delta - E_f \qquad \therefore \quad I_d = \frac{V \cos \delta - E_f}{X_d}.$$

Resolving these current components along V to get the power:

$$\text{power per phase} = P = V I_a \cos \varphi = V(I_q \cos \delta + I_d \sin \delta)$$

$$= V \left[\frac{- V \sin \delta}{X_q} \cos \delta + \left(\frac{V \cos \delta - E_f}{X_d} \right) \sin \delta \right]$$

$$\text{total power} = 3P = -3 \left[\frac{V E_f}{X_d} \sin \delta + \frac{V^2}{2 X_d} \left(\frac{X_d}{X_q} - 1 \right) \sin 2\delta \right]. \quad (8.9)$$

This equation for power is plotted on Fig. 8.26b. The two components in the equation are indicated separately. The first one is of the same form as eqn. (8.7). Equation (8.9) is in fact a general expression, since

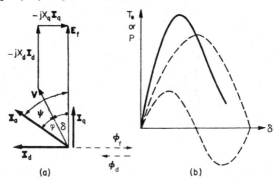

FIG. 8.26. Power/angle curve for salient-pole motor.

510

when applied to the cylindrical-rotor machine, $X_d = X_q = X_s$ and the second term is zero. The second component, which is due to the change of gap reluctance, twice every pole-pair pitch round the periphery, causes the slope of the power/angle curve to be steeper, so that the synchronising torque per radian is increased. This additional stiffness improves the transient stability.

It will be noticed that the second term is independent of the excitation. It is due to the different reluctances on direct- and quadrature-axes as a result of which the poles tend to line up with the armature field. This reluctance torque, already discussed in Chapters 1 and 3, is exploited on certain smaller motors with ratings up to about 100 kW which run synchronously without a field winding. The rotors however, are usually provided with a squirrel cage so that they can run up to speed on induction torque and pull into synchronism when the slip is small enough. Before synchronous speed is achieved, the synchronous torque is superimposed on the induction torque, but will pulsate as δ changes due to the poles slipping relative to the rotating field. The pull-in torque, near synchronism, is rather less than the peak of the torque/angle curve and must be sufficient to accelerate the rotor up to synchro-

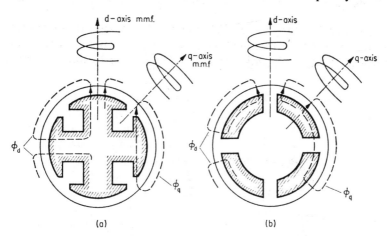

FIG. 8.27. Four-pole synchronous-reluctance-motor flux paths: (a) salient-pole construction; (b) segmental rotor construction.

nous speed against the rotating inertia and damping torques. Figure 8.27 compares a conventional salient-pole construction with one of the many recently developed forms of reluctance motor[23] designed to give convenient control over the X_d/X_q ratio which determines the reluctance torque magnitude, eqn. (8.9). Figure 8.27b shows how ϕ_q has to cross an interpolar air-gap as well as the radial air gap, thereby reducing X_q which is proportional to ϕ_q/I_q. These machines offer advantages for variable-frequency drives because of their simplicity and synchronous characteristics. A thyristor variable-frequency inverter can have its frequency controlled by a crystal oscillator to an accuracy better than 0·1%. The speed of a synchronous machine could therefore be maintained within the same accuracy without the complications of an external feedback-control circuit, providing the rate of frequency changes is restrained to avoid dynamic instability.

By considering Fig. 8.26, it can be understood that as excitation is reduced, the power factor changes to lagging. For a reluctance motor, $E_f = 0$ and the magnetising current must all be supplied from the line. The same equations hold as for the excited salient-pole machine but with $E_f = 0$ and remembering again that δ is negative for a motor:

$$I_d = \frac{V \cos \delta}{X_d} \qquad I_q = \frac{-V \sin \delta}{X_q}$$

and
$$\psi = \tan^{-1} \frac{I_d}{I_q} \qquad I_a = \sqrt{[I_d{}^2 + I_q{}^2]}$$

ψ is the angle by which $\mathbf{I_q}$ leads $\mathbf{I_a}$ and is equal to $\delta - \varphi$. The phasor diagram is drawn on Fig. 8.28 and the sign of both ψ and I_d are always positive for the reluctance motor and the power factor is always lagging. On no load, δ will be zero and the current will be just $V/X_d = I_d = I_a$, lagging 90° and magnetising. See Reference 23 for exact expressions including the effect of resistance.

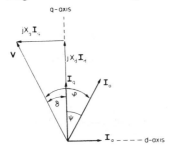

FIG. 8.28. Reluctance motor.

8.4 CIRCLE DIAGRAMS

When connected to "infinite busbars" there are two variables which may be controlled; the excitation E_f and the mechanical torque. In the following expressions, the term mechanical power P, will refer to that converted electromagnetically. It corresponds to P_g for the induction machine, though for the synchronous machine it is also equal to P_m, because the field copper losses are not supplied by induction. For a generator, P will exclude the torque losses which must be provided mechanically, but for the motor these losses will be part of the power converted electromagnetically and will be included as part of the mechanical load.

Excitation E_f Constant; Load Varying

For a cylindrical-rotor machine the locus of current can be determined, as when considering Fig. 8.22, from the motor equation (8.4):

$$\mathbf{I_a} = (V/Z_s) - (E_f/Z_s).$$

Taking $\mathbf{V}$ as the reference, the phasor $\mathbf{E_f}$, although constant in magnitude for the conditions stated, varies in phase with respect to $\mathbf{V}$ in accordance with the necessary changes in load angle. Since a constant E_f will trace out a circular locus, the same must be true of $\mathbf{E_f}/Z_s$. Since $\mathbf{V}/Z_s$ is fixed, the current phasor has the same circular locus as $\mathbf{E_f}/Z_s$. Although Fig. 8.29 shows the case where E_f is numerically greater than V, the same construction applies for reduced excitations but with a smaller circle radius. It can be seen that for generating operation $\mathbf{I_G}$ has a negative real part because the motor equation has been used. As in the case of the induction machine, the circle is merely a continuation of that for the motor with the same excitation. As shown, the machine changes from the condition of a sink taking a leading current, to a source supplying a lagging current.

Note: the current locus for a salient-pole machine departs somewhat from a simple circle and its derivation is a little more complicated.

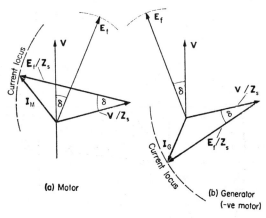

(a) Motor

(b) Generator
(-ve motor)

Fig. 8.29. Circle diagrams with constant excitation.

Constant Mechanical Power P; Varying Excitation

On steady state, the mechanical power for a motor can be obtained from the expression $P = VI_a \cos \varphi - I_a^2 R_a$ per phase. The same expression holds for a generator but with the negative sign reversed. If P is maintained constant, the equation can be satisfied for many different values of I_a and $\cos \varphi$. It can be shown that the current phasor under these conditions traces out a circular locus, the "O" curve, though this is not of great practical interest because the radius is so large. If resistance is neglected, the radius is infinite since for constant P, the active component of current, $I_a \cos \varphi$, will be constant. The locus of $\mathbf{I}_a$ is obviously a straight line in this case, variation of field current causing the kVAr to vary, with kW maintained constant.

Combined Diagram

Figure 8.30 shows the current loci for typical constant values of E_f and also for typical constant values of $VI_a \cos \varphi$ which very nearly correspond to constant mechanical power. For any particular value of load and excitation, the current phasor is readily determined; it must terminate at the intersection of the two appropriate loci. A

514

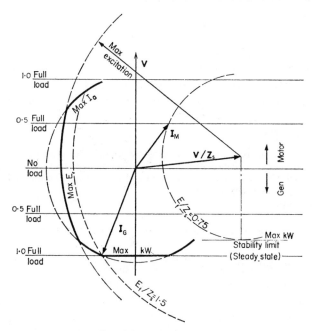

Fɪɢ. 8.30. Load diagram.

motoring current-phasor for 0·5 *per-unit* load and 0·75 *per-unit* excitation
is shown; also shown is a generating current-phasor for 1 *per-unit* load
and 1·5 *per-unit* excitation. Increase of load with constant excitation or
decrease of excitation with constant load will eventually lead to a
situation where the loci do not meet. This is the steady-state stability
limit where the machine pulls out of step. It corresponds to a power
$\hat{P} = VE_f/X_s$ per phase, neglecting resistance.

The diagram is very useful for remembering the effects of excitation
and load on the power factor. An increase of excitation causes a
generator power-factor to lag more, whereas the converse is true for the
motor. An increase of generator load causes the power factor to lead
more, in general, and again the converse is true for the motor. The

515

permissible operating region is bounded by limits determined by maximum armature current, maximum field current and maximum kW as indicated. At low excitations, the possible positions for the current phasor are restricted well within the heating limits purely from transient-stability considerations which demand a steady-state operating load angle less than 90° to avoid pull-out during rotor oscillation.

EXAMPLE E.8.5

A 3-phase, 350-kVA, 3·3-kV, 50-Hz, 6-pole, star-connected synchronous generator has a synchronous impedance per phase of $1 + j10$ Ω, and operates at a lagging power-factor of 0·9. Construct the circle diagram and hence determine the excitation E_f, the load angle and the input torque.

Determine also:

(a) the current and power factor when motoring at maximum electromagnetic torque;

(b) the current and torque when operating as a motor at 0·6 power-factor leading;

(c) the power factor when running as a generator delivering 80 A and with the excitation adjusted to the same value as the terminal voltage.

$$\text{Full-load current} = \frac{350 \times 10^3}{\sqrt{3} \times 3300} = 61\cdot1 \text{ A}.$$

$\alpha = \tan^{-1} R_a/X_s = \tan^{-1} 1/10 = 5\cdot7°$.

$Z_s = \sqrt{(1^2 + 10^2)} = 10\cdot05$ hence, $V/Z_s = 3300/(\sqrt{3} \times 10\cdot05) = 189$ A.

On the figure below, the phasor V/Z_s is set off at an angle $90° - \alpha = 84\cdot3°$ behind the V reference phasor. The current phasor of magnitude 61·1 A is set off at the power factor angle, $\cos^{-1} 0\cdot9 = 25\cdot9°$. The closing phasor E_f/Z_s is measured from the diagram to be 230 A.

Hence, $E_f = 10\cdot05 \times 230 \times \sqrt{3} = \underline{4000}$ V in line terms.

The load angle δ by measurement $= \underline{13°}$.

(a) From eqn. (8.6) the maximum torque occurs when $\delta - \alpha = -90°$ i.e. $\delta = -84\cdot3°$. By continuing the E_f/Z_s locus until this angle of separation occurs, the closing current phasor has a magnitude measured at $\underline{282}$ A and a power-factor angle of 31°. Power factor $= \underline{0\cdot856}$ lagging.

(b) A current phasor is set off at an angle $\cos^{-1} 0\cdot6 = 53\cdot1°$ till it intersects the E_f/Z_s circle. The measured value is $\underline{49\ A}$.

The gross torque can be calculated from $VI_a \cos \varphi - I_a^2 R_a$,

(i) in synchronous watts $= \sqrt{3} \times 3300 \times 49 \times 0\cdot6 - 3 \times 49^2 \times 1$

$= 168,000 - 7200 = 160,800$ watts,

(ii) in lbf ft $= (7\cdot04/1000) \times 160,800 = \underline{1130 \text{ lbf ft}}$.

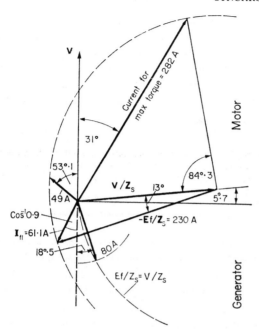

(c) A new E_f/Z_s circle is constructed of radius equal to V/Z_s. A circle of radius 80 A struck off from the origin determines the operating power factor where it intersects the excitation circle.

Power factor $= \cos 18 \cdot 5° = \underline{0 \cdot 95 \text{ leading.}}$

Note: All the above answers could have been calculated analytically, but the graphical solution is much quicker, although less accurate. Quite a close estimate of the steady-state electromechanical performance can be obtained by simple analytical solutions, if resistance is neglected. This gives the approximate equivalent circuit, consisting of an e.m.f. E_f in series with the synchronous reactance X_s, which should be the saturated value, if known. Figure 8.31a shows the phasor diagram. This represents eqn. (8.4), (simplified by neglecting R_a) and the electromechanical relationship is in the form of eqn. (8.7). V, ω_s and X_s being given, the general method of solution for various types of problem can be shown as a block-diagram sequence like the flow diagram for

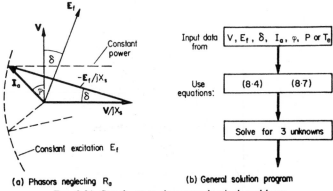

(a) Phasors neglecting R_a (b) General solution program

FIG. 8.31. Steady-state electromechanical problems.

a computer program, Fig. 8.31b. The result of neglecting the losses is that the mechanical power for a 3-phase machine, $3P = T_e \omega_s$ is also given by $3 \cdot V . I_a . \cos \varphi$. It is possible to solve for three unknowns with this additional equation and also, eqn. (8.4) has both real and imaginary parts. It may be necessary to solve either eqn. (8.4) or (8.7) first, depending on the terms of the problem, and if for example I_a and δ were specified, it would be necessary to invoke the sine rule or construct the phasor diagram to find φ first. Such a situation could arise if the problem specified a certain overload margin which would therefore fix the normal value of δ to permit this overload at $\delta = 90°$. It would be a useful exercise for the student, to formulate the program for various types of problem; for example, to compute the values of current, power factor and load angle as the excitation E_f is varied at constant power, or as the load P is varied with constant excitation. Example E.8.5 could be recalculated with resistance neglected (after considering the appropriate block diagram sequence), to assess the errors involved in this simplifying assumption, see Problem 10, p. 646 amd Ref. 33.

V-curves

These are the curves of armature current against field current for various constant mechanical loads. They could be estimated from Fig.

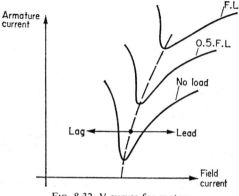

FIG. 8.32. V-curves for motor.

8.30 and the open-circuit curve relating E_f to field current. Any errors in the circle diagram due to assuming constant synchronous impedance would be included, of course, though the major allowance for saturation is made on using the o.c. curve. The V-curves are usually used only for motors, and typical ones are shown in Fig. 8.32. The broken-line locus corresponds to the minimum armature current at various loads; i.e. excitation adjusted to give unity power factor.

8.5 TESTING AND EFFICIENCY

Synchronous "Open-circuit" Test (Unity Power-factor Curve)

The o.c. and s.c. tests have already been discussed in connection with the determination of synchronous impedance. They can be taken simply, running as a generator, if a driving motor is available. The coupled exciter, run as a motor, is sometimes large enough to provide the loss energy. As an alternative, a close estimate of the magnetisation curve and the torque losses can be made by motoring the machine itself on no load. Assuming that it can be run up to speed and synchronised to a variable voltage supply, the test is similar to the no-load test on the induction motor, with the advantage that the speed is constant and the armature current and its effects can be kept to a minimum by adjusting

519

the excitation at each voltage setting, to give unity power-factor. Plotting the net power input against applied voltage will separate the mechanical from the iron loss as in Fig. 7.25. The curve of applied voltage against field current is virtually the o.c. curve.

Synchronous "Short-circuit" Test (Wattless Curve)

This is also designed for the single, unloaded synchronous machine without external drive. The machine is supplied at a low voltage so that the iron loss is very small, though it could be read off from the previous curve if desired. The operating power factor will be very low as the excitation is varied to draw armature currents up to say 25% overload. The phasor diagrams are the same as on Fig. 8.37. The relationship between field and armature currents is close to the short-circuit curve. After deducting the known friction and windage loss from the power input, the remainder, apart from the small iron loss, is the copper loss including stray load loss, though not at normal flux.

Determination of F_a and x_{a1} from Zero-power-factor Tests

From Fig. 8.6, it is known that on zero power-factor, the armature m.m.f. F_a is aligned along the same axis as the field m.m.f. F_f. By taking advantage of this feature, information can readily be obtained about the magnitude of F_a and the associated reactances x_{a1} and X_m. A greater degree of stability will result from operation at the high-field condition, i.e. with F_a demagnetising. A generator would have to be loaded inductively to give a low lagging power-factor whereas a motor would have to be run on no-load, over-excited as on the wattless curve, to draw a low leading-power-factor current. Because of the relatively small effect of the $I_a R_a$ drop, the power factor can be as high as perhaps 0·25 without introducing significant errors. Two tests are required, one for the unsaturated condition at as low a terminal voltage as possible (which can be zero for a generator with the terminals short-circuited) and the other for the saturated region beyond the air-gap line, taken at rated terminal voltage.

Figure 8.33 shows the generator phasor diagrams neglecting R_a, for two such tests, both taken at the same value of armature current for

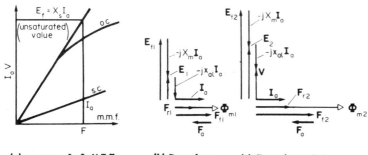

(a) o.c curve $I_a = 0$ $V = E = E_f$ (b) Zero p.f. s.c.   (c) Zero p.f. normal V

s.c. curve $V = 0$ $E_f = X_s I_a$ $E_{f1} = E_1 + X_m I_a$ $E_{f2} = E_2 + X_m I_a$

FIG. 8.33. Open-circuit and zero power-factor conditions.

convenience. From the current and voltage phasors alone, it can be seen that the voltages due to armature current, $jX_m\mathbf{I_a}$ and $jx_{al}\mathbf{I_a}$, are in direct opposition to the field produced voltage $\mathbf{E_f}$. This is a reflection of the demagnetising action of F_a which is indicated by the space phasor $\mathbf{F_f}$, due to the field being greater than the resultant $\mathbf{F_r}$ producing ϕ_m and E. For the conditions shown therefore, the phasors can be combined algebraically to give, for the terminal voltages:

$$0 = E_{f1} - X_m I_a - x_{al}I_a = E_1 - x_{al}I_a \qquad (8.10)$$

$$V = E_{f2} - X_m I_a - x_{al}I_a = E_2 - x_{al}I_a \qquad (8.11)$$

and for the m.m.f.s:

$$F_{f1} - F_a = F_{r1} \qquad (8.12)$$

$$F_{f2} - F_a = F_{r2}. \qquad (8.13)$$

Since at zero terminal voltage, F_{r1} and ϕ_{m1} are small, the magnetic circuit is virtually unsaturated so that:

$$E_1 = k_f . F_r \qquad (8.14)$$

and

$$E_2 = f(F_{r2}) \qquad (8.15)$$

for the saturated region, $f(F_{r2})$ being the non-linear relationship of the o.c. curve. Note that since I_a and F_a are both zero when this curve is taken, the gap e.m.f. E is measured directly by the terminal voltage and the resultant m.m.f. F_r is measured directly by the field current.

For steady-state calculations, it is only necessary to extract from the above equations the values of leakage reactance x_{a1} and magnetising reactance X_m. Having deduced the unsaturated value of synchronous reactance $X_s = X_{mu} + x_{a1}$, as described on p. 481 and illustrated on Fig. 8.33a for the case where resistance is negligible, X_{mu} can be found if x_{a1} is known. X_{mu} can then be corrected for the saturation level corresponding to the desired operating condition as described on p. 480. During the zero-power-factor tests, the terminal voltage V, current I_a and the field excitations F_{f1} and F_{f2} will have been recorded. Equations (8.10) to (8.15) can then be solved for the remaining eight unknowns and though only x_{a1} is required, the armature m.m.f. F_a would be needed to construct the full phasor diagram as on Fig. 8.11b.

Combining eqns. (8.10)–(8.14) gives:

$$E_2 = V + k_f(F_{f1} + F_{r2} - F_{f2}). \tag{8.16}$$

The solution of eqns. (8.15) and (8.16) simultaneously, can be achieved by trial and error. A value of E_2 is chosen and the corresponding value of F_{r2} read from the o.c. curve. These two values are then substituted in eqn. (8.16) to see if they can satisfy this equation too. If not, a new value of E_2 must be chosen and the iteration process repeated until both equations are satisfied. Substitution of E_2 in eqn. (8.11) will then give x_{a1} and substitution of F_{r2} in eqn. (8.13) will give F_a in terms of the field turns.

A simple graphical solution, not unlike that of Fig. 6.25, is possible and is left as an exercise. The same graphical solution virtually, but referred to as the Potier construction, is normally employed and is shown on Fig. 8.34. All of the equations (8.10) to (8.15) can be discerned on the figure so the solution should be readily understood. The two points P and p corresponding to rated terminal voltage and zero terminal voltage respectively are plotted separately on Fig. 8.34b. The operating conditions under which these points are obtained have two features in common as a result of current and power factor being unchanged, viz. both the internal leakage reactance drop $x_{a1}I_a$ and the demagnetising armature m.m.f. F_a remain constant. If a magnetisation curve (V/I_f) were to be taken at the same constant current and zero-power-factor, it

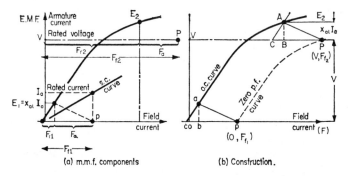

(a) m.m.f. components   (b) Construction.

FIG. 8.34. Zero power-factor test. Potier construction.

would be displaced from the o.c. curve by the sloping line $ap =$ AP, see Fig. 8.34b. AP $= ap$ is made up of the two perpendicular components $x_{al}I_a =$ AB $= ab$, and $F_a =$ BP $= bp$.

The points A and a are not known, but may be obtained as described below.

Reading from the s.c. curve, draw PC $= pc$, the total excitation to neutralise F_a on short circuit and generate $x_{al}I_a$. This may include a small portion due to remanent ampere-turns, co.

Draw CA parallel to the initial slope of the o.c. curve and ca, to intersect the curve at point A.

The line AP is thus defined and the zero power-factor curve could be constructed if desired. More important, by dropping the perpendicular AB $= ab$, the leakage reactance and armature reaction effects are separated; AB $= ab = x_{al}I_a$ and BP $= bp = F_a$. For any other value of armature current, each of these two quantities is increased or decreased in proportion.

The construction depends on the fact that the initial portion of the o.c. curve is a straight line. The test value of leakage reactance (referred to as the Potier reactance x_p) tends to be an overestimate. This is because F_{f_2} is higher than obtained at power factors nearer unity under more normal conditions with an internal e.m.f. E_2. As a result, the additional field leakage flux is sufficient to cause appreciable saturation in parts of the rotor body so that P is farther to the right than it would be otherwise. Note that, if the abscissa is in units of field current, F_a will be in terms of field turns but this is quite convenient.

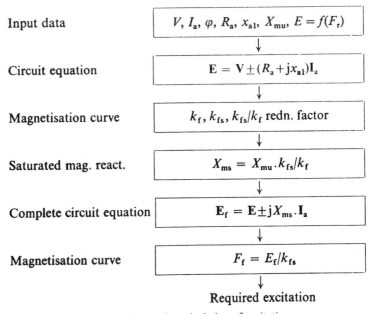

| Input data | $V, I_a, \varphi, R_a, x_{al}, X_{mu}, E = f(F_r)$ |

$\downarrow$

| Circuit equation | $\mathbf{E} = \mathbf{V} \pm (R_a + jx_{al})\mathbf{I}_a$ |

$\downarrow$

| Magnetisation curve | $k_f, k_{fs}, k_{fs}/k_f$ redn. factor |

$\downarrow$

| Saturated mag. react. | $X_{ms} = X_{mu} \cdot k_{fs}/k_f$ |

$\downarrow$

| Complete circuit equation | $\mathbf{E}_f = \mathbf{E} \pm jX_{ms} \cdot \mathbf{I}_a$ |

$\downarrow$

| Magnetisation curve | $F_f = E_f/k_{fs}$ |

$\downarrow$

Required excitation

FIG. 8.35. Flow diagram for calculation of excitation.

To determine the required excitation at any voltage, load current and power factor, the phasor diagram can be constructed, but as described on p. 480, it is really only necessary to use the equivalent circuit equations and eqn. (8.3c), the m.m.f.s being reflected in the values of the voltage components. The method has already been carefully explained but it might be helpful to summarise it in the form of a flow diagram, Fig. 8.35.

For the salient-pole machine, the unsaturated values of X_d and X_q obtained as described on p. 486, can be used to find the excitation

524

after applying the saturation correction as explained on p. 488. Alternatively, saturated values of the reactances could be used, though for the quadrature-axis, saturation is not so extensive and could be neglected. For X_d the reduction factor can be obtained in the same way as for X_{ms} and in fact a normal o.c./s.c. test analysis will give X_d directly because the power factor is nearly zero and therefore I_q is negligible. Round-rotor theory would give a close estimate of the excitation, using X_{ds} for X_{ms}. Finally, Rothert's *ampere-turn method* might be mentioned. This considers the leakage reactance drop in terms of the m.m.f. cb; i.e. the total armature m.m.f. is taken as cp so that only the s.c. and o.c. curves are necessary.

EXAMPLE E.8.6

The test results on a 3-phase, 10-MVA, 11-kV, star-connected synchronous generator are as follows:

Open-circuit test.

Line voltage	4	6	8	10	11	12	13	14 kV
Field current	60	90	126	171	203	252	324	426 A

Short-circuit test.

Armature current 566 A, field current 200 A.

Zero power-factor test.

Armature current 396 A, field current 480 A, line voltage 11 kV. Calculate the field current:

(a) when supplying 6000 kVA at zero leading power-factor to an unloaded transmission line (line charging).

(b) when generating at full load 0·8 p.f. lagging.

(c) when motoring at the same current and power factor as (b).

(d) when running as an unloaded motor taking 6000 kVA at zero p.f. leading.

Neglect armature resistance.

The Potier construction is carried out on the accompanying figure after correcting the short-circuit information to the same armature current as the zero p.f. point, i.e. at 396 A on s.c. the field current will be $(396/566) \times 200 = 140$ A.

At 396 A and 11 kV the field current is 480 A for zero p.f. lagging. The simple correction to the s.c. information is possible because the relationship between field and armature s.c. current is linear. Note that the o.c. curve is plotted in terms of line voltage for convenience so that the Potier reactance drop per phase at 396 A is scaled up by a factor of $\sqrt{3}$.

From curve $\sqrt{3}x_{al}I_a = 2·6$ kV at 396 armature amperes.
$$F_a = 102 \text{ A at } 396 \text{ armature amperes.}$$

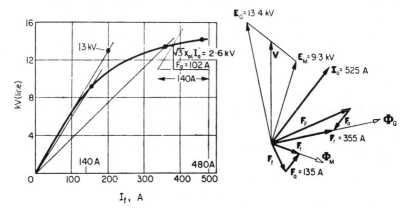

Parts (a) and (d) for zero-power-factor will be solved using m.m.f.s since only algebraic combination is necessary for this condition. For parts (b) and (c), the equivalent circuit will be used but the phasor diagrams are also shown on the figure. The same answers would be obtained by either method if worked out carefully and it would be instructive to check the answers in each case using the alternative method.

(a) At 6000 kVA, $I_a = \dfrac{6000}{\sqrt{3} \times 11} = 315$ A.

$F_a = (I_a/396) \times 102 = (315/396) \times 102 = 81$ A. $\sqrt{3}x_{al}I_a = (315/396) \times 2\cdot6 = 2\cdot06$ kV. At zero p.f. leading, see Fig. 8.16c, $E = V - \sqrt{3}x_{al}I_a = 11 - 2\cdot06 = 8\cdot94$ kV.

From the o.c. curve, this requires a net excitation current:

$$F_r = 145 \text{ A.}$$

The armature reaction ampere-turns are completely magnetising, Fig. 8.6c, so:

$$F_f = F_r - F_a = 145 - 81 = \underline{64 \text{ A.}}$$

(b) Full-load current at 10 MVA $= (10/6) \times 315 = \underline{525 \text{ A.}}$

Unsaturated synchronous reactance determined, arbitrarily, at the s.c. test currents

$$= \dfrac{13000/\sqrt{3}}{566} = 13\cdot2 \; \Omega \text{ per phase.}$$

k_f from the same condition $= 13000/200 = 65$ line V/field A.

Leakage (Potier) reactance $= \dfrac{2600/\sqrt{3}}{396} = 3\cdot8 \; \Omega$ per phase.

Unsaturated magnetising reactance $X_{mu} = 13\cdot2 - 3\cdot8 = 9\cdot4 \; \Omega.$

Air-gap e.m.f. $E = 11000 + \sqrt{3} \times j3\cdot8 \times 525(0\cdot8 - j0\cdot6)$

$= 11000 + 2080 + j2760$

$= 13400$ line V,

giving, from o.c. curve $k_{fs} = 13400/355 = 37\cdot8$ line V/field A.

Hence, saturated magnetising reactance $X_{ms} = 9\cdot4 \times 37\cdot8/65 = 5\cdot4\,\Omega$.

$$E_f = 13080 + j2760 + (2080 + j2760) \times 5\cdot4/3\cdot8$$
$$= 16040 + j6680 = 17400 \text{ line V}/\underline{+22\cdot6°}\ (\delta)$$
$$F_f = 17400/37\cdot8 \quad = \underline{460\text{ A}}; \text{ eqn. (8.3c).}$$

If the load was removed, the terminal voltage would rise to 14·2 kV with this field current maintained; see o.c. curve.

(c) The calculation procedure is the same as for (b) except for the sign reversal of the impedance drop, which has the same magnitude.

Hence
$$E = 11000 - 2080 - j2760 = 9300 \text{ line V}$$
$$k_{fs} = 9300/160 = 58 \text{ line V/field A}$$
$$X_{ms} = 9\cdot4 \times 58/65 = 8\cdot4\,\Omega$$
$$E_f = 8920 - j2760 - (2080 + j2760) \times 8\cdot4/3\cdot8$$
$$= 4320 - j8860 = 9850 \text{ line V}/\underline{-64°}\ (\delta)$$
$$F_f = 9850/58 = \underline{169\text{ A.}}$$

(d) In this case, the armature m.m.f. is completely demagnetising, see p. 484 and Fig. 8.6b. Further, $E = V + x_{a1}I_a$, see Fig. 8.37 with resistance neglected. Substituting the same values of F_a and $\sqrt{3}x_{a1}I_a$ as in (a):

$$E = 11 + 2\cdot06 = 13\cdot06 \text{ kV.}$$

From the o.c. curve this requires $F_r = 328$ A

$$\text{so } F_f = F_r + F_a = 328 + 81 = \underline{409\text{ A.}}$$

This is the case of the synchronous capacitor discussed in the next section.

Back-to-back Test

If two synchronous machines are mechanically coupled so that their poles and stator phase windings are in line, their e.m.f.s can only be in phase or in antiphase. If the mechanical coupling is modified to permit a small relative angular displacement between the rotors, or a less convenient adjustment is made to give angular displacement between the stators, a situation very similar to that of Fig. 8.19b arises. By connecting the two stator windings together in the correct phase sequence, machine 1, say, can be made to generate on to machine 2 as a motor. The terminal voltage and power factor are adjusted by varying E_1 and E_2, and the load transferred depends on the angular shift, which is rather critical. As usual on back-to-back tests, only the energy for the losses need be provided.

EXAMPLE E.8.7

A 3-phase, star-connected alternator has a synchronous impedance per phase of $1+j20\,\Omega$. For the purposes of a back-to-back test it is to be mechanically coupled to a machine having a synchronous impedance of $2+j30\,\Omega$ per phase. The alternator is to be loaded at 2000 V, 15 A and 0·8 p.f. lagging. Calculate the necessary excitation for each machine, and the angular displacement between the rotors which must be provided at the coupling. The machines are of 8-pole construction.

Taking the generator terminal voltage $\mathbf{V}$ as reference phasor then:

$$\mathbf{I_a} = 15(0\cdot8-j0\cdot6) = 12-j9 \text{ A}$$
$$\mathbf{E_1} = \mathbf{V}+\mathbf{I_aZ_1} = (2000/\sqrt{3})+(12-j9)(1+j20)$$
$$= 1347+j231 \text{ V}$$

in terms of line voltage $= 1368 \times \sqrt{3} = \underline{2370 \text{ V}}$ for the generator.

For the motor:
$$\mathbf{E_2} = \mathbf{V}-\mathbf{I_aZ_2}$$
$$= (2000/\sqrt{3})-(12-j9)(2+j30)$$
$$= 861-j342$$
$$= 926 \text{ V} \quad \text{or} \quad \underline{1600 \text{ line V}}.$$

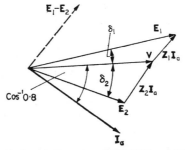

For the generator, the load angle $\delta_1 = \tan^{-1} 231/1347 = 9\cdot75°$.
For the motor, $\delta_2 = \tan^{-1}-342/861 = -21\cdot65°$.
∴ total displacement between the rotors $= 9\cdot75+21\cdot65 = 31\cdot4$ electrical degrees.
Since there are 4-pole pairs, the mechanical displacement must be $31\cdot4/4 = \underline{7\cdot85°}$.

Wattless Testing

This is the most common way in which a large synchronous machine is temperature tested while carrying full current at rated voltage, though at low power-factor. The smaller machines are over-excited as motors and supplied from a suitable test machine which can provide the leading kVA. Larger machines are run as generators supplying suitable reactors. The mechanical input is not much greater than the machine loss, but must of course include the reactor loss. In either case, the field temperatures will be higher than normal due to the over-excitation and allowance must be made for this.

Efficiency

Once the losses have been determined, the efficiency can be calculated in the usual way, the field and the total exciter loss being different at each load since field currents vary so widely. On the very large units, see Fig. 8.3c, efficiencies can approach 99%, hydrogen cooling being of considerable assistance not only in permitting higher copper loading, but in reducing the windage loss, which on an air-cooled turbo-alternator of 50 MVA is several hundred kW. Direct water-cooling of stator conductors has been in use for many years and a 500-MW unit with water-cooled rotor coils has been built and tested. Development work is proceeding also on superconducting generators[24] and proto-types have been built. These machines raise very considerable engineer-ing problems concerned with insulating the rotor—in its cryostat (refrigerator)—not only thermally but also electromagnetically. Under unbalanced fault conditions for example, negative-sequence currents (p. 81) arise, giving a backwards rotating field and double-frequency induced currents in the rotor which would destroy the super-conductivity. A rotor electromagnetic screen would therefore be necessary. In ad-dition, because of the very large magnetic field, the whole machine would have to have a conducting screen surrounding it to shield the immediate environment from its influence. The use of air-gap windings mentioned earlier, p. 215, is possible and even without superconduc-tivity, such windings are claimed to have potential maximum rating of 6,000,000 kVA! Superconducting machines might have even greater ratings. It is of course debatable whether the enormous investment required for each single unit and all this entails would justify the improvement in efficiency which is expected.

Sudden Short-circuit Test

The envelope of the current/time curve following a sudden short-circuit test from a voltage E_f on open-circuit is shown on Fig. 8.36. The d.c. component, which depends on the switching instant, has been extracted by measuring the peak-to-peak values of the wave from the oscillogram and these have been replotted to a scale reduced by $1/\sqrt{2}$. In Sections 3.7 and 3.10, the transient reactance of two coupled coils 1 and 2 was shown to be $\omega L_{11}(1 - k_{12}^2)$. This test is concerned with the measurement of such parameters.

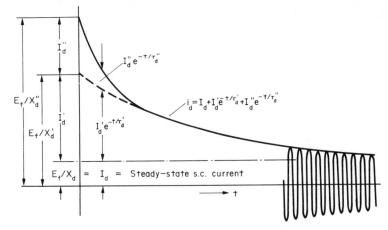

FIG. 8.36. Sudden short-circuit test; analysis of oscillogram.

In the steady state, a synchronous machine rotor is at rest with respect to the stator m.m.f. wave, but if the *peak* stator current is changing with time, transformer-induced currents in the coupled field and damper windings resist the mutual-flux change. Decay of these induced currents is governed by the effective circuit time-constants $\tau_d' = L_{ff}(1 - k_{af}^2)/R_f$ for the field (typically 1 second) and $\tau_d'' = L_{dd}(1 - k_{ad}^2)/R_d$ for the damper (typically 0·03 seconds). Consequently, as a near approximation, the reactions of these two circuits on the stator current can be considered separately and superimposed,

530

since the damper reaction, during the *subtransient* region, is completed before the field reaction decays appreciably.

The effective reactance of stator and damper windings together to a changing value of I_d is called the *direct-axis subtransient reactance* X_d'' and the initial current is thus limited to E_f/X_d''. If no damping action were present, the initial current would be limited to E_f/X_d' where X_d', the *direct-axis transient reactance*, is due to stator and field windings together; i.e. as for a transformer with two s.c. secondaries, damper and field *leakage* impedances are effectively in parallel across the magnetising reactance. The final steady-state short-circuit current is limited by the *direct-axis synchronous reactance* to E_f/X_d. By analysing the oscillogram as indicated, these various reactances and time constants can be calculated. The transient reactances are rather higher in value than the leakage reactance and typically in the range 0·1–0·25 *per-unit*.

Since on "dead short circuit" the power factor is very low, I_q is negligible and the quadrature-axis reactances are not involved in the behaviour. It is otherwise on higher resistance fault conditions and X_q' can then be taken as X_q since normally there is no quadrature-axis field winding. However, the *quadrature-axis subtransient reactance* X_q'' will not be equal to X_q if there are paths for quadrature-axis damper currents.

Other tests are necessary to determine the reactance to zero-sequence (in-phase) currents and to negative-sequence currents which establish a backwards rotating stator m.m.f. wave. The most obvious way of measuring the *zero-sequence impedance* would be to connect all the phases in series and supply them with single-phase current. The machine would be run at synchronous speed and the impedance calculated from the measured input voltage, current and power. For the *negative-sequence impedance*, the phases would have to be connected to a 3-phase supply. The machine would then be driven at synchronous speed, though in the opposite direction to the rotating field produced by the stator currents. The field winding should be short-circuited so that its reaction in terms of the induced double-frequency currents, would be present. The impedance could then be calculated from suitable input measurements, though for these tests

531

there is usually a considerable harmonic content in both current and voltage so that precise treatment of the problem is rather complicated.[2]

Sudden-short-circuit tests are not usually carried out on induction machines and d.c. machines though, like the synchronous machine, they could also feed high currents into a supply system fault due to the internal e.m.f. generated by the flux existing during the short circuit. Their characteristics under those conditions must therefore be calculated if their contribution to the fault current is likely to be significant. For the induction machine the transient, in general, is less severe and dies out very quickly (in a few cycles) because there is no independent source of excitation. For the d.c. machine, the current peaks can be much higher than for a.c. machines; up to 15 *per-unit* or more on compensated machines.[9]

8.6 OTHER MODES OF OPERATION

In the synchronous mode, there is one speed of operation, corresponding to $n_s = (f_{stator} \pm f_{rotor})/p$. Although this general expression includes the possibility of having both windings connected to a.c. systems, set independently at any particular frequencies (doubly-fed machines); such machines are inherently unstable and require extraordinary arrangements to provide some damping. The possibility of incorporating thyristors in the machine windings may lead to special forms of the synchronous mode. Already, brushless machines, see Fig. 8.3b, using a.c. excitation sources with rotating rectifiers, have become generally accepted for standard synchronous machine duties. Other types of brushless motor will be discussed shortly. The application of more refined control-techniques can lead to improved transient behaviour of normal machines during faults or overloads. The excitation can be regulated as some individual or combined function of voltage, current, power factor or load angle to get maximum performance from a given machine. An interesting development is the provision of two field windings with their axes usually but not necessarily in space quadrature. It will be recalled from p. 485, that the direct-axis component of excitation affects the reactive power, whereas the quadrature-axis component of excitation governs the torque. For a normal machine, if the operating power factor demands a low excitation, stability margins deteriorate, see Fig. 8.30, but if the quadrature excitation can be boosted this need not occur. By suitable control of the two windings, therefore, stability is improved and operation at load angles in excess of 90° is possible. Some of the more common operational modes will now be discussed.

Starting of Motors

The synchronous machine as generally understood, develops synchronous torque only at speed $n_s = f_{supply}/p$ so there is a problem of producing a torque for starting and running up to speed. It may, for special reasons, be coupled to a machine which could act as a motor during the starting period. In general however, a synchronous motor is designed so that it can develop enough induction torque to bring it from standstill, nearly up to synchronism. Thereafter, the application of excitation to the field must be sufficient to develop the necessary pull-in torque to accelerate the rotor inertia up to synchronous speed. In some cases, eddy currents in the face of a solid-iron pole are sufficient for damping and starting, but if a pole-face winding *is* provided, it must be of high enough resistance to give the necessary starting torque with normal starting-voltage applied. Starting methods are generally the same as for induction motors, though a resistor is usually connected across the field winding to limit its terminal p.d. resulting from the induced voltage, when there is armature-flux/field-coil relative motion. If the motor field is normally excited through a rectifier circuit, this induced voltage could produce a d.c. component of current and a pulsating torque which would interfere with the starting process. Extra precautions must then be taken in the circuit to minimise this component.[11] On small machines, the exciter armature may be left in circuit and as its d.c. voltage builds up with speed, the corresponding d.c. component of the motor field-current eventually becomes large enough to give smooth synchronisation. When very high starting-torques are required, the damping windings are phase-wound in the pole-face and brought out to slip rings. External resistance is inserted for starting and the rings are later short circuited for damping.

Electrical Braking of Synchronous Machines

The three usual methods are available for synchronous machines. *Reverse current braking* is not used very often and would depend on the induction characteristics. *Regenerative braking* occurs automatically if the mechanical torque on a motor shaft reverses and causes the load angle to change from retard to advance. Operation is confined to synchronous speed so cannot be used to slow down a motor unless frequency- or pole-changes are made. *Dynamic or rheostatic braking* is commonly used. The stator is disconnected from the supply and connected instead to a three-phase bank of resistors. The stored energy in the rotating parts is thus converted to electrical energy and heat, as the machine slows down. The rate of conversion and the deceleration depend on the excitation and the external resistance value.

Operation at Various Voltage and Frequencies

The equations and phasor diagrams already developed can be applied even when V and f are changed, over a moderate range. The V/f ratio must be such as to avoid excessive saturation or reactive-current changes and in the calculations all the frequency-sensitive parameters must have their correct values. Synchronous motors are sometimes used on large turbo-electric marine drives and in this case, with constant flux on the variable-speed supply generators, the voltage and frequency rise and fall together. For reduced frequencies, since the reactance falls, the resistance drops can no longer be neglected as in some of the performance equations developed. Operation under variable-frequency conditions is becoming more common with the availability of static, variable-frequency supplies which have already been discussed on pp. 427 and 512. Aircraft electrical systems provide another example of non-standard frequency operation, usually 400 Hz. From eqn. (6.1a), p. 291, it can be seen that operation at high speeds improves the power/weight ratio, figures of about 2 kW/kg being achieved. Such machines necessitate special design, of course—to control the iron losses and, for generators, to deal with the problem of the variable-speed engine-shaft drive; see p. 431.

534

The Synchronous Capacitor (Compensator)

This machine is a synchronous motor running on no load and designed for a wide variation of field current. When over-excited, Fig. 8.37a, the machine takes a leading current like a three-phase capacitor but of controllable rating. When under-excited, Fig. 8.37b, within the stability limit, the motor behaves as an inductor of controllable rating. Figure 8.25 has illustrated that the transient stability is dependent on E_f and hence on F_f so that the lagging kVAr capacity will be less than the leading kVAr capacity. These machines in units up

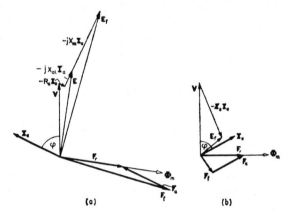

FIG. 8.37. Synchronous compensator; (a) leading power-factor; (b) lagging power-factor.

to 1000 MVA or more, are used frequently on long transmission lines to control the voltage at points remote from the power station. They also act to increase the power capacity of the transmission system by controlling the effective line reactance.

Power-factor Correction

Most items of electrical equipment in industry, e.g. inductive coils, transformers and induction motors, require lagging power-factor current for normal operation. To improve the power factor for any given installation, equipment taking leading kVAr is often installed.[11] The result is a reduction in total current taken and in the losses and monetary charges. It may be economically sound to use static capacitors for this purpose but the synchronous capacitor has the added advantage that it can also provide mechanical power and power-factor control is available by changing the excitation. As a simple illustration, consider an installation where there are just two induction motors, IM_1 and IM_2 operating with an overall power-factor angle φ_M, see Fig. 8.38. A synchronous motor taking a current I_S at a power-factor-angle φ_S is now provided to correct the overall power factor to $\cos \varphi$. If the power ratings, efficiencies and lagging power-factor angles of the induction motors are: P_1 and P_2, η_1 and η_2, φ_1 and φ_2 respectively, the currents $\mathbf{I}_1$ and $\mathbf{I}_2$ assuming star connection are:

$$\mathbf{I}_1 = (P_1/(\eta_1 . \sqrt{3} . V))(\cos \varphi_1 - j \sin \varphi_1),$$
$$\mathbf{I}_2 = (P_2/(\eta_2 . \sqrt{3} . V))(\cos \varphi_2 - j \sin \varphi_2).$$

Total current before correction $= \mathbf{I}_1 + \mathbf{I}_2 = I_M \cos \varphi_M - j I_M \sin \varphi_M$
$$= \underline{\qquad I_P - j I_Q \qquad}$$

If the synchronous motor provides a mechanical power P_S with efficiency η_S, then the power component of its current is:

$$I_{PS} = P_S/(\eta_S . \sqrt{3} . V).$$

The phasor diagram shows these relationships and it follows that the overall power-factor angle is defined by:

$$\tan \varphi = \frac{I_Q - I_{QS}}{I_P + I_{PS}}.$$

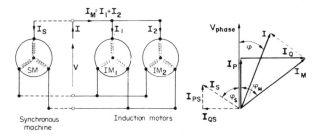

FIG. 8.38. Power-factor improvement.

536

This equation can be solved for any one unknown, typically I_{OS} and the synchronous machine kVA rating; $\sqrt{3} . V . \sqrt{(I_{PS}^2 + I_{OS}^2)} . 10^{-3}$, given the required value of φ. The power component I_{PS} could of course be zero, the motor then running on no-load as a synchronous capacitor. Note that on the phasor diagram, if all current components are multiplied by V, we have watt, voltampere, VAr (or kW, kVA, kVAr) triangles. See Appendix E, Problem 22 for Chapter 8.

The Synchronous Induction Motor (Auto-synchronous Motor)

This machine is similar to an induction motor with a wound secondary, so that it has good starting characteristics. When d.c. is then applied to the secondary, it pulls into synchronism and thereafter is used as a synchronous motor with its power-factor control features.

If an induction motor is driven at synchronous speed, its equivalent circuit is the same as for a synchronous machine with zero excitation. This follows from the discussion in Section 3.9 and later in Appendix B. At $n = n_s$, the secondary current is zero ($R_2/s = \infty$). The equivalent circuit and the phasor diagram for this condition, neglecting iron loss, are shown in Fig. 8.39. The sum of $x_1 + X_m$ is, in fact, the reactance presented at the motor terminals at synchronous speed, i.e. the synchronous reactance. The magnetising m.m.f., $\mathbf{I}_m N_1$ in induction-machine terms, is, in synchronous-machine terms, $\mathbf{F}_a$, due to the current $\mathbf{I}_a = \mathbf{V}/\mathbf{Z}_s$ which is the same as $\mathbf{I}_m$ for this condition. If now the secon-

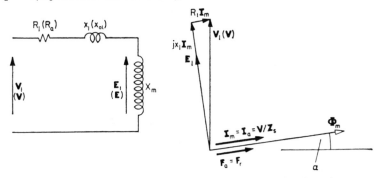

FIG. 8.39. Synchronous and induction motors on no load and synchronous speed.

537

dary is excited with direct current, for example in accordance with Fig. 5.30a, so that it establishes its own leakage flux ϕ_{f1} and mutual flux ϕ_f, there is an additional e.m.f. in the equivalent circuit of value E_f. The equivalent circuit is now identical with Fig. 8.10, and the behaviour will be substantially in accordance with cylindrical-rotor theory; the air gap being uniform, apart from slotting effects, and the m.m.f. waves being nearly sinusoidal.

At this stage it is instructive to compare the behaviour of synchronous and induction machines in physical terms. For this purpose it will be assumed that the same stator winding is being used and is connected to the supply line so that the resultant mutual-flux axis of ϕ_m is rotating at synchronous speed as indicated on Fig. 8.40. For the synchronous machine, the rotor poles are excited by d.c. currents whereas for the induction machine they are produced by induced currents though still rotating synchronously with the stator poles; see p. 373. For the synchronous machine on no load, the load angle δ_{fr} is nearly zero, the stator and rotor poles being virtually in line. For the induction machine on no load, however, the load angle δ_{ar} is near its optimum value of 90°, i.e. $\varphi_2 = 0°$; see Fig. 7.15. It cannot be quite equal to 90° for a motor with a short-circuited secondary, because the secondary reactance and therefore the slip would have to be zero; i.e. the speed would be synchronous, with no rotor induced e.m.f. or current to provide the no-load loss torque.

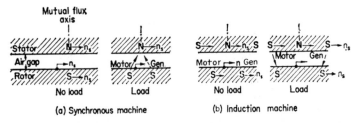

(a) Synchronous machine   (b) Induction machine

FIG. 8.40. Comparison of synchronous and induction modes of operation.

On load as a motor, the rotor of a synchronous machine is retarded by the load, giving an electromagnetic force in the direction of rotation

538

which increases to a maximum as the load angle increases to rather less than 90°, Figs. 8.23 and 8.26. Figure 8.40a shows the situation both for motoring and generating, the rotor poles being advanced in the second case so that the electromagnetically produced torque opposes the applied mechanical driving torque. The rotor m.m.f., as shown by F_f on Fig. 8.41a, is always less than 90° displaced from F_r and Φ_m because of stability considerations limiting the load angle. For a motor, the power factor is leading if F_f is large enough. For the induction machine, Figs. 8.40b and 7.15 show that the rotor poles still move backwards for motoring and forwards for generating but the load angle then increases beyond 90° and I_2 is always at a lagging power-factor, $\cos \varphi_2$. Consequently, the power factor of a normal induction motor is always lagging. Maximum torque occurs at a load angle of approximately 135°, see pp. 384 and 393. For leading power-factor operation as a motor, or lagging power-factor as a generator, it would be necessary to inject an externally supplied secondary voltage of appropriate phase; Fig. 7.38c. The rotor-pole position would then correspond to a load angle less than 90° on no load, though increasing with motoring torque as before.

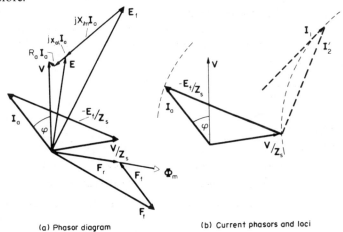

(a) Phasor diagram   (b) Current phasors and loci

FIG. 8.41. Synchronous induction motor on load.

539

The calculation of d.c. excitation for a synchronous induction motor can be made in the same way as described on p. 524, since synchronous operation is the same in the two cases. An alternative method is sometimes used which takes advantage of the fact that X_m is not very different in magnitude from Z_s. From the development of eqns. (8.3a, b and c), it follows that:

$$\frac{F_f}{F_a} = \frac{E_f}{I_a X_m} \simeq \frac{E_f}{I_a Z_s} \quad \therefore \quad F_f \simeq \frac{E_f}{Z_s} \cdot \frac{F_a}{I_a} \text{ ampere turns.} \quad (8.17)$$

Similarly, $F_r \simeq (V/Z_s)(F_a/I_a)$ since $V \simeq E$.

E_f/Z_s is therefore the required excitation in terms of the effective armature turns F_a/I_a. Consequently, by constructing the current-component phasor diagram of Fig. 8.41b for any desired current and power factor, E_f/Z_s can be measured directly and with the connection of Fig. 5.30a for example, the d.c. field current to produce F_f ampere turns is:

$$\frac{E_f}{Z_s} \times \frac{\text{Effective stator turns per phase}}{\text{Effective rotor turns per phase}} \times \sqrt{2}.$$

Allowance for saturation would require further information as already discussed in Sections 8.2 and 8.5.

Up to now, the stator iron loss has been neglected but can be taken into account conveniently, in the same way as for the induction motor, i.e. by adding $\mathbf{I_p}$ to the $\mathbf{I_m}$ phasor to give $\mathbf{I_0}$ as the total current at synchronous speed. The phasor $-\mathbf{E_f}/Z_s$ is then drawn from the end of the $\mathbf{I_0}$ phasor in the same way as the phasor $\mathbf{I_2'}$ compensating the rotor m.m.f. of the induction motor. For asynchronous operation however, since the rotor ampere-turns are produced by induction, the 90° lagging phase shift means that $\mathbf{I_2'}$ always lags $\mathbf{V}$. $-\mathbf{E_f}/Z_s$ always leads $\mathbf{V}$.

Figure 8.42 shows the synchronous and induction circles corresponding to the two modes of operation. A current phasor is drawn for each condition when the machine is just providing its own mechanical loss. To produce useful mechanical power would necessitate the current having a larger in-phase component. The value of E_f has been chosen so that the machine operates at unity power-factor on full load when

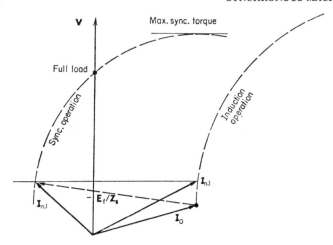

FIG. 8.42. Current loci for synchronous and induction motor operation.

running synchronously. By controlling the excitation manually or automatically, the power factor can be adjusted as desired at any load. When the motor reaches its maximum torque as a synchronous machine, it pulls out of synchronism, but continues to run as an induction motor, with a superimposed alternating torque if the excitation is left on. When the load falls, the machine will resynchronise in a similar manner to that following the normal starting process as an induction motor.

For an ordinary induction motor, the value of I_m is kept as small as possible to ensure a good power factor. The small air gap required is a disadvantage for synchronous-motor operation because the field m.m.f. will then be relatively small. This means that the armature m.m.f. tends to predominate and govern the performance, bringing about excessive changes of power factor as the load changes. With a larger air gap, the required field m.m.f. must be increased, but the armature m.m.f. is unchanged and the effect of armature reaction is reduced. The power factor as an induction motor is impaired by this

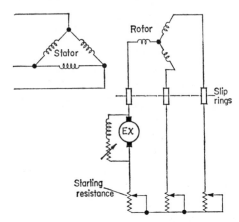

FIG. 8.43. Typical circuit for synchronous induction motor.

change, but since the induction characteristics are required only for starting and perhaps for excessive overloads, this fault is not serious. The larger air gap and the increase in E_f/Z_s also gives rise to a higher peak synchronous torque, see eqn. (8.7). The machine must of course be somewhat larger, to accommodate the extra field ampere-turns.

A typical connection diagram is shown in Fig. 8.43. The exciter may be left permanently in series with one of the leads as indicated, providing the induced voltage at standstill is not too high.

Synchronous induction motors are very often installed along with other induction motors so that they may improve the overall power factor of the plant. Their leading kVAr capacity is designed to offset the lagging kVAr demand from the induction motors. They have been made for ratings up to 30,000 kW.

EXAMPLE E.8.8

The induction motor for which data is given in Examples E.7.4 and E.7.5 is to be run as a synchronous motor at unity power-factor and rated primary current. The delta-connected rotor is supplied with d.c. across two slip rings only, so that one phase will carry 2/3 of the total d.c. and the other two phases, which are in series and connected across the first phase, carry 1/3 of the total d.c. Note that the stator/rotor turns ratio is 3/1.

Estimate, utilising the solution of Example, E.7.5, p. 425:

(a) the useful mechanical output, neglecting the losses associated with the provision of d.c. excitation,

(b) the total direct current required and the exciter voltage,

(c) the ratio of d.c. to normal a.c. in the heavily loaded rotor phase,

(d) the approximate ratio of maximum induction torque to maximum synchronous torque; i.e. pull-out torque.

The accompanying figure shows the synchronous circle and construction.

(a) Mechanical output $= \sqrt{3}VI_a \cos\varphi -$ stator copper loss $-$ n.l. loss

$$= \sqrt{3} \times 6600 \times 120 - 3 \times 120^2 \times 1\cdot31 - 15000 - 23700$$

$$= 1274 \text{ kW} = \underline{1710 \text{ horse power.}}$$

(b) The rotor current in primary a.c. terms (E_t/Z_s) by measurement $= 123$ A.

Actual d.c. in heavily loaded phase $= 123 \times 3 \times \sqrt{2} = 523$ A.

Total d.c. required $= 523 \times 3/2 = \underline{785 \text{ A.}}$

Exciter voltage $= 523 \times 1\cdot31 \times (1/3)^2 = 76$ V.

The d.c. excitation power is therefore $76 \times 785 = 59\cdot5$ kW. If this was provided by a mechanically-coupled exciter, this power, together with the total exciter losses would have to be deducted from the mechanical output calculated for (a).

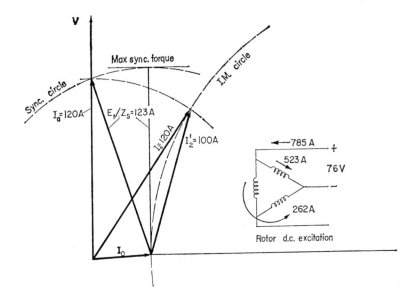

(c) Normal rotor a.c. = 100×3 = 300 A. Ratio = 523/300 = 1·75.

(d) Maximum synchronous torque $\sqrt{3} \times 6600 \times 123$ A = 1410 syn. kW.
   Ratio = 1740/1410 = 1·24.

Note that with normal stator current and heating, the output would be improved over the induction-motor rating, but only at the cost of prohibitive rotor heating due to the 75% current overload, in one phase, so in practice the rating would have to be reduced. Also a larger air gap would be used to improve the synchronous characteristics and the extra rotor heating would impose further limitations on the rating.

Brushless Synchronous-motor and Inverter Drive Systems

This section is not concerned with the brushless excitation of synchronous machines via a.c. exciter and diodes, Fig. 8.3b, but mainly with machines based on the same principle explained on Fig. 6.54 when discussing commutatorless d.c. motors. When the arrangement is in accordance with Fig. 6.54f, with the field on the rotor, it becomes a matter of definition as to whether the machine is regarded as a d.c. type with a static commutator, or a variable-frequency synchronous machine with the frequency controlled by switching, activated from shaft position. Even at the stage of considering Figs. 3.27 and 3.28, the only obvious difference between the synchronous and d.c. machines was the insertion of a rectifier system in the armature winding. The original idea, which dates back even before the days of transistors, was to try and get d.c.-machine characteristics without the penalty of a commutator and brush system, and many small machines have since been built with this intention in mind. The ostensible attempt in Fig. 6.54 to maintain field and armature m.m.f.s in quadrature, on the average, by using a shaft position detector, is clearly performing the function of a mechanical commutator with the same number of segments as of bilateral switches. In this sense the arrangement could with reason be called a brushless d.c. machine. However, once the power-electronic equipment is provided, it is a simple matter to incorporate control of this torque angle.[25,26] In this sense it could be considered as a variable-frequency synchronous machine with the possibility of load-angle control i.e., angle between direct-axis m.m.f. and resultant m.m.f. (between E_f and V) or the angle between any two m.m.f. axes. The sinusoidal variation of this angle with load, as defined by eqn. (8.7), would no longer apply and the speed would no longer be constant. The analysis based on variable frequency synchronous-machine theory offers a more straightforward explanation in detail of the machine behaviour. This is especially so when considering the differences in performance between round-rotors and salient-pole rotors.

If the field winding is d.c. excited, the machine is not really brushless, but permanent magnets could be used, p. 612, or if the supply is from an inverter with appreciable harmonics in the waveform, see Fig. 7.31, it is possible to provide field-winding excitation by induction. The harmonic fields are not stationary with respect to the field winding like the fundamental rotating field. Consequently, if the field winding is short-circuited by diodes, the induced field voltages from the harmonics will be rectified and produce a unidirectional field current; the harmonics, which would otherwise cause additional rotor losses, are harnessed to provide the excitation. [27]

There are many variations of the inverter-supplied brushless synchronous-motor and quite large ratings have been constructed, [26] with the supply from the 3-phase mains through a rectifier and d.c. link. Special circuit arrangements are required to force the switch-off of the inverter thyristors in sequence (forced commutation). If the synchronous motor is overexcited so that it operates at leading power factor, the circuit voltages in the d.c. link, which are derived from the a.c. supply, are in such a sense, relative to the thyristor current, that they apply, in sequence, a reversing voltage across the appropriate thyristors which switch off naturally (natural commutation). This leads to considerable simplification in the power-electronic circuits. Control of speed is invested in the voltage of the d.c. link, as for a d.c. machine; the armature–coil frequency being automatically adjusted to match the speed (as on the d.c. machine) by position control of the switching.

Perhaps it should be pointed out here, that inverter drives can impart other characteristics as a result of the control facilities available, even on standard synchronous motors. By controlling the excitation so that the power-factor is held constant, instead of allowing it to vary as assumed in the derivation of eqn. (8.7), the torque/load-angle variation becomes approximately tangential instead of sinusoidal. A wide range of speed control with minimum inverter rating becomes possible if the power factor is held at unity. The unusual torque characteristic permits a faster speed response.[14]

These and other possible developments resulting from the control, power-electronic and microprocessor revolutions mean that conventional theory must not be taken as the final word in the delineation of machine types. This is not to say that conventional theory is wrong, since the vast majority of machines will continue to be built on this basis, but that it must be adapted to allow for these new, somewhat special developments which have taken place and will continue to do so.

Single-phase Synchronous Machines

Small-power single-phase synchronous drives can be provided utilising either reluctance torque or hysteresis torque.[18] Reluctance torque has already been discussed briefly in Sections 1.3, 3.7 and 8.3. Hysteresis torque on a rotor arises because the change of induced magnetism B_i lags the change of magnetising force H, giving rise to a constant torque angle, as already described in Section 2.1. In either case, a rotating field must be provided by a phase-splitting circuit as for the single-phase induction motor. The rotor of a hysteresis motor is a smooth steel cylinder having magnetic properties such that the hysteresis loss is high. The torque between the stator field and the induced rotor field is present at all speeds from zero up to synchronous speed and then the rotor gives behaviour like a permanent magnet machine, see Appendix A, with adaptable load angle. The motor is thus self-starting.

Single-phase generators of conventional construction are occasionally used in small sizes but as discussed on p. 509 the damping windings must also be designed to carry the currents induced by the backwards component of the pulsating stator field. Interference from this component on the flux waveform is thereby minimised.

Another special case of single-phase generator is the *inductor alternator* used to provide high-frequency supplies in the range 1000 to 10,000 Hz for use in induction furnaces. Here, all the windings are on the stator including the d.c. excitation coils and a homopolar construction (see p. 20) is sometimes employed. The rotor construction is different from that shown on Fig. 1.12a, being made of laminated steel and having a large number of teeth but no winding. When the rotor is stationary, the air-gap flux density in any position is steady, though it has a spatial variation from maximum to minimum over a tooth pitch. When the rotor is driven, the permeance in any particular spatial position is time variant and the local flux pulsations induce high-frequency voltages in suitably placed a.c. stator coils. Heteropolar types of high-frequency generator, which employ a rather different form of rotor and stator slotting are also used. [21]

Stepper Motors

These motors,[18,28,29] important in the machine-tool and computer industries and other position-control applications, are not synchronous motors in the conventional sense. They have reluctance-type rotors and toothed stators provided with coils, often 3-phase or 4-phase. Over a rotor-slot pitch, the rotor can occupy several stable positions depending on which of the stator coils are excited. The supply is pulsed d.c.

Figure 8.44 shows a 2-pole, 3-phase arrangement but a 4-pole construction (with twice the number of coils and teeth) is more common. The positions of the stator and rotor are shown for the three different slot orientations. Clearly there are a total of 12 stable positions per revolution, so that it can be moved in 30° steps, any desired number of

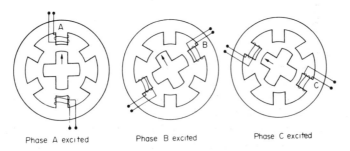

Phase A excited   Phase B excited   Phase C excited

FIG. 8.44. Stepper motor.

547

times and at a speed depending on the pulse frequency. The system lends itself to simple digital control, though in practice there are design problems to be overcome, mostly concerned with providing adequate damping.

The figure is a very simplified version of the many configurations now available. For example, round rotors with permanent magnets are sometimes used and the stator poles may have multiple slotting to reduce the angular movement per step. Another way of achieving this is to have multiple stack rotors and stators, with a small angular displacement between the stators so that the angle per step can be reduced to that of a single stack, divided by the number of stacks connected in parallel mechanically. The maximum number of steps per revolution can be designed to be as large as 200.

Switched Reluctance (SR) Motors[29,36]

The magnetic geometry of these machines is virtually the same as the Stepper Motors, but in addition, there is a rotor-position detector which governs the switching sequence of the d.c. pulsed, stator-pole windings. If this switching is in accordance with a fixed rotor position, the behaviour is similar to the brushless d.c. machine represented in Fig. 6.54f, though the rotor poles are not excited, nor need they be permanent magnets. The behaviour is also determined by the supply and circuit impedance and the switching angle can be manipulated with the supply, to give a variety of characteristics. The motor is structurally very simple, as are the power electronic circuits. The complication is in the control of switching and supply but this is readily achieved with modern electronic techniques, to produce an integrated, flexible, overall design. Noise and torque pulsations may give rise to problems in certain circumstances, as indeed occur with some inverter drives.

CHAPTER 9

ALTERNATING-CURRENT COMMUTATOR MACHINES: LABORATORY MACHINES

9.1 VOLTAGE, FREQUENCY AND SPEED RELATIONSHIPS

Commutator Action

Figure 9.1 is a schematic drawing of a d.c. machine. The commutator brushes are shown, for convenience, making direct contact with the armature coils. The coils are assumed to be moving at constant speed in a steady flux. The coil e.m.f.s in either top- or bottom-half sections are unidirectional as indicated by the arrows. With the brushes on the quadrature axis, the e.m.f.s are additive between the brushes. A d.c. potential difference arises across the brushes. Although individual coils are continually changing their position, the brushes, being stationary with respect to the poles, are always related to coils in a particular region of the magnetic field.

If the brushes are now rotated clockwise say, to the direct axis, through an angle of 90°, the e.m.f.s between brushes will cancel so that the brush voltage is zero. Another 90° of brush rotation causes a potential reversal at the brushes, full reverse voltage appearing across them. A further 90° and the brush voltage falls to zero again and if a complete revolution of the brushgear is made, the brush polarity and voltage will return to the original condition. A voltmeter connected across the brushes would have recorded one cycle of alternation. Therefore, if the brushes were to be rotated at n revolutions per second,

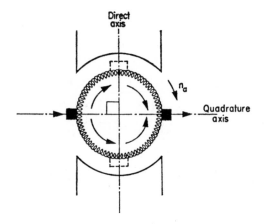

FIG. 9.1. Commutator action.

an a.c. voltage of frequency $f = n$ cycles/sec ($f = pn$ cycles/sec for a 2p-pole machine), would appear across them. The amplitude of this voltage would correspond to the maximum d.c. voltage available with the brushes fixed in the quadrature axis, see eqn. (5.9), p. 250.

The same behaviour of brush voltage would be observed if the brushes were stationary and the field instead were rotated *counter-clockwise* at the same speed. *Relative to the field system*, the "movement" of the brushgear would still be clockwise. It is the relative brush/field speed which determines the frequency at the brushes. For example, if the brushgear and field were rotating at the same speed and in the same direction, the brush voltage would be d.c. again. Note, although the speed of the armature n_a has been assumed to remain constant, the actual value of this constant speed does not affect the fundamental *frequency* of the brush voltage, though of course, the conductors must cut the flux to produce an e.m.f. However, the *magnitude* of this e.m.f. *does* depend on the relative conductor/field speed.

In a practical case, the brushgear is nearly always fixed in space and will be assumed so in the following discussion. Under these conditions, it is the *speed of the field in space n_f* which determines the *brush frequency.* The field can readily be made to "travel" around the periphery by using polyphase windings supplied with polyphase currents.

Slip-ring Action

Permanent connections to particular conductors brought out to slip rings as indicated in Fig. 9.2 will experience the same potential variations as the conductors to which they are attached. As before, the conductor voltages will be proportional to the relative conductor/field speed. The conductor and slip-ring frequency will also be proportional to this relative speed.

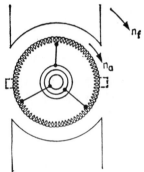

FIG. 9.2. Slip-ring action.

Assuming the brushes are stationary and the flux per pole is constant, the relationships just developed can be summarised with reference to Fig. 9.2 as follows.

Conductor and slip-ring frequency $= p(n_a - n_f)$ for a $2p$-pole machine.

Brush frequency $= pn_f$ for a $2p$-pole machine.

All induced voltages proportional to $(n_a - n_f)$.

These statements can be checked for example against the known behaviour of the d.c. machine by substituting $n_f = 0$. For this special

551

case, the brush frequency is always zero, independent of the armature rotational speed. However, in the general case, the commutator acts as a frequency changer from $p(n_a - n_f)$ to pn_f or vice versa. Numerous types of variable speed a.c. machines are based on this frequency changing action, and in the next section, the more common types will be described briefly. Developments in Power-electronic circuits have diminished the importance of many of these machines—apart from the single-phase (universal) series commutator-motor.

It should be noted that if the flux per pole is not constant, i.e. if the amplitude of the flux wave varies with time as in the case of the single-phase pulsating field, there will be a transformer e.m.f. at the brushes in addition to the motional e.m.f. Transformer e.m.f.s arise in all machines during transient operation.

The a.c. commutator machines to be described do not in general have a salient-pole construction, but follow the practice on induction machines, with which they have close affinity.

9.2 PRACTICAL EXAMPLES OF COMMUTATOR MACHINES

The Rotary Converter

In normal operation this device is an a.c./d.c. converter though it can be made to operate satisfactorily as a d.c./a.c. "inverter". The machine is basically the same as the one shown in Fig. 9.2 with the

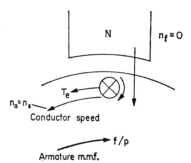

FIG. 9.3. Rotary converter.

552

field d.c. excited, $n_f = 0$. An a.c. voltage is applied to, or extracted from the slip rings, which are connected to symmetrical tappings round the armature winding, usually six per pole pair to give six-phase working. A rotating field, moving at a speed $n_s = f/p$ revs/sec relative to the armature conductors, is established. If by some means the armature can be brought up to the same value of speed but in the opposite direction relative to space, i.e. $n_a = -f/p$, then:

$$\begin{bmatrix} \text{speed of armature} \\ \text{field in space} \end{bmatrix} = \begin{bmatrix} \text{speed of armature field} \\ \text{relative to conductors} \end{bmatrix}$$

$$+ \begin{bmatrix} \text{speed of conductors} \\ \text{relative to space} \end{bmatrix}$$

$$= f/p + (-f/p)$$

$$= 0$$

i.e. the armature field is stationary in space and will thus lock in synchronously with the stator field if this is d.c. excited, and a uniform torque will be produced. Figure 9.3 shows the electromagnetic picture. To give a driving torque in the same direction as the rotation, the conductor current must be flowing into the paper, but the induced e.m.f. is in the opposite sense. The machine is acting as an *inverted* synchronous motor with the stator and rotor windings interchanged from the normal arrangement. However, the field is stationary in space, so a d.c. voltage and load current can be taken from the commutator brushes. This is a generated current and will be in opposition to the motoring current shown in Fig. 9.3. When these are super-imposed, the resultant current and therefore the copper loss, will be relatively low.

The rotary converter as such is rarely made nowadays, but the Autodyne, which is the same machine basically, has been developed as a high-gain rotary power-amplifier which, supplied from a.c., gives a variable-voltage d.c. output with fast response. It is neces-sary, however, to have additional field windings and a built-in feedback loop to achieve this performance.

The Commutator Frequency Changer

This again is similar to the device shown in Fig. 9.2 but there is no winding in the stator. The stator iron is present merely to provide a low-reluctance path for the working flux. Polyphase alternating current is supplied to the slip rings and establishes an armature field which rotates at speed f/p *relative to the conductors*. This speed is not affected by physical rotation of the armature because the slip-ring

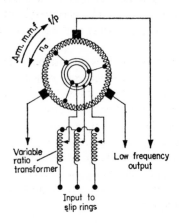

FIG. 9.4. Commutator frequency changer.

brushes remain connected to the same conductors irrespective of rotor movement, unlike the brushes on a commutator. However, when the armature is rotated, the *speed of the field in space, n_f,* is affected. Rotation is normally in the opposite sense to the direction of the rotating field at a speed n_a. The value of the field speed in a clockwise sense is therefore $n_f = (f/p) - n_a$; see Fig. 9.4. Consequently, the commutator-brush frequency is $pn_f = f - pn_a$ Hz. If, for example, it is desired that this should have a value s times the slip-ring frequency, then n_a must be adjusted to $(1-s)f/p$. This immediately suggests a means of obtaining a slip-frequency source for control of induction-

554

motor speed and power factor. A three-phase commutator frequency changer is directly coupled to the shaft of an induction motor, its slip rings are supplied from the same a.c. source and the winding is arranged for the same number of poles as the motor. The speed is automatically $(1-s)f/p$ and so the output frequency at the brushes is sf, no matter what the speed of the induction motor. Voltage control can be achieved by supplying the frequency-changer slip rings through a three-phase tapped transformer. For phase control, the commutator brushes must be "rocked" round to a new spatial position, such that the space wave of flux travelling around the periphery, begins to traverse the winding section between the brushes at the desired instant in the supply-voltage cycle.

The device can also be used as an independent source of variable frequency, usually a low frequency, in which case it is driven by a variable-speed motor, just large enough to provide the mechanical losses. The armature speed affects the brush frequency directly, since this is equal to $f - pn_a$ Hz. The relative field/conductor speed, as already explained, is f/p so that the generated voltage is independent of the rotational speed n_a. This permits the generation of low frequency with adequate output voltage. The induction frequency changer described on p. 453, is at a disadvantage for very low frequencies because the output voltage falls with frequency.

Schrage or Rotor-fed Motor

The commutator frequency changer offers even better possibilities in that it can be developed into a self-contained variable-speed motor, generating its own slip-frequency voltage for injection to the secondary windings, see Fig. 7.38. To do this the stator is provided with secondary windings, which will experience the flux changes associated with the speed of the flux in space, i.e. at a speed $n_f = n_s - n_a = sn_s$ in a clockwise sense. The electromagnetic picture of Fig. 9.5 shows a North pole moving at this speed and inducing a current in the stator winding, such that the torque is clockwise on the stator conductor. Since this is not free to move, the reaction torque on the rotor causes counter-clockwise rotation. A little thought will reveal that this is the action of an induction motor, but with the primary winding on the rotor, i.e. "inverted" operation.

If now this secondary winding, which also experiences flux changes at speed $n_f = sn_s$, is connected to the commutator brushes, Fig. 9.5b, it receives an injected e.m.f. of the correct slip frequency. With the

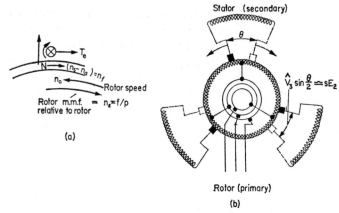

FIG. 9.5. Schrage motor.

secondary star-connected, and one end only of each phase taken to the commutator, a fixed voltage would be obtained and could be phase shifted by brush rocking to give power-factor improvement. More usefully, if additional brushes are provided as indicated, with controllable movement of the brush pairs in contrary directions and both secondary phase ends are connected, the magnitude of the injected voltage can be changed by altering the brush separation θ, see eqn. (5.10), to give speed control. Crossing over the brushes reverses the phase of the injected voltage, giving super-synchronous speeds. Power-factor control can be obtained by moving all the brushes together round the commutator away from the "neutral" position, where the injected and induced secondary voltages are in time phase. The expression for speed follows from eqn. (7.13), p. 436, as:

$$n = n_s(1 - s) \simeq n_s\left(1 - \frac{\hat{V}_3 \sin \theta/2}{E_2}\right),$$

where $\hat{V}_3$ is the maximum available commutator voltage ($\theta = 180°$ electrical). E_2 is the induced secondary voltage at standstill. The last term of eqn. (7.13) which is not included here, leads to a slight fall of speed with load, Fig. 7.37.

As a practical point, it is not possible, for commutation reasons, to tap directly on to the primary winding for the commutator voltage, so a lower voltage winding (fewer turns) is used, this being magnetically coupled to the primary winding by being placed in the same rotor slots. Schrage motors can only be made for ratings up to a few hundred kW. Restrictions are imposed, because of commutation problems associated with movable brushgear and because the primary power must be supplied through sliding contacts at the slip rings, voltages being limited to about 660 V.

Stator-fed Motors

These are variable-speed induction motors having the rotor winding tapped to a commutator which converts the mains-frequency injected voltage to slip frequency for the rotor coils. The speed of the field in space corresponds to both the brush frequency and the stator frequency since the brushes and the stator winding are both connected to the same supply. A step-down variable-ratio transformer, which may take the form of an induction regulator (Fig. 7.3), is usually interposed between the line and the commutator to give voltage and/or phase control. The stator/commutator connection may be either in parallel, Fig. 9.6a, giving substantially constant flux and shunt-type speed/torque characteristics, or in series, Fig. 9.6b, giving a machine flux dependent on load and hence producing series-type speed/torque characteristics. Ratings may be in excess of 2000 kW, with stator voltages as for induction motors. For use of a thyristor commutator for this machine, see Reference 4.

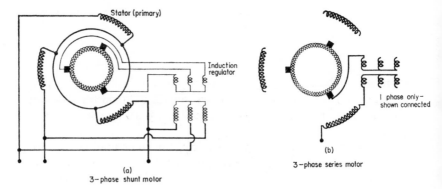

Induction
regulator

I phase only−
shown connected

(b)
3−phase series motor

(a)
3−phase shunt motor

FIG. 9.6. Stator-fed motors.

Scherbius Machine

This has a rotor winding provided only with a commutator, as on a d.c. machine and the stator-fed motors. It is driven at any convenient speed. The stator is excited from a low-frequency source and hence produces a field rotating in space at $n_f = f/p$. The brush frequency is $pn_f = f$ Hz, i.e. the same as the input frequency. The field may be self, or separately excited. The machine is used as a high-power slip-frequency source for speed control of large induction motors.

Leblanc Phase Advancer

This has the same type of rotor as the Scherbius machine but the stator core is unwound. The excitation is at low frequency through the brushes, and is usually obtained from the slip rings of an induction machine for which the power factor is to be improved. If p is the number of pole pairs on the advancer, the flux established must rotate in space at a speed sf/p. Irrespective of the rotor speed and the coil movement, the brushes provide ingress to a delta-connected circuit in which currents flow with a fixed spatial orientation, Fig. 9.8a. The rotor speed does, however, affect the induced e.m.f., since the voltage is

558

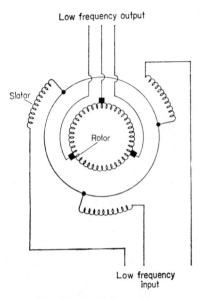

FIG. 9.7. Scherbius machine.

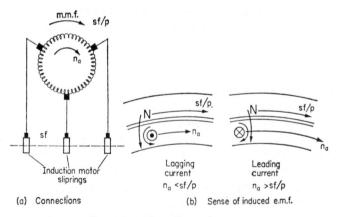

(a) Connections   (b)   Sense of induced e.m.f.

FIG. 9.8. Leblanc phase advancer.

559

determined by the relative conductor/field movement. Increasing the speed from zero in the same direction as the rotating field, Fig. 9.8b, causes the e.m.f. to fall from a value corresponding to the behaviour of the device as a stationary 3-phase inductor, to zero, when $n_a = sf/p$. At this speed, due to the absence of relative flux/conductor movement, the only voltage across the brushes is due to IR drop. A further increase of rotor speed, causing the conductors to move faster than the field and cut it in the reverse sense, Fig. 9.8c, will reverse the phase of the induced e.m.f. for the same current. The device now behaves as a negative inductance taking a current leading the voltage drop. Such a voltage, in series with the secondary winding of the induction motor, to which the brushes are connected, will improve the motor power-factor. The magnitude of the voltage can be controlled by the speed of the advancer and is of course dependent on the flux produced by the brush currents, i.e. on the motor secondary current. Stator windings have been added in certain practical cases to modify the characteristic.

Single-phase Series Motor

Many ingenious forms of single-phase commutator machines have been developed. Only a few have survived the test of time, however, and the series motor is by far the most important of these. The basic circuit is the same as the d.c. series motor, Fig. 6.39. The stator construction is different, being similar to the induction motor or the amplidyne. It is laminated to reduce iron losses and permit the flux to be in time-phase with the field current, as eddy currents in the iron would other-wise delay the flux rise and fall. It will be remembered that for maximum torque, the stator and rotor m.m.f.s should be in space quadrature and this is easily arranged by placing the brushes at 90 electrical degrees to the stator field-axis. The torque is then a direct function of the product of these m.m.f.s. There is a further condition to observe which becomes more obvious when the currents are alternating; viz. to get maximum torque, it is necessary that the m.m.f.s should be in time phase as can easily be seen by considering the product of two sine waves which are not in time phase; see Fig. 3.8. Connecting the armature and field windings in series ensures that the m.m.f.s rise and fall together, but in addition, eddy-current effects must be minimised by lamination of the iron paths.

As in all single-phase circuits, the power pulsates at twice the supply frequency and this manifests itself as a torque pulsation[33]. Because the torque is a function of (current)^2, this being insensitive to polarity, the series machine can be used either on d.c. or a.c. (but with lower efficiency) and is known as the *universal motor* in the smaller sizes where it is very popular for driving small machine tools and domestic equipment. The construction is more expensive than the d.c. series motor and many special features, including compensating windings and carefully designed inter-pole circuits are necessary in motors with a rating of a few hundred kW, such as are required for a.c. traction drives. Operation at reduced frequency, $16\frac{2}{3}$ Hz or 25 Hz, is usual for such applications, to ease the commutation difficulties.

A corresponding development of the d.c. shunt motor for a.c. use has not led to the same degree of success. The current in the shunt excited field would lag the armature voltage and current by an angle approaching 90° and little net torque would be produced. This is because the field inductance is much higher relatively, than that of the armature. Time-phase correction could be, and has been applied; for example, by means of a series capacitor, but in practice, additional equipment is necessary to produce a single-phase commutator motor with shunt characteristics and speed control.

Single-phase Repulsion Motor

One form is shown schematically in Fig. 9.9. The brushes are short circuited and the a.c. supply is taken only to the field. The brush axis must be displaced from the field axis in order that the induced e.m.f. and current in the rotor winding will give rise to an m.m.f. having a component in quadrature with the field m.m.f. so that a torque is produced. Limited speed control can be effected by angular brush shift. The motor is not made in very large sizes because of commutation difficulties but the principle is utilised on repulsion-start, single-phase induction motors.

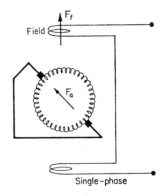

FIG. 9.9. Single-phase repulsion motor.

9.3 LABORATORY MACHINES AND EQUIPMENT

Generalised Laboratory Machines

The application of tensor analysis to the circuit equations of electrical machines has not only given an impetus to machines studies, but has also high-lighted the underlying similarity in their basic electromagnetic action. Kron's mathematical concept of the "primitive machine" and its build-up to any particular kind of machine has been matched practically by the production of a "generalised machine". Several variations of this are now available commercially.

The rotating armature arrangement is similar to that of the commutator frequency changer. Both stator and rotor windings can be connected for d.c., single-phase or polyphase excitation. The commutator is provided with a multiplicity of brushes to give d.c., metadyne or a.c. commutator machine arrangements, and in some cases one set of brushes can be rotated at high speed by means of a separate driving motor. The slip rings can be connected to various tappings at the non-commutator end of the armature. Numerous other facilities are available for measuring temperature, speed, load angle and torque, in

562

both steady and transient states, and induced voltage measurements in various parts of the magnetic circuit by means of search wires and coils.

A salient-pole stator or rotor is necessary to demonstrate saliency effects satisfactorily. However, with the uniform air-gap arrangement, most other kinds of machine described in this book, together with many other variants, can be achieved on this one device. Although the performance of a machine designed specifically for one mode of operation cannot be represented accurately, the generalised machine offers a convincing demonstration of the underlying unity in machine theory and operation. The degree of inaccuracy is quite small for a.c. commutator machines, relatively small for induction motors and would be quite appreciable in some respects for salient-pole machines.

The facilities available vary with the different makes of machine and Fig. 9.10 shows only one type. In this case, there is a small d.c.

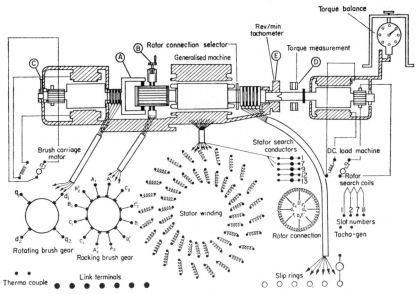

FIG. 9.10. Schematic arrangement of Generalised Laboratory Machine. (By courtesy of Mawdsley's Ltd.)

motor to drive the rotating brush carriage and a large, coupled d.c. load machine (dynamometer) which can be used as a main driving motor or a loading generator. The dynamometer stator is free to move in bearings against the restraint of a spring balance to register the torque reaction. The generalised machine stator has a 4-pole, 2-layer winding with each coil brought out to terminals at which the various phase, compensating winding and field coil connections are made. A 2-pole winding is readily formed although the coils which span 90° mechanical, will then only be half full-pitch. It is even possible to go up to 14 poles in demonstrating certain new types of pole-changing windings, see p. 431.

The rotor has a symmetrical 4-pole full-pitch lap winding brought out to the commutator at one end and tapped at the other end for various phase connections to the slip rings. With only two diametral commutator brushes down, a 2-pole commutator winding is formed with half pitch coils.

Three sets of commutator brushgear give:

(a) a symmetrical 4-brush rotatable carriage;
(b) contra-rotating brushgear for Schrage motor operation;
(c) search brushes with variable span.

Apart from commutator machine operation, either stator or rotor can be used as the primary, though the closed commutator winding imposes some restriction on the polyphase connections available on the rotor. The commutator and slip rings can be shorted out with a heavy-section brass ring mounted concentrically. This gives an approximation to a squirrel-cage rotor.

For transient-torque measurements, a special section of the shaft between the two main machines is magnetised at high frequency by a group of stationary coils. That portion of the torque which is not absorbed in the generalised machine inertia and in its mechanical loss when motoring, causes a torsional strain, giving rise to permeability variations, in this shaft section. Another group of stationary coils act as secondary windings in which a resultant voltage appears under load conditions having an average value which is nearly proportional to this torque.

The possibilities of demonstration and experiment are very considerable indeed. At an elementary level, the device can be arranged to show the general performance of most machines described in this book and other machines also. At a more advanced level, the two-coil concept described briefly in Section 3.8 can be checked for different winding arrangements by measuring the variation of mutual inductance with angular position and correlating this with torque and e.m.f. variations. A salient-pole member is necessary to demonstrate satisfactorily the production of reluctance torque due to self-inductance variations and some generalised machines provide for this.

At an even more advanced level, the various mathematical transformations involved in generalised machine analysis, as discussed in the next chapter, can be verified. The rotating brushgear is useful in this regard for the purpose of checking the transformation to d–q axes. By rotating the brushes at the same speed as that of the air-gap field in space, the slip-ring voltages appear as d.c. voltages at the commutator brushes. The 3-phase to 2-phase transformation is readily checked by suitable arrangement of the winding connections. The rotating brushgear also lends itself to various novel modes of operation and the verification of the propositions discussed in Section 9.1.

Alternative General-purpose Laboratory Equipment

The first general-purpose machine for use in teaching was brought out several years ago in the form of the Schrage motor, with facilities for running as a synchronous, induction or commutator machine and also in single-phase modes. Other students' sets include those with interchangeable rotors, machine assembly kits and special machines with some or all of the coils brought out to external terminals. Another system is to provide a group of normal machines with adequate facilities for investigating the natural characteristics of the main machine types particularly, by the methods described in the earlier chapters. This approach is well suited to the needs of the increasing number of students who, while not taking up the subject of electrical machines as a specialised study, nevertheless require a full knowledge of their potentiality for use in industrial drives and power systems, for example. This would necessitate a consideration not only of the machine itself, but also of the machine as an element in a larger system.

Figure 9.11 shows a teaching console designed by the author with the specific intention of meeting the requirements of this last-mentioned approach to laboratory courses. The machines, though rather smaller than is usual in traditional laboratory practice, are nevertheless large enough to satisfy the same criterion of giving satisfactory correlation of theory with practical measurement. One of the main advantages of this equipment is the possibility of running lecture and laboratory courses in step, with all students performing the same experiment at the same time. The console is completely self-contained, with transformers and machines; instruments, rheostats and stroboscope; speed and torque measurement, changeover switches, circuit breakers and fuses to give complete protection, and d.c. and a.c. voltage-control facilities. Simple plug-in wiring permits the rapid connection of any desired circuit configuration on a mimic diagram.

The 3-machine set consists of a d.c. dynamometer and a d.c. series/shunt/compound motor on which all the d.c. work can be conducted. The 3-phase a.c. machine can be used in synchronous and induction modes and even as a transformer, though specially designed, 3-coil tapped transformers are also provided. The machines are readily uncoupled and replaced by other types though this would normally be for specialised courses only.

An important advantage of using small machines is that most of the ancillary equipment is reduced in size, leading to economies in space and power requirements and giving virtually a fully equipped laboratory for one group of students at a cost appreciably less than the larger generalised laboratory machines. Possibilities of damage to the equipment by careless use are remote and would not in any case be very serious in replacement costs. This means that students can explore the behaviour of machines with considerable freedom of experiment. Overloads, electromechanical transients, pull-out, starting, stalling and reversal tests are readily performed. The equipment lends itself to the development of new experiments, for example with special thyristor supplies, again of low power. Thus, after verifying basic machine

Fig. 9.11. Electrical Machines Teaching Console. (By courtesy of Nickerson Electrical Controls Ltd., and Normand Electrical Company Ltd.)

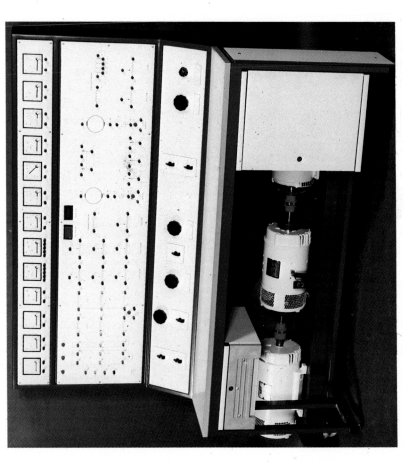

Fig. 9.12. Electrical Machines Teaching Console for 1500-watt machines. (By courtesy of Nickerson Electrical Controls Ltd., Manchester, M11.)

theory [preceded if necessary by training in the use of the various types of instrument in resistive and inductive circuits and in the use of the dynamometer for measuring the torque at the mechanical "terminals" (the coupling), p. 357], a systematic investigation might be undertaken to test the validity of the various machine equations developed. The machine parameters can be measured by the methods described in the Testing Sections, 4.7, 6.9, 7.4 and 8.5; see also Appendix C. This will lead to a physical understanding of the machine equations which, as a useful exercise, could profitably be listed, as already done in part for the d.c. machine; pp. 318 and 319. The behaviour of machines when considered as elements in a larger system forms an interesting laboratory project. If several consoles are interconnected, they can represent a miniature power system for load flow, stability and fault studies.

Thus the equipment permits considerable flexibility without the complications sometimes associated with a single, general-purpose machine. Although the emphasis is on the machine terminal characteristics and equations, these are of more general import to the majority of students than a study of specialised machine-design features. In conjunction with a closely integrated lecture course,[30] the laboratory work can therefore be made interesting and instructive as theory and practice are immediately interrelated.

Alternative equipment is available using 1500-watt machines, but designed with the same basic philosophy in mind as far as practicable. Figure 9.12 shows the console, of somewhat similar size but provided with the larger machine set and loading equipment and digital read-out of speed and torque.

CHAPTER 10

AN INTRODUCTION TO GENERALISED CIRCUIT THEORY OF ELECTRICAL MACHINES

10.1 THE "PRIMITIVE" MACHINE IN DIRECT- AND QUADRATURE-AXIS TERMS

The Circuit Approach

For each winding of an electrical machine, it is possible to write down a circuit equation in the form $V = \Sigma Z . I$, balancing the terminal voltage against the sum of the winding resistance drop and the self-induced and mutually induced voltages. Given the terminal voltages, the simultaneous solution of all the winding equations will yield the currents, from which the torque and the component voltages can be derived. Apart from the standstill condition, the windings are in relative motion so that at least some of the inductances will vary with angular position. In general therefore, the equations have time-varying coefficients. In this form, the equations cannot be solved analytically by any routine procedure like the use of Laplace transforms, but, as already discussed in Section 3.8, it is possible to apply another mathematical transformation which replaces the time-varying inductances with related, but constant inductance-coefficients. In this way the generalised-circuit-theory approach permits a direct solution for all kinds of steady-state and transient electrical circuit conditions; each case starting from similar circuit equations, for all machine types. For transient mechanical conditions, other than for small oscillations and for relatively long mechanical time constants, a numerical digital-computer solution or an analogue-computer solution becomes necessary.

An important consequence of this method is that one so-called primitive machine, to be described shortly, can be used as a basis for deriving the performance of the large majority of electrical machine types and their control circuits by a series of routine mathematical manipulations. This unified circuit theory of electrical machines has been highly developed, and, quite obviously, such a powerful method cannot be satisfactorily explained without a detailed study. However, it might be useful, particularly for those students who are proceeding to work on the power-system and control-system fields, to give an outline of the method, to relate it to the classical theory based on equivalent circuits and their associated equations and phasor diagrams. It is unavoidable, in such a brief review, that certain expressions will have to be quoted without full proof. However, the notation and presentation generally is derived from the companion volume,[2] so if this is consulted, a more detailed explanation will be found. In the following sections an elementary knowledge of matrix algebra will help in understanding the arrangement of the equations and the few simple manipulations which are carried out.

As can be confirmed by reference to Section 3.8, the equation for coil 1 of a 2-coil system, expressed in terms of self and mutual inductances as for eqn. (3.10a) is:

$$v_1 = R_1 . i_1 + \frac{\mathrm{d}(L_{11} . i_1)}{\mathrm{d}t} + \frac{\mathrm{d}(M . i_2)}{\mathrm{d}t}$$

$$= R_1 . i_1 + L_{11} . \frac{\mathrm{d}i_1}{\mathrm{d}t} + M . \frac{\mathrm{d}i_2}{\mathrm{d}t} + \left[i_1 . \frac{\mathrm{d}L_{11}}{\mathrm{d}\theta} + i_2 . \frac{\mathrm{d}M}{\mathrm{d}\theta} \right] \frac{\mathrm{d}\theta}{\mathrm{d}t}. \quad (10.1)$$

This expression includes the resistance drop, the transformer voltages and the motional or speed voltages which, being associated with mechanical movement and power, are related to the torque produced. Figure 10.1 shows a typical variation of mutual inductance with angle. In passing, it might be mentioned that the aim of the a.c. machine designer is to produce a sinusoidal variation by suitable arrangement of

569

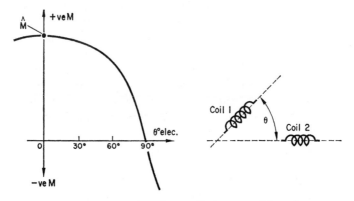

FIG. 10.1. Variation of mutual inductance with angle.

the magnetic-circuit geometry. Note that maximum positive mutual inductance occurs at $\theta = 0$ when the coils are in line and magnetising in the same sense. The inductance is zero at $\theta = 90°$ but here, $dM/d\theta$ has its maximum negative value. For a sinusoidal variation and with maximum mutual inductance occurring at $\theta = 0$, the inductance at any angle would be expressed as $M = \hat{M} \cos \theta$. In this case, the maximum value of $dM/d\theta$ would then be the same as the maximum mutual inductance $\hat{M}$. Note that when θ is used in the equations, it refers to electrical angles.

The D.C. Machine

The d.c. machine can be taken as an illustration, based on the information given and the equations derived in Chapter 6. A generalised derivation would be the usual procedure, starting from the "primitive" machine. This has a commutator-type rotor with two brush arms at right-angles, one of them being in line with the stator salient poles if present, or, in the case of a uniform air-gap machine, in line with the axis of one of the stator windings. On the basis of work done in Chapter 6, it will be realised that if commutation phenonema are neglected, the

570

effects of currents through a brush axis via the armature winding, give rise to an m.m.f. along the brush axis. Hence these effects can be represented by a single, "pseudo-stationary" coil, so called because only the m.m.f. is really stationary. With two brush axes in quadrature as on the amplidyne of Fig. 6.16b, there must, of course, be two such rotor coils and these will be designated d and q. On the stator of the primitive machine, there are in general D and Q coils in line with the d and q coils respectively. If on the actual machine, there are multiple

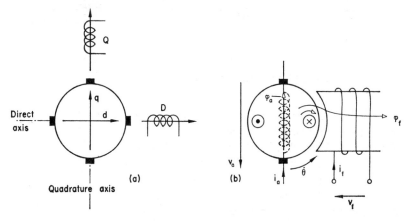

Fig. 10.2. (a) Primitive d–q machine; (b) simple d.c. machine.

windings on rotor and stator, each must be represented by its own set of direct- and quadrature-axis coils on the primitive machine. In the case of the stator, where there is a winding which is actually on the direct or quadrature axis, a single D or Q coil will suffice for this winding. On the primitive machine, the windings are not interconnected; an operation which is performed later in the solution if necessary. Figure 10.2a shows such a machine for the case where there is only one rotor winding and one stator D and one stator Q coil.

Figure 10.2b shows a simple d.c. machine which will be used as an introduction to the method. On the basis of the above, the single field winding would be designated D and with only one brush axis in quadrature, the armature would be represented by a q coil. For the moment, suffices a and f will be retained for armature and field windings and these in turn will correspond with, or be related to, coil 2 and coil 1 respectively in eqn. (10.1). Figure 10.2b also shows the positive conventions to be used and these have been chosen to give the most convenient form of the equations. They are summarised below.

Positive i_f and ϕ_f magnetise in the normal positive X direction.

Positive i_a and ϕ_a magnetise in the normal positive Y direction.

Positive angles are measured in the normal positive anticlockwise direction.

Positive speed, in electrical radians per second, $d\theta/dt$ $(\dot{\theta})$, will also be in the same direction as positive θ.

Positive electromagnetic torque will turn out to be in the same direction as positive θ, as in motoring operation.

Terminal voltages will be taken as positive if they would drive positive currents and all other voltages will be taken as if the element concerned was a positive sink, i.e. the voltage rise across the element opposes positive current-flow or current-change; see Section 3.2.

As shown by Fig. 5.22, for example, the voltage developed at the quadrature brushes of a d.c. machine corresponds to the maximum voltage occurring between diametral slip-ring tappings. This voltage is due to a series of terms for each coil as shown by eqn. (10.1) but because of the continuous changes of connection resulting from brush/commutator action, some of these terms give a zero resultant as detected at the brushes. The overall magnetising effect of the d.c. armature winding can be represented by a coil fixed in space at $\theta = 90°$, along the brush axis. The self inductance L_a of this "pseudo-stationary" coil is not time variant and can in fact be measured at the brush terminals. Thus there is an armature-voltage component $L_a \cdot di_a/dt$. For the third term of eqn. (10.1), there is no corresponding $M \cdot di_f/dt$ voltage because any transformer-type voltage in the armature coils due to field-flux changes, would cancel between brushes on the quadrature axis. For the speed-voltage terms, since L_a is not time variant, there is no voltage due to self inductance changes and the final term leads to the summed effect of all the $i_f . dM/dt$ induced e.m.f.s in individual armature coils.

From Fig. 10.1 it can be seen that $dM/d\theta$ has a maximum negative value at $\theta = 90°$. An alternative explanation of this negative sign follows from consideration of the armature-coil e.m.f. sense. For positive speed and positive i_f, the direction of the conductor e.m.f. under the pole, as indicated on Fig. 10.2b, is seen to be supporting positive i_a, magnetising upwards, and hence this voltage is negative when considered as a back e.m.f. or as a voltage drop in the direction of current flow. For a d.c. machine, the maximum voltage between brushes occurs for a brush angle of 90° to the field axis and the overall effect of the $dM/d\theta$ terms for individual armature coils can be considered as due to the one pseudo-stationary coil at this angle. The value of this $dM/d\theta$ will be designated $-M_\phi$ and it has already been shown on p. 294 to have a magnitude of $(z_s/\pi) . (\phi/i_f)$. For a sinusoidal distribution of flux, this would in fact be equal to the mutual inductance between the field and the armature, if, for measurement purposes, the brush axis was aligned along the field axis. The speed voltage has previously been expressed as $E = k_\phi . \omega_m$, where ω_m here is the mechanical angular velocity. In the new terms, the speed voltage is, from $i_2 . (dM/d\theta) . (d\theta/dt)$, equal to $i_f . (-M_\phi) . (d\theta/dt)$, where $d\theta/dt$ is now the electrical angular velocity. M_ϕ is directly related to k_ϕ as explained on p. 294, so it is a simple matter to obtain its value and its variation with the saturation level, from an open-circuit test.

It is now possible to write down the armature-winding and field-winding circuit equations. There are no speed voltages or mutually induced voltages in the field winding because the armature circuit and its m.m.f. are in quadrature with the field, irrespective of rotor movement. In the equations, $\dot{\theta}$ represents $d\theta/dt$ and p is the symbol for d/dt. Hence:

$$v_a = (R_a + L_a . p) . i_a - M_\phi . \dot{\theta} . i_f$$

$$v_f = \qquad\qquad (R_f + L_f . p) . i_f.$$

It is a considerable advantage in the application of generalised theory if the techniques made available by the use of matrix algebra are employed. In matrix form, the equations become:

$$\left[\begin{array}{c} v_a \\ v_f \end{array}\right] = \left[\begin{array}{cc} R_a + L_a p & -M_\phi \cdot \dot{\theta} \\ & R_f + L_f p \end{array}\right] \left[\begin{array}{c} i_a \\ i_f \end{array}\right]$$

where the coefficients have been separated from the currents in an impedance array. The method of reading the individual equations can be deduced by comparing with the original form of these equations. The first term in any particular row of the impedance matrix is associated with the first term in the column of the current matrix. The second term in any particular row of the impedance matrix similarly is associated with the second term in the column of the current matrix and so on, if there are more than two rows and columns. The matrix equation will be written in bold type and is in the form:

$$\mathbf{v} = \mathbf{Zi,}$$

where each element here, is itself a matrix array. If the voltages are known and it is required to find the currents, the differential equations can be solved by algebraic manipulation using Laplace transforms, which express the current transforms $\bar{\mathbf{i}}$, in terms of the voltage transforms $\bar{\mathbf{v}}$ and an inverse transformation then gives the actual currents. In matrix form, the solution is expressed as:

$$\bar{\mathbf{i}} = \mathbf{Z}^{-1} \cdot \bar{\mathbf{v}},$$

where $\mathbf{Z}^{-1}$ is the inverse of the impedance transform matrix. For many practical problems however, as will be shown later on in the chapter, it is possible to obtain solutions without having to resort to Laplace transformation.

The impedance matrix can be divided into parts, separating the resistive terms into an **R** matrix, the inductance terms into an **L** matrix and the motional voltage terms (which may be related to the inductance

574

terms), into an array usually called the **G** matrix. Hence the voltage can be written:

$$\mathbf{v} = (\mathbf{R} + \mathbf{L}p + \mathbf{G}.\dot{\theta})\mathbf{i}. \tag{10.2}$$

Now power in matrix form is written down as $\mathbf{i}_t.\mathbf{v}$, where $\mathbf{i}_t$ is the transpose of $\mathbf{i}$, being written as a row matrix instead of a column matrix. Hence, for example, the terminal input power for the d.c. machine equations is:

$$
\begin{array}{|c|c|}\hline i_a & i_f \\\hline\end{array}
\begin{array}{|c|}\hline v_a \\\hline v_f \\\hline\end{array}
= i_a v_a + i_f v_f.
$$

Multiplying eqn. (10.2) by $\mathbf{i}_t$ gives:

$$\mathbf{i}_t.\mathbf{v} = \mathbf{i}_t\mathbf{R}.\mathbf{i} + \mathbf{i}_t.\mathbf{L}p.\mathbf{i} + \mathbf{i}_t.\mathbf{G}.\dot{\theta}.\mathbf{i}.$$

These components of the total power may be recognised as the copper loss, the rate of storing inductive energy and the mechanical power respectively. The torque can be obtained from the last term on dividing by the mechanical speed ω_m, which is equal to $\dot{\theta}/$pole pairs in mechanical radians per second. Therefore the torque is:

$$\mathbf{i}_t.\mathbf{G}.\mathbf{i} \times \text{pole pairs} \quad \text{or} \quad \mathbf{i}_t.\mathbf{G}.\mathbf{i} \text{ per pole pair}.$$

For the d.c. machine being considered, the **G** matrix consists of only one term, there being only one speed-voltage component. Forming the $\mathbf{i}_t.\mathbf{G}.\mathbf{i}$ product gives the torque per pole pair as:

$$
\begin{array}{|c|c|}\hline i_a & i_f \\\hline\end{array}
\begin{array}{|c|c|}\hline & -M_\phi \\\hline & \\\hline\end{array}
\begin{array}{|c|}\hline i_a \\\hline i_f \\\hline\end{array}
=
\begin{array}{|c|c|}\hline i_a & i_f \\\hline\end{array}
\begin{array}{|c|}\hline -M_\phi.i_f \\\hline \\\hline\end{array}
= -i_a.M_\phi.i_f.
$$

Since it has already been shown that $dM/d\theta$ is equal to $-M_\phi$, this new expression for torque gives the same value as eqn. (3.13), p. 103. It also gives the same value as eqn. (6.4), p. 316, i.e. using also eqn. (6.2):

$$T_e = k_\phi . i_a = i_f . (k_\phi/i_f) \qquad . i_a$$

$$= i_f . \left\{ \frac{\text{pole pairs}.}{\pi} z_s \frac{.\phi}{i_f} \right\} . i_a$$

$$= -i_f . \qquad M_\phi \qquad . i_a \qquad \text{per pole pair.}$$

The negative sign can be confirmed by physical reasoning from Fig. 10.2b since with positive i_a and i_f the torque is clockwise, opposing the positive sense of θ which is counterclockwise.

When the machine has windings on both axes, as on the amplidyne, for instance, it is convenient to use lower-case letters for the rotor, d and q, and use upper-case letters for the stator, D and Q. The elements in the impedance matrix can be filled in, following a standard routine. The resistance and self-inductance terms occupy the main diagonal. All windings on the same axis have mutual inductance and the appropriate spaces in the matrix must have M_{Dd} and M_{Qq} inserted, each in two places, for the two windings. Rotational voltage terms occur only in the d and q coils and are due to currents flowing in the axis windings which are in quadrature. Hence, for example, with only one stator winding, D say, the equations are:

$$v_d = (R_d + L_d p) . i_d + \qquad K_1 . \dot{\theta} . i_q + \qquad M_{dD} p . i_D$$

$$v_q = \qquad K_2 . \dot{\theta} . i_d + (R_q + L_q p) . i_q + \qquad K_3 . \dot{\theta} . i_D$$

$$v_D = \qquad M_{Dd} p . i_d \qquad + (R_D + L_D p) . i_D$$

from which the primitive machine matrix is readily formed and is, in fact, usually written down directly without first writing the equations above, see eqn (10.3). Note that on steady state, the exciting function

$i_f = i_D$ is a steady d.c. current so that $p = d/dt$ becomes zero and with it the associated inductance terms. The coefficients K_1, K_2 and K_3 must have the dimensions of inductance since they multiply current/time to produce a voltage component and it can be understood from the previous explanations that K_2 and K_3 are negative. By similar reasoning, K_1 can be shown to be positive. If each winding is assumed to produce a sinusoidal flux distribution, or if only the fundamental component of the actual distribution is considered, then the coefficients can be shown to be equal respectively to the corresponding inductance term in the adjacent d or q row; i.e. $K_1 = L_q$, $K_2 = -L_d$. It has already been explained when dealing with the simple d.c. machine, that K_3 is equal to $-M_\phi$ which for a sinusoidal distribution has the same magnitude as the maximum mutual inductance $\hat{M}$ between field and armature, if these were aligned along the same axis. Since the armature is symmetrical, this value of $\hat{M}$ is equal to M_{dD} in the equations above. It follows also, that for a symmetrical armature winding, $R_d = R_q = R_a$.

It is now possible to give some simple applications of the equations. On open circuit, the voltage produced at the q brushes v_q, is obtained by putting $i_d = i_q = 0$ since there is no rotor current in this condition. Therefore, v_q is $K_3 . \theta . i_D$. Since $K_3 = -M_\phi = k_\phi/(i_f \times$ pole pairs) see p. 295; $\theta = \omega_m \times$ pole pairs and $i_f = i_D$, then $v_q = k_\phi . \omega_m$, which is the same as eqn. (6.2a), p. 294.

As a further example, consider the torque expression. The **G** matrix can be extracted from the above equations and consists of the $K_1 . K_2$ and K_3 terms. Substituting the appropriate values for a sinusoidal distribution, the torque per pole pair $\mathbf{i_t . G . i}$ is:

i_d	i_q	i_D						i_d
					L_q			i_d
			$-L_d$				$-M_{dD}$	i_q
								i_D

577

which, when multiplied out, gives:

$$-i_q.L_d.i_d + i_d.L_q.i_q - i_q.M_{dD}.i_D.$$

$$= -i_d i_q(L_d - L_q) - i_q.M_{dD}.i_D.$$

The first term in this torque expression is a reluctance torque and is zero if the armature inductance on the quadrature axis is the same as that on the direct axis, which would be true for a uniform-air-gap machine like the amplidyne but not for a salient-pole type of d.c. machine provided with d and q brush sets. If the d-axis brushes are not present, as on the simple d.c. machine, then $i_d = 0$ and only the final term has a value, which is the same as derived earlier with $i_q = i_a$, $M_{dD} = M_\phi$ and $i_D = i_f$.

The presence of additional windings is easily accounted for in the impedance matrix by including further rows and columns as appropriate. With an additional Q winding, like the interpoles or compensating winding, the transient-impedance matrix equation is written down

			d	q	D	Q		
d	v_d	d	$R_d + L_d p$	$L_q \dot{\theta}$	$M_{dD} p$	$M_{qQ}.\dot{\theta}$	d	i_d
q	v_q	q	$-L_d.\dot{\theta}$	$R_q + L_q p$	$-M_{dD}.\dot{\theta}$	$M_{qQ} p$	q	i_q
D	v_D	D	$M_{Dd} p$		$R_D + L_D p$		D	i_D
Q	v_Q	Q		$M_{Qq} p$		$R_Q + L_Q p$	Q	i_Q

with $=$ between the first two matrices.

$$(10.3)$$

directly as shown here in eqn. (10.3), following the rules described. The letters added at the sides of the rows and the tops of the columns help in identification when numerical values have been inserted in the matrices.

It should be noted at this stage, that the matrix is applicable whatever the nature of the exciting function, d.c. or a.c. or any other time variation. For sinusoidal excitation, p is replaced by $j.2\pi f$, see p. 315.

578

10.2 THE "PRIMITIVE" FOR A.C. MACHINES

Transformations from 3-phase to d–q Axes

The impedance matrix Z_{dq} just presented in eqn. (10.3), has a one-to-one physical correspondence with the arrangement of a d.c. machine having stator and rotor direct-axis and quadrature-axis windings. The same matrix might not be expected to have such a relationship with say a 3-phase a.c. machine. However, by a suitable series of transformations, the impedance matrix derived from the winding-circuit equations of this machine, Z_{ABC}, can be reduced to exactly the same form as Z_{dq}. In general, for any a.c. machine, the direct- and quadrature-axis coils do not then correspond directly to those of the actual machine, though they represent, by their resultant m.m.f., the resultant m.m.f. of the windings they replace. It is not essential that the transformed winding axes are fixed with respect to the stator; they may alternatively be aligned with a moving reference frame. If they are so chosen that they move at the same speed as the air-gap field itself, then their currents and voltages under steady-state a.c. conditions will, in fact, turn out to be steady d.c. values, giving the steady field pattern of this reference axis, with stator and rotor m.m.f.s stationary with respect to one another as described in Section 1.3. For the synchronous machine in d–q axes, the D coil will then be the same as the actual machine field winding; as on the d.c. machine. If instead, the reference axis is so chosen that it is fixed with respect to the armature winding, then the currents and voltages in all transformed winding-circuits will be at the same frequency as those in the armature itself.

To get the machine impedance matrix into the form of eqn. (10.3), for what is called the stationary reference frame (which implies that the D and d axes are coincident), two transformations are performed to give this equivalent machine. Although in purely mathematical terms, transformations generally need have no direct physical significance, in electrical machine analysis they are so chosen that the new system produces the same m.m.f. and the same total power as the original. Further, by a suitable choice of constant, not only is the power invariant in the transformation so that $\mathbf{v}_t . \mathbf{i}$ is the same in the old and new systems, but the transformations for voltage and current are the same. The transformation for impedance is different, though it is simply related to the voltage and current transformations, as in the case of referring secondary parameters to the primary on a transformer. The situation is best explained in physical terms by equating the m.m.f.s in the two systems.

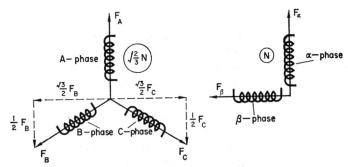

FIG. 10.3. Transformation from 3-phase to 2-phase axes.

For the major class of machines which are wound with three phases and indeed for any polyphase machine, a change to an equivalent 2-phase system greatly simplifies the subsequent working. Figure 10.3 shows a 3-phase and a 2-phase set of coils and for convenience, the α-phase of the 2-phase system is taken to be in line with the A-phase of the 3-phase system. The particular choice of turns ratio shown gives rise to equal voltage and current transformations and also invariant power. Resolving m.m.f.s along the α- and β-axes:

$$F_\alpha = N.i_\alpha = \sqrt{(2/3)}N.i_A - \tfrac{1}{2}.\sqrt{(2/3)}N.i_B \quad -\tfrac{1}{2}.\sqrt{(2/3)}N.i_C$$

$$F_\beta = N.i_\beta = \qquad \sqrt{3}/2.\sqrt{(2/3)}N.i_B - \sqrt{3}/2.\sqrt{(2/3)}N.i_C$$

In matrix form and dividing throughout by the number of 2-phase-machine turns N:

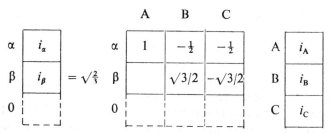

which is of the form:

$$\mathbf{i}_{\alpha\beta} = \qquad \mathbf{C} \qquad \mathbf{i}_{ABC},$$

C being the required transformation matrix from 3-phase to 2-phase currents. The same transformation applies for 3-phase to 2-phase voltages. The inverse transformation from 2-phase to 3-phase, $\mathbf{C}^{-1}$, can be shown in this instance to be equal to the transpose of $\mathbf{C}$; i.e. the first row in $\mathbf{C}$ becomes the first column of $\mathbf{C}_t$ and hence of $\mathbf{C}^{-1}$. Similarly the second row of $\mathbf{C}$ becomes the second column of $\mathbf{C}_t$ and $\mathbf{C}^{-1}$. For the case where the sum of the three phase-currents is not equal to zero, with a 3-phase, 4-wire system, zero-sequence currents are present; see Fig. 3.15. The broken-line portions of the matrices above would then include the zero-sequence current i_0, which is a function of the sum of the three phase-currents. Zero-sequence currents arise only under special conditions and will be neglected subsequently.

A second transformation converts the 2-phase system into a d–q system. Figure 10.4a shows the time variation of the 2-phase m.m.f.s and Fig. 10.4b shows the spatial configuration of a 2-phase armature system which is rotating counterclockwise at synchronous speed as on an inverted synchronous machine. The m.m.f. established by the phase currents will also rotate at synchronous speed relative to the rotor, but in the opposite direction to the movement of the rotor; see Fig. 9.3. The resultant armature m.m.f. $\mathbf{F}_a$ will therefore be stationary in space and for the case of an inverted synchronous machine it would be locked to the field produced by the stator field winding, at an angle dependent upon the machine loading conditions. Suppose that at time angle $\omega t = 90°$, when F_α has its maximum value and F_β is at a zero

581

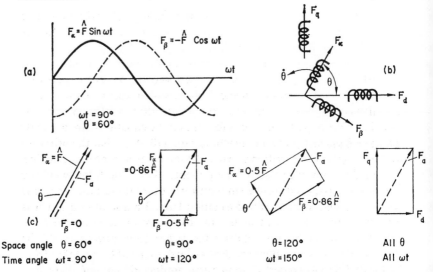

Fig. 10.4. Transformation from rotating to stationary axes:
(a) time variation of phase m.m.f.s; (b) α-β and d-q axes;
(c) m.m.f. space phasors at various instants.

value, the space angle θ of the α-phase is 60° to the horizontal. The space phasors of m.m.f.s shown on Fig. 10.4c are drawn to follow the subsequent time and space variations of F_α, F_β, and F_a. For each example shown and in fact at every instant, F_a is of constant magnitude cf. Fig. 5.33, and is fixed in space. It could equally well be produced by two coils d and q in quadrature, having the same turns N and giving F_d and F_q of appropriate magnitude. For any angle θ, from Fig. 10.4b:

$$F_d = i_d . N = \cos \theta . i_\alpha N + \sin \theta . i_\beta N$$
$$F_q = i_q . N = \sin \theta . i_\alpha N - \cos \theta . i_\beta N$$

from which:

				α	β			
d	i_d		d	$\cos \theta$	$\sin \theta$	α	i_α	
q	i_q	$=$	q	$\sin \theta$	$-\cos \theta$	β	i_β	.

This transformation, $\mathbf{C}$, say, which converts $i_{\alpha\beta}$ to i_{dq}, has the interesting and useful characteristic that $\mathbf{C}^{-1} = \mathbf{C}_t = \mathbf{C}$ itself so that it can be used to transform from d–q to α–β also.

The relationships between the 3-phase and 2-phase and between the 2-phase and the d–q axis systems are now known so that by successive substitution, the equations in the 3-phase system can be converted to equations expressed in terms of direct and quadrature axes. When these substitutions are performed, it is found that the equations are in exactly the same form as for the d.c. machine, eqn. (10.3), so that the primitive for an a.c. machine is written down in the same way as for a d.c. machine, making due allowance also for the presence of any additional windings other than those used on the simplest form of each machine type. It is possible to find a direct transformation from 3-phase to d–q but it is convenient for the present to use the two-stage process. Further, the use of matrix methods formalises the substitution procedure. If the transformation from 2-phase to 3-phase is designated $\mathbf{C}_1$ and that from d–q to α–β is designated $\mathbf{C}_2$ and their product $\mathbf{C}_1\mathbf{C}_2$ as $\mathbf{C}$ (which is then the d–q to A–B–C transformation), then for the current:

$$\mathbf{i}_{ABC} = \mathbf{C}.\mathbf{i}_{dq} \quad \text{and} \quad \mathbf{i}_{dq} = \mathbf{C}^{-1}.\mathbf{i}_{ABC}.$$

For the voltage, the same transformations hold.

For the impedance, the appropriate transformation is obviously going to be related to those for voltage and current. It can be shown to be:

$$\mathbf{Z}_{dq} = \mathbf{C}_t.\mathbf{Z}_{ABC}.\mathbf{C}.$$

When these matrix multiplications are carried out, a procedure which requires a considerable amount of algebraic manipulation, the same result is obtained as by direct substitution; i.e. the equations of the a.c. machine have the same form of impedance matrix as in eqn. (10.3). For the synchronous machine referred to stationary axes fixed with respect to the single field winding, there would just be one stator coil, D, which would be the same as the field winding itself, the transformation leaving this particular winding impedance unchanged. Even though the transformation of impedance may be simplified a little if C is applied in two stages as C_1 and C_2, it is still a very laborious process. Fortunately, it is not usually necessary to carry out this work once its validity has been proved to the prospective user for the general case, using impedance symbols. Thereafter, the result may be assumed and the calculations can be started from Z_{dq} written down in a routine fashion to get the primitive for the particular machine being considered. Unless there are further transformations to be applied to account for the interconnection between windings for example, this matrix for the primitive machine is also the final transient impedance matrix expressed in d–q axes. Because of the transformations chosen, the total power and torque derived from the d–q equations are in fact the power and torque for the actual machine, so if these are the requirements of the solution, no further transformations are necessary. If currents or voltages, whichever were the unknowns, are required in 3-phase terms, then the appropriate transformation must be applied as stated above to convert i_{dq}, say, to i_{ABC}. This transformation is not so tedious to perform as that for impedance and also, because of the nature of C, it is unnecessary to calculate the inverse because C^{-1} is equal to C_t, which can be written down directly.

The routine analytical solution of simultaneous differential equations demands that they are linear and that all the coefficients are constant. Examination of eqn. (10.3) shows that these conditions will be fulfilled if there is no change in saturation level, so that all the inductances are constant and further, that the rotational speed $\dot{\theta}$ is constant. Strictly, then, it is only possible to solve for *electrical circuit transients* by straightforward methods, and it must be permissible, therefore, to assume that the electrical transient will die out before the speed has time to change. If on the other hand the mechanical time constants are not significantly longer than the electrical-circuit time constants, then a computer step-by-step solution or its equivalent is necessary to deal with any *electromechanical transients*. The solution of a.c. machine problems proceeds in a similar manner to that for d.c. machines. Having formed the transient-impedance-matrix and replaced $p = d/dt$ by zero for d.c. excitation, by $j.2\pi f$ for sinusoidal excitation and by the Laplace transform operator for any other form of excitation, the solution for currents, given the 3-phase voltages, follows as below:

Transform v_{ABC} to v_{dq}.

Invert Z_{dq} matrix.

Currents found from $\bar{I}_{dq} = Z^{-1}.\bar{v}_{dq}$.

Power, torque and 3-phase currents follow if required.

The Synchronous Machine

For simplicity, only steady-state operation will be discussed. For this case, the exciting function is a steady d.c. field current so that p can be replaced by zero in the impedance matrix of eqn. (10.3). Further, with only one direct-axis field winding, the Q row and column will be omitted and the D coil will be given the symbol f as for the d.c. machine. On steady state, the speed $\dot{\theta}$ is the synchronous value equal to $2\pi f$ electrical radians per second. In eqn. (10.3) therefore, $R_d = R_q = R_a$; $L_d p = L_q p = L_D p = M_{dD} p = 0$; $L_q \dot{\theta} = 2\pi f L_q = X_q$; $L_d \dot{\theta} = X_d$; $M_{dD} \dot{\theta} = X_{fd}$ say and $R_D = R_f$. The matrix equation is therefore:

				d	q	f			
d	v_d	=	d	R_a	X_q		d	i_d	
q	v_q		q	$-X_d$	R_a	$-X_{fd}$	q	i_q	(10.4)
f	v_f		f			R_f	f	i_f	

It is now possible to relate the parameters in the equation above to those discussed in Chapter 8, with particular reference to Fig. 8.15. Since the transformations for voltage and current are the same, then the voltage/current ratios are unchanged in the transformation. Consequently, X_d is the same as the direct-axis synchronous reactance and X_q is the same as the quadrature-axis synchronous reactance; see p. 486. R_a is the resistance per phase and R_f is the field-winding resistance. i_d and i_q are closely related to the I_d and I_q of 2-axis theory. It will be shown in Section 10.3 that r.m.s. ABC currents and voltages are changed numerically by a factor of $\sqrt{3}$ when transformed to d–q axes. Consequently, the value of the phase current I_a which is equal to $\sqrt{(I_d^2 + I_q^2)}$, see Fig. 8.15, can be found from the values of i_d and i_q, as $(1/\sqrt{3}).\sqrt{(i_d^2 + i_q^2)}$. On steady state, v_d, v_q, i_d and i_q are d.c. values.

If it is required to find the open-circuit voltage from the equations above, then as on the d.c. machine, $i_d = i_q = 0$ and substituting, neglecting R_a, gives:

$$v_d = 0 \qquad v_q = -X_{fd}.i_f \quad \text{and} \quad v_f = R_f i_f.$$

Transforming v_q back to 3-phase quantities will give the r.m.s. open-circuit voltage which is equal in magnitude to E_f as: $v_q/\sqrt{3} = X_{fd}.i_f/\sqrt{3}$. X_{fd} can therefore be determined from the o.c. curve in the same way as M_ϕ for the d.c. machine. X_{fd} could be replaced by $\sqrt{3}.E_f/i_f$.

The torque of the synchronous machine is obtained by extracting the **G** matrix giving the torque per pole pair from:

i_d	i_q	i_f

	X_q		i_d
$-X_d$		$-X_{fd}$	i_q
			i_f

which when multiplied out and divided by the mechanical speed $\omega_m = \dot{\theta}/\text{pole pairs} = \omega_s$, gives:

$$T_e = [-i_d i_q (X_d - X_q) - i_q i_f X_{fd}]/\omega_s.$$

When i_d and i_q have been determined as described from $Z^{-1}.v_{dq}$ and substituted in this expression, the result is exactly the same as eqn. (8.9), p. 510.

The terminal power can be obtained from $\mathbf{v_t i}$ in d–q terms.

The Induction Machine

For the normal induction machine, some simplification in the terms of the impedance matrix is possible because all the stator windings are identical, all the rotor windings are identical and the air gap is uniform. Therefore, all the mutual inductances are the same, of value M, say, and there are only two self inductances, L_s for the stator primary, say, and L_r for the rotor secondary. The suffices have been changed here to bring out a point about the inductances used in the equivalent circuits of 3-phase machines, see also p. 381. Assuming for the moment that all the flux produced by stator or rotor crosses the air gap, the total induced voltage in stator phase A say, with all three phases excited and the rotor open circuited is:

$$L_{AA}.p.i_A + L_{AB}.p.i_B + L_{AC}.p.i_C.$$

Since there is a 120° electrical spatial displacement between the phases, the mutual flux due to currents in B and C will be demagnetising on A with positive B and C currents. For sinusoidal distributions, this fact can be expressed more precisely by stating that the mutual flux with neighbouring phases will be cos 120° times the self flux so that:

$$L_{AB} = -\tfrac{1}{2}L_{BB} = -\tfrac{1}{2}L_{AA} = L_{AC};$$

since all the phases are identical. Hence, for a 3-wire system for which $i_B + i_C = -i_A$, the total induced voltage in phase A can be written:

$$L_{AA}.p.i_A - \tfrac{1}{2}L_{AA}.p.(-i_A) = \tfrac{3}{2}L_{AA}.p.i_A = L_s.p.i_A.$$

With all stator (or rotor) phases excited, therefore, the apparent inductance detected at the phase terminals would be 3/2 times the value measured with only that particular phase excited, see also Fig. 7.2.

In practice, not all the stator-produced flux does cross the air gap and there is leakage with respect to the rotor. This gives the effect of a series leakage inductance l_s, say, increasing the value of L_s. Similarly, for the rotor, L_r must include the leakage l_r. Unlike the stator/rotor mutual flux crossing air the gap and linking each phase, the phase leakage flux is not so much affected by the presence of excitation on the other phases because a good portion of this leakage takes a direct path across the slots. When measuring the equivalent-circuit parameters by s.c. and o.c. tests therefore, the mutual inductance so determined will be 3/2 times that of one phase alone since all phases are excited in these tests.

For the balanced 2-phase machine, there is no such increase of the phase mutual-flux component because the α- and β-phases are in space quadrature. Nevertheless, when a 3-phase system is transformed to 2-phase, this effect is not lost and the ratio of phase terminal voltage to phase current is unchanged. For the condition with all phases excited, the terminal impedance $V_A/I_A = V_\alpha/I_\alpha$. This occurs because of the particular transformation chosen and which is applied identically to both voltage and current, preserving their ratio unchanged.[31] Thus the parameters required for the α–β and hence for the d–q impedance matrices, can be derived directly from the normal equivalent circuit. For example, with rotor parameters referred to the stator, $L_s = l_s + M$ or, in terms of reactances, $X_s = x_s + X_m$, where x_s is the same as x_1, the leakage reactance of the equivalent circuit and X_m is the magnetising reactance of the same circuit. A similar treatment applies to L_r and X_r and to the reactances of the synchronous machine. It is interesting to note that the equivalent-circuit method almost assumes the result of the transformation procedure which converts time-varying inductances into constant parameters.

Note that the machine equations are now being expressed in terms of self and mutual inductance instead of leakage and mutual inductance as in the earlier chapters. Since for the induction machine and the transformer it is really the leakage reactance which dominates the performance, the calculations from the impedance matrix would involve the differences of the relatively large self- and mutual-inductance terms so that a high degree of computational accuracy is called for. Any independent measurements of self and mutual inductance must be precise and taken at the same saturation level so that their difference is equal to the normally unsaturated leakage-inductance. Alternatively, the equations can be rewritten in terms of this difference, or, to be more correct, in terms of the effective inductance in the form $L_s - M^2/L_sL_r$;

see eqn. (3.11), p. 100. It is possible, of course, by further manipulation of the equations, to derive the normal equivalent circuits for polyphase and single-phase induction machines and proceed by the methods of Chapter 7. For balanced, normal operation, there is little point in using any other method than the equivalent-circuit solution, but generalised theory has the advantage that it also covers transient, unbalanced and special forms of operation following a single, routine, logical approach.

The same transformations from A–B–C axes to d–q axes already explained for the synchronous machine, can be used for the induction machine also and the impedance matrix of the primitive machine can be written down straight away as in eqn. (10.3). Direct- and quadrature-coils are required for both stator and rotor. In a stationary reference frame, fixed to the member excited by line-frequency currents, all the currents in the solution will be at line frequency. In this case, for steady-state operation, p is replaced by $j.2\pi f$ (see p. 315) and $\dot\theta$ is replaced by $(1-s).2\pi f$ which is the actual mechanical speed only for a 2-pole machine. Hence the transient-impedance matrix of eqn. (10.3) after making the substitutions described, becomes, on steady state:

	d	q	D	Q
d	R_r+jX_r	$(1-s)X_r$	jX_m	$(1-s)X_m$
q	$-(1-s)X_r$	R_r+jX_r	$-(1-s)X_m$	jX_m
D	jX_m		R_s+jX_s	
Q		jX_m		R_s+jX_s

R_s and R_r are the stator and rotor resistances per phase respectively and the reactances are derived from the normal equivalent circuit as already explained. As it stands, this 4×4 matrix is inconvenient to

589

handle by manual methods but can be simplified by the routine techniques of matrix algebra for example, if only the stator currents or the rotor currents are required.

It is not necessary to use referred rotor impedances in the solution but it is usually convenient. If this is done, then the equations can be represented by a modification of the standard equivalent circuit.[32] This involves one circuit for the direct axis and another for the quadrature axis. There is effective mutual coupling between the two circuits if

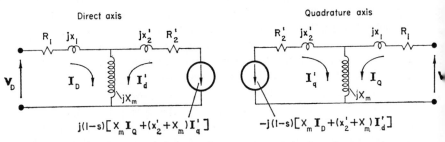

FIG. 10.5. Equivalent circuit from induction-machine impedance-matrix.

rotation takes place, as can be seen by the presence of the speed-voltage terms which incorporate $(1-s)$. Figure 10.5 shows the two circuits and to bring out the close relationship to the circuit of Fig. 7.11, the usual symbols are employed as far as possible. $R_s = R_1$, $R_r = R_2'$, $X_s = x_1 + X_m$ and $X_r = x_2' + X_m$. Upper case letters have been used for voltages and currents to indicate that they have a direct numerical relationship with the r.m.s. values of the actual phase voltages and currents. The speed-voltage terms take the place of the modification from R_2' to R_2'/s is the normal equivalent circuit. The rotor terminals are assumed to be short circuited so that $v_d = v_q = 0$.

The torque can be obtained from the **G** matrix as before and for steady state can be reduced to the usual expression; $3 . I_2'^2 R_2'/s\omega_s$ Nm. For the transient state, the **G** matrix can be read from the matrix on p. 589, extracting only the terms including the speed $(1-s)$. Hence the instantaneous torque per pole pair is:

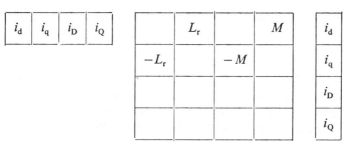

which when multiplied out gives

$$i_d L_r i_q + i_d M i_Q - i_q L_r i_d - i_q M i_D$$
$$= M(i_d \cdot i_Q - i_q \cdot i_D), \qquad (10.5)$$

there being no reluctance torque since the L_r terms cancel.

The treatment of the unbalanced machine is carried out by formulating the primitive machine equations with the appropriate changes to the winding impedances. The single-phase machine is a special case where the current in one stator winding is zero and the last row and column are therefore omitted. Since an unbalance involves the presence of both positive- and negative-sequence currents, quite apart from zero-sequence currents in certain cases, another transformation is desirable from the primitive in d–q axes, to give the equations in terms of positive and negative sequence voltages, currents and impedances. When feeding in the voltage or current in this P–N matrix, the actual machine terminal voltages, or currents must, of course, have this additional dq/PN transformation applied.

10.3   GENERAL SOLUTION PROCEDURES AND EXAMPLES

Since there is rather a large number of operations required before the solution of a numerical problem can be obtained, a flow diagram has been drawn, Fig. 10.6, showing the steps in the solution for the various machine types discussed. C is used as a general symbol for a transformation matrix. The following examples will demonstrate the validity of the method by using material which should be familiar, from the earlier chapters.

A few additional comments about the d.c. machine are appropriate. This was used as a simple introduction to the method and no transformations were involved since the d.c. machine equations are already expressed in terms of stationary axes. So far it has been assumed that the field current is independently controlled and that the brushes are in fact on the quadrature and direct axes. Sometimes, these conditions do not obtain in which case it is possible to derive a suitable transformation to convert the primitive matrix into one which represents the machine in its real situation.

If the brushes are not on d and q axes but on g and h axes, say, which are set at different angles, a transformation matrix can be obtained by equating the d and q m.m.f.s to the components produced by g and h axes resolved along the d and q axes. The method is similar to that used to convert α–β currents to d–q currents. The primitive machine impedance Z_{dq} is then transformed in the usual manner by $C_t . Z . C$ to give Z_{gh} as the actual machine transient-impedance-matrix. This brush-shift transformation will not be pursued further here. A more common form of transformation matrix concerns the interconnection of the machine windings or connection to any external system. Consider, for example, the series machine. When the connections are completed, then $i_a = i_f = i$, say. In matrix form:

$$\begin{bmatrix} i_a \\ i_f \end{bmatrix} = \begin{bmatrix} 1 \\ 1 \end{bmatrix} \begin{bmatrix} i \end{bmatrix} \quad \text{so } C = \begin{bmatrix} 1 \\ 1 \end{bmatrix} \quad \text{and } C_t = \begin{bmatrix} 1 & 1 \end{bmatrix}.$$

The transient impedance matrix from $C_t . Z . C$ is therefore:

$$Z_{\text{machine}} = \begin{bmatrix} 1 & 1 \end{bmatrix} \begin{bmatrix} R_a + L_a p & -M_\phi . \dot{\theta} \\ & R_f + L_f p \end{bmatrix} \begin{bmatrix} 1 \\ 1 \end{bmatrix}$$

$$= \boxed{\begin{array}{c|c} 1 & 1 \end{array}} \; \boxed{\begin{array}{c} R_a + L_a p - M_\phi . \dot{\theta} \\ \\ R_f + L_f p \end{array}} = \boxed{R_a + L_a p - M_\phi . \dot{\theta} + R_f + L_f p}$$

The new voltage is:

$$\mathbf{C_t V} = \boxed{\begin{array}{c|c} 1 & 1 \end{array}} \; \boxed{\begin{array}{c} v_a \\ \\ v_f \end{array}} = \boxed{v_a + v_f} = \boxed{v} \quad \text{say.}$$

The machine equation follows as $v = Zi = [R_a + R_f + (L_a + L_f)p - M_\phi . \dot{\theta}]i$.

This example, though rather trivial, does show how the transformation works to bring about the final machine equation, though in this case it could have been derived directly by normal methods. The field circuit impedance has now been brought into the armature circuit as a result of matrix manipulation. A similar effect would occur if a load impedance was connected in series with the machine and the appropriate connection matrix derived and applied as above. By this technique, it is possible to incorporate the whole system into one matrix. There may be several field windings, for example, supplied from an external exciter or provided with excitation as some function of the machine voltage or current. Their interaction, particularly under transient conditions, might be difficult to visualise without the aid of a formalised method for including their interconnection in a logical manner. Simplification of the transient impedance matrix may be possible and could yield the transfer functions between pairs of variables to show their interdependence. This method of representing system elements by block diagrams is very useful in control-system analysis and some simple examples are given in Appendix C. The mechanical system too can be incorporated in the matrix as an additional row and column, by invoking the mechanical system equation between the machine torque and the mechanical load- and inertia-torques; eqn. (6.9).

The flow diagram of Fig. 10.6 shows that for both d.c. and a.c. machines, the starting point can be the $\mathbf{Z_{dq}}$ matrix, written down in a

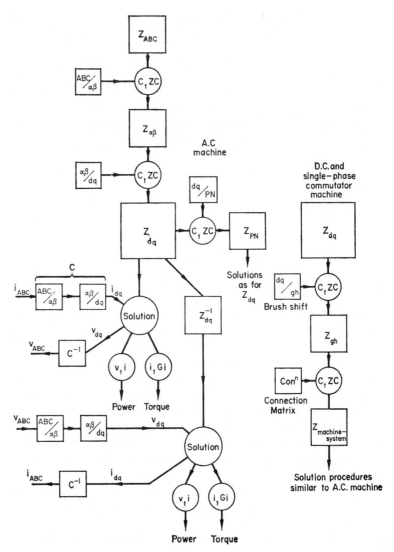

FIG. 10.6. Flow diagram for generalised-circuit-theory solutions.

594

similar way for all machines but making due allowance for the presence of each machine winding which is fitted, and which in general requires to be represented by a separate direct-axis and quadrature-axis coil. For a.c. machines, it is not necessary to go through the A–B–C/d–q impedance transformation each time because the validity of this transformation has already been proven. The sequence is only shown on Fig. 10.6 for record purposes. Input data, current or voltage, is applied to the final transient impedance matrix after being transformed from the actual machine currents or voltages and the solution for various quantities proceeds as shown. Transformation back to the machine windings may be accomplished if desired. For unbalanced machine conditions the $\mathbf{Z}_{dq}$ matrix is first transformed to P–N axes. For d.c. and single-phase commutator machines, the procedure is only slightly different in that the primitive $\mathbf{Z}_{dq}$ matrix is identified immediately with the actual machine arrangement. If required, transformations can be applied to allow for brush shifting or interconnection to the machine windings or to external machines and circuits. The same procedure applies to a.c. machines if they are connected to external circuits or when self excited, as with some brushless alternators.

The next example uses the data of Exercise E.8.1 p. 489, for a synchronous-machine illustration. A considerable amount of algebraic manipulation is involved, not all of it shown in detail. Though this makes the solution very long, it must be remembered that it follows routines which are readily adaptable to digital or analogue computation. Actually, this is not the type of problem for which generalised analysis is normally called into use. Nevertheless it is a useful exercise because for transient and unbalanced conditions, the same basic procedures are followed though there may be additional transformations involved. The flow diagram will help in following the solution.

The problem is specified by the following data, all in *per-unit*.

$$V = 1, \quad X_d = 0{\cdot}9, \quad X_q = 0{\cdot}6, \quad I_a = 1$$

at $0{\cdot}8$ p.f. lagging; machine generating. It is required to find the e.m.f. due to the field current E_f and the load angle δ.

Now the conventions adopted in this chapter, see Fig. 10.2b, assume that positive torque and speed are in the same direction and all windings are treated as power sinks. Therefore, since the equations are motor equations, the generator will appear as a "negative" motor. Either the current or voltage will therefore be reversed. Reversing the voltage gives

$$v_A = -\sqrt{2} \sin \omega t; \qquad v_B = -\sqrt{2} \sin(\omega t - 2\pi/3);$$

$$v_C = -\sqrt{2} \sin(\omega t - 4\pi/3);$$

taking the r.m.s. value of phase voltage as 1 per unit. For the currents, the power factor angle is $\cos^{-1} 0.8 = 36.9°$; hence:

$$i_A = \sqrt{2} \sin(\omega t - 36.9°); \qquad i_B = \sqrt{2} \sin(\omega t - 156.9°);$$

$$i_C = \sqrt{2} \sin(\omega t - 276.9°).$$

For the synchronous machine, the rotor electrical angular velocity θ relative to the stator D winding is the same as the frequency of the voltage and current, $2\pi f = \omega$. When θ is integrated, the actual rotor angle is therefore equal to $\theta = \omega t + \psi$, where ψ is a constant angle. By suitable choice of time zero, ψ can be made equal to the load angle δ and in any case, it is related to this load angle, being a displacement of the rotor windings superimposed on the steadily increasing angle ωt. For this particular problem, with the above choice of voltage and current expressions, $\psi = \theta - \omega t$, will be equal to δ.

It is necessary first to transform voltages and currents to d–q axes, and this will be done in two stages as described in Section 10.2.

				A	B	C			
α	v_α	$= \sqrt{\frac{2}{3}}$	α	1	$-\frac{1}{2}$	$-\frac{1}{2}$	A	$-\sqrt{2} \sin \omega t$
β	v_β		β		$\frac{\sqrt{3}}{2}$	$\frac{-\sqrt{3}}{2}$	B	$-\sqrt{2} \sin(\omega t - 2\pi/3)$
							C	$-\sqrt{2} \sin(\omega t - 4\pi/3)$

$$
\begin{array}{c|c}
\alpha & v_\alpha \\
\hline
\beta & v_\beta
\end{array}
=
\frac{-2}{\sqrt{3}}
\begin{array}{c|c}
\alpha & \sin \omega t - \tfrac{1}{2}\sin(\omega t - 2\pi/3) - \tfrac{1}{2}\sin(\omega t - 4\pi/3) \\
\hline
\beta & (\sqrt{3}/2)\sin(\omega t - 2\pi/3) - (\sqrt{3}/2)\sin(\omega t - 4\pi/3)
\end{array}.
$$

By expanding, using the sin (A–B) formula and collecting terms:

$$v_\alpha = -\sqrt{3}\,\sin \omega t$$

$$v_\beta = \sqrt{3}\,\cos \omega t.$$

This first stage in the transformation has produced a 2-phase system, with v_α in phase with v_A, leading v_β by 90° and increased in magnitude by $\sqrt{(3/2)}$. This is the 2-phase to 3-phase turns ratio to give identical voltage and current transformations with invariant power. The second stage is:

$$
\begin{array}{c|c}
 & \quad\alpha\qquad\quad\beta \\
\end{array}
$$

$$
\begin{array}{c|c}
d & v_d \\
\hline
q & v_q
\end{array}
=
\begin{array}{c|c|c}
d & \cos\theta & \sin\theta \\
\hline
q & \sin\theta & -\cos\theta
\end{array}
\begin{array}{c|c}
\alpha & -\sqrt{3}\,\sin\omega t \\
\hline
\beta & \sqrt{3}\,\cos\omega t
\end{array}
$$

$$
\begin{array}{c|c}
d & v_d \\
\hline
q & v_q
\end{array}
=
\begin{array}{c|c}
d & \sqrt{3}(-\cos\theta\,\sin\omega t + \sin\theta\,\cos\omega t) \\
\hline
q & \sqrt{3}(-\sin\theta\,\sin\omega t - \cos\theta\,\cos\omega t)
\end{array}
=
\begin{array}{c}
\sqrt{3}\,\sin(\theta - \omega t) \\
\hline
-\sqrt{3}\,\cos(\theta - \omega t)
\end{array}.
$$

As explained before, $\theta - \omega t$ is equal to the load angle δ.

The current transformations can be made in the same way but since the instantaneous current expressions contain an additional angle because the power-factor is not unity, the transformations will be even more lengthy. The reader will be spared and given the answer directly as:

597

$$
\begin{array}{c|c}
d & i_d \\ \hline
q & i_q
\end{array}
=
\begin{array}{|c|}
\hline
-\sqrt{3}\sin(\theta-\omega t+36\cdot9^\circ) \\ \hline
\sqrt{3}\cos(\theta-\omega t+36\cdot9^\circ) \\ \hline
\end{array}
=
\begin{array}{c|c|}
d & -\sqrt{3}\sin(\delta+36\cdot9^\circ) \\ \hline
q & \sqrt{3}\cos(\delta+36\cdot9^\circ) \\ \hline
\end{array}.
$$

It can be seen that there is a considerable amount of work involved even up to this stage but like the impedance transformation to d–q axes from A–B–C axes, the work for subsequent problems could be reduced since the result is now known in terms of the original data. Note that there have been changes of sign as a result of the transformation and that the magnitude of $\sqrt{(i_d^2+i_q^2)}$ will be $\sqrt{3}$ times the r.m.s. current in 3-phase terms. Similarly for the voltage there is a factor of $\sqrt{3}$, a result which was anticipated in Section 10.2. The equations for the 3-phase machine in d–q axes can now be written down for steady-state operation as in eqn. (10.4). The equation for the field winding can be added directly since this is already on the D axis and does not require transformation. X_{fd} will be replaced by $\sqrt{3}E_f/i_f$ as explained in Section 10.2, and R_a is being neglected.

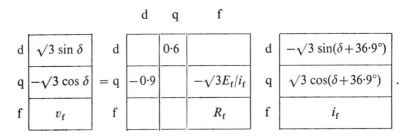

From the d-axis equation:

$$\sqrt{3}\sin\delta = \sqrt{3}\cdot0\cdot6[\cos\delta.\cos36\cdot9^\circ - \sin\delta.\sin36\cdot9^\circ]$$

$$\sin\delta = 0\cdot48\cos\delta - 0\cdot36\sin\delta$$

$$\tan\delta = 0\cdot48/1\cdot36 \quad \text{and} \quad \delta = \underline{19\cdot4^\circ}.$$

From the q-axis equation:

$$-\sqrt{3}\cos\delta = -\sqrt{3}[(-0.9)\sin\delta\times\cos 36.9° + \cos\delta\times\sin 36.9°$$
$$+(E_f/i_f)\times i_f]$$

$$\cos\delta = -0.72\sin\delta - 0.54\cos\delta + E_f$$

$$0.942 = -0.72\times 0.33 - 0.54\times 0.942 + E_f$$

from which $E_f = 1.69\ per\ unit.$

$$i_d = \sqrt{3}\sin(19.4° + 36.9°) = \sqrt{3}\times 0.833$$

$$i_q = \sqrt{3}\cos(19.4° + 36.9°) = \sqrt{3}\times 0.555.$$

Since i_d and i_q are $\sqrt{3}$ times the r.m.s. values of I_d and I_q of 2-axis theory, then: $I_d = 0.833$ per unit and $I_q = 0.555$ per unit.

The answers are the same as obtained in Example E.8.1.

The final worked example, for an induction machine, will be based on the data of Example E.7.1, p. 389. To avoid another extended series of calculations, only the torque will be calculated, using the expression derived from the impedance matrix. Following the routine procedure, the values of i_d and i_q and of i_D and i_Q should be obtained from $\mathbf{i} = \mathbf{Z}^{-1}.\mathbf{\bar{v}}$ after transforming the primary supply voltages in a similar way to the procedure of the last example. Fortunately there is a shorter way which will nevertheless check that the rather different expression for torque, eqn. (10.5), in terms of the reaction between components of stator and rotor m.m.f.s which are in space quadrature, will give the same answer as the equivalent-circuit method,

In terms of the r.m.s. values of stator and referred-rotor currents I_1 and I_2' and their corresponding power-factor angles φ_1 and φ_2, the instantaneous currents can be expressed as $i_1 = \sqrt{2}I_1\ \sin(\omega t - \varphi_1)$ and $i_2' = \sqrt{2}I_2'\ \sin(\omega t - \varphi_2)$. Now on a 3-phase machine under balanced steady-state conditions, the total power is a steady quantity, see p. 66. This means that the torque too is a constant quantity and so it does not matter which instant of time is chosen to calculate it. The choice should

599

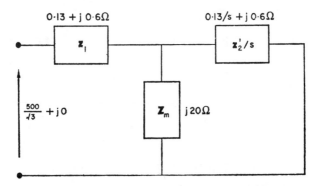

obviously be made to give the simplest expression. From the previous example, the effect of the transformation from A–B–C to d–q axes is known and this permits the values of i_d and i_q as well as i_D and i_Q to be written down from the A-phase values of i_1 and i_2'. By appropriate choice of instant, making $\psi = 0$, the angle in the final expressions at the top of p. 598 becomes equal to the lagging power-factor angle ($-36°\cdot9$) taken as positive. Hence:

$$i_D = -\sqrt{3}I_1 \sin \varphi_1 \qquad\qquad i_Q = \sqrt{3}I_1 \cos \varphi_1$$

$$i_d = -\sqrt{3}I_2' \sin \varphi_2 \qquad\qquad i_q = \sqrt{3}I_2' \cos \varphi_2.$$

Substituting in the expression $M(i_d . i_Q - i_q . i_D)$

$$T_e = M . (\sqrt{3})^2 . I_1 . I_2'(-\sin \varphi_2 . \cos \varphi_1 + \cos \varphi_2 . \sin \varphi_1)$$

$$= \quad 3 . M , I_1 . I_2' \sin(\varphi_1 - \varphi_2) \text{ per pole pair.}$$

It now remains to find the values of M, I_1 and I_2' from the data of Example E.7.1. The equivalent circuit is shown on the attached figure, the magnetising resistance being omitted since it has not been allowed for in this chapter and in this problem it has very little effect anyway. The values of I_1 and I_2' will be recalculated from the slightly modified circuit though the numerical working will not be shown. The following expressions deduced directly from the equivalent circuit are useful alternatives to those used previously, particularly if computational facilities are available:

$$I_1 = \frac{V_1}{z_1 + \dfrac{Z_m \cdot Z_2'}{Z_m + Z_2'}} = \frac{V_1 \cdot (Z_m + Z_2')}{z_1 \cdot Z_m + z_1 \cdot Z_2' + Z_m \cdot Z_2'}.$$

$$I_2' = \frac{Z_m}{Z_m + Z_2'} \cdot I_1 = \frac{V_1 \cdot Z_m}{z_1 \cdot Z_m + z_1 \cdot Z_2' + Z_m \cdot Z_2'} = \frac{V_1}{z_1 + Y_m \cdot z_1 Z_2' + Z_2'}$$

where $Z_2' = z_2'/s = R_2'/s + jx_2'$, and $Y_m = 1/Z_m$

Substituting $z_1 = 0{\cdot}13 + j0{\cdot}6 \, \Omega$, $Z_2' = 2{\cdot}6 + j0{\cdot}6 \, \Omega$ and $Z_m = j20 \, \Omega$ and with V_1 as reference phasor $= 288{\cdot}5 + j0$ gives:

$$I_1 = 85{\cdot}2 - j49{\cdot}7 \text{ A} = \underline{98{\cdot}6} \quad \underline{/-30{\cdot}2^\circ}$$

$$I_2' = 87 \quad -j37 \quad \text{A} = \underline{94{\cdot}4} \quad \underline{/-23^\circ}.$$

Since the machine has 8 poles, the torque is:

$$T_e = 4 \times \frac{20 \, \Omega}{2\pi \times 50} \times 3 \times 98{\cdot}6 \times 94{\cdot}4 \times \sin(30{\cdot}2^\circ - 23^\circ)$$

$$= 4 \times 0{\cdot}0636 \times 3 \times 98{\cdot}6 \times 94{\cdot}4 \times 0{\cdot}125$$

$$= \underline{894 \text{ Nm}}.$$

The answers are within slide-rule accuracy of the previous calculations on p. 390. The nature of the above calculation, which involves the difference of large quantities, requires a good pocket calculator for satisfactory accuracy.

The examples used in this section have not been chosen to demonstrate the power of the generalised-theory method, but only to illustrate its unifying features and to show that in numerical problems, the rather different expressions give rise to the same answers as by classical methods. Indeed, for balanced steady-state operation and for transient operation which does not involve a greater-than-2nd-order differential equation, the earlier chapters have shown that rather simpler methods are usually adequate for analysing and predicting performance. What generalised theory does, is to provide a systematic method of analysing a machine system, no matter how complex, and which covers the various possible operating modes. With the increasing use of thyristor/ machine combinations, for example, such a method is of great value in setting up computer programs for studying performance. For those who intend to specialise in engineering systems involving electrical machines, this chapter can only be regarded as giving an indication of the analytical tools required and, it is hoped, some little encouragement to study them in more detail with the aid of a text written with an appreciation of the deeper implications of such a study.

At this stage, the scope for further practical examples is not very extensive unless a considerable familiarity with matrix algebra and with Laplace transforms can be assumed. However, the reader could well check his understanding of this introductory chapter by attempting the following exercises.

(1) Formulate the transient impedance matrix for the transformer having the data of Example E.4.2, p. 152. Neglect R_m.

(2) Formulate the transient impedance matrix for the induction machine having the data of Example E.7.1, p. 389. Neglect R_m and take the speed as 712·5 rev/min.

(3) From the data of Example E.6.4, p. 329 and using the answers to part (c)(ii) converted to S.I. units; formulate the torque equation in matrix notation as on p. 575 and check that this gives the correct electromagnetic torque when multiplied out. Assume the machine has 4 poles.

(4) Show that for a cylindrical-rotor synchronous machine, the torque expression given by matrix multiplication as on p. 586, can be deduced from eqn. (8.7), p. 502. The phasor diagram of Fig. 8.26a will be helpful and note also that $X_d = X_q = X_s$ for a cylindrical rotor machine.

APPENDIX A

PERMANENT MAGNETS FOR ELECTRICAL MACHINES

FOR many years, permanent magnets have been providing loss-free excitation for very small d.c. motors and also providing supply-independent excitation for large hydro-electric generating plant, see Section 8.1. Continued advances in the development of new materials have increased the usage of permanent-magnet machines considerably. Magnets can now be smaller, lighter, readily moulded into desired shapes and can retain their magnetism under much more severe conditions. This is fortunate, since it coincides with the explosive demand for very-low-power drives in a variety of electronic, control and computer equipment using stepper motors and d.c. motors—of both conventional and brushless types.

Although the latest rare-earth compounds with ferromagnetic materials, e.g. samarium-cobalt, are still relatively expensive, they have increased further the power/weight ratio of permanent-magnet machines, quite apart from the improvement in efficiency, due to the absence of excitation losses. This is especially important in low-speed, high-torque, direct-coupled drives where efficiencies are very low, but the energy saving and the reduction in heat to be dissipated are always attractive design considerations. Even normal-speed d.c.- and synchronous permanent-magnet machines up to 10 kW or more have become commercially viable in certain cases, so the subject of PM machines is now of greater importance than hitherto.

Permanent magnet materials have progressed from the early alloys of carbon steel through several other alloys of iron, until the breakthrough in the early 1930s with the use of aluminium, nickel and cobalt in various proportions with 50-60% iron, to form the Alnico series, permitting

603

much smaller sizes. Further work continued with powdered materials and the ferrites (like naturally occurring magnetite), which are mixtures of iron with other metallic oxides. These are very resistant to demagnetisation, having high coercivity and also have high electrical resistivity. Although the volume required is rather larger than for Alnico, the high H_c leads to a much shorter length. Rare-earth compounds combine very high coercivity with fairly high remanence B_{res}. Permanent magnets operate in Quadrant 2 and clearly, the values of B_{res} and H_c are relevant to the suitability of PM material.

To understand the above points, consider the typical part-hysteresis loop of Fig. A.1, with emphasis particularly on the upper region of Quadrant 2, where, after being saturated with an intense magnetising

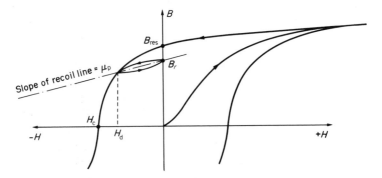

FIG. A.1. B/H characteristic.

field, this has been reduced to zero and is being reversed. The magnetic domains will be coming out of alignment with the forward field but will not have reversed their inelastic rotations which occurred during the rising part of the characteristic in Quadrant 1, see Section 2.1. If demagnetisation continues to a negative intensity H_d, say, but is then brought back to zero, the B/H, characteristic will return along the lower half of a minor hysteresis loop and, if the magnet is "short circuited" through a soft-iron keeper, it will intersect the B axis, at a lower value B_r. However, reapplying H_d will bring the flux density back along the upper half of the loop to the same value as before. Several excursions of this kind are used to stabilise the magnet so that in service it will continue to operate along this loop providing the negative intensity H_d is not exceeded. On a good material, the loop is very narrow and the mean line, called the recoil line, has a slope virtually the same as that at B_{res}. It has the dimensions of permeability and is called the permanent permeability μ_p. On the newer materials, μ_p is only slightly greater than μ_o, see Fig. 2.1, and the recoil line is almost coincident with the demagnetisation curve, extending in some cases down to zero flux density and below. In this case, it does not need stabilising nor a keeper to prevent loss of magnetisation, which is a useful feature when considering the need to assemble and occasionally to dismantle a magnetic circuit.

A selection of demagnetisation curves is shown on Fig. A.2 for comparison. It would not be possible to see clearly the curve for 1%

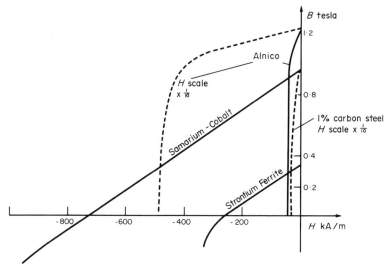

FIG. A.2. Demagnetisation curves.

carbon steel if plotted on the full H scale so this has been divided by 10 and the Alnico curve for one of the best of the family has been plotted twice, using both H scales. Comparing the broken-line curves shows an improvement similar to that between Alnico itself and samarium-cobalt. There are many other materials available to meet the variety of technological needs. The ferrite shown is often used for machines and is relatively inexpensive by comparison with samarium-cobalt.

Magnet Dimensions

In practice, of course, a permanent magnet is used to provide a field in a "working region" which is usually an air gap. Pole pieces may be necessary to complete the magnetic circuit and distribute the flux in a suitable manner. These will be of soft iron and absorb only a small m.m.f. which will be ignored in the following equations. Referring to the arrangement of Fig. A.3, the magnet has a cross-sectional area A_m, length l_m and flux density B_m. The corresponding features for the air gap are A_g,

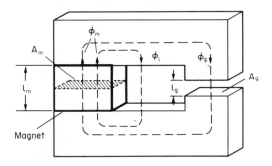

FIG. A.3. Permanent magnet with pole pieces and air gap.

l_g and B_g. A leakage flux is indicated as ϕ_l, but will be neglected for the moment.

Applying Ampere's circuital law, eqn. (2.4):

$$\int H \cdot dl = \text{current enclosed} = \text{zero} = H_m \cdot l_m + H_g \cdot l_g$$

from which the length of the magnet must be:

$$l_m = \frac{-H_g}{H_m} \cdot l_g.$$

This expression explains why the length of modern magnets can be quite surprisingly short, due to the possibility of working at very high negative field intensities.

Another equation can be formed by considering the identical fluxes in magnet and air gap, viz.:

$$\phi_m = B_m A_m = B_g A_g = \phi_g,$$

606

neglecting leakage, from which the cross-sectional area of the magnet:

$$A_m = \frac{B_g}{B_m} \cdot A_g.$$

Hence the magnet volume required is:

$$|A_m l_m| = \frac{B_g}{B_m} A_g \frac{B_g}{\mu_o H_m} l_g = \frac{B_g^2}{\mu_o} \frac{\text{(Air-gap volume)}}{B_m H_m}.$$

The effect of leakage is to require a greater magnet flux so that its volume must increase in proportion to the leakage factor: (useful flux + leakage flux)/useful flux. The magnet volume is seen to be a minimum when the product $B_m H_m$ is maximum, which occurs at a point on the curve somewhat higher than $B_{res}/2$. It has been noted in Section 2.2 that $B \times H$ has the dimensions of energy density in joules/m^3; now it is shown that the maximum BH value is also an important factor in assessing permanent-magnet-material suitability. It varies from about 2 kJ/m^3 for 1% carbon steel, to about 40 kJ/m^3 for Alnico, rather less than this for the ferrites and as much as 160 kJ/m^3 for samarium-cobalt. Taking this last figure and referring to Fig. A.2, substitution will show that the required magnet volume must be about the same as the air-gap volume (a B_g of about 0·4 would be possible, allowing say, 20–25% for leakage). This is only to be expected since μ_p for samarium cobalt is only a little high than μ_o, so $B_g^2/\mu_o = B_g H_g \simeq B_m H_m$, if operating at the optimum point $B_r/2$. Here, $B_m = -\mu_p(H_o - H_m) = -\mu_p H_m$. Magnet lengths may be little greater than the air-gap length, but increase if $B_m > B_r/2$ or as leakage increases. This can be high on very small machines.

Magnetic Equivalent Circuit

For a permanent magnet which has been stabilised and has its operation confined within this limit, the recoil line defines a simple linear equation. Converting the B/H to a flux/m.m.f. relationship, eqns. (2.5) or (2.6) form the basis for a magnetic equivalent circuit as already seen in Fig. 2.11. On Fig. A.4a, the magnet flux at zero m.m.f., $\phi_r = B_r \times A_m$. The m.m.f. at zero flux, $F_o = H_o \times l_m$, where H_o is the value of H at the intersection of the recoil line extended back to zero flux density. The

ratio:
$$\frac{\text{m.m.f.}}{\text{flux}} = \frac{F_o}{\phi_r} = \frac{H_o l_m}{B_r A_m}$$

has the dimensions of a reluctance $\mathscr{R}_o$, the slope of the recoil line being the permanent permeability $\mu_p = B_r/H_o$. The magnet therefore behaves as if it were a source of m.m.f. F_o behind a reluctance $\mathscr{R}_o$ corresponding to a permeance Λ_o and gives a flux $\phi_r = F_o/\mathscr{R}_o$ if "short circuited" by a soft-iron keeper.

Referring to Fig. A.3 and adding the air-gap reluctance $\mathscr{R}_g = l_g/(\mu_o A_g)$ into the equations, still neglecting leakage, the flux is $\phi_m = F_o/(\mathscr{R}_o + \mathscr{R}_g)$ $= \phi_g$. The source m.m.f. is absorbed in the two components $\phi_m \mathscr{R}_o$ and $\phi_g \mathscr{R}_g$ as indicated on Fig. A.4a, although the actual m.m.f. absorbed by the iron really extends only to the point $H_c \times l_m$, the point F_o being a mathematical abstraction for the purpose of formulating a linear model. The effect of the "load" (the air gap) is to move the operating point along the recoil line. Point "P" does not therefore correspond exactly to the optimum value of BH. Note also that the air gap would have to be "short-circuited" in order to bring the flux to ϕ_r.

We can now extend the model to include the effect of leakage flux and any m.m.f. acting in the circuit as a result of an energised coil or armature-reaction m.m.f. F_a. Figure A.4b is the equivalent circuit. F_a is shown in a demagnetising sense, i.e. opposing F_o. The circuit is topologically the same as Fig. 4.28b for two paralleled transformers. The corresponding terms are:

Fig. A.4b	F_o, F_a	$\mathscr{R}_o, \mathscr{R}_g, \mathscr{R}_1$	ϕ_m, ϕ_g, ϕ_1
Fig. 4.28b	E_A, E_B	Z_A, Z_B, Z	I_A, I_B, I

The solution follows directly from eqns. (4.7) and (4.9) but note that in the present case, the flux ϕ_g must be negative to the direction shown, if it is to represent a flux produced by the magnet in opposition to the demagnetising m.m.f. F_a. Hence, from eqn. (4.7):

$$\phi_m = \frac{F_o \mathscr{R}_g + (F_o - F_a)\mathscr{R}_1}{\mathscr{R}_1(\mathscr{R}_o + \mathscr{R}_g) + \mathscr{R}_o \mathscr{R}_g}.$$

ϕ_g follows the same equation but with the numerator suffices changed and from eqn. (4.9):

$$\phi_1 = \frac{F_o \mathscr{R}_g + F_a \mathscr{R}_o}{\text{(denominator as above)}}.$$

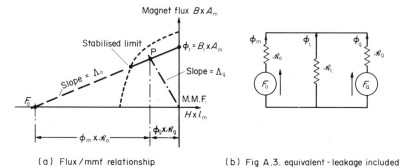

(a) Flux/mmf relationship   (b) Fig A.3. equivalent - leakage included

FIG. A.4. Equivalent circuit development.

The equations will be illustrated by calculating the various flux components using the following rounded-off circuit parameters, at various settings of F_a.

$F_o = 10,000$ At, $\phi_r = 1$ mWb. Reluctances in At/weber, with leakage having a value equal to 20% of the air gap flux; $\mathscr{R}_o = 10^7$, $\mathscr{R}_g = 10^6$, $\mathscr{R}_l = 5 \times 10^6$.

Condition		ϕ_m	ϕ_l	ϕ_g	ratio ϕ_m/ϕ_g
$F_a = 0$	$\mathscr{R}_l = \infty$	0·91	0	−0·91	1·0
$F_a = 0$	$\mathscr{R}_l = 5 \times 10^6$	0·92	0·15	−0.77	1·19
$F_a = 1000$	$\mathscr{R}_l = 5 \times 10^6$	0·85	0·31	−0·54	1·57
$F_a = 2000$	$\mathscr{R}_l = 5 \times 10^6$	0·77	0·46	−0·31	2·48
$F_a = 3000$	$\mathscr{R}_l = 5 \times 10^6$	0·69	0·62	−0·07	9·86

Permanent-magnet D.C. Machines

The behaviour of permanent-magnet d.c. machines is similar to the separately-excited case with constant excitation. The recoil line has a very low apparent permeability, $\mu_p \simeq \mu_o$, so that the cross-magnetising component of armature m.m.f. will not cause loss of flux with load due to pole-tip saturation, providing that on the demagnetised side of the pole

609

arc the m.m.f. does not take the magnet beyond its stabilised limit. This must, of course, be checked for the worst condition which can arise, making due allowance for any direct-axis component of m.m.f., F_{ad}. The numerical example above has shown that leakage increases considerably as the directly opposing m.m.f. increases, and this also has the effect of limiting the change of magnet flux. Although machines in the kW sizes are made, the vast majority are in the small-motor range, e.g. motor-vehicle auxiliary drives, robot actuators, office and computer equipment, tape recorders and record players. Even where the universal series motor is dominant, in the mains-supplied domestic-appliance field, a challenge is being offered; a simple single-phase, power-electronic circuit being used to give rectified-voltage (and speed) control, e.g. by varying the firing-angle (α) delay. Many permanent-magnet d.c. machines are brushless, sometimes with the power transistors incorporated in the machine itself and with mounting and external connections similar to the conventional machines they are replacing. PM d.c. tachometer generators giving voltage proportional to speed are in common use.

Permanent-magnet Synchronous Machines

For synchronous machines, the situation is rather different in that F_a may have a large direct-axis demagnetising component, e.g. on short circuit at zero power factor. Again, sufficient margin must be allowed to ensure that the worst demagnetising condition does not take the magnet beyond the stabilised limit. The equivalent circuit could be deduced from o.c. and s.c. tests, the latter conducted with a limiting inductor if necessary to prevent excessive short-circuit current. From these tests, the effective e.m.f. 'E_f' behind the effective synchronous reactance 'X_s' follow and performance as a motor or a generator could then be calculated from these fixed parameters, which include some allowance for saturation, corresponding to the condition at which the tests were taken. The approximate circuit of Fig. 8.31 could easily be supplemented to include the armature resistance. A more precise analysis would have to allow for the appreciable saliency effect of most practical permanent-magnet synchronous machines. Figure A.5a is a typical rotor arrangement for a modern machine,[18] and it shows that the d-axis flux has to negotiate the magnet-material path with its low value of μ_p, whereas the q-axis flux is only in the soft iron. This leads to the interesting effect that

610

$X_q > X_d$ and reference to eqn. (8.9) and Fig. 8.26 will explain why the power/angle curve is of different shape, with a maximum value occurring where $\delta > 90°$ (Fig. A.5b). Measurements of axis reactances would have to use the slip test (p. 486) or its equivalent, involving a 90° shift of the stator m.m.f. to align with either the maximum or minimum rotor permeance.

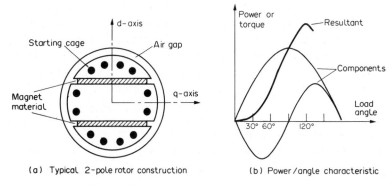

(a) Typical 2-pole rotor construction   (b) Power/angle characteristic

FIG. A.5. Permanent-magnet synchronous machine.

Permanent-magnet synchronous machines are also available in sizes up to and beyond 10 kW. Though they are more efficient than coil-excited machines of equivalent rating, the flux density is much lower (see Figs. A.2 and A.4a). The active-rotor volume $\pi d^2 l$, will have to be larger (see eqn. (6.1a)), and this is mostly achieved by increasing the length. A ferrite magnet would require about a 50% increase in core length compared with a rare-earth magnet. However, the soft-iron thicknesses would be smaller with the longer cores and in addition, the required thickness of magnet is so small that overall, the outside diameter of the machine carcass is reduced. The improved efficiency is achieved at the cost of some increase in weight, volume and price. It is possible, however, to operate at power factors much closer to unity than induction motors of equivalent rating, quite apart from the improved efficiency and without any loss in ruggedness and simplicity of the cage rotor. The reduction in kVA requirements leads to appreciable savings in inverter costs, should this be the mode of supply. A requirement, as in the fibre-processing industry, for multi-motor, variable-speed drives locked in synchronism, is an obvious application for this type of machine.

611

Brushless-machine Circuit with Regenerative Facility

It has already been explained on p. 544 that brushless machines can be regarded either as d.c. machines with static commutators, or synchronous machines with a variable frequency automatically governed by the shaft speed, sensed in each case by a shaft-position detector. Figure 6.54 showed the development from the conventional d.c. machine and with two switchings per pole pair, the torque angle varies between $0° \pm 90°$, giving a large torque pulsation (Fig. 6.54e). If the number of switchings is increased to six, the m.m.f. moves forward in steps of $60°$ instead of $180°$ so that the torque-angle swing is only $\pm 30°$. An appreciable reduction of the pulsation follows and the salient-pole theory of synchronous machines becomes a reasonable analytical approach. The d.c. supply, either from a battery or rectified source is switched sequentially through an inverter to the three tappings per-pole-pair of the stator winding. The general arrangement looks like a three-phase a.c. machine supplied through an inverter. The operation will now be described, since it could be applied also to conventional synchronous and induction machines. This particular system was for a brushless machine and devised for an 11 kW vehicle drive with a short-term overload of 26 kW.[35]

The 'power-conditioner' circuit is shown in Fig. A.6a and can be regarded in two sections, the d.c. link and the inverter. The d.c. link regulates the current through the inductor to a preset value by means of the d.c. chopper power-transistors. T_M is ON when the current is below this value causing it to rise. Above the setting, T_M is switched OFF and the current then starts to decrease through the diode D_M. There is a small ripple in the current due to this action as shown on the line current i_A of Fig. A.6b. The power-transistor/diode pair T_R and D_R act for reverse current through the inductor—opposite to the direction shown—i.e. when power flow is reversed and the d.c. link accepts regenerated power. This changeover was actual achieved by switching the one transistor by mechanical relay from top to bottom position, to avoid accidental short circuit of battery supply. The diode D_B provides a path for return of inductor energy, through D_M to the battery, should the inverter be blocking during its switching operations.

To run the machine as a motor, the transistors in the inverter, controlled by shaft position, are switched sequentially as indicated by

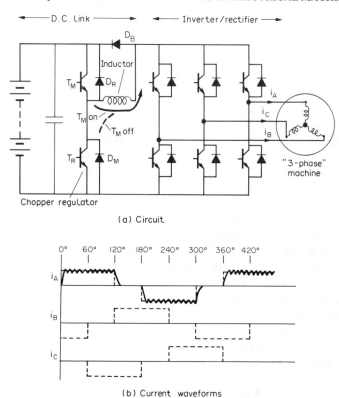

(a) Circuit

(b) Current waveforms

FIG. A.6. D.C./3-phase power conditioner.

the waveforms of Fig. A. 6b. This shows only two conducting together except during overlap (120° conduction). This again is a protection against d.c. link short circuit. If conduction was for 180° the inverter would always have at least two elements conducting on one side and one on the other, so that short circuit, though limited by the inductor, would be more likely on a switching changeover. Nevertheless 180° conduction is often used, with careful protection arrangements, because it gives a rather higher r.m.s. value, the line waveforms corresponding to those of Fig. 7.31b. For regeneration, the diodes in the inverter act like a three-phase bridge rectifier (see Problem E.6.7, p. 341), and the inductor current is reversed from the direction shown. With T_R switched ON, 'short-circuiting' the bridge, the inductor is being charged and at a predetermined current, T_R is switched OFF. The inductor can then discharge through D_R, feeding energy into the battery. This mechanism of regeneration is basically the same as discussed previously on p. 362 in connection with the chopper-controlled d.c. motor (see also Ref. 33).

Survey of Small-motor Options

The range of permanent-magnet motors[18] starts with the very special, minute, microwatt, stepper motor for driving quartz wrist-watches. Shaded-pole construction continues the range through the milliwatt sizes—with hysteresis motors at the lower end. Beyond about 1 watt, stator windings with phase-splitting circuits are also used, with permanent-magnet rotors at the lower speeds and cage rotors for synchronous speeds requiring less than 8 poles. For higher speeds, or where large speed-variation is required, conventional d.c. or brushless motors, often with permanent magnets, tend to be used when the supply is from batteries. Static variable-frequency inverters bring in competition here with either permanent-magnet synchronous machines or induction motors. For mains-voltage supplies, the universal series motor is dominant, though as mentioned earlier, this position may be challenged by developments in permanent-magnet motor systems. Low efficiencies are often regarded with toleration in the low-power range, but it is worth pointing out that even small savings of power per motor by the use of permanent-magnet machines may well become decisive when the vast numbers of motors in use is considered. A good example of this is the shaded-pole induction motor used in millions for driving small fans and where efficiencies of the order of 20% are typical.

The above remarks cover the range of small motors—once referred to as fractional horse-power motors. Beyond about 1 kW, there are many other considerations in choosing a drive which are discussed briefly in Appendix D. The whole range of criteria must be taken into account as far as possible and this includes energy savings, simplicity of maintenance, absence of sliding contacts and rotor supplies. It may be said in summary that the availability of the new permanent-magnet machines, while not leading to immediate takeovers in the range where other motor types have dominated, nevertheless need to be appreciated as serious competitors when the overall schemes are considered.

APPENDIX B

EQUIVALENT CIRCUITS DERIVED FROM FLOW DIAGRAMS

THE idea of representing a d.c. machine by a transfer function and a block diagram was introduced in Section 6.6. The block diagram approach is very useful for reducing a complex interconnected system of elements to a simpler form, once the transfer functions of the components are known. To combine the elements in this way requires a suitable algebra to be devised and the one in common use in Control System Theory will be outlined briefly. The purpose of this exercise is to permit the derivation of the steady-state equivalent circuits of electrical machines in a very concise manner. The method has been suggested[1] on the basis that in an electrical machine, there is a cause and effect sequence involving voltages, currents, m.m.f.s and fluxes so that it can be treated in a similar way to a simple control system.

Consider Fig. B.1, which shows a series LR circuit and its block diagram representation. The induced e.m.f., $e = L\,di/dt$, opposes the change of current so that this is limited to $(v-e)/R$. The interpretation of the top element in the block diagram of Fig. B.1b is that it gives the ratio between the output i, and the input $(v-e)$. Multiplying the input by the transfer function $1/R$ gives the output, i.e. $i = (v-e).1/R$. The induced e.m.f. behaves like a negative feedback and is a function of the output current. The feedback transfer function is Lp, where p is the operator d/dt. The output e from the feedback circuit is given on multiplying its input i by the transfer function; i.e. $e = Lp.i$.

It is now necessary to consider some basic elements of block-diagram algebra before proceeding further. Consider Fig. B.2, which shows a general feedback loop with a reference input R, and the controlled output C. C is fed back negatively through a transfer function H and the difference, or error $E = R - HC$ is applied to

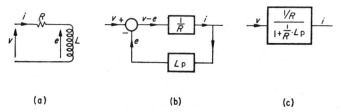

(a)   (b)   (c)

FIG. B.1. Inductive circuit as a feedback system: (a) circuit; (b) block-diagram representation; (c) overall transfer function.

615

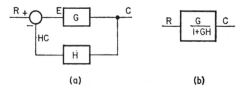

(a)   (b)

FIG. B.2. Reduction of feedback loops: (a) single feedback loop;
(b) reduced system.

the forward transfer function G to give the output C. By considering the loop, and multiplying the transfer functions by their appropriate inputs, the following equations can be written:

$$C = GE = G(R - HC)$$

Hence $$C = R \cdot G/(1 + GH)$$ by rearranging the equation.

Thus the block diagram can be simplified to one block which is a function of the original two blocks. The output is now obtained on multiplying the input R, by the modified transfer function which represents the overall behaviour.

Another configuration which arises frequently is when two transfer functions are in cascade, or series. This situation has already been met in connection with Fig. 6.33 where it was found that the combined effect is obtained by multiplying the two transfer functions together. One further rule concerns the situation where there is more than one input to the system and which may in fact be at a different point from the first input. In this case it is necessary to apply the theorem of superposition. Each input is considered separately with the others assumed to be zero. The total output is then obtained by linear combination of the individual responses of the system to the different inputs; see Fig. B.3. An example occurs in connection with Fig. B.5.

Referring back to Fig. B.1b, it is now possible to simplify this to one block. In the

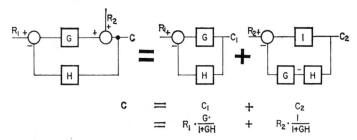

FIG. B.3. Superposition of two inputs.

616

notation of Fig. B.2, G is $1/R$ and H is Lp. Consequently, the ratio between the output i, and the input v, which is the overall transfer function is given by:

$$\frac{1/R}{1 + \frac{1}{R}.Lp} = \frac{1}{R(1 + \tau p)} \quad \text{where } \tau = L/R.$$

A similar transfer function has been met before when dealing with the d.c. generator in Section 6.6. It corresponds to the ratio between field current and field voltage. Also in Section 6.6 was explained the condition for replacing the operator p by $j\omega$. This procedure is adopted in the following, because the steady-state, constant-frequency equivalent circuits are being derived.

TRANSFORMER EQUIVALENT CIRCUIT

The flow diagram of Fig. B.4 for the 2-winding transformer, shows the intereactions in a step-by-step sequence, which is almost self explanatory. The primary current I_1 follows in a similar manner to i in Fig. B.1, though E_1 is now due to the resultant mutual flux ϕ_m and z_1 is the leakage impedance $R_1 + jx_1$. I_1 is now multiplied by N_1 and at the next summing point, the resultant m.m.f. $I_1 N_1 - I_2 N_2$, $(F_p + F_s)$, is obtained and applied to the mutual circuit permeance Λ, to get the mutual flux ϕ_m. The e.m.f.s follow from the e.m.f. equation (4.1), which at constant frequency reduces to $KN\phi_m$. One feedback loop produces E_1 at the first summing point. The second feedback loop, after producing E_2, and applying this to the total of secondary leakage and load impedances $z_2 + Z_L$ to get the secondary current I_2, gives the secondary m.m.f. $I_2 N_2$ required at the m.m.f. summing point.

The diagram is simplified by first removing the m.m.f. feedback loop. The forward transfer function G is equal to Λ and the feedback transfer function is $H = kN_2{}^2/(z_2 + Z_L)$. The diagram then becomes rearranged as Fig. B.4b. Now the transformer equivalent circuit is really the ratio V_1/I_1 expressed as the apparent impedance viewed from the primary terminals. The ratio between any other two quantities could be obtained by a different treatment of Fig. B.4b, but in this case the forward transfer function which applies is that between V_1 and I_1, i.e. $G = 1/z_1$. The feedback function H is equal to the product of the remaining terms. The overall transfer function, giving the ratio between the output I_1 and the input V_1 follows from the appropriate combination of G and H. This ratio is in fact the admittance of the equivalent circuit.

Drawing on previous knowledge of the magnetising reactance X_m which is equal to E_1/I_m, the expression can be further simplified since

$$\frac{E_1}{I_m} = \frac{KN_1\phi_m}{I_m N_1/N_1} = KN_1{}^2 \frac{\phi_m}{I_m N_1} = KN_1{}^2 \Lambda.$$

The final expression for the admittance is shown below Fig. B.4b and this is clearly equal to the inverse of an impedance consisting of z_1 in series with the parallel combination of jX_m and $(z_2 + Z_L)(N_1/N_2)^2$. The latter term is the referred value of

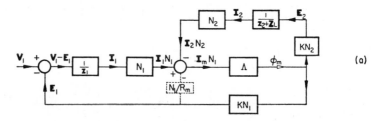

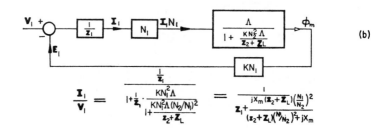

$$\frac{\mathbf{I}_1}{\mathbf{V}_1} = \frac{\frac{1}{\mathbf{Z}_1}}{1 + \frac{1}{\mathbf{Z}_1} \cdot \frac{KN_1^2 \Lambda}{1 + \frac{KN_1^2 \Lambda (N_2/N_1)^2}{\mathbf{Z}_2 + \mathbf{Z}_L}}} = \frac{1}{\mathbf{Z}_1 + \frac{jX_m (\mathbf{Z}_2 + \mathbf{Z}_L)(\frac{N_1}{N_2})^2}{(\mathbf{Z}_2 + \mathbf{Z}_L)(\frac{N_1}{N_2})^2 + jX_m}}$$

$$\frac{\mathbf{V}_1}{\mathbf{I}_1} = \quad\quad\quad jX_m \quad\quad \mathbf{Z}'_L \quad (c)$$

FIG. B.4. Two-coil transformer: (a) flow diagram; (b) reduction to one transfer function; (c) equivalent circuit.

the total secondary circuit impedance, and the equivalent circuit shown will be recognised from previous work; Figs. 3.25 and 4.7. The magnetising resistance $R_m = E_1/I_p$ could be included as shown by adding a subsidiary negative feedback loop with transfer function N_1/R_m, between E_1 and the m.m.f. summing point to give:

$$\mathbf{I}_m N_1 = \mathbf{I}_0 N_1 - \mathbf{I}_p N_1 = (\mathbf{I}_1 N_1 - \mathbf{I}_2 N_2) - (\mathbf{E}_1/R_m) N_1.$$

THE INDUCTION MACHINE EQUIVALENT CIRCUIT

The flow diagram for the induction machine is practically the same as for the transformer except that the secondary e.m.f. and reactance must be multiplied by the slip s. The equivalent circuit can be derived in just the same way as for the transformer and the only change will be in the value of the secondary circuit impedance which is divided throughout by the slip s and becomes $(R_2 + jsx_2 + Z_{ext})/s$.

THE SYNCHRONOUS MACHINE
EQUIVALENT CIRCUIT

Figure B.5a shows the flow diagram which has many similarities to that for the transformer. However, the secondary m.m.f. (field m.m.f.) is no longer due to a transformer-type feedback reaction, but is independently controlled by the field

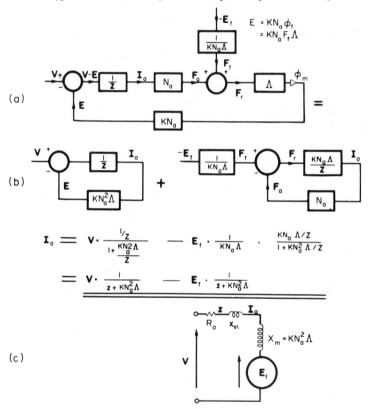

FIG. B.5. Synchronous machine: (a) flow diagram; (b) reduction by superposition; (c) equivalent circuit.

619

current applied, which gives rise to a flux ϕ_f and e.m.f. E_f. From the e.m.f. equation (5.8), at constant speed, this e.m.f. can be expressed as:

$$E_f = KN_a\phi_f = KN_aF_f\varLambda \quad \text{from which} \quad F_f = (1/KN_a\varLambda).E_f.$$

This equation explains the transfer function interposed between the voltage E_f and the m.m.f. summing point. Note that $\mathbf{E}_f$ is fed in with a negative sign in order that the resulting circuit equation shall be in line with the usual form; i.e. $\mathbf{I}_a = (\mathbf{V}-\mathbf{E}_f)/\mathbf{Z}_s$. Note also that $\mathbf{F}_a$ is *added* to $\mathbf{F}_f$ at the summing point in line with the convention adopted throughout the book.

The simplification of the diagram in this case involves the consideration of the two inputs $\mathbf{V}$ and $\mathbf{E}_f$ separately, so the diagram is broken down into two separate figures which have different forward- and feedback-transfer functions, Fig. B.5b. In each case, the forward transfer function is equal to the product of all the elements between the input voltage, $\mathbf{V}$ or $-\mathbf{E}_f$, and the armature current $\mathbf{I}_a$. When the two individual responses are examined, it is seen that they each consist of the voltage divided by the same impedance $R_a + jx_a + KN_a^2\,\varLambda$. By analogy with the transformer case, the expression $KN_a^2\varLambda$ is equal to the magnetising reactance X_m. The equivalent circuit is shown in Fig. B.5c, cf. Fig. 8.10. It corresponds, as does the flow diagram, to the motoring condition but it can be used for the generator by treating this as a negative motor or alternatively by redrawing the motoring flow-diagram with both $\mathbf{V}$ and $-\mathbf{E}_f$ having opposite signs to give $\mathbf{I}_a = (\mathbf{E}_f - \mathbf{V})/\mathbf{Z}_s$.

THE D.C. MACHINE EQUIVALENT CIRCUIT

From the equivalent-circuit viewpoint, the d.c. machine can be regarded as a synchronous machine with a rectifier in the armature circuit. This was the method used in the previous general derivation of equivalent circuits given in Section 3.8, though some refinements were also discussed. Consequently, for the d.c. machine, Fig. B.5c applies if the reactance $X_s = x_a + X_m$ is omitted, F_a is changed to F_a' and E_f is changed to E.

It would be instructive, on the flow diagram of Fig. B.5a, to consider how the provision of the field m.m.f. F_f could be shown for a shunt- or series-excited machine, tapping off V or I_a to give the necessary feedback function. The diagrams can also be used to represent transient conditions if inductance is retained as for Fig. 6.33, and it is worth a temporary digression at this stage to demonstrate, in a simple fashion, the action of a feedback control system, by modifying the circuit of Fig. 6.32 slightly. On Fig. B.6a, the generator field excitation is now provided by the difference between a reference voltage v, and the actual output equal to e_g on open circuit, which is fed back in a negative sense. The difference $v - e_g = v_f$ is a direct measure of the error between input and output voltages and tends automatically to correct any discrepancy. To explain this analytically, the block diagram of the system has been drawn on Fig. B.6b, the forward transfer function G, between v_f and e_g being the same as already derived from Fig. 6.32. τ is the field-circuit time constant L/R and K is the steady-state gain equal to the output voltage divided by the field terminal

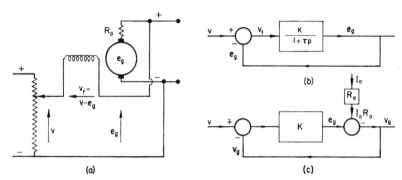

FIG. B.6. Closed-loop feedback control of generator voltage: (a) circuit, (b) block diagram on open circuit; (c) block diagram on load; steady state.

voltage. The feedback transfer function is unity: $H = 1$ so that the overall transfer function is:

$$\frac{e_g}{v} = \frac{G}{1 + GH} = \frac{K/(1 + \tau p)}{1 + K/(1 + \tau p)} = \frac{K}{1 + K + \tau p} = \frac{K/(1 + K)}{1 + \tau p/(1 + K)}.$$

Two effects of the negative feedback are now shown up. Firstly, the transient response is improved since the time constant is reduced by a factor $(1 + K)$ and secondly, the overall gain is reduced by the same factor. This loss of gain must normally be made up by inserting an amplifier in the field winding circuit which effectively increases the value of K. A further result of the feedback becomes obvious if the load condition is considered. The voltage drop $R_a I_a$ can be regarded as an additional input to the loop and the total output voltage obtained by superposition. On steady state, with $p = 0$, Fig. B.6c shows the block diagram on load from which the total output due to the two "inputs" v and $R_a I_a$ is given by:

$$v_g = \left(\frac{K}{1 + K}\right).v - \left(\frac{1}{1 + K}\right).R_a I_a.$$

The second term follows from the fact that the forward transfer function from $R_a I_a$ to v_g is unity and so the effective resistance drop is seen to be reduced by the factor $(1 + K)$. This means that the regulation is very small and is determined by the magnitude of the gain K. The overall effects of negative feedback give rise therefore to a substantially constant-voltage generator output which follows changes in field voltage with rapid response.

More elaborate systems can be dealt with in a similar manner; for example, the block diagram for a Ward–Leonard control system would require the motor transfer function to be cascaded with that of the generator and the feedback would normally be from a tachometer generator mounted on the motor shaft so that the system would act to maintain constant motor speed. Problem 9 for Chapter 6 in Appendix E, gives a further exercise in the use of block diagrams to obtain the appropriate motor trans-

621

fer function. When the system has more elements the overall differential equation is, naturally, more complex. The effects of gain and feedback still improve the response and reduce the regulation but, in addition, the introduction of more time lags accentuates the tendency to oscillatory behaviour; see Fig. 3.35. The corrective action can be upset because of the delays in processing the signal through the system and unstable oscillations of increasing magnitude could arise and would require suitable damping and stabilising circuits to be incorporated. The possibility of machines themselves to develop oscillatory behaviour can therefore be explained from the block diagrams. These show that there are inherent feedback loops, which, with the mechanical system included as in Problem 9 just referred to, reflect the presence of two energy stores. Oscillatory energy interchange between the magnetic field and the mechanical inertia could occur under conditions of light damping.

GENERAL EQUIVALENT CIRCUIT

On Fig. B.7 is shown a general equivalent circuit from which all the previous equivalent circuits can easily be derived if the appropriate operating conditions are taken into account. For the transformer and induction machine, E_f is omitted and the slip s must be taken as unity for the transformer. For the synchronous and d.c. machines, $s = 0$ which leaves, for the synchronous machine, the circuit of B.5c if the suffices on the primary quantities are omitted. For the d.c. machine, the reactance terms are removed as before. The magnetising resistance, if included, is connected across the air-gap e.m.f. as shown.

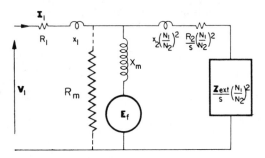

Fig. B.7. General equivalent circuit.

APPENDIX C

THE MEASUREMENT OF POWER IN 3-PHASE CIRCUITS: THE 2-WATTMETER METHOD

To determine the equivalent circuit parameters of machines by test requires the measurement of power, often in a polyphase circuit in which the number of phases is usually three. If three separate wattmeters were to be used, each measuring the phase power, it would be necessary, for the delta connection, to bring out six leads from the circuit in order to supply all the wattmeter current-coils. It is possible however to devise a measuring circuit which only requires the three line leads to be accessible and which employs only two wattmeters. In fact, only one wattmeter is necessary if suitable changeover arrangements are made, Fig. C.1b.

Consider Fig. C.1a. The total instantaneous power is:

$$v_A i_A + v_B i_B + v_C i_C.$$

For a 3-wire system:

$$i_A + i_B + i_C = 0.$$

It is possible therefore to eliminate one of the currents in the power equation so that it does not have to be measured. Substituting for:

$$i_C = -i_A - i_B$$

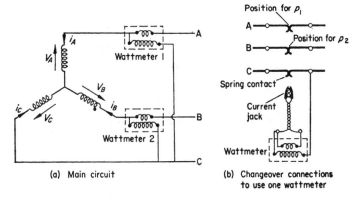

(a) Main circuit

(b) Changeover connections to use one wattmeter

Fig. C.1. Two-wattmeter method for measuring 3-phase power.

623

in the power equation, the total power is:

$$(v_A - v_C).i_A + (v_B - v_C).i_B$$
$$= \text{reading of wattmeter } 1(P_1) + \text{reading of wattmeter } 2(P_2).$$

The wattmeters can of course only respond to the average value of the instantaneous voltage × current product, see Fig. 3.8, so that the sum of $P_1 + P_2$ gives the average of the total circuit power. The equations show that this is true irrespective of whether or not the load is balanced. For a 3-phase balanced load, the three pulsating components of the phase power cancel one another, so that in this case, the total average power is the same as the total instantaneous power. Note that although a star-connected load is shown, the 2-wattmeter method is still valid with a delta connection. This can be demonstrated readily by transforming the delta circuit to an equivalent star-connected circuit.

BALANCED LOAD CONDITIONS

This is a particularly interesting case because the variations of the wattmeter readings with power factor follow simple curves from which the power factor itself can easily be determined. The phasor diagram for a lagging power factor is shown on Fig. C.2a and taking φ as positive under this condition the wattmeter readings are:

$$P_1 = V_L I_L \cos(30 - \varphi) \quad \text{and} \quad P_2 = V_L I_L \cos(30 + \varphi),$$

the line voltage and current V_L and I_L being the same amplitudes for each wattmeter.

$$P_1 + P_2 = V_L I_L \times 2.\cos 30.\cos \varphi = \sqrt{3} V_L I_L \cos \varphi$$
$$P_1 - P_2 = V_L I_L \times 2.\sin 30.\sin \varphi = V_L I_L \sin \varphi$$

from which
$$\tan \varphi = \sqrt{3}\,\frac{P_1 - P_2}{P_1 + P_2} = \sqrt{3}\,\frac{(P_1/P_2) - 1}{(P_1/P_2) + 1}.$$

Hence the ratio of the wattmeter readings is related to $\tan \varphi$ and therefore to $\cos \varphi$.

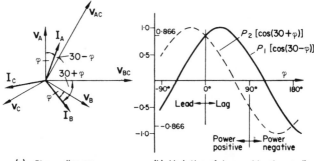

(a) Phasor diagram

(b) Variation of two wattmeter readings ($V_L.I_L$ constant)

FIG. C.2. Balanced load conditions.

The variations of the wattmeter readings with power factor, can be plotted from the expressions for P_1 and P_2. It must be remembered that the reading of one wattmeter in isolation gives no straightforward indication of the total power because it is measuring a line voltage and a phase current. This introduces a 30° time-phase displacement which is additional to that between phase voltage and current. A wattmeter reading reverses when the time-phase angle between the current in the current coil, and the voltage across the voltage coil exceeds 90°, cosines being negative beyond this angle. For a single-phase circuit, a wattmeter reversal means that the energy flow has reversed, the load becoming a source or vice versa. With the 2-wattmeter method however, it is the algebraic sum of P_1 and P_2 which must become negative before a reversal of energy flow is indicated. Figure C.2b shows that there are some easily remembered ratios of P_1/P_2 at the 30° points. For example, at 0·866 p.f. (30°), the ratio is 2:1. At a power-factor angle greater than 90° it can be seen that the sum of the wattmeter readings is negative; the power flow has reversed.

As a practical point it should be noticed that machine loads particularly are never perfectly balanced and the "average" power factor should be obtained from:

$$\frac{P_1 + P_2}{\sqrt{3} \cdot V_L \cdot I_L}.$$

V_L and I_L may themselves have to be averaged from two or more slightly different readings. However, near unity power factor, instrument errors may accumulate so that a value slightly greater than unity may result. In this case, the ratio of the wattmeter readings will be more accurate and give an "average" power factor which is always less than, or equal to unity.

APPENDIX D

MOTOR APPLICATION NOTES

THERE are many factors which must be considered when choosing a motor for a particular drive. In the following notes, the main points are highlighted but the final choice is usually a matter for the specialist. Not all the factors can be dealt with here, but the summary notes, together with the relevant sections of the main text, will be sufficient to give a good appreciation of the general problem.

The solution of this problem is not usually unique and opinions may differ widely in some cases as to the most suitable motor-drive system. However, it may be said in brief that the correct motor is the one which meets the required specification at minimum cost, but this is not a simple matter to estimate. It must include not only the capital cost of the motor-drive system itself but also the running costs. The capital cost includes the provision of any special supplies and control gear to run the motor. The running cost includes interest charges on capital equipment and buildings and charges for the power losses consumed in the machine and control circuits; obviously power factor and efficiency are of significance here. Maintenance is also a regular running cost and would normally be higher with complex control gear and with machines having rubbing contacts at slip rings and commutators. Installation costs too may be decisive; for example, special foundations are required for motor-generator sets but not for static conversion equipment. Such equipment also makes less demands on space and is less noisy than rotating machinery. On the other hand, there is considerable harmonic generation which poses a suppression problem.

Some motors are excluded from a particular application because the operating environment is "hostile" in some way, as in conditions of high temperature, high vacuum, high speed or in the presence of corrosive liquids. A brushless type of machine then becomes essential. Induction motors are generally the cheapest type of machine, especially if a single-cage rotor is satisfactory. The price increases as more demands are made on the control side and which might necessitate a wound-rotor machine. Even then the power factor remains poor unless correction is applied, and the brushless synchronous machine may therefore become competitive. If simple, infinitely-variable speed-control is required, then a.c. or d.c. commutator machines are called for, unless a variable-frequency supply with induction or synchronous motors can be justified. This special supply could be partly offset against the cost of d.c. or variable-voltage a.c. apparatus required for commutator machines.

SUMMARY OF FACTORS AFFECTING THE CHOICE OF MOTOR

Speed. Maximum speed at 50 Hz = 3000 rev/min for normal a.c. machines. Frequency changers required for higher speeds. Commutator machines can run at higher speeds though not at the higher ratings. For low speeds, the possibility of using a gearbox with a high-speed, and therefore a smaller and cheaper motor must be considered against the more expensive, but also more reliable low-speed direct drive.

Speed variation.
Constant speed—Synchronous or closed-loop control of d.c. or induction motor.
Approximately constant speed—Induction motor, d.c.-shunt or-compound motor.
Discrete speed variation—pole-changing on a.c. machine.
Natural variation with load—d.c. and a.c. series motors.
Small speed range—Induction motors with voltage or resistance control.
Wide speed range—d.c. or a.c. commutator motors, or variable-frequency supplies with induction or synchronous motors.

Load requirements.
Low or high starting torque.
Accelerating time—high torque required from zero to maximum speed.
Decelerating time—convenient and effective braking circuits.
Constant torque; constant power; torque = $f(\omega_m)$; overloads in service; duty-cycle of load variation; reversal of rotation—fast or slow.

Supplies. Provision of special supplies—d.c. or variable frequency.

Environment. Suitable motor; motor enclosure and protection; accessibility.

Running costs. Efficiency; power factor; maintenance; interest and depreciation considerations.

Capital costs. Must include all necessary equipment from main supply point to output shaft. D.C. and variable-frequency supplies usually start from H.V. a.c. mains, apart from battery-supplied systems.

Referring now to table D.1. it should be noted that it can only be used as an approximate guide to the relative performance of the different motor types in their natural state. Special design and control can often improve weak natural features, though some sacrifice in other aspects of performance may result. With the growing concern for energy conservation, more attention is now being given to improving efficiencies. The recent tendency to reduce costs by improving power/weight ratios, e.g. with better insulation permitting higher temperatures, has meant that efficiencies have been falling on the lower-kW sizes of machine. This trend, to overemphasise low capital costs could be reversed, e.g. by increasing core lengths to reduce flux densities. Attempts are now being made also to optimise running conditions, e.g. by voltage and/or speed reductions to reduce losses when operating at light-load.

The number of motors in use, in the U.K. for example, runs into millions and the great majority of these are in the small-power (< 1 kW) range. In terms of energy consumption, approximately two thirds of all power generated, (virtually all of it by large synchronous machines), is consumed by electric motors. About half of this is due to sizes between 1–100 kW, most of which are 3-phase induction motors of about 10–15 kW rating. D.C. and synchronous motors each account for about 10% of the energy used in electrical drives; single-phase and other small-power machines consuming about another 10%.

TABLE D.1. SUMMARY AND COMPARISON OF MOTOR FEATURES
Machine types listed very approximately in order of increasing first cost.

Type	Efficiency	Power Factor	Starting Torque	Overload Capacity	Speed Control	Speed Response	Maintenance
Induction machines							
Single-cage	Very good	Poor	Poor	Good	Poor	Poor	Excellent
Double-cage (or deep-bar)	Good	Poor	Good	Good	Poor	but good with variable-frequency supplies, as for Speed control	Excellent
Wound rotor	Very good	Poor	Very good	Good	Moderate		Good
Synchronous machines							
Reluctance	Moderate	Poor	Poor	Poor			Excellent
D.C. excited	Excellent	Excellent	Poor	Good			Good
D.C. series shunt and compound	Good		Excellent	Excellent	Excellent	Very good	Moderate
A.C. comm. motors	Good	Moderate	Very good	Good	Very good	Poor	Moderate

628

Synchronous motors are mostly in the larger sizes so that their numbers are much less than d.c. machines which cover a wider range and are themselves much smaller in numbers than 3-phase induction machines. See Reference 37.

Although the more interesting drives are those requiring special features and control, e.g. in the metal forming, paper, mining, machine-tool, transport, food processing and textile industries, the great majority of drives are for pumps, compressors, e.g. for refrigerators, and fans, all of these much used by the chemical and petroleum industries, which in total probably consume most of the energy used in motor drives. The much larger numbers of small-power drives are associated with leisure activities and games as major consumers, and with office and computer equipment and domestic appliances for refrigeration, washing, cleaning, etc., see p. 614.

Clearly the subject of motor applications is too vast to do other than indicate the general pattern. The text has mostly concentrated on describing the special features of the main machine types so that the applications for which they would be suitable can be appreciated. Reference back to the later sections of Chapters 6, 7 and 8 will fill out the rather concise statements made in this Appendix. In the list of References on page 653, the Conference Proceedings produced by the Institution of Electrical Engineers are full of useful and relatively-easily-digestible information on Electrical Machines, Applications and the Power Electronic circuits associated with them. The most recent ones[14,18,26,34,36] are especially interesting. See also References 29, 35 and 38.

APPENDIX E

PROBLEMS WITH ANSWERS

N.B. Many of these problems and several others with special emphasis on the topic of Electrical Drives, have been solved in the Author's companion volume, *Worked Examples in Electrical Machines and Drives*, Pergamon Press, 1982.

CHAPTER 2

(1) A sample of iron consists of laminations 0·4 mm thick and is built up into a square core of uniform cross section for testing purposes. The total weight of the core is 15 kg. The magnetising coil is supplied from a controllable sinusoidal source of voltage, which permits tests at various flux densities and frequencies to be carried out. Calculate the eddy current loss at 50 Hz when the peak flux density is 1·2 teslas. Take the effective iron resistivity as 0·22 $\mu\Omega$-m and the density as 7800 kg/m^3.

The power input to the coil after deducting its resistance loss is taken at two different flux densities with the frequency kept at 50 Hz. When the maximum flux density is 1·2 teslas the net input is 25 watts and this falls to 18·1 watts when the maximum flux density is reduced to 1·0 tesla. What is the hysteresis loss at 50 Hz and the value of Steinmetz index?

(2) In the armature slot of Fig. 2.9 are two conductors of rectangular cross section laid on top of one another, each conductor being one side of a single-turn coil. The top-layer conductor is 5 mm from the slot opening and the bottom-layer conductor is 2·5 mm from the bottom of the slot. Each conductor is 10 mm deep. Calculate:

(a) the *self* inductance per metre of axial length for each conductor due to that portion of its flux crossing within the slot;

(b) the flux density at the top of the slot when each conductor carries 250 A flowing axially in the same direction.

(3) From the answers to Example E.2.1 derive the electrical analogue of the magnetic circuit in the form shown in Fig. 2.11.

CHAPTER 3

(1) Using the data of Example E.3.1, and considering case (a)(i), express the voltage V_{CA} and the current I_{CA} as complex numbers and check that the phase kVA is given by $V^*_{CA} I_{CA}$. Take V_{AB} as reference phasor.

Two wattmeters are used to measure the total power input to the load. One wattmeter has its current coil in line A and its voltage coil measuring V_{AB}. The other wattmeter measures the current in line C and the voltage V_{CB}. Express these voltages and currents as complex numbers and hence calculate the readings of the two wattmeters from V^*I for each wattmeter and check that the sum of the readings gives the total power dissipated. See Appendix C.

(2) Using the data given on Fig. 3.19a write down the equation for the external characteristic for each battery.

Calculate the common terminal voltage and the current delivered by each battery when the total current drawn by an external load is 5 A.

(3) The temperature rise of a certain machine may be taken to follow a simple exponential law, of time constant τ hours, and to reach a maximum value $\hat{\theta}$ if full load is maintained. The machine is subjected to a repetitive duty cycle in which full load is alternately switched on for a period of τ hours and then off for τ hours. After several cycles, the temperature rise varies regularly between an upper limit θ_u and a lower limit θ_1. By considering the laws of the heating and cooling curves deduced from eqn. (3.21), calculate the temperature rise θ_u reached at the end of an "on" period and the reduced temperature rise θ_1 at the end of an "off" period, the temperature then tending to zero. Express the temperature rise in terms of $\hat{\theta}$.

CHAPTER 4

(1) A transformer core has a square cross-section of 20 mm side. The primary winding is to be designed for 230 V, the secondary winding for 110 V and a further, centre-tapped, 6/0/6-V winding is to be provided. If the flux density is not to exceed 1 tesla, find a suitable number of turns for each winding. Neglect all transformer imperfections.

(2) A 20-kVA, 3810/230-V, 50-Hz, single-phase transformer operates at a maximum flux density of 1·25 teslas. The mean length of the magnetic circuit is 1·4 m and its cross sectional area is 16,000 mm². The half hysteresis-loop is given by the following readings.

B	0	0.3	0.6	0.8	1.0	1.1	1.2	1.25	tesla
H	44	69	100	128	168	209	297	356	At/m
	−44	−17	15	48	109	162	275		At/m

The primary winding has 860 turns. The eddy current loss in the iron is 10^4 watts/m³. Plot the no-load current wave of the transformer assuming the back e.m.f. is sinusoidal. Note that the component of I_p due to the eddy-current loss must be drawn in time-phase with the e.m.f. and added to the hysteresis and magnetising components of I_0.

(3) Assuming instead that the no-load current is sinusoidal and of the same peak value as found in the last question, estimate and plot the approximate variation of flux density which this would produce.

(4) The transformer of question 2 is to be used as an inductor employing one winding. To keep the reactance substantially constant with current, the core is sawn through transversely and packed with brass to give an effective total air gap of 5 mm. Calculate the voltage rating at 50 Hz if the low voltage winding is used and the current is at rated value. The winding resistance and the m.m.f. absorbed by the iron may be neglected.

(5) The peak value of the mutual flux for a transformer operating on a constant-

631

voltage, constant-frequency supply is itself substantially constant over the operating range. Explain why this is so by reference to Fig. B.4a, p. 618, and by considering how the feedback effects of m.m.f. and e.m.f. react against any tendency for $\bar{\phi}_m$ to change.

(6) If a coil of N turns carries a current i, the flux change produced per ampere-turn change is Λ, the permeance of the magnetic circuit. If a second coil is mutually coupled, however, and has N_e turns with a circuit impedance z_e, the induced current i_e will oppose the change of flux so that the effective permeance is reduced to $| z_e/(jX_m + z_e) | \times \Lambda$, neglecting the small effect of iron loss. By considering the second coil m.m.f. as a negative feedback $i_e N_e$, opposing the input iN, see Fig. B.4, prove that this expression is correct. Note that this permeance reduction factor could also be obtained from the equivalent circuit and that it may be regarded as a general damping factor explaining, for example, the effect of an eddy-current m.m.f. resisting any change of flux.

(7) The mean length of turn (total copper length/number of turns) is the same for both the primary winding (N_1 turns) and the secondary winding (N_2 turns) for a particular transformer. Both windings also operate at the same current density. Neglecting magnetising current show that $R_2' = R_1$ and $R_1' = R_2$. If the leakage permeances for the two windings are the same show that $x_2' = x_1$ and $x_1' = x_2$.

(8) In a certain transformer the winding leakage reactances are four times larger than the winding resistances. Estimate the power factor for which the voltage regulation is zero.

(9) The primary and secondary impedances of a single-phase transformer are:

$$z_1 = 1 \cdot 4 + j5 \cdot 2\ \Omega \quad \text{and} \quad z_2 = 0 \cdot 0117 + j0 \cdot 0465\ \Omega.$$

The primary/secondary turns ratio is 10·6/1. If the primary voltage is 6600 V, calculate the kW output when the secondary load takes 300 A at 0·8 p.f. lagging.

(10) A 3-phase, 11,000/660-V, star/delta transformer is connected to the far end of a distribution line for which the near-end voltage is maintained at 11 kV. The effective leakage reactance and resistance per phase of the transformer are respectively 0·5 Ω and 0·25 Ω referred to the low-voltage side. The reactance and resistance of each line are respectively 2 Ω and 1 Ω.

It is required to maintain the terminal voltage at 660 V when a line current of 260 A at 0·8 lagging power-factor is drawn from the secondary winding. What percentage tapping must be provided on the high-voltage side of the transformer to permit the necessary adjustment? The transformer magnetising current may be neglected and an approximate expression for regulation may be used. Neglect the changes to the impedances due to the alteration of the turns ratio.

(11) An 800-kVA transformer at normal voltage and frequency requires an input of 7·5 kW on open circuit. At reduced voltage and full-load current it requires 14·2 kW input when the secondary is short circuited. Calculate the all-day efficiency if the transformer operates on the following duty cycle:

6 hours	500 kW	0·8 p.f.
4 hours	700 kW	0·9 p.f.
4 hours	300 kW	0·95 p.f.
10 hours	No load	

(12) On a certain 60-kVA, 50-Hz transformer the losses at normal rating are distributed as follows:

Eddy-current loss	300 W
Hysteresis loss	400 W
Copper loss	650 W

An open-circuit test at half voltage and normal frequency requires an input of 205 W. Calculate Steinmetz index. What would be the efficiency if run from a 60-Hz supply at normal voltage and current and at unity power-factor?

(13) For what purposes are tertiary windings used on 3-phase transformers? Explain how they can assist in unbalanced loading conditions if suitably connected.

On open-circuit, a 3-phase, star/star/delta, 6600/660/220-V transformer takes 50 kVA at 0·15 p.f. What is the primary input kVA and power factor when for balanced loads the secondary delivers 870 A at 0·8 p.f. lagging and the tertiary delivers 260 line amperes at unity power-factor? Neglect the leakage impedances.

(14) Two single-phase transformers operate in parallel to supply a load of 24 + j10 Ω. Transformer A has a secondary e.m.f. of 440 V on open circuit and an internal impedance in secondary terms of 1 + j3 Ω. The corresponding figures for transformer B are 450 V and 1 + j4 Ω. Determine the current and *terminal* power-factor of each transformer. What is the terminal voltage?

(15) Briefly discuss the essential and ideal conditions to be observed when operating two 3-phase transformers in parallel.

Two 11,000/660-V 3-phase transformers, one 200 kVA and the other 500 kVA are supplying a total load of 700 kVA at a power factor of 0·9 lagging. Calculate the output current of each transformer. The internal impedances referred to the secondary are respectively 0·014 + j0·06 Ω and 0·007 + j0·043 Ω per phase.

(16) The following are the light-load test readings on a 3-phase, 100-kVA, 400/ 6600-V, star/delta transformer:

Open circuit: supply to 400-V side	1250 W, 400 V
Short circuit: supply to 6600-V side	1600 W, 314 V, full-load current

Calculate the efficiencies at full-load 0·8 p.f. and at half full-load unity power-factor. Calculate also, the maximum efficiency and the *per-unit* resistance and reactance.

(17) The open-circuit test on a transformer rated at 230 V, 50 Hz, was taken over a range of voltages and frequencies, the following readings being noted:

Applied voltage	250	200	150	100	V
Frequency	54·4	43·5	32·6	21·7	Hz
Input power	130	93·5	62·4	36·2	W

If the copper loss can be taken as negligible during this test, calculate the eddy-current loss and the hysteresis loss at normal voltage and frequency.

(18) Describe economical methods of performing a temperature rise test on:

      (a) two identical 3-phase transformers,

      (b) a single 3-phase transformer.

Determine the supplies required for the first case for 2000-kVA, 33/6·6-kV, delta/star transformers which each have 20 kW copper loss at full load, 6·5 kW iron loss, a leakage reactance of 5% and take a no-load current of 4 A when supplied from the 6·6 kV side. In the test, the iron loss is to be supplied from the 6·6 kV side.

(19) A 50-Hz, Scott-connected transformer supplies an unbalanced 2-phase load at 200 V per phase. For the leading phase (phase "a") the load has a resistance of 10 Ω and an inductance of 42·3 mH. For the other phase, the load consists of a resistor of 13·3 Ω and a capacitor of 318 μF in series. Neglecting magnetising current and the internal impedance of the transformers, calculate the line currents on the 3-phase side. The main transformer primary/secondary turns ratio is 12/1.

(20) A 2-phase system supplies, through a Scott-connected transformer, a 3-phase star-connected load in which the line to neutral voltages are balanced but the phase currents are as follows:

Phage A; (connected to teaser transformer) $I_A = 50$ A at 0·866 p.f. leading,

Phage B; $I_B = 70·7$ A at 0·966 p.f. leading,

Phage C; $I_C = 50$ A at unity power-factor.

Neglecting the magnetising current and internal impedance, calculate the supply currents and power factors. The 2-phase line voltage is 3300 V and the 3-phase line voltage is 400 V.

CHAPTER 5

(1) A 4-pole, 250-kW, 300-V, 1000-rev/min, d.c. generator has an armature diameter of 0·56 m and core length of 0·28 m. There are 264 conductors in 66 slots forming a lap winding with single-turn coils. The length of one coil is 2 m and the copper cross section is 35 mm². Calculate the armature resistance between brushes (see p. 354) and the necessary e.m.f. at full load, neglecting the voltage absorbed at the brush contact. The copper resistivity at the working temperature can be taken as 0·021 μΩ m. Calculate the average flux density required to generate this e.m.f. and also the armature m.m.f. per pole at full load.

If 11 equalisers are used, to which conductors should they be connected? (See Fig. 5.9 for numbering system.)

(2) A 6-pole d.c. motor has an armature diameter of 0·9 m and a core length of 0·35 m and is to operate at an average flux density of about 0·66 tesla when supplied at 200 V and running at 200 rev/min. The internal voltage drop at full load may be taken as 10 V. If the slot pitch must not be less than 20 mm nor greater than 35 mm choose a number of slots to accommodate: (a) a wave winding, (b) a lap winding. Specify the essential coil particulars in each case, see Examples E.5.2 and E.5.4. What will be the value of B in each case?

Note. In this case either a lap or wave winding would be satisfactory. This is not generally the case though the situation does arise on intermediate sizes of machines.

(3) A 6-pole d.c. machine is required to generate an e.m.f. of 260 V when running at 450 rev/min. The flux per pole is not to exceed 0·02 Wb. If the armature has 74 slots determine the particulars of a suitable wave winding.

(4) A 3-phase, star-connected, 50-Hz, 6-pole generator has a fundamental flux per pole of 0·015 Wb. The stator has 90 slots and 4 conductors per slot and each stator coil spans 12 slots.

If the flux in the air gap contains a 30% third harmonic and a 20% fifth harmonic with respect to the fundamental, determine the r.m.s. values of the phase and line induced e.m.f.s.

(5) Derive an expression for the generated e.m.f. of a d.c. machine and explain the modifications necessary so that it is suitable for calculating the e.m.f. of an a.c. machine.

A 3-phase, 50 Hz, 6-pole, star-connected alternator has 972 conductors distributed in 54 slots. The coils are chorded by one slot. Calculate the fundamental-frequency line voltage on no load, when the fundamental flux per pole is 0·01 Wb.

If the r.m.s. current per conductor is 96 A, what is the peak value of the fundamental, 5th and 7th harmonic components of the armature m.m.f. wave in ampere-turns per pole?

(6) A 3-phase star-connected winding has a resistance of R Ω per phase. Show that the d.c. excitation power for any given value of ampere-turns per pole is the same for both of the connections shown in Fig. 5.30.

CHAPTER 6

N.B. Conversion factors for "engineers" units; 1 hp = 746 watts, 1 lbf ft = 746/550 Nm.

(1) By considering the average force and torque on the conductors of a d.c. armature winding each of active length l, carrying the current per circuit $I_a/2a$ in an average flux density B_{av}, show that the total torque due to the Z conductors in an armature of diameter d is:

$$\frac{pz_s}{\pi}.\phi.I_a \text{ Nm}, \quad \text{where} \quad z_s = Z/2a.$$

(2) A 4-pole d.c. armature wave winding has 294 conductors.

(a) What flux per pole is necessary to generate 230 V at 1500 rev/min?

(b) What is the electromagnetic torque at this flux when the armature current is 120 A?

(c) How many interpole ampere-turns are required with this current if the interpole gap density is to be 0·15 tesla and the radial air gap is 8 mm? Neglect the m.m.f. absorbed by the iron.

(d) Through what mechanical angle must the brushes be moved if it is required to produce a direct-axis magnetisation of 200 At/pole?

(3) It is found that the voltage of a certain d.c. shunt generator will not build up. List the possible causes of failure.

(4) A d.c. shunt-wound generator rated at 220 V and 40 A armature current has the following o.c. characteristic when running at 500 rev/min with its shunt field excited:

E.M.F.	71	133	170	195	220	232	V
Field current	0·25	0·5	0·75	1·0	1·5	2·0	A

The armature resistance is $0.25\,\Omega$. The shunt field resistance is $110\,\Omega$ and there are 2500 turns per pole. Calculate:

(a) the range of external field-circuit resistance necessary to vary the voltage from 220 V on full load to 170 V on no load when the speed is 500 rev/min.

(b) the maximum voltage on o.c. if the speed is reduced to 250 rev/min and all the external field circuit resistance is cut out.

(c) the series-winding m.m.f. required to give a level-compound characteristic at 220 V, 500 rev/min.

Armature reaction and brush drop may be neglected.

(5) The shunt field inductance (unsaturated) of the machine in the last question is 23 henrys. Derive the voltage transfer function of the machine when running as a separately excited generator at 1000 rev/min on open circuit, operating on the unsaturated part of the magnetisation curve and without external field resistance.

If the field is excited from a sinusoidally alternating voltage at what frequency will the phase angle between output armature voltage and input field voltage be 45° and what will then be the voltage gain?

(6) The machine of Question (4) is to be run as a motor from 220 V. A speed range of 2/1 by field control is required. Calculate, neglecting F_a' and brush drop:

(a) the range of external field circuit resistance required as a shunt motor to permit speed variation from 500 rev/min on no load to 1000 rev/min with the armature carrying its rated current of 40 A.

(b) The value of series-field ampere-turns required to cause the speed to fall from 500 rev/min on no load, by 6% when full-load current is taken.

(c) The speed regulation, (no load to 40 A load), with this series winding in circuit and the shunt field set to give 1000 rev/min on no load.

(d) By how much would this series winding increase the torque at 40 A compared with condition (a) at the minimum field-current setting?

(7) The motor of the last question is to be started from a 220 V supply which is also connected directly across the shunt field circuit with its external resistance cut out. The series winding is not in circuit. An external armature resistance is used to limit the starting current to 80 A and is left in circuit while the motor runs against a torque T_m corresponding to this field flux and rated armature current. Calculate the steady-state speed and show that the torque developed during the build up of speed is given by the expression $T_e = 352 - 7.03\,\omega_m$ Nm.

Show that the response of speed is of simple exponential form when this load torque T_m is constant and develop the differential equation when the total coupled inertia is 13.5 kg-m². What is the electromechanical time-constant?

(8) The shunt motor of Question (6) when running on no load (mechanical and iron loss only) takes a total current of 5 A when both field and armature are directly connected across the 220 V supply. What is the speed, efficiency and output hp when the armature current has increased to 40 A?

(9) Show that the differential equation relating d.c. motor speed to terminal voltage for a pure-inertia load and including the armature inductance is:

$$k_\varphi(\tau_m \cdot \tau_e \cdot p^2 + \tau_m \cdot p + 1)\omega_m = V,$$

where $\tau_m = JR/k_\phi{}^2$, is the electromechanical time-constant and $\tau_e = L/R$ is the electrical-circuit time-constant. This equation should be derived from the transfer

636

function ω_m/V, formed by considering the d.c. motor as a feedback system in a similar manner to Fig. B.5a, p. 619, but with the mechanical system in place of the magnetic circuit; i.e. using the relations $T_e = k_\phi \cdot I_a$; $\omega_m = T_e/J.p$ and for the feedback, $E = k_\phi . \omega_m$. Note that from the quadratic characteristic equation, see p. 128, it follows that the speed response will be oscillatory of natural frequency,

$$\frac{1}{2\pi}\sqrt{\left(\frac{k_\phi^2}{JL} - \frac{R^2}{4L^2}\right)} \text{ Hz}$$

if $4\tau_e > \tau_m$, which would only occur on a high-performance, low-inertia system. Note also that the quadratic reduces to the same equation as on p. 314 if L is neglected.

(10) A d.c. series motor has a *per-unit* resistance of 0·05 based on the full-load $R_a I_a$ drop as a fraction of the e.m.f. at rated load; i.e. $V = 1·05$ p.u. and 1 *per-unit* speed then becomes the rated value; see p. 321, noting the change of voltage base. Assuming that the machine is unsaturated, i.e. $k_{\phi p.u.} = I_{ap.u.}$ calculate:

(a) the *per-unit* speed and current when the torque is 0·5 p.u.

(b) the *per-unit* speed and torque when the current is 0·5 p.u.

(c) the *per-unit* current and torque when the speed is 0·5 p.u.

(11) A 500-V d.c. series motor has an armature circuit resistance of 0·8 Ω. The magnetisation curve at 550 rev/min is as follows:

Field current	20	30	45	55	60	67·5	A
Generated e.m.f.	309	406	489	521	534	545	V

The motor drives a fan, the total mechanical torque being given by the expression $T_m = 10 + (\text{rev/min})^2/2250$ lbf ft.

Plot the speed/torque curves and hence find the steady-state speed and torque under the following circuit conditions:

(a) when an external starting resistance, used to limit the starting current at full voltage to 60 A, is left in circuit;

(b) when all the external resistance is cut out;

(c) when only 2/3 of the series winding turns are used, a field tapping being provided at this point in the winding. The armature circuit resistance may be taken as unchanged at 0·8 Ω.

(12) The magnetisation curve of a d.c. series motor is taken by running it as a separately excited generator in accordance with Fig. 6.21b, field and armature currents being adjusted together to the same value. The following readings were obtained when the speed was maintained constant at 400 rev/min:

Terminal voltage	114	179	218	244	254	V
Field current	30	50	70	90	110	A

(a) If the motor armature and field resistance are each 0·08 Ω find the current and speed at which a torque of 350 lbf ft will be developed as a series motor running from 250 V.

(b) If the motor is driving a hoist and the load can "overhaul" the motor so that the operating point is in the positive-torque, negative-speed quadrant, how much external armature circuit resistance will be necessary to hold the speed at 400

637

rev/min when the torque is 350 lbf ft? This will be the same current as in part (a) but note that the e.m.f. is now reversed, supporting current flow; the machine is generating and the resistance is absorbing $V + E$ volts.

(13) A 220-V d.c. series motor runs at 700 rev/min when operating at its full-load current, 20 A. The motor resistance is 0·5 Ω and it may be assumed unsaturated. What will be the speed if:
(a) the load torque is increased by 44%?
(b) the motor current is 10 A?

(14) A 250-V d.c. shunt motor has $R_a = 0·15$ Ω. It is permanently coupled to a constant-torque load of such magnitude that the motor takes an armature current of 120 A when running at a speed of 600 rev/min. For emergency, provision must be made to stop the motor from this speed in a time not greater than 0·5 sec. The peak braking current must not exceed twice the rated value and dynamic braking is to be employed with the field excited to give rated flux corresponding to rated speed running. Determine the value of the external armature-circuit resistance required and the maximum permissible inertia of machine and load which will allow braking to standstill within the specified time.

(15) A 500 V, 60 hp, 600 rev/min d.c. shunt motor has a full-load efficiency of 90%. The field resistance is 200 Ω and the armature resistance is 0·2 Ω. Calculate the rated *armature* current and hence find the speed under each of the following conditions at which the machine will develop an electromagnetic torque equal to the rated value.
(a) Regenerative braking; no limiting resistance.
(b) Plugging, (reverse current) braking—external limiting resistance of 5·5 Ω inserted.
(c) Dynamic braking—external limiting resistance of 2·6 Ω inserted.

The field current is maintained constant and armature reaction and brush drop may be neglected.

The machine is braked from full-load motoring, using the circuit configurations of (b) and (c). What time does it take in each case to bring the machine to rest? The inertia of the machine and coupled load is 4·6 kg m² and the load, which is coulomb friction, is maintained.

(16) A d.c. shunt motor has its supply voltage so controlled that it produces a speed/torque characteristic following the law: Rev/min = $1000\sqrt{[1 - (0·01T_e)^2]}$, where T_e is in Nm. The total mechanical load, including the machine loss-torque has the following components: coulomb friction 30 Nm; viscous friction (α speed), 30 Nm at 1000 rev/min; and a fan-load torque (α(speed)²), 30 Nm at 1000 rev/min. The total coupled inertia is 4 kg m². By plotting the motor and the load speed/torque characteristics, determine the balancing speed and also calculate the time to reach 98% of this speed starting from rest. Note that the answer given corresponds to a graphical integration with increments taken between the following speeds:

0, 100, 200, 300, 400, 500, 600, 650, 685, and 715 rev/min.

(17) When the motor of Question (16) is supplied instead, from its rated voltage, at rated flux and field current, it runs at 1000 rev/min when driving the same mechanical load and it then draws rated armature current. Take these rated values as 1 *per unit*, except for speed for which, with $V_R/k_{\phi R} = 1$ *per unit* speed, the equations

developed on p. 319 *et seq.* can then be used directly with *per-unit* quantities. Assuming that the armature resistance has a value of 0·06 *per unit* and that the *per-unit* field current is given by the expression: $I_f = 0.6\phi/(1-0.4\phi)$ where ϕ is the *per-unit* flux (k_ϕ *per unit*), calculate:

(a) the magnitude of 1 *per-unit* speed in rev/min;

(b) the required terminal voltage in *per unit* if the resistance is increased to 0·2 *per unit*, the field current is reduced to 0·6 *per unit* and the speed is to be at the rated value;

(c) the required I_f and the value of I_a *per unit* if the speed is to be 600 rev/min with the terminal voltage set at 0·5 *per unit*;

(d) the required values of I_f and I_a when the terminal voltage is equal to the rated value and the speed is to have magnitudes of (i) 0·8 *per unit* and (ii) 1200 rev/min.

Note that the answers to (d) will show that field control of speed is unsatisfactory for this type of mechanical load because of the excessive currents.

(18) Three d.c. generators are operating in parallel with excitations such that their external characteristics, which may be taken as straight lines over the working range, are defined by the following pairs of points:

| | Terminal Voltage | | |
Load current	Generator A	Generator B	Generator C
0	492·5	510	525
2000 A	482·5	470	475

Determine the terminal voltage and the current in each generator when:

(a) the total load current is 4350 A;

(b) the load is completely removed without change of excitation currents.

(19) In the circuit of Fig. 6.50, machine 1 is a motor and machine 2 is a generator, the supply being 250 V. The armature currents are respectively 50 A and 40 A and the field currents are respectively 2 A and 2·4 A. Taking the armature resistance and the brush drop to be 0·3 Ω and 2 V respectively for each machine, calculate the sum of iron loss, mechanical loss and stray loss per machine, and hence estimate the operating efficiencies of motor and generator under these conditions.

(20) An electrically driven automobile is powered by a d.c. series motor rated at 72 V, 200 A. The motor resistance is 0·04 Ω and its inductance is 0·006 henry. Power from a 72 V battery is supplied via an ON/OFF controller having a fixed frequency of 100 Hz but with a variable mark/space ratio. When the machine is running at 2500 rev/min, the generated e.m.f. per field ampere is 0·32 V, which may be assumed as a mean "constant" value over the operating range of current. Determine the maximum and minimum currents and the mean torque and power produced by the motor when operating at a speed of 2500 rev/min with a pulsed d.c. supply as described and a ratio of t_{ON} to t_{OFF} of 3:2. The mechanical losses may be neglected.

(21) A d.c. shunt motor is being considered as a drive for different mechanical loads having the following characteristics: (a) constant power ($T_m\omega_m$), (b) constant torque and (c) torque proportional to speed. It is desired to know the effect on armature current and speed of making various changes on the electrical side. Taking as a basis that rated voltage, armature current and field current give rated speed and

torque, express armature current and speed in *per unit*, when the following changes are made: (i) field current reduced to give half flux; (ii) armature voltage halved and (iii) armature voltage and field flux both halved. Consider loads (a), (b) and (c) in turn and neglect all machine losses.

CHAPTER 7

N.B. Conversion factors for "engineers" units; 1 hp = 746 watts, 1 lbf ft = 746/550 Nm.

(1) A 440-V, 3-phase, 50 Hz, 8-pole, star-connected induction motor has the following equivalent circuit parameters per phase:

$$R_1 = R_2' = 0.1\,\Omega \quad x_1 = x_2' = 0.7\,\Omega \quad R_m = 100\,\Omega \quad X_m = 25\,\Omega.$$

Calculate the rotor current referred to the stator, the stator current, the input power-factor, the torque and efficiency at 4% slip and the starting torque:
(a) using the exact circuit;
(b) using the approximate circuit;
(c) neglecting the stator impedance $R_1 + jx_1$.
Take the mechanical loss as 1 kW.

Note. It will be found that method (c) is quite inaccurate even at the slip of 4% where the torque is nearly 90% of the maximum. The full-load slip on this motor would be about 2% and then the errors would be less noticeable, R_2'/s being the dominant parameter. In subsequent problems, the approximate circuit, Fig. 7.11b, is to be used unless otherwise stated.

(2) A 3-phase, 200 hp, 3300-V, star-connected induction motor has the following equivalent circuit parameters per phase:

$$R_1 = R_2' = 0.8\,\Omega \quad x_1 = x_2' = 3.5\,\Omega.$$

Calculate the slip at full load if the friction and windage loss is 3 kW. How much extra rotor resistance per phase would be necessary to increase the slip to three times this value with full-load torque maintained? How much extra stator resistance per phase would be necessary to achieve the same object without changing the rotor circuit resistance and what loss of peak torque would result?

(3) In a certain 3-phase induction motor the leakage reactance is five times the resistance for both primary and secondary windings. The primary impedance is identical with the referred secondary impedance. The slip at full load is 2%. It is desired to limit the starting current to three times the full-load current. By how much must:
(a) R_1 be increased?
(b) x_1 be increased?
(c) R_2 be increased?
What loss of maximum torque would result in each case if the extra impedance was left in circuit?

(4) A 440-V, 3-phase, 6-pole, 50-Hz, delta-connected induction motor has the following equivalent circuit parameters at normal frequency:

$$R_1 = 0.2\,\Omega; \quad R_2' = 0.18\,\Omega; \quad x_1 = x_2' = 0.58\,\Omega \text{ per phase.}$$

640

The machine is subjected in service to an occasional fall of 40% in both voltage and frequency. What total mechanical load torque is it safe to drive so that the machine just does not stall under these conditions?

When operating at normal voltage and frequency, calculate the speed when delivering this torque and the power developed. Calculate also the speed at which maximum torque occurs.

If now the machine is run up to speed from a variable voltage, variable-frequency supply, calculate the required terminal voltage and frequency to give the "safe" torque calculated above: (a) at starting; (b) at 500 rev/min. Repeat the calculation for the machine to develop a torque equal to the maximum value occurring at rated voltage and frequency. In each case, adjust the air-gap flux (E_1/f) so that this is constant for any particular torque, see Fig. 7.24 and the associated text.

By how much would the starting torque be increased over that occurring at normal voltage and frequency if both V and f are halved? Note; use the $I_2{}^2 R_2/s\omega_s$ ratios to avoid unnecessary duplication of calculations.

(5) A 3-phase, 50-Hz, 50-hp, 500-V, 6-pole, delta-connected induction motor has the following equivalent circuit parameters, all referred to the primary:

$$R_1 = R_2' = 0.5\,\Omega \text{ per phase}; \quad x_1 = x_2' = 2\,\Omega \text{ per phase}.$$

The phase turns ratio is 2/1, stator to rotor, but the rotor is star connected. Find the torque developed at 971 rev/min.

How much resistance will have to be added to the rotor circuit to reduce the speed to 900 rev/min with the torque maintained?

(6) The primary and secondary resistances per phase of a 3-phase, 50-Hz, 1460 rev/min induction motor are respectively $0.25\,\Omega$ and $0.017\,\Omega$. The effective primary/secondary turns ratio per phase is $3.6/1$. When started at reduced voltage with the slip rings short-circuited the power factor is 0.22.

Neglecting magnetising current, estimate the leakage reactance and hence calculate the value of maximum torque in lbf ft and the horsepower at normal speed with 500 V per phase applied.

(7) In what ways is a 3-phase induction motor similar to, and in what ways is it different from a 3-phase transformer? Explain how the differences and similarities affect the parameters and treatment of the equivalent circuit.

Determine the *per-unit* values of the starting torque and the maximum torque using full-load torque as the base quantity, for an induction motor having a full-load slip of 3%. The primary and referred secondary impedances are identical and the leakage reactance is five times the resistance.

(8) A 3-phase, 50-Hz, 4-pole slip-ring induction machine has the following equivalent circuit parameters per phase:

$$R_1 = 1\,\Omega, R_2 = 4\,\Omega, x_1 = 4\,\Omega \quad \text{and} \quad x_2 = 12\,\Omega.$$

The magnetising admittance referred to the primary is:

$$Y_m = 0.005 - j\,0.025\,\Omega^{-1} \text{ per phase}.$$

Both the primary and secondary windings are star connected and the primary/secondary phase turns ratio as $1:\sqrt{3}$.

Using the exact circuit, and neglecting the mechanical losses, determine the magnitude and sense of the power flow at the electrical terminals and at the mechanical coupling when the speed is 1560 rev/min, the line voltage is 660 V and the slip rings are short-circuited.

(9) Using the approximate circuit with the data of problem (1), show on a power-flow diagram like Fig. 6.34, the electrical terminal input, the iron loss, the stator copper loss, the air-gap power, the rotor copper loss, the electromechanical output (negative in this case), and the total mechanical input when the machine is running at 720 rev/min in the *opposite* direction to the rotating field; i.e. plugging. Show that the *total* input to the machine is equal to the sum of all the losses. Note that the rotor copper loss will appear in the control lead, the control terminals being short-circuited in this case.

(10) A 6-pole induction generator operates at 50 Hz, 88·5 kVA, 0·77 p.f. leading and at a line voltage of 500 V. The stator is star-connected and the rotor runs at 1,030 rev/min to generate this power. Calculate the constants of the approximate equivalent circuit given that the machine takes 55·5 A at 0·1 p.f. when running at synchronous speed from 500 V. The stator and referred rotor resistances may be assumed identical.

Note. Problems (11) to (15) are intended to be solved by using a circle diagram.

(11) On no load at 500-V, a certain 3-phase, 50-Hz, 4-pole induction motor takes 16 A at a power factor of 0·2. With the rotor locked and 125 V applied, it takes 37·5 A at 0·3 p.f.

If, when the motor is running at full load and supplied at 500 V, the power factor is at its maximum, estimate the full-load current, power factor, torque, speed, output and efficiency. Also estimate the starting torque and maximum torque at full line voltage. Assume $R_2' = 1·3 R_1$.

By considering the circle diagram and its basis, how would you expect the performance to be affected if the motor was run from a 500-V, 60-Hz supply?

(12) A 3-phase, 4-pole, 440-V, 50-Hz slip-ring induction motor is started at full voltage uncoupled from its load. When a fixed external rotor resistance of value four times the rotor resistance itself is inserted at starting, the starting current and power factor are respectively 200 A and 0·6. The motor runs up to speed and the slip rings are then short circuited. On steady conditions being achieved the current and power factor become 12·5 A and 0·2. Assuming $R_1 = R_2'$,

(a) Estimate the maximum torque and the total external resistance, expressed in terms of the rotor resistance, required to get this torque at standstill with full voltage applied.

(b) What would be the line current if started from full voltage without external rotor resistance?

(c) How much external rotor resistance would be required to double the natural starting torque?

(13) For a certain polyphase induction motor, which at full load operates at its maximum power factor, the starting current and the no-load current at full voltage are in the ratio of 10/1 and the power factor in each case is 0·25.

Assuming the rotor and stator copper losses are equal, determine the percentage reduction in voltage which would just permit stable operation at a torque not exceeding the full-load value. What would be the power factor and slip at full-load torque with the voltage thus reduced?

(14) A 400-hp, 1100-V, 3-phase induction motor gave the following data during

642

light-load tests, the figures representing line values.

No load   1100 V 100 A 22·2 kW total input; mechanical loss 8 kW
Locked rotor   230 V 200 A 18 kW total input
Resistance between any two stator terminals 0·144 Ω.

Using this resistance, satisfy yourself that the stator copper loss for any given line current is the same whether the stator is assumed to be star or delta connected.

(a) Estimate the stator line current and the ratio of stator to rotor m.m.f. at full load.
(b) Estimate the increase of stator resistance per phase required to reduce the starting current to three times the full-load value and what reduction in starting torque will this cause? Express the additional resistance in terms of R_1.
(c) Estimate the output kW, power factor and the efficiency if the machine is driven as an induction generator by a water turbine delivering 400 hp.

(15) Two identical 3-phase, 500-V induction machines are mechanically coupled for a back-to-back test. One machine is motoring from a 500-V, 50-Hz supply and the other is generating on to a 3-phase, 500-V synchronous machine for which the frequency can be controlled. It is required to run the motoring induction machine at 40 hp. Neglecting the slight change in the generating machine parameters due to operating at a reduced frequency, estimate the rating of the synchronous machine for this duty and the frequency at which it must be run.

Line input readings for each machine

No load	500 V	8·5 A	1·6 kW
Locked rotor	100 V	30 A	1·6 kW

Friction and windage loss per machine 0·8 kW. $R_1 = R_2'$

(16) An 8-pole, 50-Hz induction motor coupled to a flywheel, drives a load which requires a torque when running light of 110 Nm. For an intermittent period of 8 sec, a pulse load is to be supplied such that there is an instantaneous rise of load torque to a total value of 550 Nm. What must be the combined rotational inertia of the system to ensure that the peak motor torque does not exceed 400 Nm? The motor speed/torque characteristic may be taken as linear and it gives a torque of 350 Nm with a slip of 5%.

What would be the peak motor torque if the coupled inertia was 200 kg m² ?
Note—it will be helpful if reference is made to Fig. 6.47 and the associated text.

(17) Using the data from the exact-circuit calculation of Problem (1), determine the following particulars when operating in the dynamic braking mode and using the 2-lead connection of Fig. 5.30b. In each case, the changeover to braking is to be made from motoring at 4% slip and the circuit adjustments must retain the rotor current initially at the same value as when running at this slip.

(i) Find the required d.c. excitation current and voltage when the rotor resistance is unchanged. What is then the initial torque on changeover and the maximum torque when running down to rest if this d.c. current is maintained. Neglect the change of X_m due to saturation.

(ii) Repeat the calculation for the condition where the excitation is adjusted so that the air-gap flux (E_1), is maintained instead, at the 4% slip value. Extra rotor-circuit resistance is now required to limit the rotor current to rated value. What is the value of this extra resistance? Note that the speed for maximum torque for this circuit

643

condition occurs when $R_2 = Sx_2$, since E_1 and I_m, not I_1, are being maintained constant.

Saturation may be neglected throughout but show that the effect would be to reduce the peak torque approximately in proportion to the reduction of X_m.

(18) The values of x_2' and E_1 for a particular induction motor are known at a frequency f_{base}. Show by developing the following expressions, that the rotor current and torque are independent of the supply frequency f, but depend upon the slip frequency f_2 Hz, if flux/pole, (E_1/f_{base}), is constant.

$$\mathbf{z_2}' = f(R_2'/f_2 + jx_2'/f_{base})$$

$$\mathbf{I_2}' = \frac{E_1}{R_2'/(f_2/f_{base}) + jx_2'},$$

$$T_e = \frac{3}{2\pi . f_{base}/p} \; I_2'^2 . \frac{R_2'}{f_2/f_{base}}.$$

(19) An induction motor, the data for which is the same as in Example E.7.1, p. 389, is to be run from a constant-current source at normal frequency. Calculate:
   (a) the slip for maximum torque;
   (b) the maximum torque when the stator current is set at the rated value (105 A);
   (c) the rotor and magnetising currents I_2' and I_m for this condition;
   (d) the input impedance for this condition;
   (e) the required stator line-voltage to sustain this condition.
   (f) What supply frequency will be required to develop the maximum possible torque at standstill, if the starting current is to be 5 times the rated value. What is the input impedance and the voltage required at starting? What will then be the rotor current, the magnetising current, and the torque?

   Use the "exact" equivalent circuit, excluding R_m, and neglect saturation except for part (f) when it can be assumed that X_m falls to 10% of its normal value.

(20) Using the "exact" equivalent circuit of the induction motor and noting the relationships:

$$\mathbf{I_0} = \mathbf{E_1 Y_m}, \quad \mathbf{I_1} = \mathbf{I_2' + I_0} \quad \text{and} \quad s\mathbf{E_1} = \mathbf{I_2' z_2' + V_3'},$$

where $\mathbf{z_2}' = R_2' + jsx_2'$, show that the general expression for $\mathbf{I_2}'$ including the effect of an injected secondary voltage $\mathbf{V_3}$, if present, is given by the following modification to eqn. (7.12):

$$\mathbf{I_2}' = \frac{\mathbf{V_1} - (1/c)(\mathbf{V_3'}/s)}{c[(R_2'/s) + jx_2'] + (R_1 + jx_1)} = \frac{\mathbf{V_1} - (1/c)(\mathbf{V_3'}/s)}{\mathbf{Z_2' + Y_m z_1 Z_2' + z_1}},$$

where c is the ratio $\dfrac{\text{terminal voltage}}{\text{air-gap voltage on no load}} = \dfrac{\mathbf{V_1}}{\mathbf{E_1}} = 1 + \mathbf{Y_m z_1}$,

and $\mathbf{Z_2}' = \mathbf{z_2'}/s$.

CHAPTER 8
N.B. Conversion factors for "engineers" units; 1 hp = 746 watts, 1 lbf ft = 746/550 Nm.

(1) The water turbines for a hydro-electric station have an optimum economical running speed of about 78 rev/min. Find the number of poles and the nearest practicable speeds for direct-coupled alternators to generate 50 Hz, 25 Hz and 60 Hz.

(2) Justify, explaining the limitations, the use of a series reactance to represent the effects of armature reaction in a synchronous machine.

Two 3-phase, 3·3 kV, star-connected alternators are connected in parallel and supply a load of 800 kW at 0·8 p.f. lagging. The prime movers are so set that one machine delivers twice as much power as the other. The more heavily loaded machine has a synchronous reactance of 10 Ω per phase and its excitation is adjusted so that it operates at 0·75 p.f. lagging. The synchronous reactance of the other machine is 16 Ω per phase.

Calculate the current, e.m.f., power factor and load angle of each machine. The internal resistances may be neglected.

(3) Two star-connected, non-salient-pole synchronous generators of identical rating operate in parallel to deliver 25,000 kW, 0·9 power-factor lagging current at 11 kV. The line induced e.m.f. of machine 1 is 15 kV and the machine delivers 10 MW, the remaining power being supplied by machine No. 2.

Determine for each machine:

(a) the load angle in electrical degrees,
(b) the current,
(c) the power factor,
(d) the kVA.

Find also the line induced e.m.f. of machine No. 2. Take the synchronous reactance for each machine as 4·8 Ω per phase and neglect all power losses.

(4) A 3-phase, 8-pole, 50-Hz, 6600-V star-connected synchronous motor has a synchronous impedance of $0·66 + j6·6$ Ω per phase. When excited to give a generated e.m.f. of 4500 V per phase it takes an input of 2500 kW. Calculate the electromagnetic torque in lbf ft, the input current, power factor and the load angle.

(5) The motor of the last question is operating at a current of 180 A at unity power-factor. What is the value of the generated e.m.f. (E_f)? Under these conditions, what is the mechanical output and the efficiency if the mechanical, excitation and iron losses total 50 kW?

(6) A 3-phase synchronous motor is paralleled on an infinite system. Explain by means of phasor diagrams what happens when:

(a) the machine excitation is increased whilst it is uncoupled from the mechanical load;
(b) the mechanical load is applied with the excitation maintained constant.

A 3-phase, 5000-kVA, 11-kV, 50-Hz, 1,000-rev/min, star-connected synchronous motor operates at full load 0·8 p.f. leading. The synchronous reactance is 0·6 *per-unit* and the resistance may be neglected. Calculate for these conditions the synchronising torque per mechanical degree of angular displacement.

(7) Deduce an expression for the pull-out torque of a synchronous motor based on synchronous impedance.

A 3300-V, 3-phase, 50-Hz, star-connected synchronous motor has a synchronous

645

impedance of $2+j15\,\Omega$ per phase. Operating with an excitation corresponding to an e.m.f. of 2500 V between lines, it just falls out of step at full-load. To what open-circuit e.m.f. will it have to be excited so that it will just remain in synchronism at 50 % excess torque?

(8) Show that for any given excitation the armature current phasor of a synchronous machine has a circular locus if saturation is neglected, the synchronous impedance being constant.

In a certain synchronous generator running at rated load-current, the e.m.f. behind synchronous reactance is 1·5 *per-unit* based on the terminal voltage, and the power factor is 0·866 lagging. If the armature resistance may be neglected and the excitation is left unchanged determine from the circle diagram and check analytically:

(a) the maximum torque in *per-unit* based on full-load torque;
(b) the current in *per-unit* based on full-load current when operating:

    (i) as a generator at 0·866 p.f. leading,
    (ii) as a motor at 0·5 p.f. leading.

What will the power factor be when operating as a generator at full-load current and with the e.m.f. behind synchronous reactance adjusted to 1 *per-unit*?

(9) Deduce the shape of the current locus of a non-salient-pole synchronous motor for constant excitation and variable load. The synchronous reactance may be assumed constant.

A 3-phase, 220-V, 50-Hz, 1500 rev/min, mesh-connected synchronous motor has a synchronous impedance of $4\,\Omega$ per phase. It receives an input line current of 30 A at a leading power-factor of 0·8. Find the line value of the induced e.m.f. and the load angle expressed in mechanical degrees.

If the mechanical load is thrown off without change of excitation, determine the magnitude of the line current under the new conditions. Neglect losses.

(10) Check analytically the answers obtained from the circle diagrams of Example E.8.5.

Note that the next four problems involve the approximation of neglecting R_a. It may be useful, therefore, to refer back to the solution program and phasor diagram of Fig. 8.31.

(11) A synchronous machine has a synchronous reactance of $X_s = 1$ *per unit*. Find the *per-unit* values of current, power, torque and the power factor and state the machine function when the e.m.f. due to field current and the load angle δ have the following values; (e.m.f.s in *per unit* based on $V = 1$, so that $E_f = E_{f.p.u.}(\cos\delta + j\sin\delta)$.

(a) 0·5, 0°; (b) 1·5, 0°; (c) 0·5, −30°; (d) 1·5, −30°; (e) 0·5, +30°; (f) 1·5, +30°.

Derive the *per-unit* equation required, i.e. $I_a = (\pm 1 \mp E_r)/jX_s$, and sketch the phasor diagram for each case. Note that either the generator or motor equation could be used. The sign of the real part of I_a will then show whether the choice has been correct for a positive motor or generator.

(12) A 3-phase, 4-pole, 400-V, 200-hp, star-connected synchronous motor has a synchronous reactance of $0.5\,\Omega$ per phase, and the resistance may be neglected. Calculate the load angle and the input current and power factor when the machine is working on full load with the e.m.f. adjusted to 1 *per-unit*, based on the terminal voltage. The mechanical loss is 10 kW.

(13) A 6-pole, 3-phase, star-connected synchronous motor has an unsaturated synchronous reactance of 12·5 Ω per phase, 20% of this being due to leakage flux. The motor is supplied from 11 kV at 50 Hz and drives a gross mechanical torque of 50×10^3 Nm. The field current is so adjusted that the e.m.f. E_f read off the air-gap line is equal to rated terminal voltage. Calculate the load angle, input current and power factor and also the maximum possible output power with this excitation, before pulling out of step. Resistance and saturation may be neglected and E_f assumed to be unchanged when the power increases to the maximum.

With the current calculated as above, to what value would the excitation have to be adjusted in terms of E_f, if the power factor was to be unity? What would then be the output?

To what value would E_f have to be adjusted so that the machine could operate as a synchronous capacitor at the same armature current?

Repeat all the calculations allowing for saturation by reducing all those components which would be affected, by a factor of one-third; e.g., the new load angle would correspond to an e.m.f. reduced to two-thirds of its previous value and X_m would also be reduced to two-thirds of its unsaturated value.

(14) A 1000-kVA, 6·6-kV, 50-Hz, 3-phase, 6-pole, star-connected synchronous machine is connected to an infinite system. The synchronous impedance per phase can be taken as constant at j50 Ω. The machine is to operate as a motor taking full-load current and with the excitation so set that a 50% overload is possible before it pulls out of synchronism. Calculate the necessary voltage E_f behind synchronous reactance to permit this overload margin. What will be the output power and the input power factor at full-load current, neglecting all the machine power losses?

If with the same mechanical load, E_f was reduced by 25%, what would be the change in power factor and the new overload margin?

What excitation, in terms of E_f would be necessary to operate at 0·8 lag and then at 0·8 lead with the same output power as above, and what would be the corresponding armature currents?

(15) For the machine for which data was given in Example E.8.1, p. 489 calculate for the same load angle, the ratio between the torque produced when the salient poles are unexcited to that produced with $E_f = 1$ per unit.

(16) Show that if the armature resistance of a polyphase salient-pole synchronous machine is neglected, the load angle at which maximum torque occurs is given by: $\cos^{-1} -A + \sqrt{(A^2 + \frac{1}{2})}$ where

$$A = E_f X_q / 4V(X_d - X_q).$$

A 6·6-kV, 5-MVA, 6-pole, 50-Hz, star-connected synchronous generator has $X_d = 8·7$ Ω per phase and $X_q = 4·35$ Ω per phase. If the excitation is so adjusted that $E_f = 11$ kV (line) and the load angle is 30 electrical degrees determine:

(a) the power factor, output current and power in *per unit*,
(b) the load angle at maximum torque,
(c) the ratio between maximum torque and that occurring with $\delta = 30°$,
(d) the stiffness ($dP/d\delta$) in newton metres per mechanical radian for a load angle of 30°,
(e) the frequency of small undamped oscillations if the total coupled inertia is 8200 kg m², the mean load angle being 30°,

647

(f) the stored-energy constant.

(17) The induction machine for which data is given in Problem (8), p. 641, is operated as a synchronous generator with a rated output of 5 kVA at 1000 line volts. The primary winding is used as the field, two stator terminals being supplied with d.c. The secondary is used as the star-connected armature winding. A magnetisation curve taken with the slip-rings open-circuited and the rotor running at 1500 rev/min gave readings as follows:

Field current	4	6	8	10	12	14	16	A
Armature line-voltage	490	735	900	1000	1070	1115	1160	V

The field current required to circulate rated current through the armature windings with the slip rings short-circuited was 7 A.

Determine the range of d.c. excitation voltage and current required to enable the generator to operate from no load to full load at 0·8 power factor lagging, with a constant terminal voltage of 1000 V. Allow for the saturation of synchronous reactance.

(18) For Example E.8.6, p. 525, and with particular reference to the saturated conditions corresponding to part (b), compare the saturated ·value of synchronous reactance as given by $x_{a1} + X_{ms}$, with that based on the short-circuit ratio V/I_a'', see p. 482. What is the steady current if the machine is short-circuited from an open-circuit condition with the voltage set at the rated value?

Calculate also the load excitation for part (b) using the ampere-turn method, p. 525, and noting that F_r then corresponds to $V = 11$ kV on the o.c. curve.

(19) The test results on a 5-MVA, 6·6-kV, 3-phase, star-connected synchronous generator are as follows:

Open-circuit test:

Generated line e.m.f.	3	5	6	7	7·5	8·0	8·4	8·6	8·8	kV
Field current		25	44	57	78	94	117	145	162	181	A

Short-circuit test at rated armature current requires 62 field amperes.
Zero p.f. lagging test at 6·6 kV and rated current requires 210 field amperes.

If the field resistance is 1·2 Ω cold, 1·47 Ω hot, calculate for normal machine voltage the range of exciter voltage and current required to provide the excitation from no load up to full load at 0·8 p.f. lagging. The armature resistance may be neglected.

(20) A 3-phase, 500-kVA, 3·3-kV, star-connected synchronous generator has a resistance per phase of 0·3 Ω and a leakage reactance per phase of 2·5 Ω. The excitation when running at full-load 0·8 p.f. lagging is 72 A. The o.c. curve at normal speed is as follows:

Line voltage	2080	3100	3730	4090	4310	V
Field current	25	40	55	70	90	A

Estimate the value of the full-load armature ampere-turns per pole in terms of the field turns, and hence calculate the range of field current required if the machine has to operate as a synchronous motor at full kVA from 0·2 leading to 0·8 lagging power-factor.

(21) A star-connected synchronous motor rated at 600 kVA, 440 V unity power-factor is to be used to improve a works power-factor by running on no load as a

synchronous capacitor. From tests as a generator on open circuit, short circuit and zero power-factor the following information was obtained:

Potier reactance drop 33·5 V per phase.

Field equivalent of full-load armature reaction 6·6 A (F_a in terms of field turns).

Open-circuit curve:

Terminal voltage	210	330	430	505	560	V
Field current	9	14	20	28	40	A

Plot a curve of machine leading kVAr against field current for values up to 800 kVAr with the machine operating as a synchronous capacitor at 440 V. Stator resistance and mechanical losses may be neglected.

Hence, determine the permissible rating of the machine in this mode of operation if the field current can be increased by 30% over the value required to operate the machine as a motor at its full-load rating and unity power-factor.

(22) A 6·6-kV, 50-Hz synchronous motor is to be used to improve a works power factor by running as a synchronous capacitor at zero power-factor leading. The works load is 400 kW at 0·8 lagging. Calculate the necessary kVAr rating if the power factor is raised to: (a) unity; (b) 0·95 lagging; (c) 0·95 leading. For condition (a) find the e.m.f. behind synchronous reactance required (E_t).

If for condition (a), the motor is to drive a load and the motor rating is 400 kVA, what maximum continuous power output is possible without exceeding this total kVA rating for the machine? What would then be the required excitation E_t? In all cases, assume a synchronous impedance of 1 *per unit* and neglect all machine losses.

CHAPTER 9

(1) A 6-pole, 50-Hz Schrage motor with brushes in the "neutral", (in-phase) position gives a speed variation on no load from 200 rev/min to 1800 rev/min when the maximum brush separation corresponding to a pole pitch is used. What mechanical angle of brush separation is necessary to give:

600 rev/min on no load?

1600 rev/min on no load?

(2) The induced voltage across one pole pitch of a 50-Hz, 6-pole Schrage-motor commutator is 36 V r.m.s. The stator secondary winding can be arranged in parallel or series to give an induced voltage at standstill of either 30 V or 60 V. Find in each case the speed range on no load. Also find, with the paralleled stator winding, the mechanical angular separation of the brushgear to give speeds of 1800, 400 and −100 rev/min.

(3) A 4-pole stator winding is excited from a 50-Hz supply. The rotor, which also has a 4-pole winding, is provided with tappings to slip rings at one end and to a commutator at the other end. If the rotor is driven at 1750 rev/min calculate the frequency of the voltages appearing at the commutator and at the slip ring brushes under the following conditions if both sets of brushgear are stationary:

(a) Rotor driven in the same direction as the stator field.

(b) Rotor driven in the opposite direction to the stator field.

At what speeds must the commutator brushgear itself be rotated so that the commutator frequency is the same as the slip ring frequency for condition (a)?

ANSWERS

Chapter 2

(1) 8·3 W; 16·7 W, 1·65.

(2) (a) 1·05 μH/m, 2·62 μH/m; (b) 0·06 tesla.

(3) Reluctance of various branches in At/Wb

Yoke = 8670, Pole = 2555, Pole leakage = 305,000, Air gap = 45,600, Teeth = 6930, Armature core = 267. Total "supply" m.m.f. = 4580 At per pole.

Chapter 3

(1) 11·4 kW, 88·6 kW.

(2) $I_1 = 3$ A, $I_2 = 2$ A, $V = 1·8$ V.　　　(3) $\theta_u = 0·731\ \hat{\theta}, \theta_1 = 0·269\hat{\theta}$.

Chapter 4

(1) 2607 turns, 1247 turns, 2 × 68 turns.

(2) Peak current = 0·58 A. Peak eddy-current component = 0·083 A.

(4) 297 V.　　　　　　　　　(15) $253/-22°·6$ A, $358/-27°·5$ A

(8) 0·97 leading.　　　　　　　　　　　　　　　　(E reference).

(9) 600 V, 144 kW.　　　　　(16) 96·56%, 96·81%; 97·25% at 88·5 kW.

(10) 89·4%.　　　　　　　　　$Z = 0·016+j0·0446$ p.u.

(11) 95·86%.　　　　　　　　(17) 55·2 W, 59·5 W.

(12) 1·62, 97·87%.　　　　　(18) Mag losses 6·6 kV, 8 A, 13 kW.

(13) 1110 kVA, 0·818 p.f.　　　　　Copper losses. Booster transformer

(14) $I_A = 8·45$ at 0·97 p.f.　　　or equivalent, in series with each

　　　$I_B = 8·13$ A at 0·85 p.f.　　line to supply 1680 V, 35 A, 13·3 kW

　　　$V = 425$ V.　　　　　　　　per phase.

　　　　　　(19) $I_A = 1·15$ A, $I_B = 1·58$ A, $I_C = 1·422$ A.

　　　　　　(20) $\mathbf{I}_a = 5·25/+30°$　$\mathbf{I}_b = 6·75/+3°·5$ A.

Chapter 5

(1) 0·0099 Ω, 308·2 V; 0·57 tesla, 6,860 At.

Equalisers to conductors 1–67, 7–73, ..., 61–127.

(2) (a) 86 slots, coils and commutator bars and conductors in series. $y_c = 29$, progressive winding. Coil span 1–15. $B = 0·666$ tesla.

(b) 132 slots, 264 coils and commutator bars, $z_s = 88$, $B = 0·654$ tesla. Coil span 1–22 for 1 slot chording.

With 22 equalisers connect 1–89–177, 5–93–181, ..., 85–173–261.

(3) 8 conductors per slot, 296 single-turn coils, $\phi = 0·0195$ Wb $y_c = 99$, progressive winding, coil span 1–13.

(4) $k_{d1} = 0·956, k_{d3} = 0·645, k_{d5} = 0·2, k_{p1} = 0·95, k_{p3} = 0·588, k_{p5} = 0$. R.M.S. phase e.m.f. = 183 V, R.M.S. line e.m.f. = 316 V.

(5) $k_{d1} = 0·957, k_{p1} = 0·984$; 587 V; 6600 At, 196 At, 60 At.

Chapter 6

(2) (a) 0·0156 Wb, (b) 176 Nm, (c) 3160 At per pole, (d) 4·1°.

(4) (a) I_f 1·86 − 0·75 A, 8 − 116 Ω, (b) 88·5 V, (c) 840 At per pole.

(5) 4·78/(1 + 0·2p) armature volts/field volt; 0·8 cycles/sec, 3·38 V/V.

(6) (a) I_f 1·5 − 0·38 A, 36·7–469 Ω; (b) 250 At pole; (c) 22%; (d) + 23·8%.

(7) 25 rad/s; $1·92\dfrac{d\omega_m}{dt} + \omega_m = 25$, $\tau_m = 1·92$ seconds.

(8) 453 rev/min, 84·1%, 10·4 hp.

(10) 1·435 and 0·707; 2·05 and 0·25; 1·91 and 3·65.

(11) (a) 475 rev/min, 109 lbf ft; (b) 615 rev/min, 177 lbf ft;
(c) 665 rev/min, 206 lbf ft.

(12) (a) 82·5 A, 390 rev/min; 5·82 Ω.

(13) (a) 578 rev/min; (b) 1433 rev/min.

(14) $k_\phi = 3·7$; 0·82 Ω, 6·4 kg m².

(15) 97 A, 648 rev/min, 66 rev/min, 339 rev/min; 0·167 sec, 0·223 sec.

(16) 730 rev/min, 10·3 sec.

(17) (a) 1064 rev/min; (b) 0·95; (c) $I_f = 0·706$, $I_a = 0·817$; (d)(i) $I_f = 1·376$, $I_a = 0·717$;
(d)(ii) $I_f = 0·715$, $I_a = 1·504$.

(18) (a) $I_A = 1500$ A, $I_B = 1250$ A, $I_C = 1600$ A, $V = 485$ V;
(b) $I_B = -1500$ A, $I_B = 500$ A, $I_C = 1000$ A, $V = 500$ V.

(19) 545 W, 85·42%, 85·43%.

(20) 134 A, 105 A; 17·75 Nm, 6·25 hp.

(21) For (i), (ii) and (iii), ω_m is respectively 2, 0·5 and 1 *per unit*.
For I_a; (i) 1, 2 and 4; (ii) 2, 1 and 0·5, (iii) 2 *per unit* throughout.

Chapter 7

(1) (a) 83·8 A,　　89·2 A,　0·83,　　671 Nm,　87·6%,　119·5 Nm.
(b) 86 A,　　　93·3 A,　0·835,　706 Nm,　87·5%,　123　Nm.
(c) 97·5 A,　103 A,　　0·935,　910 Nm,　91·6%,　492　Nm.

(2) 0·0117; 1·6 Ω referred to stator; 15·8 Ω, 77·3% reduction.

(3) (a) Increase to 13·15 R_1　　62·7% reduction in maximum torque
(b) Increase to 2·45 x_1　　39·6% reduction in maximum torque
(c) Increase to 13·15 R_2　　no reduction in maximum torque.

(4) 1800 Nm; 909 rev/min, 230 hp, 847 rev/min; (a) 70 V, 4·53 Hz for "safe" torque; 276 V, 29·53 Hz for max torque. (b) 102 V, 7·65 Hz and 300 V, 32·65 Hz for max torque. 1·57 times starting torque.

(5) 374 Nm; 0·306 Ω per phase.

(6) $x_1 + x_2' = 2·09$ Ω per phase; 380 lbf ft, 106 hp.

(7) 0·37 p.u., 1·75 p.u.

(8) Machine is generating 8·36 kW at 0·556 p.f. leading; mech input 11·3 kW.

(9) Powers in kW; 15·49, 0·76, 9·75, 4·98, 9·75, −4·77, −5·77.

(10) $R_1 = R_2' = 0·105$ Ω. $x_1 = x_2' = 0·202$ Ω, $R_m = 52$ Ω, $X_m = 5·22$ Ω.

(11) 50 A, 0·84, 151 lbf ft; 1415 rev/min, 40·5 hp, 83·3%;
97 lbf ft, 245 lbf ft.

(12) (a) 376 lbf ft, 6·7 R_2; (b) 245 A; (c) 1·2 R_2.

(13) 25%; 0·707, 0·13.

(14) (a) 330 A, 1·89/1; 6·6 R_1, 50% reduction; (c) 258 kW 0·73 p.f. 86·5%.
(15) 31·2 kVA, 0·88 p.f., 46 Hz.
(16) 660 kg m^2, 537 Nm.
(17) (i) 104·5 A, 20·9 V, 28 Nm, 3370 Nm; (ii) 2·4 Ω, 106·5 A, 21·3 V, 700 Nm, 1280 Nm.
(19) (a) 0·00631; (b) 4088 Nm; (c) 72·1 A, 74·2 A; (d) 9·84 + j10·29 Ω; (e) 2589 V! (f) 2·5 Hz, 0·1685 + j0·0915 Ω, 174 V, 285·5 A, 381 A, 8098 Nm.

Chapter 8

(1) 76 poles, 78·9 rev/min; 38 poles, 78·9 rev/min; 92 poles, 78·2 rev/min.
(2) Machine No. 1, 124·3 A, 4·99 kV, 0·75, 18·9°;
Machine No. 2, 52 A, 4·13 kV, 0·9, 19°.
(3) Machine No. 1, 16·9°, 660 A, 0·8, 12·6 MVA;
Machine No. 2, 27·6°, 820 A, 0·96, 15·6 MVA; 14·6 kV.
(4) 22,400 lbf ft, 236 A, 0·93 p.f. leading, −19·5°.
(5) 6·72 kV; 2599 hp, 94·2%.
(6) E_f = 1·44 p.u., $δ$ = −19·4°; 5,650 Nm/degree.
(7) 4 kV.
(8) (a) $\hat{T}_e$ = 2·39 p.u., (b) (i) 2·38 p.u. (b) (ii) 0·76 p.u.; 0·93 leading.
(9) 268 V, −6°, 20·6 A
(11) I_a in per unit; (a) −j0·5 reactor; (b) j0·5 capacitor; (c) 0·25−j0·567 motor; (d) 0·75+j0·3 motor; (e) 0·25+j0·567 gen; (f) 0·75−j0·3 gen. Real part of I_a = Torque and Power in *per unit*.
(12) −14·9° mech., 238 A, 0·965 p.f. lagging.
(13) −32·8°(−36·6°), 286 A at 0·96 p.f. lagging (422 A at 0·65 p.f. lagging), 13000 hp (11800 hp); 12·6 kV line (12·9 kV), 7300 hp (10800 hp) 17·2 kV (17·7 kV). Saturated values in brackets.
(14) $δ_{fl}$ must be −41·8°, E_f = 6403 V/phase, 976 kW, 0·976 p.f. leading; $δ_{fl}$ = −62·7°, 0·925 p.f. lagging, 12%, 4280 /−81·7°, 8190 /−31·2°, I_a = 85·3±j64 A.
(15) 0·32.
(16) (a) 0·987 lag, 1·282 p.u. 1·265 p.u. (b) 66·2°. (c) 1·496/1.
(d) 2·78 × 10^5 Nm (e) 0·92 Hz (f) 8·95 secs.
(17) x_a=12 Ω, X_{mu}=162 Ω, X_{ms}=119 Ω, k_{fs}=91·5 V/A, 10·1/16·4 fld A 20·2/32·8 fld V.
(18) x_{al} + X_{ms} = 11·6 Ω = 0·96 *per unit* at E = 11 kV (line); V/I_a'' = 11 Ω = 0·91 *per unit*. Steady-state short-circuit current = 577 A = 1·1 *per unit*. F_f by ampere-turn method = 346 A.
(19) 83–226 V, 69–154 A.
(20) F_a = 27·6 A, F_f 80–34 A.
(21) F_f max = 30·2 A giving 450 kVA leading.
(22) 300, 168 and 432 kVAr, E_f = 2 per unit. 265 kW, I_a = 0·66+0·75 *per unit*. E_f = 1·75−j0·66 *per unit* (12300 V line).

Chapter 9

(1) 20°; 32·4° mechanical.
(2) −200/2200 rev/min, 400/1600 rev/min, −28°, 20°, 44°.
(3) (a) 50 Hz, 8·33; (b) 50 Hz, 108·33 Hz, 1250 or 1750 rev/min.

REFERENCES

THE following list of references, mentioned in the text, is only a selection from the vast amount of literature on the subject of Electrical Machines and Their Applications. The references are mostly concerned with recent developments and several of them are in fact Conference Proceedings. Each of these contains many short papers on various topics and though only a few are noted, many of the remaining papers are relevant to the subject matter of the book.

1. HARRISON, H. The similarities of electrical machines when operating in the balanced steady-state condition. *International Journal of Electrical Engineering Education (I.J.E.E.E.)*, Vol. 3, No. 1, 1965.
2. HANCOCK, N. N. *Matrix Analysis of Electrical Machinery*, 2nd Edition. Pergamon Press (1974).
3. APPLETON, A. D. Superconducting Machines. *Science Journal*, April 1969.
4. EM 70, *Electrical Machines in the Seventies Conference at the University of Dundee*, July 1970.
   APPLETON, A. D. The role of superconducting machines in industry.
   CHATTOPADHYAY, A. K., HINDMARSH, J. and LIPO, T. A. Performance and analysis of thyristor commutator motor.
5. LAITHWAITE, E. R. *Linear Electric Motors*. Mills and Boon Technical Library (1971).
6. INSTITUTION OF ELECTRICAL ENGINEERS (I.E.E.) CONFERENCE No. 120, 1974. *Linear Electric Machines.*
7. LAITHWAITE, E. R. *Propulsion Without Wheels*. English Universities Press (1966).
8. DAVIES, E. J. Airgap windings for large turbogenerators. *Proceedings I.E.E.* Vol. 118, March 1971.
9. HINDMARSH, J. Large d.c. machine performance and design. Series of articles in *Electrical Times*, Vols 134/135, 1958/59.
10. GRIMSHAW, K. P. The electromagnetic principles of commutation in d.c. machines. *I.J.E.E.E.* Vol. 2, No. 1, 1964.
11. I.E.E. CONFERENCE, No. 10, 1965. *The Application of Large Industrial Drives.*
    JAHN, M. H. Implication of the application requirements on large d.c. rolling-mill drives.
    STEPHEN, D. D. Industrial applications for brushless synchronous motors.
    GROVE, P. F. Power-factor improvement and the use of synchronous motors.
12. I.E.E. CONFERENCE, No. 17, 1965. *Power Applications of Controllable Semiconductor Devices.*
    SCHOFIELD, J. R. G., SMITH, G. A. and WHITMORE, M. G. The application of thyristors to the control of d.c. machines.
    BOWLER, P. The application of a cycloconverter to the control of induction motors.
13. HINDMARSH, J., MUKHTAR, E. S. and GUPTA, R. K. A new commutation detector. *International Conference on Electrical Machines at the Technical University*, Vienna, September 1976.
14. I.E.E. CONFERENCE NO. 93, 1972. *Electrical Variable-speed Drives.* (See also No. 179, 1979.)

653

REFERENCES

HENDER, B. S. Variable-speed drives for electric vehicles and trucks.

BISHOP, K. W. J. Variable-frequency inverter drives.

AYERS, P. J. Eddy-current couplings as a general purpose industrial drive.

ALSTON, I. A. and HAYDEN, J. T. Wide-range synchronous motor drive.

15. BATES, J. J. Thyristor-assisted commutation in electrical machines. *Proceedings I.E.E.* Vol. 115, June 1968.

16. WOLFENDALE, E. A d.c. motor with s.c.r. commutator and synchronous speed-control. *Control*, September 1965.

17. BHAGWAT, P. G. Speed control of d.c. shunt motor using thyristor bridge. *Control*, May 1967.

18. I.E.E. CONFERENCE No. 136, 1976, and No. 202, 1981. *Small Electrical Machines:* WERNINCK, E. H., The methodology of motor selection; BERMOND, G. and FLYNN, J. B., The Isosyn motor, a new generation of permanent-magnet synchronous motors.

19. CHALMERS, B. J. and WILLIAMSON, A. C. Stray losses in squirrel-cage induction, motors. *Proceedings I.E.E.* Vol. 110, October 1963.

20. RAWCLIFFE, G. H. and FONG, W. Close ratio, single-winding induction motors. *Proceedings I.E.E.* Vol. 110, May 1963.

21. HANCOCK, N. N. *Electric Power Utilisation.* Pitman (1967).

22. CHALMERS, B. J. and WOOLLEY, I. General theory of solid-rotor induction machines. *Proceedings I.E.E.* Vol. 119, September 1972.

23. LAWRENSON, P. F. and AGU, L. A. Theory and performance of polyphase reluctance machines. *Proceedings I.E.E.* Vol. 111, August 1964.

24. APPLETON, A. D., ANDERSON, A. F. and ROSS, J. S. H. A discussion on large superconducting a.c. generators. *International Conference on Electrical Machines at The City University London*, September 1974.

25. CHALMERS, B. J., PACEY, K. and GIBSON, J. P. Brushless d.c. traction drive. *Proceedings I.E.E.* Vol. 122, July 1975.

26. I.E.E. CONFERENCE No. 123, 1974. *Power Electronics, Power Semiconductors and Their Applications.* (See also No. 154, 1977 and No.   , 1984.)
McLEOD, B. D., RENFREW, A. and SHEPHERD. An inverter drive suitable for traction.
PUTZ, U. The converter-fed brushless synchronous motor.

27. CHALMERS, B. J., MAGUREANU, R. M. and HINDMARSH, J. General principle for brushless synchronous machines and its application in an inverter-fed drive. *Proceedings I.E.E.* Vol. 119, November 1972.

28. BELL, R., LOWTH, A. C. and SHELLEY, R. B. *The Application of Stepper Motors in Machine Tools.* Machinery Publishing Co. (1970).

29. FIRST EUROPEAN CONFERENCE ON ELECTRICAL DRIVES/MOTORS/CONTROLS. Peter Peregrinus Conference Publication No. 19, 1982.

30. HINDMARSH, J. An integrated lecture and laboratory course for electrical machines. *I.J.E.E.E.* Vol. 14, No. 1, 1977.

31. HANCOCK, N. N. The parameters used in matrix and tensor analysis of electrical machines. *I.J.E.E.E.* Vol. 5, No. 4, 1967.

32. KRAUSE, P. C. and THOMAS, C. H. The simulation of symmetrical induction machinery. *American I.E.E.E. Transactions*, Vol. PAS-84, November 1965.

33. HINDMARSH, JOHN. *Worked Examples in Electrical Machines and Drives.* Pergamon Press. 1982.

654

34. I.E.E. Conference No. 213, 1982. *Electrical Machines—Design and Applications.* Chalmers, B. J., Ward, F. S. and Pride, A. A. A 500-kW eddy-current dynamometer.
35. Proceedings of the I.E.E.E. Industry Application Society Conference 1982. Miller, R. H., Nehl, T. W., Demerdash, N. A., Overton, B.P. and Ford, C. J. An electronically controlled permanent-magnet synchronous-machine-conditioner system for electric passenger vehicle propulsion.
36. I.E.E. Drives/Motors/Controls, Conference Proceedings 1983.
37. Energy Efficiency and Electric Motors. U.S. Dept. of Energy Report HCP/M50217–01, 1978.
38. Electric Motor Handbook. Werninck, E. H. (Ed.). McGraw-Hill (U.K.) Ltd. 1978.

INDEX

The following points are made to assist in the location of a particular topic:

(1) Many subjects are not included in the Index because they are listed separately in the Contents, p. vii, and have separate sections devoted to them. This applies particularly to the material of the first three chapters, much of which is revisory. Reference should first be made, therefore, to the various sections under the individual chapter headings in the Contents.

(2) Each machine type has its own section in the Index, but subjects which are common to two or more types of machine, e.g. load sharing, stability, torque angle, variable-frequency supplies, etc., may be found separately listed.

656

ALSO BY THE AUTHOR . . .
WORKED EXAMPLES IN ELECTRICAL MACHINES AND DRIVES

This is a handbook to the author's companion volume "Electrical Machines and Their Applications". It can also be regarded as a text on Electrical Drives taught through worked examples. This emphasis on Drives throughout the book reflects the aspect of electrical machines of most general engineering interest. Competence in the subject requires practice in solving numerical problems to appreciate the relationship between the machine, its equations and the data. With nearly a hundred questions, many of them having multiple and subsections, a wide variety of topics is covered at undergraduate level, with introductory ideas for post-graduate studies. Steady-state and transient behaviour; operation with power-electronic circuits and an introduction to machine/semi-conductor modelling, are all illustrated. The chapter problems are arranged at increasing levels of difficulty and by revision of the equations and discussion of the engineering implications, it is hoped to reveal some of the fascination and remove some of the mystery thought by some to surround the subject.

Contents: *Introduction & review of basic theory*: Aim of the book; Foundation theory; Equivalent circuits; Power-flow diagram; *Transformers*: Solution of equations; *D.C. machines*: Revision of equations; Solution of equations; Per-unit notation; Series motors; Braking circuits; *Induction machines*: Revision of equations; Solution of equations; Constant-(primary) current operation; Improved starting performance; Single-phase operation; Speed control by slip-power recovery; *Synchronous machines*: Summary of equations; Solution of equations; Per-unit notation; Electromechanical problems; Constant-current operation; Operating charts; Multi-machine problems; Salient-pole & reluctance-type machines: synchronising power; *Transient behaviour*: Transient equations; *Power-electronic/electrical machine drives*: Chopper-controlled d.c. machine; Thyristor convertor/d.c. machine drive; Power-electronic control of a.c. machines; Comparison of D.C., Induction and Synchronous machine drives with Power-electronic control; Mathematical & computer simulation of machine systems; Index.

210 × 148 mm 290 pp. 120 illus. (1981)
0 08 026130 2 (flexicover), 0 08 026131 0 (hardcover)

Available from most bookshops or in case of difficulty directly from Pergamon Press at the most convenient address (see p. iv).

SOME INDEPENDENT OPINIONS OF PREVIOUS EDITIONS

"... this well-established text provides the student with a comprehensive and concise introduction to the subject ... will be of interest to electrical engineering students wishing to study machines in depth."

ASLIB Book List

"Those engineers who have studied electronics in the past and wish to bring themselves up to date ... will find the book extremely useful."

Institute of Marine Engineering

"... should prove invaluable to students and engineers wishing to become more familiar with the subject."

Electrical Equipment

"An excellent, eminently readable book for the understanding of machine construction, practice and performance."

Electronics and Power

"... all students interested in electrical machinery will find their needs covered up to the stage of intense specialization."

British Engineer

"This book will be welcomed by those who desire to attain a comprehensive knowledge of electrical machinery."

The Electrical Supervisor